Leveraging AI and Nanotechnology for Materials, Devices, and Manufacturing

Ashok Vaseashta
https://orcid.org/0000-0002-5649-0067
International Clean Water Institute, USA

Ioan Stamatin
3NanoSAE Technology Transfer Center, University of Bucharest, Romania

IGI Global
Scientific Publishing
Publishing Tomorrow's Research Today

Published in the United States of America by
IGI Global Scientific Publishing
701 East Chocolate Avenue
Hershey, PA, 17033, USA
Tel: 717-533-8845 | Fax: 717-533-7115
Website: https://www.igi-global.com E-mail: cust@igi-global.com

Library of Congress Cataloging-in-Publication Data

Names: Vaseashta, A. (Ashok) editor | Stamatin, Ioan, 1952- editor
Title: Leveraging AI and nanotechnology for materials, devices, and manufacturing / edited by Ashok Vaseashta, Ioan Stamatin.
Description: Hershey, PA : IGI Global Scientific Publishing, [2026] | Includes bibliographical references and index. | Summary: "This book aims to bridge the gap between fundamental nanotechnology principles and cutting-edge AI techniques, offering a holistic perspective on their synergistic applications"-- Provided by publisher.
Identifiers: LCCN 2025015504 (print) | LCCN 2025015505 (ebook) | ISBN 9798337328836 h/c | ISBN 9798337328843 s/c | ISBN 9798337328850 ebook
Subjects: LCSH: Nanostructured materials--Industrial applications | Artificial intelligence--Industrial applications | Nanotechnology--Industrial applications | Manufacturing processes--Data processing http://id.loc.gov/authorities/subjects/sh2010100514
Classification: LCC TA418.9.N35 .L47 2026 (print) | LCC TA418.9.N35 (ebook) | DDC 620.1/15--dc23/eng/20250625
LC record available at https://lccn.loc.gov/2025015504
LC ebook record available at https://lccn.loc.gov/2025015505

British Cataloguing in Publication Data
A Cataloguing in Publication record for this book is available from the British Library.

Table of Contents

Detailed Table of Contents

Artificial intelligence (AI)-driven innovations in nanomaterials are reshaping traditional research paradigms, enabling accelerated discovery and design. This chapter reviews recent progress and emerging pathways in the convergence of AI and nanomaterials science, highlighting their synergistic potential in advancing the development of novel materials. Key areas of focus include leveraging existing and new experimental datasets to elucidate structure–property relationships, improving synthesis processes by employing AI-based optimization strategies, and integrating AI with advanced characterization techniques to extract critical features for iterative materials design improvement. Several case studies illustrate how AI enhances predictive modeling, synthesis control, and property characterization across diverse classes of nanomaterials. For user-defined applications, AI facilitates reverse engineering, shifting nanomaterials research from a trial-and-error methodology toward a predictive, intelligent, and application-oriented framework. The chapter concludes with a glimpse of some forward-looking state-of-the-art and niche applications, as this convergence leads to the Fourth Science Paradigm of data-exhaustive discovery, while also addressing the ethical considerations associated with integrating AI into this rapidly evolving field.

The contamination of water and water bodies is now considered a growing challenge globally, affecting both developed and developing countries. To this effect, researchers and scientific innovators are pushing their limits to ensure that more technologies for developing efficient and adaptive materials in water treatment are actualized. Nanotechnology, particularly the use of nano-sponges, functional membranes, and advanced nanocomposites, is shown to be very promising due to their properties, surface areas, and ability to perform several functions that might seem complicated for conventional materials. One challenge this faces is that the traditional experimental approach to material discovery and synthesis typically requires a significant amount of time and involves multiple trial-and-error processes. Artificial Intelligence (AI), specifically machine learning (ML), is an emerging powerful tool that can accelerate nanomaterial discovery, synthesis, and optimization, specifically for water purification. AI algorithms can be used to predict the relationship between the parameters for synthesis, the performance of the material and in turn help to optimize pore structures, adsorption efficiency, and membrane selectivity etc. in this chapter we explore a method that proposes to integrate AI into the design, synthesis, and deployment of nano-sponges, smart membranes, and hybrid nanomaterials for water treatment, highlighting case studies, simulation techniques, and predictive models. Eclectic styles, converging AI with nanotechnology, can facilitate the emergence of autonomous labs that optimize materials in real-time. It can also lead to innovative manufacturing systems based on digital twins.

The convergence of artificial intelligence and nanomaterials science is revolutionizing the discovery, design, and optimization of advanced materials with unprecedented speed and precision. AI-driven methodologies, such as predictive modeling, generative algorithms, and automated design of experiments, enable the rapid identification of novel nanomaterials with tailored properties for diverse applications in energy storage, electronics, healthcare, and environmental sustainability. Traditional materials discovery relies on resource-intensive experimental approaches and iterative trial-and-error methods, often requiring years or even decades to bring new materials from concept to commercial use. This chapter describes how AI accelerates this process by leveraging machine learning, deep learning, and big data analytics to predict material behaviors, optimize synthesis pathways, and identify promising candidates with minimal experimental intervention.

In the pursuit of environmentally responsible innovation, sustainable bio-composites have emerged as vital materials in construction, automotive, and aerospace. However, traditional methods for material optimization, defect detection, and Life Cycle Assessment are often labor-intensive and experimentally challenging. This chapter explores the role of advanced Machine Learning (ML) models in revolutionizing the design, evaluation, and regulation of sustainable bio-composites. Emphasis is placed on Graph Neural Networks, Vision Transformers, and Physics-Informed Neural Networks. GNNs model fiber-matrix interactions to predict performance and biodegradability, while ViTs enable real-time defect detection. PINNs integrate physics-based and data-driven learning to simulate material behavior. The chapter also addresses ethical, legal, and regulatory issues in AI integration—such as data privacy in ML-based LCA tools, intellectual property, and environmental compliance—offering a roadmap for responsible AI in sustainable bio-composites.

Integration of artificial intelligence into nanomaterials research is revolutionizing the discovery, design, and optimization of advanced materials. This chapter presents a comprehensive framework for AI-driven discovery of nanomaterials, encompassing data acquisition, predictive modeling, intelligent experimentation, and iterative feedback loops. It explores sector-specific applications across energy, environment, biomedical, manufacturing, and agriculture, while addressing technical, integration, and ethical challenges that influence large-scale adoption. Emerging tools such as quantum computing, federated learning, and digital twins are discussed in the context of accelerating innovation. The proposed roadmap highlights the crucial roles of infrastructure, policy, and education in establishing a sustainable and ethically informed ecosystem for AI-enabled nanomaterials research. By bridging interdisciplinary gaps, the chapter offers strategic insights for researchers, policymakers, and industry stakeholders aiming to harness AI for transformative advancements in nanotechnology.

We explore the integration of AI and nanotechnology in STEM education, highlighting their potential to transform pedagogical approaches and prepare students for future technological progress. The chapter underscores the importance of embedding AI and nanotechnology into STEM curricula by reviewing recent scientific literature. It explores how AI can enhance teaching practices, assessment methods, and administrative tasks, thereby offering personalized learning experiences that boost student engagement and performance. Furthermore, the chapter highlights how nanotechnology can bridge the gap between theoretical knowledge and practical application, providing students with hands-on experience using advanced materials and techniques. An analysis is provided of the curriculum development frameworks, effective integration strategies, and the challenges and solutions related to implementing these technologies in education. A sizable section is dedicated to a case study at the University of Bucharest, Faculty of Physics, where AI and nanotechnology have been successfully incorporated into STEM programs, demonstrating their impact on students' skillsets.

Artificial Intelligence (AI) and Nanotechnology are two of the transforming subjects in science, innovations, and industry, where there are far-reaching consequences in education. Given that such technologies are developing faster than ever before, it is crucial to incorporate them into the learning curriculum to produce future-oriented students. The chapter provides a thorough analysis of key concepts in AI and Nanotechnology, highlighting them and making them easily approachable to maintain conceptual clarity for both students and educators. It examines the potential of digital solutions to increase engagement, develop an understanding of concepts,

Manjunatha Badiger, NMAM Institute of Technology, Nitte University, India
P. S. Sushma, NMAM Institute of Technology, Nitte University, India
Savidhan Shetty C. S., NMAM Institute of Technology, Nitte University, India
Tanya Mendez, NMAM Institute of Technology, Nitte University, India
Jose Alex Mathew, AJ Institute of Engineering and Technology, India
Varuna Kumara, Moodlakatte Institute of Technology, India

Nanoelectronics development experiences a transformation through the union of artificial intelligence (AI) with nanotechnology, which leads to highly precise and high-performing devices. The chapter demonstrates how AI approaches enhance nanoelectronic device creation through the systematic optimization of each development stage, up to fabrication and testing phases. The article introduces nanoelectronics transistors, photonic devices, and biomedical implants as major applications while describing how AI handles design optimization, defect detection, performance predictions, and multi-scale simulations. The power of AI emerges in specific cases, which show its ability to revolutionize transistor fabrication, photonic crystal structure development, and electronic defect analysis. At the end of the chapter, problems like limited data availability, scalability issues, and unreliability are talked about. Next, the authors look at the future possibilities for self-operating fabrication equipment and AI-quantum computing approaches that work together.

Shikha Chaudhary, Shobhit Institute of Engineering and Technology, India
Vineet Vishnoi, Shobhit Institute of Engineering and Technology, India
Shivam Chaudhary, Shobhit Institute of Engineering and Technology, India
Nidhi Tyagi, Shobhit Institute of Engineering and Technology, India
Prakhar Consul, IILM University, Greater Noida, India

On a global scale, issues of food security and safety are gaining attention to ensure a consistent supply of safe, nutritious food. Nanotechnology is applied in processes like developing new materials and improving food safety. Nanomaterials help detect contamination by microbes, chemicals, and insecticides. Nanosensors identify allergens and pathogens. Nanoencapsulation enhances food processing by transporting bioactive compounds, improving bioavailability, and extending shelf life. Materials such as nanoparticles, nanocapsules, nanoclays, quantum dots, dendrimers, and nanoemulsions are used in nano fertilizers and pesticides, as well as for promoting plant growth, genome alteration, and transgene expression. These possess antimicrobial properties, boost plant immunity, and transport active substances. Nanocomposites offer acid resistance, recyclability, thermal stability, and improved storage properties. This chapter points out the importance of nanotechnology and artificial intelligence (AI) in transforming agriculture, enhancing food packaging, detecting pathogens, and improving crop productivity and global food security.

Peptide-based therapeutics offer high specificity and safety but face challenges such as instability, rapid clearance, and limited bioavailability. Traditional discovery and formulation methods are slow, costly, and constrained by the vast peptide sequence space. Artificial intelligence (AI) is revolutionizing this field by enabling de novo peptide design, predicting bioactivity, and reducing reliance on trial-and-error approaches. At the same time, nanotechnology provides advanced delivery platforms (such as liposomes, lipid nanoparticles, polymeric nanoparticles, and micelles) that can encapsulate or functionalize peptides to enhance stability, prolong circulation, and enable targeted or controlled release. The convergence of AI and nanotechnology creates a synergistic framework: AI not only accelerates the discovery of novel peptide drugs but also optimizes nanocarrier design, improving delivery outcomes. This chapter highlights how this dual advancement in smart drug design and intelligent delivery paves the way for safer, more effective, and personalized peptide therapeutics.

This chapter explores advanced bloodstain detection and analysis using fluorescent nanoparticles, enhancing forensic investigations through improved sensitivity and

Foreword

The convergence of artificial intelligence (AI) and nanotechnology is more than a turning point in science and engineering; it is a reframing of what is possible. While nanotechnology grants us the tools to design and orchestrate matter at the atomic and molecular scale, AI provides the capacity to perceive hidden structure in vast design spaces, to reason across multivariate constraints, and to accelerate discovery cycles that once took years into weeks or days. Together, they introduce a new era of materials intelligence: one in which we not only search for better materials, devices, and manufacturing paradigms, but also systematically create them.

Leveraging AI and Nanotechnology for Materials, Devices, and Manufacturing, edited by Prof. Ashok Vaseashta and Prof. Ioan Stamatin, arrives at precisely the right moment. Around the world, researchers and practitioners are racing to develop high-performance materials for energy storage and conversion, resilient and secure electronics, adaptive biomedical interfaces, and sustainable, digitally orchestrated manufacturing. Yet progress in each of these domains has historically been constrained by the slow pace of experiment-design-characterize cycles and by the limits of human intuition in navigating combinatorial complexity. The chapters assembled here demonstrate, with clarity and conviction, how AI-guided nanoscience can overcome these bottlenecks—discovering novel chemistries, predicting properties from first principles and data-driven surrogates, and closing the loop between simulation, synthesis, and deployment. The editors, scientists with outstanding research in advanced materials and nanotechnologies, have curated a volume that is both comprehensive and forward-looking. The contributions trace a coherent arc from foundational methods to application-ready solutions, which include:

- At the materials frontier, how machine learning models accelerate the search for tunable alloys, 2D heterostructures, metamaterials, and functional nanocomposites—linking atomic-scale descriptors to macroscopic performance and reliability. Bayesian optimization, generative models, and active learning emerge as central tools to navigate immense design spaces with principled efficiency.

- In device engineering, how AI informs the co-design of architecture and fabrication, enabling sensors, photonic and plasmonic structures, neuromorphic elements, and flexible electronics whose behavior can be predicted and adapted in situ. The synergy between data-rich characterization (in optical microscopy, spectroscopy, and tomography) and physics-informed learning unlocks unprecedented visibility into failure modes and lifecycle dynamics.
- In manufacturing, how the fusion of digital twins, robotics, and nanoscale process control ushers in the era of self-optimizing production. Intelligent platforms integrate real-time metrology with reinforcement learning and control theory to achieve zero-defect fabrication, reduced waste, and verifiable sustainability metrics—essential pillars for resilient supply chains and responsible innovation.

What makes this book especially valuable is its insistence on depth without sacrificing practical application. Readers will find rigorous methodological treatments alongside case studies that connect algorithms to outcomes—whether improving cathode stability in next-generation batteries, enhancing selectivity in catalytic surfaces, tailoring interfaces for bioelectronics, or scaling nanomanufacturing with trustworthy AI-driven quality assurance. Equally important, several chapters address the governance, safety, and ethics of deploying AI-enabled nanotechnologies, including data provenance, reproducibility, model interpretability, and standards for human oversight. This responsible-innovation lens is vital: as our tools grow more powerful, so too must our frameworks for stewardship.

For students, this volume is a map of a rapidly evolving landscape. For researchers, it is a compendium of techniques and exemplars that invite replication, extension, and critique. For industry leaders and policymakers, it provides a blueprint for strategic investment, outlining where to build capabilities, how to integrate data and computing with laboratory infrastructure, and how to align incentives with societal benefits. Across all audiences, the message resonates: multidisciplinary collaboration—spanning materials science, data science, chemistry, physics, engineering, and manufacturing—is not optional; it is the engine of progress.

Prof. Vaseashta and Prof. Stamatin have made a landmark contribution. The chapters herein do not merely report advances; they outline a methodology for discovery in the age of intelligent matter—closed-loop workflows, interoperable data standards, and hybrid models that blend domain knowledge with statistical learning. If the 20th century taught us how to harness electrons and information, the 21st will teach us how to engineer matter with intention, guided by algorithms and reverse design that learn from every experiment and every fabrication run.

The present volume will inspire its readers to think bolder, design faster, manufacture smarter—and, above all, to build a future in which the reduced dimensions of matter and the most advanced forms of artificial intelligence work in concert for the greatest human good.

Nicolae-Victor Zamfir
Member of the Romanian Academy, Romania

Preface

The convergence of Artificial Intelligence (AI) and Nanotechnology is redefining the paradigm of materials research, enabling unprecedented precision in design, synthesis, characterization, and device integration. The history of materials science, typically driven by empirical discovery and incremental modeling, is now undergoing a transformative acceleration with the integration of data-driven intelligence and atomically precise engineering. In this landscape, AI functions as both a knowledge amplifier and an autonomous innovator, distilling vast, multidimensional datasets—ranging from quantum simulations to high-throughput experimental outputs—into actionable insights. Nanotechnology, on the other hand, provides structural and functional scaffolding at the atomic and molecular scales, where emergent physical, chemical, and electronic phenomena dictate device performance. The conventional discovery pipeline in nanotechnology relies on complex density-functional theory calculations, trial-and-error synthesis, and painstaking analysis of characterization results. In the new paradigm, graph neural networks and transformer-based generative models propose atomistic motifs, while reinforcement learning agents optimize multi-parameter synthetic routes in real time. Additionally, 4D-STEM or ptychographic TEM streams are interpreted on-the-fly by self-supervised computer-vision stacks. The result is a fast-feed-forward cycle capable of shrinking a decade of empirical exploration into months—or even days—of autonomous iteration.

This book seeks to illuminate how linking AI algorithms—spanning machine learning (ML), deep reinforcement learning, and generative approaches—with nano-fabrication and characterization platforms such as electron microscopy, scanning probe techniques, and synchrotron-based methods can establish a self-refining cycle of materials development. From accelerated inverse design of catalysts, semiconductors, and quantum materials to predictive synthesis routes and defect-aware device architecture, these coupled capabilities promise orders-of-magnitude reductions in design-to-deployment timelines. It offers insights into how these cutting-edge fields are revolutionizing material discovery, device fabrication, manufacturing processes, and education. In parallel, closed-loop laboratories that integrate robotics, AI orchestration, and nanoscale metrology are beginning to demonstrate the prospect of fully autonomous discovery engines.

The scope of this volume extends from the theoretical underpinnings of AI-aided material informatics to the practical challenges of nanomanufacturing fidelity, scalability, and interfacial control in nanostructured devices. It highlights not only technological progress but also the epistemological shift wherein hypothesis generation, validation, and optimization are increasingly distributed across human expertise and digital intelligence. The chapters herein aim to provide researchers, engineers, and technologists with a deep yet integrative perspective on how the fusion of AI and nanotechnology is shaping the next era of adaptive materials and intelligent devices. Ultimately, the objective is not merely to catalogue techniques but to inspire a re-imagining of the innovation pipeline—where the iterative cycle of design, synthesis, characterization, and deployment fuses into a seamless, high-throughput system governed by both human ingenuity and machine intelligence. The convergence explored in this work marks the emergence of a factual cyber-physical nexus for materials science, poised to accelerate breakthroughs across energy, healthcare, electronics, and beyond.

The focus then shifts to nanomanufacturing platforms that can respond to algorithmic decisions with angstrom-level precision. Atomic-layer deposition reactors, molecular-beam epitaxy systems, and robotic colloidal synthesis stations are now coupled to Lab-VIEW/Python APIs, permitting deep-reinforcement agents to vary temperature, precursor flux, or plasma power as they navigate composition–structure–property landscapes. Crucially, these "self-driving" labs close the loop with high-bandwidth, in-situ metrology—4D-STEM, synchrotron nanoprobes, scanning probe microscopes with real-time spectral imaging—whose data streams are parsed by convolutional vision transformers to detect defects, quantify domain orientations, and update the belief state of the optimizer.

Characterization is no longer an endpoint but a co-evolving node in a cyber-physical graph. Multimodal data fusion—linking diffraction contrast to X-ray absorption edges, Raman tensors, and pico-ampere IV curves—enables latent-space embeddings where compositional, structural, and electronic descriptors coexist. These embeddings feed uncertainty-aware decision engines that decide whether to refine a generative model, trigger a new synthesis run, or promote a candidate to device integration. For example, the pipeline can route a doped-tetragonal perovskite, predicted to exhibit a negative formation enthalpy and sub-50 meV polaron binding energy, directly into a nanophotonic LED architecture, while dynamically re-optimizing electrode patterning in the electronic-design-automation stack.

Device design itself is increasingly co-optimized with materials discovery. Physics-informed neural surrogates for Maxwell–Bloch, non-equilibrium Green's functions, and finite-element electro-thermo-mechanical solvers allow AI agents to propose nano-architectures—whether memristive crossbars, topological qubits, or plasmonic metasurfaces—whose fabrication tolerances are back-propagated into the synthesis space. The interplay of ab-initio descriptors, process-compatibility constraints, and system-level performance metrics defines a multidimensional Pareto front that would be inaccessible without machine intelligence. Throughout the volume, we emphasize data stewardship and interoperability: FAIR ontologies, Materials-Graph IDs, and streaming protocols that bridge institutional repositories such as OQMD, NOMAD, and the Materials Project with proprietary laboratory data lakes. Ethical considerations around algorithmic bias, sustainability of GPU-intensive workloads, and governance of autonomous experimentation are addressed explicitly, as they will shape how this technology scales beyond the laboratory.

The objective of this work is not merely to catalogue techniques but to articulate a cohesive framework in which stochastic generative AI, deterministic first-principles theory, and nanometer-scale fabrication coalesce into a single, recursively improving discovery engine. We envision readers employing these strategies to accelerate breakthroughs in solid-state batteries, neuromorphic hardware, quantum photonics, and heterogeneous catalytic platforms—domains where the tyranny of dimensionality once constrained progress. By intertwining algorithmic cognition with atom-level control, the field stands on the cusp of a phase transition: from human-limited exploration to autonomous, cyber-physical innovation. The target audience of this book will consist of professionals and researchers working in the fields of Materials Science, Materials Engineering, Nanoscience, Artificial Intelligence, Advanced Manufacturing, and various other related disciplines, e.g., Environment, Chemistry, Education, Administrative sciences and management, Sociology, and subjects related to Public Policies and Decision Support. Moreover, the book will provide insights into and support to executives concerned with the management of expertise, knowledge, information, and policy development. Additionally, the book will be handy to many researchers, graduate/postgraduate students, scientists, policymakers, industry professionals, and practitioners. Some chapters can be adopted as text supplements for independent study. We invite the reader to engage with the methods, adopt them critically, and help define the next epoch of materials science.

CHAPTER OVERVIEW

The book consists of twelve chapters that describe various aspects of the convergence of AI with nanotechnology. It is worth mentioning that, out of over 70 submissions for this book, only a limited number of chapters were selected to cover this rapidly evolving subject. The first chapter by Vaseashta and co-authors reviews recent progress and emerging pathways in the convergence of AI and nanomaterials science, highlighting their synergistic potential in advancing the development of novel materials. Key areas of focus include leveraging existing and new experimental datasets to elucidate structure–property relationships, improving synthesis processes by employing AI-based optimization strategies, and integrating AI with advanced characterization techniques to extract critical features for iterative materials design improvement. Several case studies are illustrated, describing how AI enhances predictive modeling, synthesis control, and property characterization across diverse classes of nanomaterials. For user-defined applications, AI facilitates reverse engineering, shifting nanomaterials research from a trial-and-error methodology toward a predictive, intelligent, and application-oriented framework. The chapter concludes with a glimpse of some forward-looking state-of-the-art and niche applications, as this convergence leads to the Fourth Science Paradigm of data-exhaustive discovery, while also addressing the ethical considerations associated with integrating AI into this rapidly evolving field.

The next chapter by Akhayere et al. presents the synthesis and designs of nanomaterials, nano membranes, and nano sponges using AI for advanced water treatment. The contamination of water and water bodies has reached an epic level globally, affecting both developed and developing countries. Researchers and scientific innovators are pushing their limits to ensure that more technologies leading to the development of efficient and adaptive materials for water treatment are actualized. Nanotechnology, particularly the use of nano-sponges, functional membranes, and advanced nanocomposites, is shown to be very promising due to their properties, surface areas, and ability to perform several functions that might seem complicated for conventional materials. One of the challenges is that the traditional experimental approach to material discovery and synthesis typically requires a significant amount of time and involves multiple trial-and-error processes. AI, specifically machine learning (ML), is an emerging powerful tool that can accelerate nanomaterial discovery, synthesis, and optimization, specifically for water purification. AI algorithms can be used to predict the relationship between the parameters for synthesis, the performance of the material and in turn help to optimize pore structures, adsorption efficiency, and membrane selectivity etc. in this chapter we explore a method that proposes to integrate AI into the design, synthesis, and deployment of nano-sponges, smart membranes, and hybrid nanomaterials for water treatment, highlighting case studies, simulation techniques, and predictive models. Converging AI with nanotechnology can facilitate the emergence of autonomous labs that optimize materials in real-time.

The next chapter by Kishore et. al., on predictive modeling for nanomaterials using AI-assisted design of experiments (DoE), discusses revolutionizing the discovery, design, and optimization of advanced materials with unprecedented speed and precision using AI-driven methodologies, such as predictive modeling, generative algorithms, and automated design of experiments, which enable the rapid identification of novel nanomaterials with tailored properties for diverse applications in energy storage, electronics, healthcare, and environmental sustainability. Traditional materials discovery relies on resource-intensive experimental approaches and iterative trial-and-error methods, often requiring years or even decades to bring new materials from concept to commercial use. However, AI accelerates this process by leveraging machine learning, deep learning, and big data analytics to predict material behaviors, optimize synthesis pathways, and identify promising candidates with minimal experimental intervention.

The next chapter by Ismail et al. is on the next generation of sustainable biocomposites through Machine Learning. This chapter presents an application of AI and nanotechnology for performance optimization and life cycle evaluation of sustainable materials. This shift is driven by global concerns over the depletion of fossil fuels, the rise in plastic waste, and the growing carbon footprint of products worldwide. Furthermore, integrating AI into life cycle analysis (LCA) processes has shown potential in identifying significant contributors to environmental impacts and quantifying uncertainties.

The next chapter by Salunkhe et al. is on the discovery of nanomaterials using AI-based technology accelerators. The chapter draws motivation from nanomaterial discovery, which has relied heavily on empirical experimentation and incremental improvement, and discusses computational models, machine learning (ML) algorithms, and automated experimentation platforms. AI-enabled platforms explore and optimize vast design spaces far more efficiently than traditional trial-and-error methods, serving as technology accelerators. The chapter describes the impact of using such technology accelerators in nanomaterials discovery.

While there exists a vast amount of knowledge, it is critical to educate and train the next generation in these areas. The next chapter by Dobica et al. discusses how to integrate AI and nanotechnology into science, technology, engineering, and math (STEM) education. The chapter presents curriculum development, case studies, and future trends. By exploring the integration of AI and nanotechnology in STEM education, the chapter highlights its potential to transform pedagogical approaches and prepare students for future technological progress. The chapter underscores the importance of embedding AI and nanotechnology into STEM curricula by reviewing recent scientific literature. It explores how AI can enhance teaching practices, assessment methods, and administrative tasks, thereby offering personalized learning experiences that boost student engagement and performance. Furthermore, the chapter highlights how nanotechnology can bridge the gap between theoretical knowledge and practical application, providing students with hands-on experience using advanced materials and techniques. An analysis is provided of the curriculum development frameworks, effective integration strategies, and the challenges and solutions related to implementing these technologies in education.

Continuing on educational aspects, a chapter by Upadhyay et al. describes several educational tools to integrate AI and Nanotechnology in future-oriented studies. The chapter provides a thorough analysis of key concepts in AI and Nanotechnology, making them easily approachable to maintain conceptual clarity for both students and educators. It examines the potential of digital solutions to increase engagement, develop an understanding of concepts, and bridge the gap between theoretical and practical knowledge, using interactive applications, virtual simulations, and adaptive learning platforms. The chapter presents a perceptive look at the issues confronting teachers and schools in teaching advanced technology-based syllabi, such as a lack of funds, uneven distribution of technological facilities, and the need to train instructors. Moreover, it also provides viable recommendations to policymakers, curriculum designers, and educators that they can use to promote equity, inclusivity, and successful learning.

Yet another application of AI is in waste reduction. A chapter by Pehlivan et al. describes the forecasting of E-Waste generation, the fastest-growing category of solid waste worldwide, using a Nonlinear Autoregressive Neural Network (NARNET). Recycling not only mitigates environmental degradation but also serves as a strategic approach to sustainable development. As a core element of the circular economy, recycling emphasizes minimizing ecological harm by recovering valuable materials from waste streams. This is reportedly the first study to estimate and forecast e-waste generation in Germany. Given the scale and complexity of e-waste generation, accurate forecasting is essential for informing policy, optimizing collection systems, and aligning recycling infrastructure with future demand.

The use of AI in nanoelectronics device fabrication is described next by Badiger et al. Traditional fabrication methods for nanoelectronics face several obstacles, including difficulties in achieving exactness, large-scale production, and managing manufacturing errors. AI-driven approaches resolve these limitations through powerful data analytics algorithms and optimized machine learning models, which upgrade device reliability and performance levels.

From an application standpoint, food safety and security are vital to almost everyone. A chapter by Chaudhary et al. describes the integration of Nano-enhanced and AI-driven Technologies for Food Security and Safety through Scientific Advances. Nanomaterials help detect contamination by microbes, chemicals, and insecticides. Nanosensors identify allergens and pathogens, and nanoencapsulation enhances food processing by transporting bioactive compounds, improving bioavailability, and extending shelf life. Materials such as nanoparticles, nanocapsules, nanoclays, quantum dots, dendrimers, and nanoemulsions are used in nano fertilizers and pesticides, as well as for promoting plant growth, genome alteration, and transgene expression. These possess antimicrobial properties, boost plant immunity, and transport active substances. Nanocomposites offer acid resistance, recyclability, thermal stability, and improved storage properties. This chapter points out the importance of nanotechnology and artificial intelligence (AI) in transforming agriculture, enhancing food packaging, detecting pathogens, and improving crop productivity and global food security.

The next chapter by Has et. al. describes peptide drug discovery with AI-assisted predictive modeling. The 21st century has witnessed a paradigm shift in peptide drug research with the rapid advances in structural biology, recombinant biologics, and modern synthetic and analytical strategies. Peptides share a significant proportion of today's pharmaceutical market. To date, more than 100 peptide-based drugs worldwide have been authorized by the Food and Drug Administration (FDA), and many more are in the pipeline. This chapter provides a comprehensive roadmap for researchers seeking to utilize nanotechnology and AI in peptide therapies. Lastly, a chapter by Rawat et al. presents a forensic application by presenting bloodstain detection using advanced fluorescent AI-assisted nanoparticles. The tool presented in this chapter has proven essential, acting as a valuable means for crime prevention and serving justice through its undeniable importance.

A book on the convergence of AI and Nanotechnology, such as this edited volume, has the potential to profoundly impact the public by demystifying two of the most transformative technologies of our time and illustrating how their convergence is reshaping everyday life. By translating complex scientific concepts into accessible language, such a book can foster greater public understanding of how intelligent nanosystems are driving innovations in healthcare, clean energy, environmental protection, and smart materials. It can inspire curiosity, promote informed dialogue about ethical and societal implications, and encourage the next generation to engage in interdisciplinary research. Ultimately, it empowers readers to appreciate how AI and nanotechnology together are not just advancing science but redefining the future of human progress and sustainability.

The editors wish to express their sincere appreciation to the University of Bucharest and its Research Institute for their invaluable support in the preparation and publication of this volume. The academic guidance, institutional resources, and collaborative spirit provided by both the University and the Institute have contributed significantly to the scholarly quality and integrity of this work. We are honored to be part of a community that fosters rigorous research and meaningful intellectual exchange.

Disclaimers: The opinions, recommendations, and propositions presented in various chapters represent the findings, views, and opinions of the editors and may not represent the position(s), views(s), and/or opinion(s) of their institutions. Also, the editors do not promote, endorse, or express their opposition against any product(s) that may have been mentioned in the chapters. Although due precautions are exercised before publication, the authors, however, do not take any responsibility for unintentional infringement of any material of either their own or from elsewhere. For all suggestions, corrections, or exchanges of information, the readers are advised to contact the authors directly.

Editors:

Ashok Vaseashta
International Clean Water Institute, USA

Ioan Stamatin
3NanoSAE Technology Transfer Center, University of Bucharest, Romania

Chapter 1
Leveraging AI Innovation in Nanomaterials, Synthesis, Characterization, and Device Design

Ashok Vaseashta
https://orcid.org/0000-0002-5649-0067
International Clean Water Institute, USA

Gor Gevorgyan
Scientific Canter of Zoology and Hydroecology, National Academy of Sciences of the Republic of Armenia, Armenia

Surik Kudaverdyan
https://orcid.org/0009-0001-5999-3399
National Polytechnic University of Armenia, Armenia

Gagik Ayvazyan
https://orcid.org/0000-0001-9245-6758
National Polytechnic University of Armenia, Armenia

Mane Khachatryan
National Polytechnic University of Armenia, Armenia

Cătălin Berlic
https://orcid.org/0000-0002-5569-6338
Faculty of Physics, University of Bucharest, Romania

Adriana Balan
https://orcid.org/0000-0003-3060-3761
Faculty of Physics, University of Bucharest, Romania

Ioan Stamatin
3NanoSAE Technology Transfer Center, University of Bucharest, Romania

DOI: 10.4018/979-8-3373-2883-6.ch001

ABSTRACT

Artificial intelligence (AI)-driven innovations in nanomaterials are reshaping traditional research paradigms, enabling accelerated discovery and design. This chapter reviews recent progress and emerging pathways in the convergence of AI and nanomaterials science, highlighting their synergistic potential in advancing the development of novel materials. Key areas of focus include leveraging existing and new experimental datasets to elucidate structure–property relationships, improving synthesis processes by employing AI-based optimization strategies, and integrating AI with advanced characterization techniques to extract critical features for iterative materials design improvement. Several case studies illustrate how AI enhances predictive modeling, synthesis control, and property characterization across diverse classes of nanomaterials. For user-defined applications, AI facilitates reverse engineering, shifting nanomaterials research from a trial-and-error methodology toward a predictive, intelligent, and application-oriented framework. The chapter concludes with a glimpse of some forward-looking state-of-the-art and niche applications, as this convergence leads to the Fourth Science Paradigm of data-exhaustive discovery, while also addressing the ethical considerations associated with integrating AI into this rapidly evolving field.

1. INTRODUCTION

Nanomaterials constitute a class of engineered materials whose structures and properties are tailored at the nanoscale, imparting functionalities that diverge significantly from their bulk counterparts (Vaseashta, 2005; 2006). Their unique mechanical, electrical, and optical characteristics—derived from nanoscale phenomena such as quantum confinement, surface-to-volume effects, and tunable morphologies—make them highly desirable for a wide range of applications. When functionalized appropriately, nanomaterials hold exceptional promises in biological systems, enabling disease detection, targeted therapy, and prevention at the cellular and molecular levels (Vaseashta & Bölgen, 2022). In research, both experimental and computational approaches are employed to optimize their properties, as experimental studies refine synthesis and processing conditions, while computational methods extract material descriptors to develop generalizable models for prediction and design (Santana et al., 2019; Al-Qattan et al., 2018; Stamatin et al., 2006).

Artificial intelligence, encompassing methods such as machine learning (ML), deep learning (DL), computer vision, and natural language processing (NLP), has emerged as a transformative tool in this domain (Brunette et al., 2009; Jordan & Mitchell, 2015; LeCun et al., 2015; Voulodimos et al., 2018; Chowdhary, 2020).

AI algorithms identify patterns in large, complex datasets, enabling automated reasoning, predictive modeling, and decision-making that traditionally relied on iterative experimentation. By combining hypothesis- and data-driven approaches, it is possible to obtain deep insights into complex structure-property relationships. In materials science, and particularly nanotechnology, AI reduces the need for trial-and-error synthesis, identifies optimal fabrication routes, and extracts hidden insights from characterization data. By conjoining ML with materials selection and optimization of synthesis parameters, material designs that were previously deemed improbable are being accelerated. Its convergence enhances control over nanomaterial size, shape, and defect formation, while simultaneously accelerating the discovery of novel material functionalities.

The convergence of AI and nanotechnology is accelerating at a faster pace and fostering new capabilities in three principal areas: material design, intelligent synthesis, and advanced characterization. In the initial design phase, AI leverages large-scale property databases to identify descriptors, correlations, and emergent patterns, while facilitating the creation of optimized nanomaterials with tailored attributes. During synthesis, AI-assisted optimization learns from experimental parameters to automate and refine production processes, significantly reducing time and cost. At the characterization stage, AI-powered computer vision and image analysis extract critical structural and morphological information from microscopy and spectroscopy data, providing deep insights into nanoscale features. These approaches accelerate material assessment, enhance predictive accuracy, and streamline the discovery-to-application pipeline.

Beyond accelerating existing workflows, AI enables reverse materials design, which involves articulating target device characteristics and reverse-engineering nanomaterials to meet those specifications. This paradigm shift supports application-driven research in areas such as biomedical devices, environmental sensing, and contamination remediation. It also aids niche areas, including strategic alternatives to rare earth elements (REE) for radar communications, unmanned aerial vehicles, and the protection of critical infrastructure. By embedding AI into every stage of the materials lifecycle—design, synthesis, and characterization—nanotechnology is poised to advance rapidly in the new data-driven era. AI-based algorithms will continue to play a pivotal role in transforming nanomaterials science from an experimental trial-and-error approach to a predictive, intelligent, and application-oriented innovation (Sha et al., 2020; Wang et al., 2019; Barnard et al., 2019). Whereas traditional research relies heavily on hypothesis-based experimentation, AI introduces a paradigm of data-assisted discovery, where patterns can reveal hypotheses rather than the other way around. Hence, the synergy highlighted above has profound implications for the scientific method. In nanotechnology, options of conjoining materials and processing conditions are nearly infinite; thus, this paradigm shift shortens the discovery cycle and democratizes access to innovation. The convergence of AI and nanotechnology, therefore, represents a synergy also known as the "fourth research paradigm" in materials science.

With the rapid expansion of data centers and the increasing deployment of AI-driven computational tools, researchers have initiated several projects to integrate AI with nanotechnology and other advanced platforms for materials design, synthesis, characterization, and device design. As conceptually illustrated in Fig. 1, such integration provides new pathways for directing AI toward user-defined applications. Table 1 highlights the current state of the art, which positions the contribution of the present work within this rapidly evolving field. The convergence of AI and nanotechnology carries implications that extend well beyond academic research, influencing global industries and societal systems. Although not the primary focus of this article, the subject raises several ethical concerns, particularly regarding the transparency of algorithmic decision-making. Limited interpretability can complicate risk assessment, especially in biomedical and environmental contexts where unintended outcomes may have profound consequences. Furthermore, disparities in access to computational infrastructure and AI expertise risk amplifying existing global inequalities in technological development. Addressing these challenges is essential for ensuring the responsible and equitable integration of AI into nanotechnology and will be discussed elsewhere.

Figure 1. Convergence of AI with nanomaterials leading to novel materials, improved synthesis, and characterization.

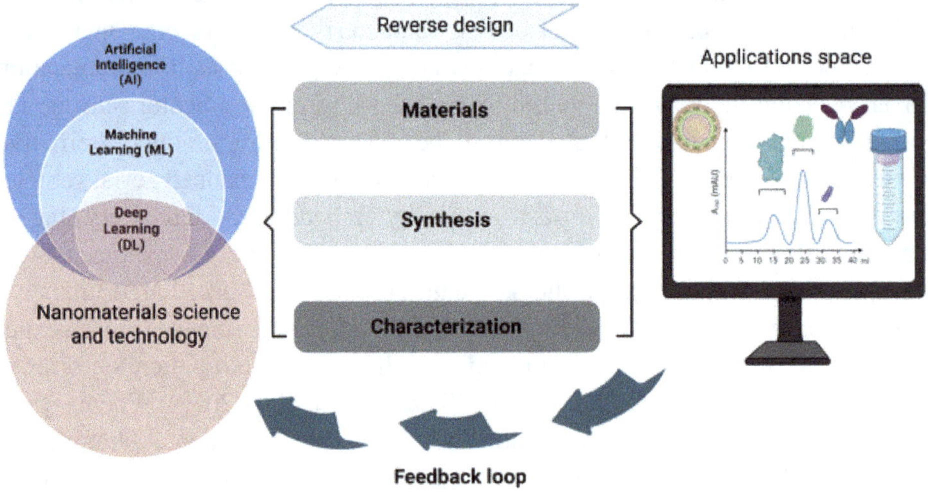

Table 1. State of the Art and Contribution of the Present Work

Work that exists already.	Contribution due to this work.
The convergence of artificial intelligence (AI) and nanomaterials research remains in its formative stage. The existing literature is sparse and primarily focuses on feature characterization and extraction, rather than broader applications. While prior initiatives such as the Materials Genome Project have demonstrated the potential of integrating AI, most efforts have been confined to augmenting or replacing traditional statistical approaches. Current datasets are relatively limited in scale and diversity, but their expansion is anticipated as high-throughput experiments, advanced simulations, and improved data-sharing frameworks continue to evolve.	This work constitutes the first comprehensive narrative review dedicated exclusively to the convergence of artificial intelligence and nanomaterials. This work outlines a structured analysis covering nanomaterials development, synthesis, and characterization, supported by illustrative case studies. The review also catalogs key databases—both publicly accessible and subscription-based studies that can serve as valuable resources for researchers. Furthermore, it highlights user-inspired applications that enable reverse engineering approaches for materials design and synthesis. Ultimately, the review addresses the ethical and societal implications associated with this rapidly evolving field.

2. LITERATURE REVIEW METHODOLOGY

This article was developed as a narrative review to provide a comprehensive and up-to-date synthesis of current knowledge on the convergence of AI and nanomaterials, emphasizing materials design, synthesis, and feature extraction through advanced characterization techniques. Within this framework, materials characterization serves as a critical feedback mechanism to refine design and material synthesis in an iterative, reverse-engineering cycle aimed at producing materials and devices with tailored properties. The methodological approach followed established international standards for narrative reviews, ensuring a systematic examination of literature on AI-enabled nanomaterials design, synthesis strategies, and characterization research.

2.1. Literature Search Strategy

A comprehensive literature search was conducted using Web of Science, Scopus, Academia, and Google Scholar, covering studies published between 2010 and 2025. Search terms combined conventional keywords and generative AI–related terminology, including nanomaterials, nanotechnology, AI, nanomaterial synthesis, nanomaterial characterization, ML, agent-based systems, and reverse engineering. Boolean operators (“AND”, “OR”) were applied to refine and expand the queries as appropriate.

Given the narrative scope of this review, a wide range of peer-reviewed publications written primarily in English were considered, with priority accorded to studies addressing AI-driven advances in materials development, synthesis, and characterization. Works that generated or utilized large datasets within these domains were also included. Applications were treated primarily as outcomes of such investigations rather than as the central focus. Publications reporting reverse-engineering approaches to improve materials design or synthesis pathways were likewise incorporated. In addition to original research articles, relevant reviews, case studies, and technical or policy reports from international agencies and expert groups were considered. Publications lacking substantive relevance to AI were excluded from core analysis. Titles and abstracts were screened independently by the authors, with full texts of potentially relevant studies retrieved for in-depth review. Extracted data included study type, analytical methods or systems investigated, and principal findings. Studies were subsequently organized into thematic categories encompassing materials development, synthesis, characterization, applications, ethical dimensions, data repositories, and emerging niche areas. Findings were synthesized narratively, with emphasis on cross-disciplinary integration. Special attention was given to works published from 2019 onward, reflecting the accelerated pace of innovation in this field.

3. AI FRAMEWORK IN NANOMATERIALS CONTEXT

Nanomaterials, with typical dimensions within the 1–100 nm size range, exhibit size-dependent physical and chemical properties that often deviate significantly from those of their bulk counterparts (Vaseashta, 2005; 2006; Vaseashta et al., 2005). Numerous examples demonstrate the remarkable functionalities that emerge from the onset of reduced dimensionality. Advances in synthesis techniques now enable the fabrication and study of diverse nanostructured systems, including nanotubes, nanowires, nanoparticles, and nanoporous materials. In such systems, electrons are strongly confined in one or more dimensions, leading to profound modifications of electronic, optical, and mechanical behavior. From a fundamental standpoint, reduced dimensions of nanosized materials corresponds to several fundamental physical quantities, viz: the electromagnetic field wavelength, the cyclotron radius in a magnetic field, the relaxation length of quasi-particles, the radius of the electron correlation length, the de-Broglie electron wavelength, the radius of Abrikosov or Josephson fluxon, and the effective path length of electrons, phonons, etc. The "phase coherence length", defined as the distance over which a particular electron in a conductor can exhibit quantum interference, is related to conductance fluctuations and localization. These quantum-mechanical, surface, and confinement effects result in properties that cannot be achieved from bulk material behavior, which explains why nanoparticles may exhibit enhanced catalytic activity, tunable behavior, high electrical conductivity, and mechanical strength (Vaseashta, 2005; 2006; Mekuye

& Abera, 2023). At the nanoscale, phenomena such as quantum confinement, high surface-to-volume ratios, and discrete electronic states govern behavior that requires extensive and complex computation. Given this complexity, AI offers a robust framework for addressing the nonlinear, multiscale, and quantum-mechanical calculations inherent to nanomaterials. Hybrid and physics-informed ML models, in particular, can integrate empirical data with fundamental domain knowledge, constraining predictions within physically realizable limits. This not only enhances model interpretability but also ensures scientifically plausible outcomes, thereby accelerating the logical design of novel nanomaterials and devices.

3.1. AI Terminology with Reference to Nanomaterials

Artificial intelligence is a field of computer science that focuses on creating systems or even machines capable of performing tasks that typically require human-level intelligence. Such tasks include learning, rationalization, problem-solving, understanding natural language, recognizing patterns, perceiving environments, and making routine to critical decisions. At its core, AI aims to simulate aspects of human intelligence through algorithms, models, and computational methods. Much of the current work is based on narrow AI (ANI), where systems are designed for specific tasks. General AI (AGI) will have the ability to understand, learn, and apply knowledge at a human conscious level, and Super AI (ASI) is believed to surpass human intelligence in virtually all aspects. Although AI can develop code for quite complex formulations in materials science, the anticipatory arrival of AGI and ASI will quite likely help reverse engineering these complex reduced-dimensional materials based on user-desired applications.

Machine learning is a subset of AI that enables computer systems to learn algorithmically from data, identify patterns, and make predictions or decisions without being explicitly programmed for every task. ML is increasingly transforming materials science by enabling the rapid analysis of complex datasets to uncover hidden patterns, predict material properties, and guide the discovery of novel compounds. By leveraging algorithms trained on experimental and computational data, ML can accelerate the design process, optimize synthesis conditions, and reduce reliance on costly trial-and-error approaches. Deep learning (DL) is a subset of ML that uses multi-layered artificial neural networks (ANN) to extract hierarchical features from large and complex datasets automatically (LeCun et al., 2015; Voulodimos et al., 2018). Unlike traditional ML methods, DL minimizes the need for manual feature extraction by learning representations directly from raw data. In materials science, DL is particularly relevant for tasks such as predicting material properties, classifying microstructures, analyzing microscopy images, and interpreting spectroscopic data. Its ability to handle large-dimensional and nonlinear data makes it a powerful tool for uncovering hidden patterns in the design, characterization, and processing of nanomaterials.

Supervised learning is a branch of ML in which algorithms are trained using labeled datasets, i.e., the input data is paired with known outputs (e.g., structure–property relations). The model learns to predict outcomes for new, unseen data based on these mappings (Anker et al., 2022). Unsupervised learning, by contrast, deals with unlabeled datasets and focuses on uncovering hidden structures, patterns, or groupings in the data without predefined outcomes (e.g., clustering nanomaterials by similarities in their features). In nanomaterials science, supervised learning is highly relevant for predicting material properties such as bandgap, Young's modulus, elasticity, or catalytic activity based on structural descriptors. Unsupervised learning is valuable for classifying nanomaterials, detecting anomalies in synthesis data, and discovering novel groupings of structures with potentially useful properties.

Convolutional Neural Networks (CNNs) are a subset of DL models designed to process grid-like data such as images. They use convolutional layers to automatically extract spatial features, making them effective for analyzing microscopy images, diffraction patterns, and spectroscopic maps in nanomaterials science. CNNs are particularly useful for tasks such as identifying defects, classifying microstructures, and automating image-based characterization of nanomaterials. Graph Neural Networks (GNNs), in contrast, are DL models designed to operate on graph-structured data, where nodes (entities) and their edges (relationships) can be explicitly represented. In nanomaterials science, atoms can be modeled as nodes and chemical bonds as edges, enabling GNNs to learn directly from material structures. This makes GNNs particularly relevant for predicting molecular or crystal properties, modeling interactions at the nanoscale, and discovering novel nanostructures with desired functionalities.

Reinforcement learning (RL) is a subset of ML in which an agent learns to make sequential decisions by interacting with an environment, receiving feedback in the form of rewards or penalties, and optimizing its strategy to maximize long-term rewards. Unlike supervised learning, RL does not require labeled datasets but instead learns through trial and error guided by reward signals. In nanomaterials science, RL is relevant for optimizing complex, multi-step processes such as material synthesis, self-assembly, and process control, where numerous variables and uncertain outcomes make traditional optimization complex. By enabling adaptive decision-making, RL accelerates discovery and facilitates the autonomous design of nanomaterials with tailored functionalities.

Lastly, Generative Adversarial Networks (GANs) are a class of DL models consisting of two neural networks—a generator that creates synthetic data and a discriminator that evaluates whether the data is real or generated. Through this adversarial process, GANs can produce highly realistic outputs. Variational Autoencoders (VAEs), in contrast, are probabilistic generative models that encode data into a latent space and then decode it back, allowing for both the reconstruction of existing data and the generation of new samples by sampling from the latent space. GANs can generate novel candidate nanostructures, microstructures, or materials images, aiding in exploring vast design spaces and augmenting training datasets, while VAEs can learn compact representations of nanomaterials' structures and properties, enabling interpolation between known materials to suggest new candidates with desired characteristics. Among several other methods, Bayesian Optimization (BO) (Frazier, 2018) is employed as an uncertainty or probabilistic optimization technique (Mukherjee et al., 2017) that balances exploration and exploitation to search for high-dimensional design spaces using limited data efficiently. Multi-fidelity Modeling (MFM) is an approach that combines data or simulations of varying accuracy and computational cost to improve predictions while reducing resource demands. Transfer Learning (TL) is an ML technique where knowledge gained from one domain or dataset is reused to enhance performance on a related task with limited data. Additionally, Digital Twins (DTs) are High-fidelity virtual replicas of physical systems that integrate real-time data, simulations, and predictive models to mimic and optimize system behavior.

4. NANOMATERIALS DESIGN

The Materials Genome Initiative (MGI) was launched as a federal multi-agency initiative to discover, manufacture, and deploy advanced materials twice as fast and at a fraction of the cost compared to traditional methods. The initiative was designed to create policies, resources, and infrastructure to support U.S. institutions in adopting strategies for accelerating the discovery, development, and deployment of advanced material systems, which are crucial for achieving global competitiveness. The MGI strategic plan outlines three goals to expand the initiative's impact in the coming years: (1) unifying the Materials Innovation Infrastructure (MII), an integrated framework of advanced modeling, computational, and experimental tools; (2) advancing the use of quantitative data to harness the power of materials information fully; and (3) educating, training, and connecting the materials research and development workforce. One of the critical pillars of the initiative is nanomaterials property optimization. A subcommittee of the initiative conducted a workshop on "Accelerated Materials Experimentation Enabled by the Autonomous MII" (AMII), and determined that the goals, viz., unify the MII framework, develop quantitative data, and educate the workforce, are the key elements in materials development, as they produce a large dataspace, which is essential to innovations. Nanomaterials research often requires several repetitive experiments and/or complex molecular dynamics simulations, while AI provides a transformative solution by enabling advanced data analysis and high-throughput screening.

Nanomaterial design typically involves three key stages: data preparation; development of mathematical models—often using complex neural networks that capture both linear and nonlinear relationships—to map features to target properties; and reverse design models that generate materials with desired characteristics, thereby streamlining and optimizing the overall design process.

4.1. Datasets Considerations in Nanomaterials Science

In nanomaterials research, high-throughput experimental data facilitates a crucial role in aiding exploration, thereby generating vast and complex datasets derived from experiments such as microscopy, spectroscopy, and bioassays, as well as modeling and simulations, often represented as large-dimensional feature descriptors. Noise and uncertainty are inherent to nanoscale measurements, and these challenges—together with limited dataset size, heterogeneity across modalities, and costly data acquisition—pose a significant challenge to traditional statistical modeling. For instance, atomic- and subatomic-scale imaging is susceptible to thermal drift, beam distortions, and operator-induced variability, leading to background interference or overlapping peaks that complicate the interpretation of materials properties and structures (Yang et al., 2022; Ji et al., 2021; Yan et al., 2020). AI addresses many of these challenges by employing denoising algorithms and multimodal data fusion, thereby improving analytical accuracy and reliability. A conceptual framework is shown in Fig. 2.

Figure 2. A conceptual framework illustrating the classification and training data sets.

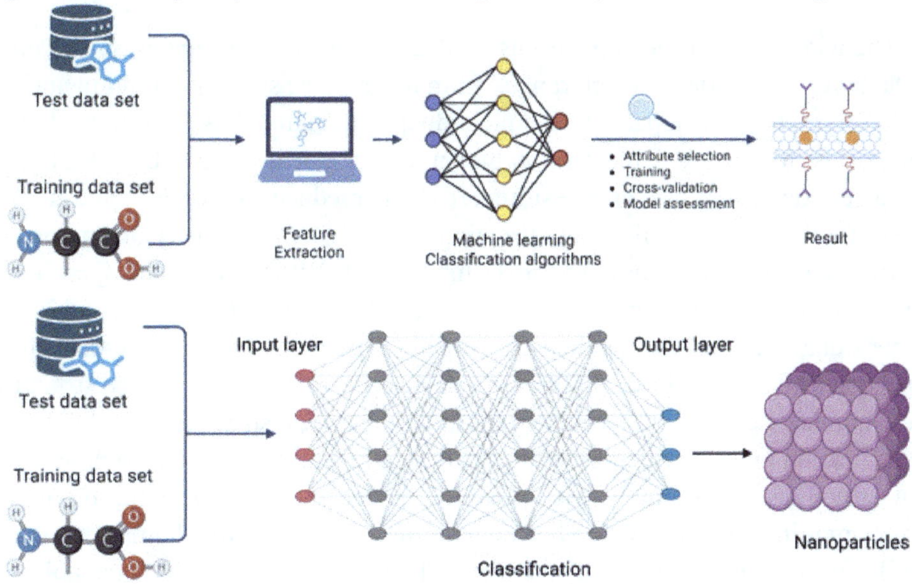

A key consideration in this field is whether data, physical principles, or a hybrid model should initiate modeling strategies. Data-based models use pattern recognition, such as CNNs, to classify nanoparticle morphologies from microscopic images—yet operate primarily on correlations without embedding physical constraints. Physics-based models, in contrast, explicitly incorporate domain knowledge, such as symmetries, conservation laws, or quantum-mechanical principles, thereby ensuring physical plausibility but sometimes at the expense of generalizability. Bridging these paradigms, physics-informed ML (PIML) integrates theoretical constraints into data-driven frameworks, producing models that combine predictive power with interpretability. This is particularly critical in nanotechnology, where purely data-driven approaches may yield nonphysical or "implausible" nanostructures. Emerging theoretical and applied perspectives highlight that both explainable AI (XAI) and PIML will be indispensable for the trustworthy deployment of AI in nanomaterials research. Several databases, specific to their applications, are mentioned in the case studies listed below. As regulatory interest in responsible innovation increases, requirements of transparency, interpretability, and physical consistency are becoming more fundamental to designing next-generation AI frameworks.

4.2: Nanomaterials Identification and Classification

The wide variety of nanomaterials produces a remarkable capability, as well as a challenge to this field, which ranges from formations, such as carbon nanotubes (CNTs), graphene, nanoparticles, semiconducting quantum dots, to complex hybrid composites. Most of these materials exhibit unique physical, chemical, and biological characteristics at the nanoscale, resulting from their reduced dimensions and atomic-level structures. This requires classification for cross-comparisons. Conventionally, classification has been done based on the morphology of these materials, specifically Zero-dimensional (0D), One-dimensional (1D), Two-dimensional (2D), and Three-dimensional (3D), as well as their functional properties. For emerging nanostructures consisting of a hybrid combination, such as partially functionalized structures with organic ligands, or hybrid organic-inorganic nanostructures, AI-facilitated classification plays a critical role due to its role in dealing with the complexity of nanostructures. Additionally, ML offers the power to extract meaningful patterns from large-dimensional data and categorize materials, tasks that humans may find challenging and not intuitive to perceive without the aid of computing capabilities. Instead of relying entirely on visual identification or categorical bins, AI-based classification is dynamic, feature-based, and scalable, ultimately supporting highly effective materials discovery platforms. Additionally, several models, particularly those using DL, are employed alongside AI modalities to enhance accuracy in analyzing their properties and structures. Such models include 2DMOINet (2D Materials Optical Identification Network), Seg Net (Segmentation Network), Mask-RCNN (Mask Regional Convolutional Neural Network), Artificial Neural Network (ANN), and ResNet (Residual Neural Network), for a variety of techniques, including analysis and characterization of nanomaterials. In addition to ML and DL, CNN models are typically trained to identify complex patterns in specialized images, primarily from scanning electron microscopy (SEM) and transmission electron microscopy (TEM).

For classification, several models are used for nanomaterials, based on the observed data. One such category of models is Supervised Learning, i.e., training these models on labeled data for classification based on learned patterns. The CNN models are trained to distinguish between different metals from TEM images and, similarly, classify graphene oxide (GO) derivatives based on CNT functionalization states using Raman spectroscopy. Supervised learning is accurate with clean datasets; however, given the noise in datasets, the models have their limitations. In contrast, unsupervised learning is exploratory in nature and can differentiate without any prior knowledge. Unsupervised learning employs various clustering modalities, including k-means clustering, hierarchical clustering, and Gaussian mixture models, which automatically classify nanomaterials based on their intrinsic similarities. This helps in categorization based on catalytic activity or chemical profiles, which may be apparent by characterization alone. Graphical neural networks (GNNs) are trained to decipher both the topological arrangement of atoms and their associated features, enabling these models to classify materials based on bonding environments and crystal families, which represents a significant success for classification.

4.3: Selected Case Studies of AI-assisted Nanomaterials Design

• In a study, GNN was used to classify 2D transition metal dichalcogenides (TMDs) by reflecting their coordination geometries, while also predicting key optoelectronic differences (Xu et al., 2024; Kumar & Kumar, 2024).

- In a study, Kajita et al. (2020) developed an ensemble descriptor method, combining ML with high-throughput screening, essentially to identify potential superionic conductors effectively from extensive material databases.
- Yao & Chen (2023) developed an AI pipeline for the classification of metallic nanoparticles using thousands of electron microscopy images. Their algorithm recognized particle shapes and also identified defect patterns and twinning that often correlate with catalytic behavior.
- Carbon nanomaterials present classification complexity, especially with fullerenes, CNTs, graphene, and amorphous carbons forming overlapping categories. Recent approaches used Raman spectral fingerprints combined with ML classifiers to distinguish between oxidized graphene, reduced graphene oxide, and pristine graphene (Mourdikoudis et al., 2018; Yang et al., 2024). These automated classifications enable high-throughput screening of carbon nanostructures.
- Mixed nanocomposites present challenges in classification since they blend multiple components. Recent studies used unsupervised clustering techniques to classify polymer–nanoparticle composites based on scattering pattern data, revealing structural motifs correlated with mechanical properties. In doing so, the classification layers could form design strategies for the next-generation nanocomposites.
- In yet another study, Yamankurt et al. (2019) identified roughly about 1,000 spherical nucleic acid (SNA) candidates by systematically varying design parameters within defined ranges and evaluating their activity using a high-throughput screening protocol. The resulting dataset was then used to train an ML model capable of predicting the activity of new SNA formulations, enabling systematic and independent optimization of design variables to enhance performance.

- Defects are critical determinants of nanomaterial properties and can be classified into various types depending on their structural nature. However, their analysis remains challenging and time-intensive, often requiring advanced characterization techniques. Interestingly, defect-engineered nanomaterials can exhibit superior optical and physicochemical properties, including enhanced redox activity. In one study, decision tree and logistic regression models were applied to defect-related datasets to elucidate their impact on metal–organic framework (MOF) materials (Wu et al., 2020). DL has further enabled the rapid identification and classification of defects from microscopic images, accelerating defect analysis protocols tremendously. Collectively, these approaches have advanced the understanding of defect formation and structural transitions in nanomaterials, providing valuable guidance for the rational design of nanomaterials with tailored functionalities.

Several models stated above improve identification and analysis. These examples demonstrate the utility of ML and high-throughput data in exploring the design of new materials. With additional experimentation, training, and descriptors, these principles can be applied to high-throughput screening of new and advanced nanomaterials, including for biomedical applications (Chen et al., 2022).

5. AI AND NANOMATERIALS INTELLIGENT SYNTHESIS

One of the crucial challenges in nanotechnology is the controlled and reproducible synthesis of nanomaterials with precise size, shape, composition, and surface chemistry. These features are susceptible to multiple interdependent parameters—temperature, pressure, reactant concentration, pH, solvent type, reaction time, and mixing dynamics—where even small perturbations can drastically alter material properties and reproducibility. Conventional optimization strategies have primarily relied on trial-and-error experimentation, guided by researchers' intuition and expertise. While effective in some cases, this approach is laborious, resource-intensive, and often incapable of revealing non-intuitive synthesis pathways. To address these limitations, statistical design-of-experiments (DOE) techniques, such as the Taguchi method (Freddi & Salmon, 2019) and response surface methodology (RSM)in DOE (Kumar & Reji, 2023), have been widely adopted. The Taguchi method employs orthogonal arrays and signal-to-noise ratios to systematically study factor interactions and identify robust synthesis conditions, even under environmental variability. RSM extends this framework by constructing empirical models to approximate parameter–response surfaces, enabling multi-objective optimization of synthesis yield, particle uniformity, and defect minimization.

Recent advances in AI offer complementary methods and a set of more powerful strategies (Sha et al., 2020; Barnard et al., 2019). ML methods, such as random forests (RF) and support vector machines (SVM) (Noble, 2006), can identify nonlinear relationships between synthesis parameters and material properties, while Bayesian optimization (BO) efficiently explores large-dimensional parameter spaces by balancing exploration and exploitation. DL approaches, such as CNNs, add further capabilities which can analyse microscopy images to correlate morphology with synthesis conditions; GNNs capture atomic- and molecular-scale structural descriptors to predict stability and reactivity; and recurrent neural networks (RNNs) can model time-dependent reaction pathways. In addition, VAEs and GANs are increasingly applied to propose novel synthesis routes and material candidates by learning latent representations of structure–property spaces. Collectively, these methodologies move beyond heuristic experimentation toward data-driven discovery. By integrating DOE frameworks with ML and DL models, researchers can iteratively refine synthesis conditions, predict physicochemical properties even before synthesis, and identify efficient, innovative pathways for nanomaterial fabrication at both laboratory and industrial scales.

For continuity, it is essential to define a few additional terms before describing the application of AI in nanomaterial synthesis. AI is used for predictive modeling of nanomaterials, while ML and DL are utilized for analyzing nanomaterial properties, synthesis, and design, which involve vast datasets that enable the estimation of functional properties of nanomaterials. Especially for biological systems, ML techniques analyze properties such as cytotoxicity based on structure, composition, surface charge, and dimension. Such techniques also include DL, Support Vector Machines (SVM), and random forest (RF) methods, which are a combination of ML models that work together as decision trees, and are trained on random samples of data and features. The final prediction is made by averaging the decisions of individual datasets, resulting in a robust and accurate outcome. These models have been used for multiple applications, including predicting the toxicity of metal oxide nanoparticles with varying compositions. SVMs, on the other hand, are highly effective and versatile learning algorithms that are used for classification and regression tasks. They classify and predict properties with high accuracy by identifying a hyperplane in the data, which aids in the design of complex materials, including complex nanocomposites.

ML is increasingly applied to synthesis optimization, where it enhances material properties by extracting insights from limited and often heterogeneous experimental datasets. For example, ML algorithms have been employed to optimize the synthesis of carbon nanotubes (CNTs) and quantum dots (QDs), improving their quality and reproducibility for use in optoelectronics and bio-imaging. Vaseashta et al. further demonstrated the application of ML-based optimization to maximize the efficiency of energy harvesters, with implications ranging from critical infrastructure protection to advanced biomedical devices (Batra et al., 2023; Palwai et al., 2022; Vaseashta & Batra, 2022; Noorali et al., 2025). Beyond individual case studies, ML accelerates the materials discovery pipeline by integrating predictive modeling with automated experimentation. Algorithms such as Gaussian process regression, Bayesian optimization, and reinforcement learning guide experimental design by efficiently navigating vast parameter spaces, drastically reducing the need for exhaustive trial-and-error. Coupled with AI-assisted robotics, these approaches enable self-driving laboratories that perform closed-loop, high-throughput experimentation—synthesizing, characterizing, and iteratively refining material candidates in real time. This synergy between ML and automation shortens discovery cycles from years to months while enabling exploration of previously inaccessible regions of chemical and structural design space.

In property prediction, AI and ML utilize diverse modeling frameworks, including SVMs, RFs, and deep neural networks, which, with architectures inspired by biological cognition, employ interconnected layers of nodes to learn complex, nonlinear relationships in large-dimensional data. Trained in extensive chemical and materials databases, these models capture nanoscale molecular interactions and emergent phenomena, providing predictive capabilities crucial for healthcare. Additional applications include drug discovery, personalized medicine, precision diagnostics, proteomics, genomics, and the prediction of adverse drug interactions.

A transformative paradigm within this domain is the inverse design methodology, which shifts materials research from empirical exploration to performance-oriented optimization. Instead of randomly screening candidates, inverse design employs generative models, evolutionary algorithms, and variational autoencoders to tailor structural, chemical, and morphological parameters toward desired functionalities. This enables the rational creation of nanomaterials with application-specific properties—for example, tuning bandgap energies for optoelectronics, optimizing surface functionalization for targeted drug delivery, or engineering catalytic sites for energy conversion. Collectively, these AI- and ML-enabled tools—ranging from synthesis optimization and predictive modeling to inverse design and autonomous experimentation—are converging to redefine nanomaterials research. By bridging theory, computation, and experiment, they provide a data-centric framework for accelerating discovery while ensuring that materials are designed with precision, scalability, and application relevance in mind.

5.1. Selected Case Studies of Nanomaterials Synthesis

AI, especially ML and reinforcement learning (RL), offers transformative potential by enabling predictive synthesis, rapid optimization, and autonomous experimentation. AI's strength lies in detecting complex, non-linear relationships embedded within multi-dimensional reaction conditions and linking them to desirable nanomaterial properties. ML models are trained on past synthesis data and hence can predict outcomes such as particle size distribution, crystallinity, or yield based on input parameters—even when experimental data are sparse or noisy. For example, regression models and neural networks forecast optimum temperature and reactant ratios necessary for growing QDs with specific optical properties. Beyond prediction, RL has been employed to create adaptive synthesis platforms where agentic AI iteratively proposes experimental conditions, receives feedback on results, and progressively recalculates strategies to optimize outcomes. When paired with automated robotic synthesis systems, these "self-driving or autonomous laboratories" reduce human intervention and dramatically shorten experimental cycles. As described later in characterization, the synthesis conducted by automated labs enhances electronic properties, primarily by controlling process parameters, and predicts parameters such as electron affinity, Fermi level energy, electronic band gap, and ionization potential of materials.

- To predict electronic properties of graphene and investigate the intrinsic relationships between nanoscale features and graphene electronic properties, an AI model quantitative structure-property relationship (QSPR) was developed using various ML methods (Huang et al., 2022; Gao et al., 2021), including the multiple linear regression (MLR), decision tree (DT), K-nearest neighbors (kNN), ANN, and SVM methods.
- Silver nanoparticles were analyzed using ML algorithms, k-means, logistic regression (LR), random forest (RF) modeling, principal component analysis (PCA), and a three-layer ANN, which successfully mapped their structural, morphological, and topographical features (Mekki-Berrada et al., 2021).
- ML can also predict the thermal conductivity of nanomaterials and is used to design high-performance thermal interface materials. ML models accurately predicted the interfacial thermal resistance between graphene and hexagonal boron-nitride using system temperature and coupling strength between layers as inputs (Zhu et al., 2024; Yang et al., 2018).
- AI models developed by Esteva et al. (2019) leveraged large datasets of nanoparticle formulations, biological targets, and therapeutic outcomes to predict effective nanocarrier designs with minimized off-target toxicity.

These predictions guided formulation strategies, which were subsequently validated experimentally, reducing development time by 30%.

- In ongoing research on energy harvesters, DL has been employed to identify optimal processing and material parameters that maximize piezoelectric coefficients, as shown in Fig. 3. Calculations of the figure of merit demonstrate that composite systems outperform single-phase materials, primarily due to their ability to achieve compositions that balance structural stability with enhanced electromechanical coupling (Batra et al., 2023; Ramprasad et al., 2017). These advances extend the applicability of piezoelectric harvesters across a broad spectrum of domains, ranging from critical infrastructure monitoring and protection to biomedical sensing and therapeutic devices.

Figure 3. Use of ML/DL in optimizing energy harvester performance.

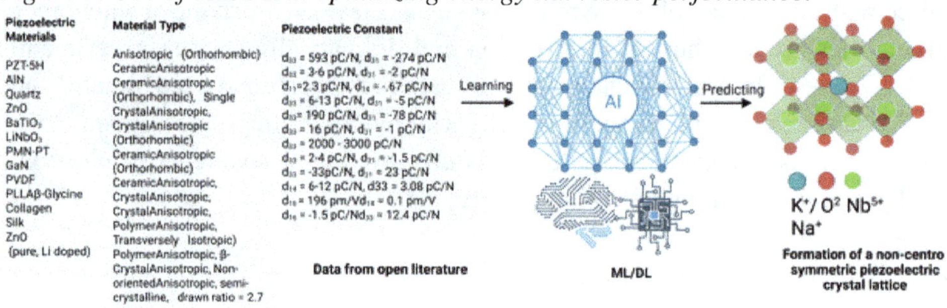

- Using ML, cell shape index (CSI) and nuclear area factors (NAFs) were used to decode the influence of nanomaterial properties on their interaction with cells, as nanotoxicity descriptors (Singh et al., 2021).
- Yao & Chen (2023) proposed a novel inverse-design framework based on RF multi-objective regression, demonstrating its capability to predict structural features from desired functional properties. By leveraging algorithms designed to handle multiple objectives concurrently, their approach systematically navigated the large-dimensional design space of multifunctional nanoparticles, enabling the identification of optimal structural parameter combinations designed for specific performance requirements.
- In a DT algorithm to analyze the immune response to nanoparticles (Thomas et al., 2014), particular emphasis was placed on physicochemical properties—such as size and zeta potential—that influence activation of the complement system. Elucidating these relationships is critical for designing safe and bio-

compatible nanomaterials, as it provides a framework for tailoring nanoparticle characteristics to mitigate adverse immune reactions.

- ANN was used in a study (Boso et al., 2011) to model the adhesion of fluorescent polystyrene nanoparticles to ampule walls as a function of shear rate and particle diameter, enabling reverse engineering of nanoparticle size. Their findings revealed the existence of an optimal particle size that maximizes surface adherence, suggesting a mechanism by which nanoparticles can be engineered to enhance accumulation at contaminated sites.
- In a study, Chan et al. (2010) proposed a reproducible and high-throughput synthesis method for forming colloidal nanocrystals, offering a systematic way to explore multidimensional parameter spaces efficiently (Chan et al., 2010). This approach enables the optimization of the synthesis process, producing nanocrystals with the desired properties and ensuring reproducibility and scalability.
- Colloidal quantum dots (CQDs) require extremely narrow size distributions to ensure uniform optical and electronic behavior, yet achieving this remains a major synthesis challenge due to the multitude of interacting parameters. A recent study demonstrated that incorporating ML in the loop can enhance monodispersity by leveraging existing experimental data to identify optimal synthesis conditions. The ML model not only suggested promising parameter combinations but also mapped regions of the synthetic parameter space that led to record-level monodispersity in PbS quantum dots (Yamankurt et al., 2019).
- In catalytic converter production, particle morphology governs activity and stability. A study presented a framework for integrating economic model predictive control (EMPC) and RL for online model parameter estimation of a class of nonlinear systems (Alhazmi et al., 2022). In this framework, EMPC optimally maintains a closed-loop system while maintaining its stability and recursive feasibility. To optimize the process, the RL agent continuously compares the measured state of the process with the model's predictions (nominal states) and modifies model parameters accordingly.
- Yoshihara et al. (2023) used ML to predict ideal chemical vapor deposition (CVD) parameters for synthesizing large area defect-free graphene sheets. Their predictive model demonstrated the formation of a large-area graphene domain on the Cu surface using the CVD conditions determined by the constructed ML model, which provided faster graphene growth compared to previously reported strategies.
- Microbial synthesis represents an eco-friendly and cost-effective approach to nanomaterial production (Fariq et al., 2017; Grasso et al., 2019; L et al., 2025; Sidorenko et al., 2022). Given the complexity of biochemical consortia

(L et al., 2025) and reaction mechanisms involved, it is critical to investigate the biosynthetic capabilities of microorganisms and establish strategies to regulate and control these processes for enhanced efficiency and reproducibility. Yu et al. demonstrated the use of ML to analyze and interpret the biosynthesis patterns of gold (Au) nanoparticles generated by different microbial species (Yu et al., 2022). By training an ML algorithm on datasets containing biosynthetic profiles and corresponding taxonomic information, the system successfully identified and classified previously unknown microbial samples based on the Au-nanoparticle synthesis process, highlighting its potential for diverse applications.

- Nikolaev et al. reported an innovative AI-based Autonomous Research System (ARES) system, leveraging ML to design and execute experiments (Nikolaev et al., 2016). The system optimized the synthesis process of single-walled CNTs by iteratively adjusting experimental parameters autonomously, demonstrating the potential of AI-driven autonomous platforms in accelerating nanomaterial discovery and fabrication.
- In a study aimed at controlling the synthesis of gold nanoparticles with defined sizes and shapes (Ma & Li, 2018; Guda et al., 2023), researchers investigated the effects of different gold precursors, reducing agents, and surfactants on nanoparticle formation. Using a modified Latin hypercube sampling strategy, experimental points were systematically selected within the response parameter space to vary the concentrations of $HAuCl_4 \cdot 4H_2O$, ascorbic acid, and hexa-decyltrimethylammonium bromide (CTAB). This approach facilitated the controlled synthesis of gold nanoparticles with either spherical or elongated morphologies. The work provides valuable insights into the complex interplay among reaction parameters and demonstrates how rational experimental design can enable the tailoring of nanomaterial synthesis for precise and reproducible outcomes.

5.2. Predictive Modeling and Optimization

Early applications of ML in nanomaterial synthesis primarily relied on supervised learning, where models predicted material properties based on reaction inputs. For instance, neural networks have been used to correlate synthesis parameters with catalyst particle sizes (Vaseashta & Batra, 2022). Similarly, SVMs and RFs have optimized chemical vapor deposition (CVD) conditions for graphene growth, outperforming linear regression and empirical heuristics by revealing hidden process variabilities. Despite these advances, predictive accuracy remains constrained by the limited availability and uneven quality of experimental data, which are often fragmented across laboratories with differing protocols. To address this challenge, recent work has adopted active learning approaches, where AI models not only predict outcomes but also identify knowledge gaps and propose targeted strategies. This iterative optimization strategy conciliates a balance between exploration and exploitation, enabling efficient data acquisition while minimizing experimental overhead.

5.3. Reinforcement Learning and Autonomous Laboratories

Reinforcement learning uses a feedback-driven framework in which an AI agent adjusts decisions based on rewards or penalties derived from experimental outcomes. This trial-and-error paradigm, executed at machine speed, has been shown to accelerate materials discovery far beyond human capability. When coupled with automated robotic platforms, RL enables "self-driving laboratories" capable of fully autonomous fabrication and testing of nanomaterials. MacLeod et al. demonstrated a system that integrates AI-controlled synthesis, in situ characterization, and real-time feedback to autonomously discover thin-film materials with superior electrical properties, reducing discovery timelines from months to a few days (MacLeod et al., 2020).

5.4. Current Challenges in AI-driven Synthesis

Despite significant advances, several challenges remain in AI-driven synthesis. First, the data limitations pose a considerable barrier, as synthesis processes are highly decentralized, with some are shared, and high-quality datasets are scarce. Variability in equipment and protocols further constrains model generalization. Scalability is yet another obstacle, as lab-scale successes often prove inadequate to directly translate to industrial-scale synthesis without additional model refinement and robustness testing. Interpretability also remains problematic, since models that appear as black boxes obscure the underlying chemical mechanisms, limiting both scientific trust and regulatory acceptance. Hybrid, physics-informed AI frameworks offer a promising pathway to bridge this gap. Among these challenges, reproducibility is particularly critical for advancing biomedical nanomaterials from laboratory research to market-ready applications. To address this, researchers are developing controllable synthesis strategies, with microfluidic platforms emerging as one of the most effective approaches (Ma & Li, 2018; Ma et al., 2017). Coupled with real-time monitoring and AI/ML-driven optimization, such methods provide powerful tools to enhance reproducibility and enable precise, scalable nanomaterial synthesis.

6. AI IN NANOMATERIALS CHARACTERIZATION

Information extraction from nanomaterials' characterization is a valuable tool that enables researchers to gain a deeper understanding of materials and uncover hidden patterns and features within them, which is critical for advancing nanoscience and engineering (Anker et al., 2022; Han et al., 2020; Lin et al., 2022; Mahjoubi et al., 2023). The information about the size, shape, surface, morphology, and other characteristics of nanoparticles is crucial for understanding structure–property relationships, though electron microscopy, scanning probe methods, spectroscopy, scattering, and tomography have produced vast volumes of multi-dimensional and heterogeneous datasets. Yet these datasets frequently contain noise, incomplete sampling, measurement artifacts, and human subjectivity in interpretation, creating a persistent problem in extracting reliable insights. AI, encompassing ML and DL, is increasingly recognized as a transformative approach to mitigate these challenges by automating feature extraction, accelerating data interpretation, and enabling predictive modeling beyond the limits of human intuition. AI's impact on nanomaterial characterization is enabled by, (a): capacity to integrate multimodal datasets—ranging from images and spectra to simulations that enables the discovery of subtle correlations often overlooked by conventional statistical methods, (b): real-time and scalable analysis by enabling adaptive data acquisition, dynamic feedback, and optimization of measurement protocols, and (c): knowledge transfer across modalities and length scales, bridging atomic-resolution measurements with mesoscale behavior, and advancing the development of digital twins of nanomaterials with predictive capability.

For example, the CNN enhances microscopy by denoising, super-resolving, and segmenting atomic-scale images, while GNNs capture the non-Euclidean topologies intrinsic to nanomaterial structures. Generative models—such as VAEs and GANs—reconstruct partial or corrupted datasets and generate hypothetical nanostructures consistent with physical constraints. RL and BO further extend these capabilities by guiding adaptive sampling strategies and optimizing instrument control, thereby focusing data acquisition on the most informative data. These methods collectively are transforming characterization from a passive, descriptive activity into an active, predictive, and increasingly autonomous paradigm. The convergence of AI with nanomaterial characterization methodologies represents a paradigm shift rather than an incremental improvement. It opens pathways to explore phenomena inaccessible to human-driven analysis, accelerates materials-by-design strategies, and lays the foundation for autonomous laboratories that integrate materials design, synthesis, and characterization. Additionally, features of nanomaterials, such as quantum confinement, surface reactivity, and morphological complexity, demand characterization with exceptional spatial and temporal precision. Techniques including TEM, SEM,

Atomic Force microscopy (AFM), scanning tunneling microscopy (STM), Raman spectroscopy, and X-ray photoelectron spectroscopy (XPS) are routinely employed to examine these properties. Due to the complexity and scale of data generated, AI is used for its capacity in handling multi-dimensional data and recognizing complex patterns.

6.1. Case Studies of AI and Optical Microscopies

SEM is a valuable tool for obtaining high-resolution surface morphology and topological information of nanomaterials, such as their structure, shape, and size (Vladár & Hodoroaba, 2020). As the volume of SEM data increases with the continuous development of technology, the analysis capabilities need to be enhanced accordingly. Several studies have leveraged DL models to automate and improve the analysis of SEM images.

- In a study (Aversa et al., 2018) researchers identified a large dataset of SEM images and used a DL model to classify them into categories, such as 0D particles, 1D nanowires, 2D films, and 3D devices.
- A study proposed an intelligent classification model using the VGG-19 deep network (Simonyan & Zisserman, 2015), and SVM model (Noble, 2006) to classify the shape types of nanoparticles in SEM images, with 97% accuracy (Dahy et al., 2023).
- A study developed an ML-based computer vision algorithm to quantitatively extract particle size, size distribution, shape, and core-shell structure information from SEM images (Kim et al., 2020). This method enables automated and high-throughput measurements, even in the presence of overlapping nanostructures.

In general, the convergence of AI with characterization results from SEM and TEM images can be achieved more quickly and accurately by conforming to a design process in steps, including data set construction, training set calibration, model training, and model verification. This creates uniformity in image interpretation.

Optical absorption spectroscopy is yet another vital characterization method for analyzing the interaction of light with materials and extracting information about electronic and molecular properties. However, the optical absorption peaks in the nanomaterials' spectrum are challenging to identify chemical constituents from absorption spectra unless special control measures, such as high-resolution separation (Chen et al., 2023) were used.

- In a study, the researchers achieved a good match between the predicted results and the experimental results by using an ML on a one-dimensional CNN model with a low mean absolute error value from the Visible absorption spectrum of metal nanoclusters (Knoppe et al., 2011). This work enables the identification of nanomaterials from their optical properties with molecular precision as a rapid, high-throughput characterization method, as conceptually shown below in Fig. 4.

Figure 4. Deep convolutional neural network-based morphology classification and feature extraction from SEM images.

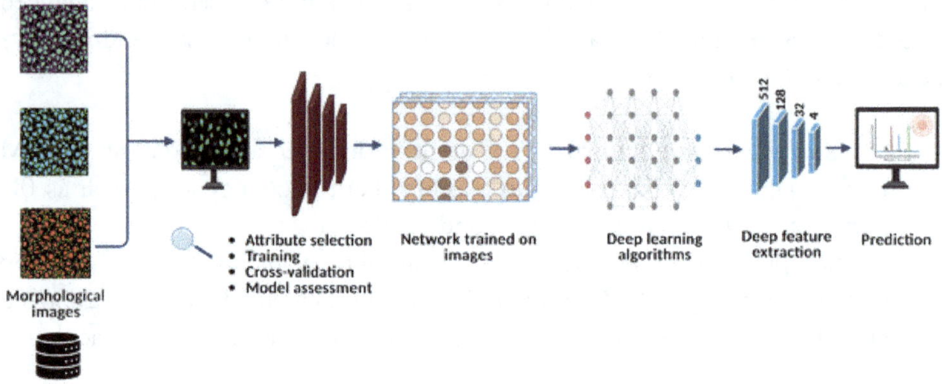

6.2. Selected Case Studies of Nanomaterials Characterization

- Groschner et al. (2021) developed a flexible two-step pipeline for the analysis of high-resolution TEM data, which uses a U-Net for segmentation followed by an RF for the detection of stacking faults. Using these segmented regions, it was possible to classify nanoparticles that contain a visible stacking fault with 86% accuracy. The combined output of the segmentation network and classifier offers a statistical distribution of features of interest, such as size, shape, and defect presence, enabling correlations between these features.
- Defects greatly influence the electronic and optical properties of 2D materials like graphene and transition metal dichalcogenides (TMDs). Several studies leveraged CNN models to analyze large sets of TEM images and to identify defects, adatoms, and grain boundaries with remarkable precision (Feng et al., 2022; Bhatt et al., 2022; Meyer, 2021). This automated defect mapping

facilitates rapid correlative studies between defect density and material performance in electronic devices.

- Several multi-modal methodologies integrate experimental imaging, spectroscopic analysis, and graph-based computational modeling to guide the rational design of nanocatalysts. These experimental descriptors are then encoded into a GNN framework with nodes and edges. The GNN is trained to learn correlations between atomic-scale structural motifs—such as step edges, vacancies, or specific alloying arrangements—and measured catalytic performance metrics (Ding et al., 2024; Price et al., 2022).
- AI-driven Design of Experiments (DoE) is transforming catalyst discovery, significantly reducing experimental workloads while improving performance and scalability. A study explores how AI-driven DoE was successfully applied to nanocatalysts discovery, explicitly focusing on the development of platinum-free hydrogen evolution reaction (HER) catalysts for green hydrogen production (Lao & You, 2024). By integrating ML, BO, and high-throughput experimentation, researchers have accelerated the identification of cost-effective, high-efficiency catalytic materials.

While AI has shown remarkable promise in materials research, several barriers remain. A major challenge is data scarcity and bias, compounded by the difficulty of data labeling and curation. Accurate annotation of microscopy and spectral datasets is labor-intensive, and inconsistent or noisy labels can significantly reduce model reliability. Generalizability is another limitation, since models trained on data from specific instruments, laboratories, or material systems often prove inadequate to perform robustly across different conditions. Furthermore, the interpretability of AI models in complex characterization tasks remains limited, undermining both scientific insight and trust in predictions. Many classification approaches are built on small, highly constrained datasets, making them vulnerable to overfitting, underperforming, and having poor transferability (Horton et al., 2025; Dunn et al., 2020; Jain et al., 2018). The reliance on black-box architecture exacerbates the problem, as models typically output classifications without transparent reasoning—this is problematic in scientific contexts where understanding causal mechanisms is essential. Compounding this challenge, nanomaterials evolve dynamically under external stimuli such as temperature or chemical environment, meaning models trained only on static 'snapshots' may miss transient yet performance-critical phenomena (Ramprasad et al., 2017). Finally, the heterogeneity of data modalities—spanning microscopy, spectroscopy, and simulation—necessitates robust multimodal fusion strategies, which remain an active area of research.

7. APPLICATIONS CONJOINING AI AND NANOTECHNOLOGY

As discussed in earlier sections, AI is increasingly integrated with nanotechnology to accelerate the design, synthesis, and characterization of nanoscale materials and devices. ML algorithms enable predictive modeling of nanostructure–property relationships, guiding the rational design of nanomaterials with tailored electronic, catalytic, and biomedical functionalities. DL enhances high-throughput analysis of microscopy and spectroscopy data, automating defect detection and feature extraction at atomic resolution. In parallel, RL coupled with autonomous robotic platforms drives closed-loop experimentation, enabling rapid optimization of reaction pathways. This convergence of AI and nanotechnology establishes a data-driven framework for scalable, precise, and adaptive engineering. Furthermore, AI-driven inverse design frameworks facilitate the discovery of nanomaterials with optimized or even desired electronic, catalytic, or biomedical properties.

7.1. Biomedical Applications

- Nano-enabled drug delivery systems leverage the unique physicochemical properties of nanocarriers—such as tunable size-dependent cellular uptake, programmable release kinetics, and surface functionalization for tissue- or receptor-specific targeting (Huang et al., 2022; Ayodele et al., 2021; Vaseashta & Stamatin, 2007). The design of optimized nanomedicines represents a complex, multi-parameter challenge involving trade-offs between efficacy, stability, and biocompatibility, with critical safety implications (Hubbell & Chilkoti, 2012; Siddique & Chow, 2020; Nance, 2019). AI models, trained on integrated biological, chemical, and physical datasets, can predict biodistribution, therapeutic performance, and toxicity with higher accuracy than conventional approaches, thereby accelerating rational design and reducing experimental burden in preclinical development. DL and ML enable the analysis of high-throughput screening datasets to uncover complex relationships between nanoparticle composition and cellular responses. For instance, ML models have been used to predict protein corona formation—the dynamic adsorption of biomolecules onto nanocarriers in biological fluids—which critically influences immune recognition, circulation time, and targeting efficiency (Huang et al., 2021). Such predictive intelligence capabilities minimize reliance on costly *in-vivo* testing and accelerate preclinical development of nanomedicines.
- AI is increasingly instrumental in image-based diagnostics that employ nanoparticle-based contrast agents. CNN facilitates automated detection and classification of malignant tissues by analyzing complex imaging data from

modalities such as magnetic resonance imaging (MRI) enhanced with targeted nanoparticles. This integration not only improves diagnostic sensitivity and specificity but also enables quantitative assessment of tumor margins and early-stage lesion detection, reducing observer variability and advancing precision oncology (Liu et al., 2021).

- Functionalized nanoparticles—such as gold nanoparticles and quantum dots—are increasingly employed to detect biomarkers linked to diseases including cancer, diabetes, and infectious infections. When combined with ML, biosensor data can be analyzed with significantly enhanced sensitivity and accuracy, even at trace biomarker concentrations. Notably, AI-driven models have demonstrated the ability to interpret complex biosensor signals from circulating tumor DNA, facilitating earlier and more precise cancer diagnoses than previously achievable (Ahmed Taha et al., 2024).
- AI-integrated nanomaterials are emerging as powerful tools in cancer therapy, particularly in targeted drug delivery and photothermal treatment. ML models leverage patient-specific data to predict optimal nanoparticle properties—such as size, surface chemistry, and drug-loading capacity—that enhance tumor accumulation, retention, and controlled release. In parallel, AI assists in designing nanoparticles for photothermal therapy, enabling precise light-induced heating that selectively eradicates cancer cells while minimizing damage to surrounding healthy tissues (Sharma et al., 2023).
- In regenerative medicine, the convergence of nanomaterials with AI is accelerating advances in scaffold-based tissue engineering. AI models can predict the influence of critical nanomaterial properties—such as porosity, mechanical strength, biocompatibility—on cell proliferation and tissue regeneration (Demir et al., 2023; Vaseashta et al., 2022). These predictions enable the rational design of nanostructures tailored for applications including wound healing, bone repair, and organ regeneration. Moreover, AI-driven optimization of scaffold architecture supports controlled cell differentiation and seamless integration with host tissues, thereby enhancing the efficacy of regenerative therapies (Demir et al., 2023; Gu, 2025).
- Recent advancements in vaccinology underscore the pivotal role of AI in optimizing lipid nanoparticles (LNPs) for mRNA vaccine delivery. The COVID-19 pandemic notably accelerated the application of AI in developing LNP formulations that enhance antigen stability and potentiate immune responses (Yuan et al., 2024). Predictive AI models were instrumental in fine-tuning LNP characteristics, significantly expediting vaccine development and reducing the timelines for formulation and testing. This AI-driven paradigm was critical to the rapid global deployment of mRNA-based COVID-19 vaccines (Baldwin et al., 2023).

- A recent study introduced EVONANO, an open-source evolutionary computational platform designed to automate the discovery of nanocarriers for cancer therapy (Stillman et al., 2021). The framework integrates multiscale, modular simulations with ML algorithms to evaluate nanocarrier behavior and therapeutic performance within biologically relevant tissue models, generated through *in-silico* virtual tumors. EVONANO's simulator can construct tumor models, extract representative microenvironments, and model nanoparticle (NP) transport dynamics to predict spatial distribution and therapeutic efficacy. The study demonstrated that ML could evolve NP designs optimized to penetrate and eradicate tumor regions while selectively minimizing systemic dosage. Results indicated that nanoparticles with high diffusivity (smaller radii) and low binding affinity (larger dissociation constants) exhibit the most significant potential for effective drug delivery (Chen et al., 2020).

7.2. Energy, Environment, Sustainable, and Green Nanotechnology Applications

- Nanomaterials have transformed energy storage, conversion, and environmental remediation through high-efficiency catalysts, lightweight battery materials, and advanced solar-harvesting platforms. AI has accelerated progress in these domains by enabling multi-scale optimization—from atomic-level catalyst design to system-level device engineering. In catalysis, AI models are particularly impactful, identifying nanoscale structural motifs that govern both activity and stability. Notably, RL has been used to design platinum-alloy nanocatalysts with enhanced oxygen reduction reaction (ORR) kinetics, achieving up to a 25% performance improvement in fuel cells (Yin et al., 2024).
- A study applied neural networks trained on experimental and computational data to predict high-capacity electrode nanomaterials for lithium-ion batteries. The models accurately forecasted performance metrics and guided synthesis efforts, driving advances in battery technology toward higher energy densities (Wang et al., 2018).
- AI is increasingly applied in the design of high-capacity battery electrodes by exploring vast compositional and structural spaces to optimize conductivity, stability, and lithium-ion transport. Predictive models narrow the search for promising nanostructured materials, reducing experimental burden and accelerating the discovery of next-generation electrodes for sustainable energy storage (Wang, 2025).

- Sustainability has become a central priority in nanotechnology development, with AI playing a critical role in advancing greener materials and processes. AI-driven design enables the development of biodegradable nanopolymers, eco-friendly nanocomposites, and energy-efficient fabrication strategies. In environmental remediation, optimized nanomaterials are engineered for selective pollutant adsorption, photocatalytic degradation of contaminants, and real-time sensing of toxic substances in air and water. By synthesizing large datasets from experiments and simulations, AI can rapidly screen and prioritize candidate nanomaterials for environmental applications (Sidorenko et al., 2022). Integration of lifecycle assessment (Vaseashta, 2015) employing AI models allows the prediction of environmental impacts, guiding the design of safer and more sustainable materials. Furthermore, AI mining of agricultural and biomass data is accelerating the discovery of bio-based nanostructures for applications in packaging, water filtration, and precision agriculture. Compared to conventional inorganic nanomaterials, these bio-derived alternatives offer lower toxicity and a reduced ecological footprint (Dave & Patel, 2023).

7.3. Electronics and Photonics

- The semiconductor industry increasingly depends on nanoscale materials—such as quantum dots (0D), nanowires (1D), and two-dimensional (2D) systems—for next-generation transistors, sensors, and photonic devices. Optimizing their electronic and optical properties presents a multi-dimensional challenge well-suited to AI. As stated above, ML models can accurately predict key parameters, including electronic bandgap, charge-carrier mobility, and optical absorption spectra, directly from atomic structures, thereby reducing reliance on computationally expensive density functional theory (DFT) simulations. AI-assisted approaches also facilitate the discovery of defect-tolerant materials, a critical factor for enhancing device performance, yield, and operational lifetime. In photonics, ML enables the inverse design of plasmonic and nanophotonic structures with tailored resonance frequencies, expediting applications in biosensing, optical communication, and integrated photonics. By efficiently navigating large design spaces, AI significantly accelerates the discovery of materials and device engineering (Butler et al., 2018; Ren et al., 2018; Malkiel et al., 2018).

Despite notable successes, several challenges reduce the impact of AI on nanotechnology applications. Notably, the complexity of biological systems limits predictive accuracy, requiring models that incorporate multi-scale physiological data (Esteva et al., 2019). Additionally, data heterogeneity and limited availability further hinder the training of robust AI models across multiple domains (Dwivedi et al., 2021).

8. CONCLUSIONS AND PATH FORWARD

The convergence of AI and nanotechnology represents a transformative "fourth research paradigm" in materials science, enabling accelerated discovery, autonomous experimentation, and intelligent design of functional materials. Unlike traditional trial-and-error methods, AI offers a data-centric framework that significantly reduces discovery cycles, enhances reproducibility, and expands the scope of exploration. By using ML-based proxy models, researchers can achieve quantum-mechanical simulation–level accuracy at a fraction of the computational cost, enabling rapid virtual screening of vast design spaces while reserving experimental validation for only the most promising candidate (Chen et al., 2019). This shift is reinforced by the integration of high-throughput screening, large-scale databases, and predictive models, which collectively reduce reliance on manual analysis and empirical intuition. AI has also advanced nanomaterial synthesis by optimizing reaction conditions and enabling precise control over size, shape, and composition. Real-time AI-assisted characterization and synthesis—where algorithmic interfaces directly with experimental platforms—are beginning to establish autonomous, closed-loop laboratories that continuously learn and adapt (MacLeod et al., 2020). These systems promise reproducibility, rapid optimization, and intelligent feedback control without human intervention, accelerating the transition toward "self-driving laboratories" powered by reinforcement learning and robotic automation. Going forward, several emerging directions highlight AI's role as a key enabler of nanoscience. Generative AI and digital twins will allow researchers to design nanomaterials virtually, predict behavior under diverse conditions, and test prototypes with unprecedented speed and accuracy (Chen et al., 2019). Federated learning and collaborative AI ecosystems will expand collective intelligence across institutions while safeguarding data privacy and sovereignty (Dwivedi et al., 2021). Equally important, sustainability considerations will be embedded within AI-driven discovery frameworks, aligning materials research with environmental and societal imperatives such as the UN Sustainable Development Goals. In biomedicine, AI is poised to advance personalized nanomedicine by predicting patient-specific biological responses to nanomaterial-based therapies, thereby improving efficacy while minimizing side effects (Esteva et al., 2019). Complementing the XAI and PIML will address concerns about transparency and trust by ensuring model predictions remain interpretable and consistent with physical laws (Ramprasad et al., 2017). Despite these advances, challenges remain in algorithm generalizability, experimental validation, and cost-effective deployment.

Progress will depend on developing robust biomedical nanomaterial databases, versatile synthesis and optimization algorithms, and integrative platforms for multimodal characterization. Ultimately, the convergence of AI and nanotechnology is reshaping the research ecosystem into one that is more autonomous, transparent, and sustainable. This paradigm shift posits AI not just as an auxiliary tool but also as a fundamental driver of next-generation nanomaterials capable of addressing pressing global challenges in energy, health, and sustainability, transforming the fourth paradigm shift to the fifth industrial revolution.

REFERENCES

Ahmed Taha, B., Kadhim, A. C., Addie, A. J., Haider, A. J., Azzahrani, A. S., Raizada, P., Rustagi, S., Chaudhary, V., & Arsad, N. (2024). Advancing cancer diagnostics through multifaceted optical biosensors supported by nanomaterials and artificial intelligence: A panoramic outlook. *Microchemical Journal*, *205*, 111307. Advance online publication. DOI: 10.1016/j.microc.2024.111307

Al-Qattan, M. N., Deb, P. K., & Tekade, R. K. (2018). Molecular dynamics simulation strategies for designing carbon-nanotube-based targeted drug delivery. *Drug Discovery Today*, *23*(2), 235–250. DOI: 10.1016/j.drudis.2017.10.002 PMID: 29031623

Alhazmi, K., Albalawi, F., & Sarathy, S. M. (2022). A reinforcement learning-based economic model predictive control framework for autonomous operation of chemical reactors. *Chemical Engineering Journal*, *428*, 130993. Advance online publication. DOI: 10.1016/j.cej.2021.130993

. Anker, A. S., Kjær, E. T. S., Juelsholt, M., Christiansen, T. L., Skjærvø, S. L., Jørgensen, M. R. V., Kantor, I., Sørensen, D. R., Billinge, S. J. L., Selvan, R., & Jensen, K. M. Ø. (2022). Extracting structural motifs from pair distribution function data of nanostructures using explainable machine learning. *npj Computational Materials, 8*(1), Article 213. https://doi.org/DOI: 10.1038/s41524-022-00896-3

Aversa, R., Modarres, M. H., Cozzini, S., Ciancio, R., & Chiusole, A. (2018). The first annotated set of scanning electron microscopy images for nanoscience. *Scientific Data*, *5*(1), 180172. Advance online publication. DOI: 10.1038/sdata.2018.172 PMID: 30152811

Ayodele, O. O., Adesina, A. O., Pourianejad, S., Averitt, J., & Ignatova, T. (2021). Recent advances in nanomaterial-based aptasensors in medical diagnosis and therapy. *Nanomaterials (Basel, Switzerland)*, *11*(4), 932. Advance online publication. DOI: 10.3390/nano11040932 PMID: 33917467

Baldwin, J., Noorali, S., & Vaseashta, A. (2023). Biology and behavior of Severe Acute Respiratory Syndrome Coronavirus contagion with emphasis on treatment strategies, risk assessment, and resilience. *COVID*, *3*(9), 1259–1303. DOI: 10.3390/covid3090089

Barnard, A. S., Motevalli, B., Parker, A. J., Fischer, J. M., Feigl, C. A., & Opletal, G. (2019). Nanoinformatics, and the big challenges for the science of small things. *Nanoscale*, *11*(41), 19190–19201. DOI: 10.1039/C9NR05912A PMID: 31397835

Batra, A., Sampson, J., Davis, A., Currie, J., & Vaseashta, A. (2023). Electrospun nanofibers doped with PVDF and PLZT nanoparticles for potential biomedical and energy harvesting applications. *Journal of Materials Science Materials in Electronics, 34*(22), 1654. Advance online publication. DOI: 10.1007/s10854-023-11066-6

Batra, A., Sampson, J., Davis, A., Currie, J., & Vaseashta, A. (2023). Electrospun nanofibers doped with PVDF and PLZT nanoparticles for potential biomedical and energy harvesting applications. *Journal of Materials Science Materials in Electronics, 34*(22), 1654. Advance online publication. DOI: 10.1007/s10854-023-11066-6

Bhatt, M. D., Kim, H., & Kim, G. (2022). Various defects in graphene: A review. *RSC Advances, 12*(33), 21520–21547. DOI: 10.1039/D2RA01436J PMID: 35975063

Boso, D. P., Ferrari, M., Decuzzi, P., & Schrefler, B. A. (2011). Optimizing particle size for targeting diseased microvasculature: From experiments to artificial neural networks. *International Journal of Nanomedicine, 6*, 1517–1526. DOI: 10.2147/IJN.S20283 PMID: 21845041

Brunette, E. S., Flemmer, R. C., & Flemmer, C. L. (2009). A review of artificial intelligence. In *2009 4th International Conference on Autonomous Robots and Agents* (pp. 385–392). IEEE. https://doi.org/DOI: 10.1109/ICARA.2009.5639918

Butler, K. T., Davies, D. W., Cartwright, H., Isayev, O., & Walsh, A. (2018). Machine learning for molecular and materials science. *Nature, 559*(7715), 547–555. DOI: 10.1038/s41586-018-0337-2 PMID: 30046072

Chan, E. M., Xu, C., Mao, A. W., Han, G., Owen, J. S., Cohen, B. E., & Milliron, D. J. (2010). Reproducible, high-throughput synthesis of colloidal nanocrystals for optimization in multidimensional parameter space. *Nano Letters, 10*(5), 1874–1885. DOI: 10.1021/nl100669s PMID: 20387807

Chen, C., Yaari, Z., Apfelbaum, E., Grodzinski, P., Shamay, Y., & Heller, D. A. (2022). Merging data curation and machine learning to improve nanomedicines. *Advanced Drug Delivery Reviews, 183*, 114172. Advance online publication. DOI: 10.1016/j.addr.2022.114172 PMID: 35189266

Chen, C., Ye, W., Zuo, Y., Zheng, C., & Ong, S. P. (2019). Graph networks as a universal machine learning framework for molecules and crystals. *Chemistry of Materials, 31*(9), 3564–3572. DOI: 10.1021/acs.chemmater.9b01294

Chen, C., Zhang, Q., Yu, B., Yu, Z., Lawrence, P. J., Ma, Q., & Zhang, Y. (2020). Improving protein-protein interactions prediction accuracy using XGBoost feature selection and stacked ensemble classifier. *Computers in Biology and Medicine*, *123*, 103899. Advance online publication. DOI: 10.1016/j.compbiomed.2020.103899 PMID: 32768046

Chen, T., Li, J., Cai, P., Yao, Q., Ren, Z., Zhu, Y., Khan, S., Xie, J., & Wang, X. (2023). Identification of chemical compositions from "featureless" optical absorption spectra: Machine learning predictions and experimental validations. *Nano Research*, *16*(4), 4188–4196. DOI: 10.1007/s12274-022-5095-7

Chowdhary, K. R. (2020). Natural language processing. In *Fundamentals of artificial intelligence* (pp. 603–649). Springer., DOI: 10.1007/978-81-322-3972-7_19

Dahy, G., Soliman, M. M., Alshater, H., Slowik, A., & Ella Hassanien, A. (2023). Optimized deep networks for the classification of nanoparticles in scanning electron microscopy imaging. *Computational Materials Science*, *223*, 112135. Advance online publication. DOI: 10.1016/j.commatsci.2023.112135

Dave, M., & Patel, N. (2023). Artificial intelligence in healthcare and education. *British Dental Journal*, *234*(10), 761–764. DOI: 10.1038/s41415-023-5845-2 PMID: 37237212

Demir, D., Bolgen, N., & Vaseashta, A. (2023). Electrospun nanofibers for biomedical, sensing, and energy harvesting functions. *Polymers*, *15*(21), 4253. Advance online publication. DOI: 10.3390/polym15214253 PMID: 37959933

Ding, R., Chen, J., Chen, Y., Liu, J., Bando, Y., & Wang, X. (2024). Unlocking the potential: Machine learning applications in electrocatalyst design for electrochemical hydrogen energy transformation. *Chemical Society Reviews*, *53*(22), 11390–11461. DOI: 10.1039/D4CS00844H PMID: 39382108

. Dunn, A., Wang, Q., Ganose, A., Dopp, D., & Jain, A. (2020). Benchmarking materials property prediction methods: The Matbench test set and Automatminer reference algorithm. *npj Computational Materials, 6*(1), Article 138. https://doi.org/ DOI: 10.1038/s41524-020-00406-3

Dwivedi, Y. K., Hughes, L., Ismagilova, E., Aarts, G., Coombs, C., Crick, T., Duan, Y., Dwivedi, R., Edwards, J., Eirug, A., Galanos, V., Ilavarasan, P. V., Janssen, M., Jones, P., Kar, A. K., Kizgin, H., Kronemann, B., Lal, B., Lucini, B., & Williams, M. D. (2021). Artificial intelligence (AI): Multidisciplinary perspectives on emerging challenges, opportunities, and agenda for research, practice and policy. *International Journal of Information Management*, *57*, 101994. Advance online publication. DOI: 10.1016/j.ijinfomgt.2019.08.002

Esteva, A., Robicquet, A., Ramsundar, B., Kuleshov, V., DePristo, M., Chou, K., Cui, C., Corrado, G., Thrun, S., & Dean, J. (2019). A guide to deep learning in healthcare. *Nature Medicine*, *25*(1), 24–29. DOI: 10.1038/s41591-018-0316-z PMID: 30617335

Fariq, A., Khan, T., & Yasmin, A. (2017). Microbial synthesis of nanoparticles and their potential applications in biomedicine. *Journal of Applied Biomedicine*, *15*(4), 241–248. DOI: 10.1016/j.jab.2017.03.004

Feng, M., Yao, W., An, J., Huang, Z., Yuan, Y., & Yao, Y. (2022). Prediction method of graphene defect modification based on neural network. *Mobile Information Systems*, *2022*, 3580762. Advance online publication. DOI: 10.1155/2022/3580762

Frazier, P. I. (2018). A tutorial on Bayesian optimization. *arXiv*. https://arxiv.org/abs/1807.02811

Freddi, A., & Salmon, M. (2019). Introduction to the Taguchi method. In *Design principles and methodologies* (pp. 159–180). Springer., DOI: 10.1007/978-3-319-95342-7_7

Gao, P., Zhang, J., Qiu, H., & Zhao, S. (2021). A general QSPR protocol for the prediction of atomic/inter-atomic properties: A fragment based graph convolutional neural network (F-GCN). *Physical Chemistry Chemical Physics*, *23*(23), 13242–13249. DOI: 10.1039/D1CP00677K PMID: 34086015

Grasso, G., Zane, D., & Dragone, R. (2019). Microbial nanotechnology: Challenges and prospects for green biocatalytic synthesis of nanoscale materials for sensoristic and biomedical applications. *Nanomaterials (Basel, Switzerland)*, *10*(1), 11. Advance online publication. DOI: 10.3390/nano10010011 PMID: 31861471

Groschner, C. K., Choi, C., & Scott, M. C. (2021). Machine learning pipeline for segmentation and defect identification from high-resolution transmission electron microscopy data. *Microscopy and Microanalysis*, *27*(3), 549–556. DOI: 10.1017/S1431927621000386 PMID: 33952372

Gu, X. (2025). Artificial intelligence for biomaterials and tissue engineering. *Journal of Intelligent Medicine*, *2*(1), 3–6. DOI: 10.1002/jim4.70000

Guda, A. A., Kirichkov, M. V., Shapovalov, V. V., Muravlev, A. I., Pashkov, D. M., Guda, S. A., Bagliy, A. P., Chapek, S. V., & Soldatov, A. V. (2023). Machine learning analysis of reaction parameters in UV-mediated synthesis of gold nanoparticles. *The Journal of Physical Chemistry. C, Nanomaterials and Interfaces*, *127*(2), 1097–1108. DOI: 10.1021/acs.jpcc.2c06625

Han, B., Lin, Y., Yang, Y., Mao, N., Li, W., Wang, H., Yasuda, K., Wang, X., Fatemi, V., Zhou, L., Wang, J. I.-J., Ma, Q., Cao, Y., Rodan-Legrain, D., Bie, Y., Navarro-Moratalla, E., Klein, D., MacNeill, D., Wu, S., & Palacios, T. (2020). Deep-learning-enabled fast optical identification and characterization of 2D materials. *Advanced Materials*, *32*(29), 2000953. Advance online publication. DOI: 10.1002/adma.202000953 PMID: 32519397

Horton, M. K., Huck, P., Yang, R. X., Munro, J. M., Dwaraknath, S., Ganose, A. M., Kingsbury, R. S., Wen, M., Shen, J. X., Mathis, T. S., Kaplan, A. D., Berket, K., Riebesell, J., George, J., Rosen, A. S., Spotte-Smith, E. W. C., McDermott, M. J., Cohen, O. A., Dunn, A., & Persson, K. A. (2025). Accelerated data-driven materials science with the Materials Project. *Nature Materials*, *24*(11), 1447–1457. DOI: 10.1038/s41563-025-02272-0 PMID: 40610673

Huang, M., Li, Z., & Zhu, H. (2022). Recent advances of graphene and related materials in artificial intelligence. *Advanced Intelligent Systems*, *4*(10), 2200077. Advance online publication. DOI: 10.1002/aisy.202200077

Huang, W., Xiao, G., Zhang, Y., & Min, W. (2021). Research progress and application opportunities of nanoparticle–protein corona complexes. *Biomedicine and Pharmacotherapy*, *139*, 111541. Advance online publication. DOI: 10.1016/j.biopha.2021.111541 PMID: 33848776

Hubbell, J. A., & Chilkoti, A. (2012). Nanomaterials for drug delivery. *Science*, *337*(6092), 303–305. DOI: 10.1126/science.1219657 PMID: 22822138

Jain, A., Montoya, J., Dwaraknath, S., Zimmermann, N. E. R., Dagdelen, J., Horton, M., Huck, P., Winston, D., Cholia, S., Ong, S. P., & Persson, K. (2018). The Materials Project: Accelerating materials design through theory-driven data and tools. In Andreoni, W., & Yip, S. (Eds.), *Handbook of materials modeling* (pp. 1–34). Springer., DOI: 10.1007/978-3-319-42913-7_68-1

Ji, Z., Guo, W., Sakkiah, S., Liu, J., Patterson, T., & Hong, H. (2021). Nanomaterial databases: Data sources for promoting design and risk assessment of nanomaterials. *Nanomaterials (Basel, Switzerland)*, *11*(6), 1599. Advance online publication. DOI: 10.3390/nano11061599 PMID: 34207026

Jordan, M. I., & Mitchell, T. M. (2015). Machine learning: Trends, perspectives, and prospects. *Science*, *349*(6245), 255–260. DOI: 10.1126/science.aaa8415 PMID: 26185243

Kajita, S., Ohba, N., Suzumura, A., Tajima, S., & Asahi, R. (2020). Discovery of superionic conductors by ensemble-scope descriptor. *NPG Asia Materials*, *12*(1), 31. Advance online publication. DOI: 10.1038/s41427-020-0211-1

Kim, H., Han, J., & Han, T. Y.-J. (2020). Machine vision-driven automatic recognition of particle size and morphology in SEM images. *Nanoscale, 12*(37), 19461–19469. DOI: 10.1039/D0NR04140H PMID: 32960204

Knoppe, S., Boudon, J., Dolamic, I., Dass, A., & Bürgi, T. (2011). Size exclusion chromatography for semipreparative scale separation of Au38(SR)24 and Au40(SR)24 and larger clusters. *Analytical Chemistry, 83*(13), 5056–5061. DOI: 10.1021/ac200789v PMID: 21609127

Kumar, A., & Kumar, M. (2024). Recent progress in emerging 2-D transition-metal dichalcogenide-based gas sensors: A review. *IEEE Sensors Journal, 24*(16), 25221–25226. DOI: 10.1109/JSEN.2024.3422972

Kumar, R., & Reji, M. (2023). Response surface methodology (RSM): An overview to analyze multivariate data. *The Indian Journal of Medical Research, 9*(3), 241–248. DOI: 10.18231/j.ijmr.2022.042

. L, S., A, S., T, G., I, R., & A, V. (2025). Micro and nanoplastics–plastisphere, bitoxicity, impact on human health, and mitigation strategies. *Nanomedicine and Nanotechnology Open Access, 10*(1), 1–17. https://doi.org/DOI: 10.23880/nnoa-16000345

Lao, Y., & You, Y. (2024). Unraveling generative AI in BBC News: Application, impact, literacy and governance. *Transforming Government: People. Process and Policy, 18*(4), 551–566. DOI: 10.1108/TG-01-2024-0022

LeCun, Y., Bengio, Y., & Hinton, G. (2015). Deep learning. *Nature, 521*(7553), 436–444. DOI: 10.1038/nature14539 PMID: 26017442

Lin, J., Liu, Y., Sui, H., Sagoe-Crentsil, K., & Duan, W. (2022). Microstructure of graphene oxide–silica-reinforced OPC composites: Image-based characterization and nano-identification through deep learning. *Cement and Concrete Research, 154*, 106737. Advance online publication. DOI: 10.1016/j.cemconres.2022.106737

Liu, H., Gao, H., & Jia, F. (2021). The value of convolutional neural network-based magnetic resonance imaging image segmentation algorithm to guide targeted controlled release of doxorubicin nanopreparation. *Contrast Media & Molecular Imaging, 2021*, 9032017. Advance online publication. DOI: 10.1155/2021/9032017 PMID: 34385899

Ma, J., Lee, S. M.-Y., Yi, C., & Li, C.-W. (2017). Controllable synthesis of functional nanoparticles by microfluidic platforms for biomedical applications – a review. *Lab on a Chip, 17*(2), 209–226. DOI: 10.1039/C6LC01049K PMID: 27991629

Ma, J., & Li, C.-W. (2018). Rapid and continuous parametric screening for the synthesis of gold nanocrystals with different morphologies using a microfluidic device. *Sensors and Actuators. B, Chemical, 262*, 236–244. DOI: 10.1016/j.snb.2018.02.001

MacLeod, B. P., Parlane, F. G. L., Morrissey, T. D., Häse, F., Roch, L. M., Dettelbach, K. E., Moreira, R., Yunker, L. P. E., Rooney, M. B., Deeth, J. R., Lai, V., Ng, G. J., Situ, H., Zhang, R. H., Elliott, M. S., Haley, T. H., Dvorak, D. J., Aspuru-Guzik, A., Hein, J. E., & Berlinguette, C. P. (2020). Self-driving laboratory for accelerated discovery of thin-film materials. *Science Advances, 6*(20), eaaz8867. Advance online publication. DOI: 10.1126/sciadv.aaz8867 PMID: 32426501

MacLeod, B. P., Parlane, F. G. L., Morrissey, T. D., Häse, F., Roch, L. M., Dettelbach, K. E., Moreira, R., Yunker, L. P. E., Rooney, M. B., Deeth, J. R., Lai, V., Ng, G. J., Situ, H., Zhang, R. H., Elliott, M. S., Haley, T. H., Dvorak, D. J., Aspuru-Guzik, A., Hein, J. E., & Berlinguette, C. P. (2020). Self-driving laboratory for accelerated discovery of thin-film materials. *Science Advances, 6*(20), eaaz8867. Advance online publication. DOI: 10.1126/sciadv.aaz8867 PMID: 32426501

Mahjoubi, S., Ye, F., Bao, Y., Meng, W., & Zhang, X. (2023). Identification and classification of exfoliated graphene flakes from microscopy images using a hierarchical deep convolutional neural network. *Engineering Applications of Artificial Intelligence, 119*, 105743. Advance online publication. DOI: 10.1016/j.engappai.2022.105743

Malkiel, I., Mrejen, M., Nagler, A., Arieli, U., Wolf, L., & Suchowski, H. (2018). Plasmonic nanostructure design and characterization via deep learning. *Light, Science & Applications, 7*(1), 60. Advance online publication. DOI: 10.1038/s41377-018-0060-7 PMID: 30863544

. Mekki-Berrada, F., Ren, Z., Huang, T., Wong, W. K., Zheng, F., Xie, J., Tian, I. P. S., Jayavelu, S., Mahfoud, Z., Bash, D., Hippalgaonkar, K., Khan, S., Buonassisi, T., Li, Q., & Wang, X. (2021). Two-step machine learning enables optimized nanoparticle synthesis. *npj Computational Materials, 7*(1), Article 55. https://doi .org/DOI: 10.1038/s41524-021-00520-w

Mekuye, B., & Abera, B. (2023). Nanomaterials: An overview of synthesis, classification, characterization, and applications. *Nano Select, 4*(8), 486–501. DOI: 10.1002/nano.202300038

Meyer, J. C. (2021). Transmission electron microscopy characterization of graphene. In *Graphene* (pp. 317–343). Elsevier., DOI: 10.1016/B978-0-12-819576-5.00012-0

Mourdikoudis, S., Pallares, R. M., & Thanh, N. T. K. (2018). Characterization techniques for nanoparticles: Comparison and complementarity upon studying nanoparticle properties. *Nanoscale, 10*(27), 12871–12934. DOI: 10.1039/C8NR02278J PMID: 29926865

Mukherjee, R., Diwekar, U. M., & Vaseashta, A. (2017). Optimal sensor placement with mitigation strategy for water network systems under uncertainty. *Computers & Chemical Engineering, 103*, 91–102. DOI: 10.1016/j.compchemeng.2017.03.014

Nance, E. (2019). Careers in nanomedicine and drug delivery. *Advanced Drug Delivery Reviews, 144*, 180–189. DOI: 10.1016/j.addr.2019.06.009 PMID: 31260712

. Nikolaev, P., Hooper, D., Webber, F., Rao, R., Decker, K., Krein, M., Poleski, J., Barto, R., & Maruyama, B. (2016). Autonomy in materials research: A case study in carbon nanotube growth. *npj Computational Materials, 2*(1), Article 16031. https://doi.org/DOI: 10.1038/npjcompumats.2016.31

Noble, W. S. (2006). What is a support vector machine? *Nature Biotechnology, 24*(12), 1565–1567. DOI: 10.1038/nbt1206-1565 PMID: 17160063

Noorali, S., Baldwin, J. C., & Vaseashta, A. (2025). Next-generation sensors using artificial intelligence as enabling technologies for biomedical and healthcare applications. In Vaseashta, A., Achour, M. E., Mabrouki, M., Tachafine, A., & Aitali, M. (Eds.), *Proceedings of the Eighth International Symposium on Dielectric Materials and Applications (ISyDMA'8)* (pp. 215–231). Springer. https://doi.org/DOI: 10.1007/978-3-031-69700-5_16

Palwai, S., Batra, A., Kotru, S., & Vaseashta, A. (2022). Electrospun polyvinylidene fluoride nanofiber membrane-based flexible capacitive tactile sensors for biomedical applications. *Surface Engineering and Applied Electrochemistry*, *58*(2), 194–201. DOI: 10.3103/S1068375522020089

Price, C. C., Singh, A., Frey, N. C., & Shenoy, V. B. (2022). Efficient catalyst screening using graph neural networks to predict strain effects on adsorption energy. *Science Advances*, *8*(36), eabq5944. Advance online publication. DOI: 10.1126/sciadv.abq5944 PMID: 36417537

. Ramprasad, R., Batra, R., Pilania, G., Mannodi-Kanakkithodi, A., & Kim, C. (2017). Machine learning in materials informatics: Recent applications and prospects. *npj Computational Materials, 3*(1), Article 54. https://doi.org/DOI: 10.1038/s41524-017-0056-5

. Ren, F., Ward, L., Williams, T., Laws, K. J., Wolverton, C., Hattrick-Simpers, J., & Mehta, A. (2018). Accelerated discovery of metallic glasses through iteration of machine learning and high-throughput experiments. *Science Advances, 4*(4), Article eaaq1566. https://doi.org/DOI: 10.1126/sciadv.aaq1566

Santana, R., Zuluaga, R., Gañán, P., Arrasate, S., Onieva, E., & González-Díaz, H. (2019). Designing nanoparticle release systems for drug–vitamin cancer co-therapy with multiplicative perturbation-theory machine learning (PTML) models. *Nanoscale, 11*(44), 21811–21823. DOI: 10.1039/C9NR05070A PMID: 31691701

Sha, W., Guo, Y., Yuan, Q., Tang, S., Zhang, X., Lu, S., Guo, X., Cao, Y.-C., & Cheng, S. (2020). Artificial intelligence to power the future of materials science and engineering. *Advanced Intelligent Systems*, *2*(4), 1900143. Advance online publication. DOI: 10.1002/aisy.201900143

Sharma, A., Sharma, A., Tselykh, A., Bozhenyuk, A., Choudhury, T., Alomar, M. A., & Sánchez-Chero, M. (2023). Artificial intelligence and internet of things oriented sustainable precision farming: Towards modern agriculture. *Open Life Sciences*, *18*(1), 20220713. Advance online publication. DOI: 10.1515/biol-2022-0713 PMID: 37854322

Siddique, S., & Chow, J. C. L. (2020). Gold nanoparticles for drug delivery and cancer therapy. *Applied Sciences (Basel, Switzerland)*, *10*(11), 3824. Advance online publication. DOI: 10.3390/app10113824

Sidorenko, A., Gutul, T., Dvornikov, D., Şeker, M. G., Arit, T., Gutul, E., Dimoglo, A., & Vaseashta, A. (2022). Synthesis of nZVI/PVP nanoparticles for bioremediation applications. *Bioremediation Journal*, *26*(2), 162–170. DOI: 10.1080/10889868.2021.1911922

Simonyan, K., & Zisserman, A. (2015). Very deep convolutional networks for large-scale image recognition. *arXiv*. https://arxiv.org/abs/1409.1556

Singh, A. V., Maharjan, R.-S., Kanase, A., Siewert, K., Rosenkranz, D., Singh, R., Laux, P., & Luch, A. (2021). Machine-learning-based approach to decode the influence of nanomaterial properties on their interaction with cells. *ACS Applied Materials & Interfaces*, *13*(1), 1943–1955. DOI: 10.1021/acsami.0c18470 PMID: 33373205

Stamatin, I., Berlic, C., & Vaseashta, A. (2006). On the computer-aided modelling of analyte–receptor interactions for an efficient sensor design. *Thin Solid Films*, *495*(1–2), 312–315. DOI: 10.1016/j.tsf.2005.08.299

. Stillman, N. R., Balaz, I., Tsompanas, M.-A., Kovacevic, M., Azimi, S., Lafond, S., Adamatzky, A., & Hauert, S. (2021). Evolutionary computational platform for the automatic discovery of nanocarriers for cancer treatment. *npj Computational Materials, 7*(1), Article 150. https://doi.org/DOI: 10.1038/s41524-021-00614-5

Thomas, D. G., Chikkagoudar, S., Heredia-Langner, A., Tardiff, M. F., Xu, Z., Hourcade, D. E., Pham, C. T. N., Lanza, G. M., Weinberger, K. Q., & Baker, N. A. (2014). Physicochemical signatures of nanoparticle-dependent complement activation. *Computational Science & Discovery, 7*(1), 015003. Advance online publication. DOI: 10.1088/1749-4699/7/1/015003 PMID: 25254068

Vaseashta, A. (2005). Nanostructured materials based next generation devices and sensors. In Vaseashta, A., Dimova-Malinovska, D., & Marshall, J. M. (Eds.), *Nanostructured and advanced materials for applications in sensor, optoelectronic and photovoltaic technology* (pp. 1–30). Springer., DOI: 10.1007/1-4020-3562-4_1

Vaseashta, A. (2006). Carbon nanotubes based devices and sensors. *Materials and Manufacturing Processes, 21*(7), 710–716. DOI: 10.1080/10426910600613595

Vaseashta, A. (Ed.). (2015). *Life cycle analysis of nanoparticles: Risk, assessment, and sustainability*. DEStech Publications.

Vaseashta, A., & Batra, A. (2022). Electrospun nanofibers of high-performance electret polymers for tactile sensing and wearable electronics. In Vaseashta, A., Achour, M. E., Mabrouki, M., Fasquelle, D., & Tachafine, A. (Eds.), *Proceedings of the Sixth International Symposium on Dielectric Materials and Applications (ISyD-MA'6)* (pp. 31–52). Springer. https://doi.org/DOI: 10.1007/978-3-031-11397-0_3

Vaseashta, A., & Bölgen, N. (Eds.). (2022). *Electrospun nanofibers: Principles, technology and novel applications*. Springer., DOI: 10.1007/978-3-030-99958-2

Vaseashta, A., Demir, D., Sakım, B., Aşık, M., & Bölgen, N. (2022). Hierarchical integration of 3D printing and electrospinning of nanofibers for rapid prototyping. In Vaseashta, A., & Bölgen, N. (Eds.), *Electrospun nanofibers* (pp. 631–655). Springer., DOI: 10.1007/978-3-030-99958-2_24

Vaseashta, A., Erdem, A., & Irudayaraj, J. (2005). Nanoparticles, nanoporous and carbon nanotube based devices for bio-molecular detection. In Vaseashta, A., Dimova-Malinovska, D., & Marshall, J. M. (Eds.), *Nanostructured and advanced materials for applications in sensor, optoelectronic and photovoltaic technology* (pp. 391–394). Springer., DOI: 10.1007/1-4020-3562-4_24

Vaseashta, A., & Stamatin, I. (2007). Electrospun polymers for controlled release of drugs, vaccine delivery, and system-on-fibers. *Journal of Optoelectronics and Advanced Materials*, *9*(6), 1606–1613.

Vladár, A. E., & Hodoroaba, V.-D. (2020). Characterization of nanoparticles by scanning electron microscopy. In *Characterization of nanoparticles* (pp. 7–27). Elsevier., DOI: 10.1016/B978-0-12-814182-3.00002-6

Voulodimos, A., Doulamis, N., Doulamis, A., & Protopapadakis, E. (2018). Deep learning for computer vision: A brief review. *Computational Intelligence and Neuroscience*, *2018*, 7068349. Advance online publication. DOI: 10.1155/2018/7068349 PMID: 29487619

. Wang, A., Kadam, S., Li, H., Shi, S., & Qi, Y. (2018). Review on modeling of the anode solid electrolyte interphase (SEI) for lithium-ion batteries. *npj Computational Materials, 4*(1), Article 15. https://doi.org/DOI: 10.1038/s41524-018-0064-0

Wang, M., Wang, T., Cai, P., & Chen, X. (2019). Nanomaterials discovery and design through machine learning. *Small Methods*, *3*(8), 1900025. Advance online publication. DOI: 10.1002/smtd.201900025

. Wang, Y. (2025). Application-oriented design of machine learning paradigms for battery science. *npj Computational Materials, 11*(1), Article 89. https://doi.org/DOI: 10.1038/s41524-025-01575-9

Wu, Y., Duan, H., & Xi, H. (2020). Machine learning-driven insights into defects of zirconium metal–organic frameworks for enhanced ethane–ethylene separation. *Chemistry of Materials*, *32*(7), 2986–2997. DOI: 10.1021/acs.chemmater.9b05322

Xu, C., Deng, X., & Yu, P. (2024). High-throughput computational study and machine learning prediction of electronic properties in transition metal dichalcogenide/two-dimensional layered halide perovskite heterostructures. *ACS Applied Materials & Interfaces*, *16*(39), 52351–52362. DOI: 10.1021/acsami.4c11973 PMID: 39361426

Yamankurt, G., Berns, E. J., Xue, A., Lee, A., Bagheri, N., Mrksich, M., & Mirkin, C. A. (2019). Exploration of the nanomedicine-design space with high-throughput screening and machine learning. *Nature Biomedical Engineering*, *3*(4), 318–327. DOI: 10.1038/s41551-019-0351-1 PMID: 30952978

Yan, X., Sedykh, A., Wang, W., Yan, B., & Zhu, H. (2020). Construction of a web-based nanomaterial database by big data curation and modeling friendly nanostructure annotations. *Nature Communications*, *11*(1), 2519. Advance online publication. DOI: 10.1038/s41467-020-16413-3 PMID: 32433469

Yang, H., Zhang, Z., Zhang, J., & Zeng, X. C. (2018). Machine learning and artificial neural network prediction of interfacial thermal resistance between graphene and hexagonal boron nitride. *Nanoscale*, *10*(40), 19092–19099. DOI: 10.1039/C8NR05703F PMID: 30288513

Yang, L., Wang, H., Leng, D., Fang, S., Yang, Y., & Du, Y. (2024). Machine learning applications in nanomaterials: Recent advances and future perspectives. *Chemical Engineering Journal*, *500*, 156687. Advance online publication. DOI: 10.1016/j.cej.2024.156687

Yang, R. X., McCandler, C. A., Andriuc, O., Siron, M., Woods-Robinson, R., Horton, M. K., & Persson, K. A. (2022). Big data in a nano world: A review on computational, data-driven design of nanomaterials structures, properties, and synthesis. *ACS Nano*, *16*(12), 19873–19891. DOI: 10.1021/acsnano.2c08411 PMID: 36378904

Yao, L., & Chen, Q. (2023). Machine learning in nanomaterial electron microscopy data analysis. In *Intelligent nanotechnology* (pp. 279–305). Elsevier., DOI: 10.1016/B978-0-323-85796-3.00009-0

Yin, P., Niu, X., Li, S.-B., Chen, K., Zhang, X., Zuo, M., Zhang, L., & Liang, H.-W. (2024). Machine-learning-accelerated design of high-performance platinum intermetallic nanoparticle fuel cell catalysts. *Nature Communications*, *15*(1), 415. Advance online publication. DOI: 10.1038/s41467-023-44674-1 PMID: 38195668

Yoshihara, N., Tahara, Y., & Noda, M. (2023). Machine learning method for determining chemical vapor deposition conditions for large-area graphene growth. *Asia-Pacific Journal of Chemical Engineering*, *18*(4), e2911. Advance online publication. DOI: 10.1002/apj.2911

Yu, T., Su, S., Hu, J., Zhang, J., & Xianyu, Y. (2022). A new strategy for microbial taxonomic identification through micro-biosynthetic gold nanoparticles and machine learning. *Advanced Materials*, *34*(24), 2109365. Advance online publication. DOI: 10.1002/adma.202109365 PMID: 34989446

Yuan, Y., Wu, Y., Cheng, J., Yang, K., Xia, Y., Wu, H., & Pan, X. (2024). Applications of artificial intelligence to lipid nanoparticle delivery. *Particuology*, *90*, 88–97. DOI: 10.1016/j.partic.2023.11.014

Zhu, C., Bamidele, E. A., Shen, X., Zhu, G., & Li, B. (2024). Machine learning aided design and optimization of thermal metamaterials. *Chemical Reviews*, *124*(7), 4258–4331. DOI: 10.1021/acs.chemrev.3c00708 PMID: 38546632

Chapter 2
AI-Inspired Synthesis and Design of Nanomaterials, Nano-Membranes, and Nano-Aponges for Advanced Water Treatment

Evidence Akhayere
https://orcid.org/0000-0002-7949-9670
Independent Researcher, Nigeria

Ashok Vaseashta
https://orcid.org/0000-0002-5649-0067
International Clean Water Institute, USA

ABSTRACT

The contamination of water and water bodies is now considered a growing challenge globally, affecting both developed and developing countries. To this effect, researchers and scientific innovators are pushing their limits to ensure that more technologies for developing efficient and adaptive materials in water treatment are actualized. Nanotechnology, particularly the use of nano-sponges, functional membranes, and advanced nanocomposites, is shown to be very promising due to their properties, surface areas, and ability to perform several functions that might seem complicated for conventional materials. One challenge this faces is that the traditional experimental approach to material discovery and synthesis typically requires a significant amount of time and involves multiple trial-and-error processes. Artificial

DOI: 10.4018/979-8-3373-2883-6.ch002

Intelligence (AI), specifically machine learning (ML), is an emerging powerful tool that can accelerate nanomaterial discovery, synthesis, and optimization, specifically for water purification. AI algorithms can be used to predict the relationship between the parameters for synthesis, the performance of the material and in turn help to optimize pore structures, adsorption efficiency, and membrane selectivity etc. in this chapter we explore a method that proposes to integrate AI into the design, synthesis, and deployment of nano-sponges, smart membranes, and hybrid nanomaterials for water treatment, highlighting case studies, simulation techniques, and predictive models. Eclectic styles, converging AI with nanotechnology, can facilitate the emergence of autonomous labs that optimize materials in real-time. It can also lead to innovative manufacturing systems based on digital twins.

1. INTRODUCTION

There is a growing demand for high-quality drinking water, largely due to the spread of waterborne diseases resulting from the introduction of various contaminants and substances into water bodies (Vaseashta et al, 2022). Among the substances introduced to water that cannot be naturally decomposed are pathogens, organic pollutants, heavy metals, industrial discharge, and pharmaceutical wastes, which then alter the characteristics of the water body (Vaseashta et. al., 2021). Several researchers have developed technologies that utilize nanotechnology to clean polluted water and wastewater, thereby upgrading outdated infrastructure and providing high-performance, low-cost treatment options. These technologies include nanoscale filtration, the adsorption of contaminants on nanoparticles (NPs), and the catalytic destruction of pollutants by nanoparticles. Numerous nanomaterials have been formed, including polymer nanomaterials, carbon nanotubes, nano zeolites, carbon, graphene, metallic NPs, graphene quantum dots, etc. Nanochemistry-based technologies are far more effective and affordable than conventional purification techniques. These techniques include creating membranes and using mixed oxides, zeolites, bimetallic nanoparticles, and carbon compounds as nano-catalysts and adsorbents to remove harmful substances in water (Yao et al., 2021). Additionally, nano-zero-valent metals and highly paramagnetically charged magnetic nanoparticles have been used as nano-sorbents to clean water. Advanced water and wastewater technical processes can be developed by integrating advanced nanotechnology with other processes. This global demand for clean drinking water is the primary driver of innovative approaches to optimizing water treatment processes. Nanomaterials, with at least one dimension on the order of 1–100 nm, offer unique size and surface-dependent properties that are ideal for water purification. For example, nano-adsorbents (carbon nanotubes, metal oxides, MOFs, etc.) can capture heavy metals and organic pollutants with high capacity, and nanoscale catalysts (e.g., TiO_2, ZnO) enable photocatalytic degradation of contaminants (Zheng et al., 2023). Similarly, engineered nano porous membranes (polymeric or inorganic) achieve high selectivity and flux for desalination and contaminant separation, while "Nano sponges" (e.g., cyclodextrin-based polymers or MOFs) can absorb organic micropollutants. However, designing such materials traditionally relies on trial-and-error synthesis and one-at-a-time testing, which is time-consuming and expensive. The integration of nanotechnology with artificial intelligence (AI), particularly through machine learning and neural networks, is driving the development of advanced nanomembranes with enhanced performance and reliability. AI-driven materials design offers a solution by learning from existing data and simulations (Vaseashta, 2022). AI can predict material properties and suggest optimal candidates before experimental testing. Recent years have seen an explosion of AI applications in materials science, which has been dubbed the "fourth paradigm" of research, but only in the past 2–3 years has this approach been widely applied to water-treatment nanomaterials. In this chapter, we demonstrate how artificial intelligence has advanced the experimental synthesis of nanomaterials, nanomembranes, and nanosponges for water purification, as well as a methodology that integrates AI into the synthesis of nanomaterials and their application in water treatment. AI algorithms enable real-time monitoring, adaptive responses to changing conditions, and proactive maintenance strategies. For instance, AI can optimize energy consumption, mitigate membrane fouling, and extend the lifespan of membranes (Chen & Lv, 2022). During these processes, the AI-enhanced systems continuously learn and improve their efficiency under diverse conditions. Although this technology can reduce the need for on-site personnel, it enables a system where water decentralization is powered by remote management. AI-driven systems can analyze real-time data and make informed decisions, ensuring consistent and sustainable operation. However, there are some existing challenges, such as developing desalination-specific AI algorithms, ensuring scalability and compatibility, and addressing data privacy and security concerns. These challenges can be further developed over time as the AI continues to learn about the specificity and nature of the systems. There have been significant improvements in the development of membranes using AI. The convergence of AI and nanomembrane technology holds immense potential for revolutionizing water treatment. Nevertheless, research in this area must continue to define the capacities and capabilities of technology fully. It is common knowledge that traditional water sources are continuously becoming contaminated, leading to rapid depletion. Hence, the need for more research on removing salt and other impurities from seawater and other contaminated water sources to reduce freshwater shortages cannot be overstated. (Gao et al., 2025). Another aspect to consider as motivation for continuous research is that the existing desalination approaches have numerous disadvantages, including high energy consumption, high production costs, and environmental damage. With the help of AI, such as machine learning and neural network algorithms, systems can learn from data, adapt to new situations, and make automatic and effective decisions. Nanomaterials have proven themselves as trusted agents for water treatment over time. For example, Nanomembranes made from graphene oxide and carbon nanotubes mixed with polymers, which have a porous nature, are popular for desalination. When we combine nanomembranes with nanomaterials and AI, we can develop new approaches to advance water treatment. Membrane operations can be studied and monitored using large datasets, enabling the continual achievement of optimal conditions, such as temperature, pressure, and flow rate, through AI-driven algorithms. By doing so, we can increase the percentage of water recovery, double the solids loading rate, and significantly boost efficiency. With AI, membranes can last up to thirty percent longer, and fouling can easily be detected at early stages. In reverse osmosis (RO) membranes, AI algorithms for reversible biofouling can help to reduce the frequency of cleaning and improve coherence across systems. There are also neural networks that predict permeate quality and develop a better process with control, ensuring that the specification of the end product is met. To conserve energy and reduce energy consumption in the treatment process, we may utilize reinforcement learning algorithms (Gao et al., 2025).This is highly efficient in reducing energy consumption. For example, when AI systems are installed, they can control the pump speed in real-time based on water quality, saving energy and managing waste. Nanomaterials and nano membranes typically exhibit unique properties, such as high surface area, high selectivity, relatively high stability, tunable pore size, and exceptional mechanical strength, which makes them applicable in various fields. The fabrication of nanomaterials and membranes often involves special techniques like layer-by-layer assembly, self-assembly, template-assisted synthesis, and other bottom-up approaches. This is because precision is required in terms of pore size, porosity, permeability, and surface chemistry. Nano materials also exhibit distinct mechanical properties, such as tensile strength, flexibility, and toughness, which enable them to survive high pressure, temperature, and chemical environments. (Akhayere et al., 2020) Nanomembrane systems are essential for various advanced technologies, providing high-performance efficiency and versatility. These must be understood and taken into account when designing a membrane.

Nanotechnology can be utilized to conserve water and the nonrenewable energy sources associated with it, thereby preventing contamination of water and other natural resources (Vaseashta, 2021). Additionally, this strategy can be employed to preserve ecosystems, habitats, and biodiversity, ensuring that subsequent generations can meet their own needs. The use of harmful chemicals and solvents has increased to the point where it is no longer practical to proceed without adopting an environmentally friendly strategy, given the rising public awareness of environmental issues. Nanotechnology holds the promise of advanced nanomaterials for treating surface water, groundwater, wastewater, and other ecological materials contaminated by hazardous metal ions, metal oxides, organic and inorganic solutes, and bio-organisms. These techniques include ultrafiltration membranes, reverse osmosis, carbon nanotubes, nanofiltration, and nanofiber filters. Several nanoremediation approaches, including nano zero-valent iron for groundwater cleanup, have been applied at large-scale cleaning sites, and their performance can be further improved with the addition of Artificial Intelligence (Wang et al., 2024).

Figure 1. AI-motivated Advanced Water Treatment Strategy using Nanomaterials, Nanomembranes, and Nano-sponges.

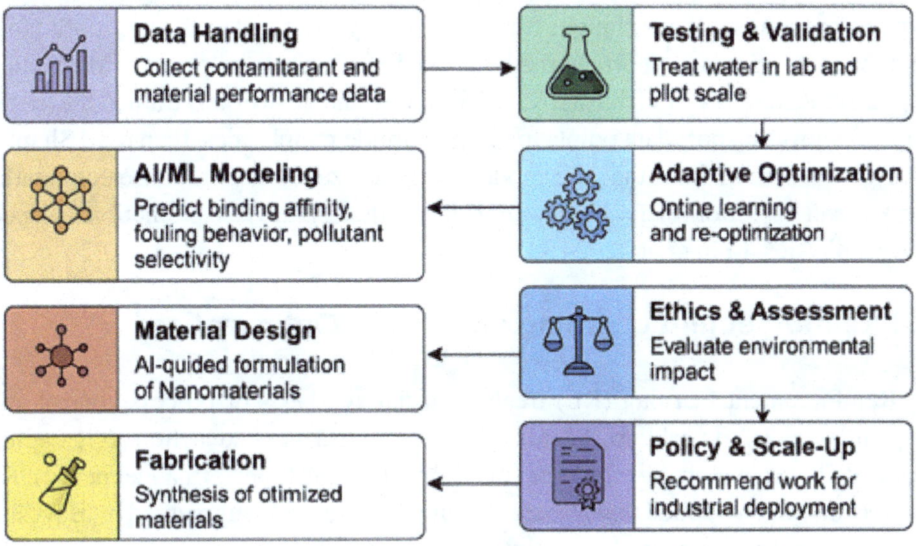

2.0 AI METHODOLOGIES FOR MATERIAL DESIGN

The application of AI technologies in creatively designing how materials function and their expected efficiency is an important part of integrating AI in nanotechnology and water treatment. The process of designing materials to suit the purpose to which they are being designed requires several trial-and-error processes. This study provides details on the AI methodologies used to achieve sustainable materials and designs for nanomaterials, aiming for efficient water treatment.

2.1. Machine Learning and Deep Neural Networks

Supervised ML forms the backbone of AI-enabled materials design. In this context, data on existing materials (compositions, structures, synthesis parameters, performance metrics) are used to train regression or classification models. For water treatment, common targets include adsorption capacity, permeability, rejection rate, or antimicrobial efficacy. Algorithms range from traditional regression trees and Support Vector Machines (SVMs) to deep neural networks and graph networks. Deep learning is more efficient when large datasets (e.g., from simulations or high-throughput experiments) are available. Graph neural networks (GNNs) are increasingly used to encode the structure of crystalline adsorbents or polymer networks, learning features that relate to adsorption or transport properties. For example, Hyunwook et al. (2022) employed atomistic machine learning (ML) to predict gas sorption in nanoporous materials, a technique that can be easily extended to aqueous pollutants. Explainable AI is also important: methods like SHAP values help ensure ML models capture real chemistry (Vishnyakov, 2025). Jeong and co-workers trained the model on 1,585 ion-transport data points for 26 polyamide membranes, then used Shapley analysis to demonstrate that the model learned size exclusion and electrostatics govern salt rejection. This builds confidence that AI-driven predictions respect known physics. (Yin et al., 2024)

2.2 Reinforcement Learning and Active Optimization

Reinforcement learning (RL) treats material synthesis or process control as a sequential decision problem. Agents (algorithms) learn to take actions (e.g., adjust reactant flows, cleaning cycles) that maximize a reward (performance metric). RL has recently been applied to water treatment processes. RL augmented by Bayesian optimization has been used to optimize activated-sludge plant operation, rapidly converging on cost-effective strategies. While RL in nanoscale synthesis is nascent, the ChatGPT Research Group study effectively demonstrated RL principles: by balancing exploration (trying new synthesis conditions) and exploitation (refining promising conditions) through Bayesian search, the AI system homed in on optimal metal organic framework MOF and covalent organic framework COF syntheses from millions of possibilities. In lab automation, RL agents could, in principle, direct robots. (Zhu et al., 2025)

2.3 Nano Adsorbents and Catalysts

A wide range of nanomaterials has been explored for adsorbing pollutants (heavy metals, dyes, pharmaceuticals) or catalyzing their degradation. For instance, iron oxide, silver, or TiO_2 nanoparticles can catalyze or disinfect, while carbon-based nanomaterials, such as graphene, carbon nanotubes (CNTs), and biochars, as well as metal-organic frameworks (MOFs), can adsorb organics and metals. AI is accelerating the discovery of such adsorbents. Large computational databases of hypothetical metal-organic frameworks (MOFs) and covalent organic frameworks (COFs), combined with deep learning, now enable the rapid screening of adsorption capacity for various contaminants. In one example, researchers have trained neural networks on millions of simulated adsorption isotherm data for multiple molecules in MOFs, enabling the prediction of adsorption for novel pollutants using the same model. Similarly, text-mining AI has been applied to extract MOF synthesis rules from the literature, and generative models are exploring vast MOF/COF chemical spaces. A recent breakthrough applied ChatGPT (a large language model) with Bayesian optimization in a multidisciplinary AI lab to optimize microwave synthesis of MOF-321, MOF-322, and COF-323, rapidly identifying conditions that improved crystallinity and porosity. By integrating AI assistants for literature review, experimental planning, and even robotic execution, Zheng et al. achieved MOF synthesis in hours that would have taken months manually. Beyond MOFs, AI has guided the design of bio-derived adsorbents and "green" nanomaterials. For example, biopolymer nanocomposites (such as cellulose and chitosan) are being tailored by machine learning (ML) algorithms to balance performance and biodegradability. Deep learning has also been used to predict the adsorption of emerging contaminants on cyclodextrin-based nano-sponges. However, these systems often require data augmentation due to the limited availability of datasets. Overall, machine learning (ML) models, including neural networks and support vector machines, trained on experimental and simulation data now routinely predict adsorption capacity, breakthrough curves, and even regeneration efficiency of adsorbents. Metaheuristic optimizers (genetic algorithms, simulated annealing) have been used to tune process parameters (pH, temperature, dosage) in adsorption experiments, effectively "automating" the search for optimal conditions. (Zheng et al., 2023)

2.4 Photocatalytic and Reactive Nanomaterials

Photocatalysts and reactive nanoparticles (e.g., TiO_2, ZnO, g-C_3N_4, zero-valent iron) play a key role in advanced oxidation processes (AOP) for water treatment. Recent studies have utilized machine learning (ML) to optimize their composition, morphology, and operating conditions. For instance, CNNs and random forests have been trained to predict photocatalytic degradation rates of dyes and pharmaceuticals based on catalyst nanostructure and light intensity. Reinforcement learning has even been applied to dynamically adjust catalyst composition in response to sensor feedback, improving long-term stability. In one case, Bayesian optimization guided the co-doping ratios of TiO_2 nanoparticles to maximize photocatalytic disinfection efficiency, reducing the number of trial combinations from hundreds to a few candidates. The synergy of AI and nanotechnology in this field is summarized by (Palansooriya et al., 2020). AI-guided nanomaterial design enhances the adsorption capacity and catalytic efficiency of nanomaterials for removing pollutants. These AI-driven approaches have opened up pathways to discover novel photocatalysts (e.g., Z-scheme composites, plasmonic-enhanced catalysts) that were previously unexplored.(Anjum et al., 2019)

3. DESIGNING MATERIALS, MEMBRANES, AND NANO-SPONGES FOR TARGETED POLLUTANT CAPTURE

Advanced nanomaterials and engineered membranes are revolutionizing the capture of pollutants from water sources. These materials provide high surface area, tunable pore sizes, and chemical functionality tailored for specific contaminants. Nano-sponges, composed of porous structures, allow selective adsorption and retention of toxic compounds. By integrating AI tools, the performance and specificity of these materials are further enhanced. This study helps us understand how the combination of surface engineering, AI predictions, and reinforcement learning is streamlining the design and manufacturing process. These innovations are crucial in the development of next-generation water purification systems.(Iravani & Varma, 2022)

3.1. Characteristics of Nanomaterials for Water Contaminants

Nanomaterials exhibit high surface-to-volume ratios and customizable chemical functionalities, making them effective in binding and degrading water pollutants. Their ability to interact at the molecular level enables the precise targeting of contaminants, such as heavy metals, pharmaceuticals, and dyes. Transition metal dichalcogenides (TMDs), such as MoS_2 nanosheets, exhibit unique adsorption and redox properties. Moreover, their layered morphology facilitates the incorporation into membranes or composite systems. These materials are also tunable through doping and functionalization for enhanced selectivity (Vaseashta, 2008). Their adaptability makes them central to modern water remediation technologies (Akhayere et al, 2021).

3.2. AI-Enhanced Pollutant-Binding Affinity Prediction

Artificial intelligence models are increasingly being applied to predict the binding affinity between nanomaterials and specific pollutants. Machine learning algorithms analyze molecular descriptors to rank or design materials with high specificity and reactivity. This computational approach reduces the need for experimental trial and error, thereby accelerating discovery and optimization. The models can incorporate quantum mechanical properties and data from material databases for enhanced prediction accuracy. These predictions also guide surface modification strategies. Hence, integrating AI streamlines the development pipeline for customized environmental nanomaterials. (Hairom et al., 2021)

3.3. Optimization of Nano-Sponge Synthesis via AI-Guided Parameters

AI can optimize the synthesis of nano-sponges by identifying ideal synthesis conditions, such as temperature, reactant ratios, and templating agents. Using unsupervised learning from materials literature, the algorithm identifies patterns and promising routes. AI also enables high-throughput virtual screening of synthesis pathways. This method reduces costs and waste associated with conventional synthesis experiments. Automated platforms driven by AI accelerate the translation from lab-scale to industrial-scale. The result is consistent, reproducible materials with tailored adsorption properties.(Iravani & Varma, 2022)

3.4. Surface Engineering for Selectivity and Anti-Fouling

Surface modifications enhance the selectivity of nanomaterials toward specific pollutants while preventing fouling. Hydrophilic coatings and functional groups can minimize biofilm formation and fouling by organic matter. Tailored surface charges allow attraction or repulsion of target contaminants. Functionalization with ligands or reactive moieties adds specificity to target pharmaceuticals or heavy metals. Anti-fouling layers also increase membrane longevity and reduce the frequency of cleaning. Surface engineering thus plays a critical role in efficient, long-lasting pollutant capture systems (Shahouni et al., 2024)

3.5. Designing Nano-Scale Pores in Sponges and Membranes

Nano-scale pore design is crucial for the selective filtration and adsorption of pollutants. The pore size determines the types of molecules that can pass through or be captured. Recent techniques include templating methods and controlled etching to achieve uniform nanoscale porosity. Pores can be engineered to allow size exclusion or specific molecular interactions. Materials like graphene and metal-organic frameworks (MOFs) provide structural integrity while supporting customizable pore networks. Accurate control of porosity enhances water flux and selectivity (Namla et al., 2025).

3.6 Optimizing Nano Membranes Using AI

AI algorithms are used to simulate membrane performance under different operating conditions and compositions. These simulations help in designing membranes with optimized thickness, porosity, and functional groups. Neural networks can also correlate structure-property relationships from experimental datasets. AI-guided optimization reduces the need for expensive prototyping and testing. The technology facilitates rapid innovation and scaling up of nano-membrane technologies. As a result, tailored membranes for specific pollutants can be produced more efficiently. (Krishnan et al., 2024)

3.7 Reinforcement Learning for Manufacturing Optimization

Reinforcement learning (RL) algorithms are used to optimize the manufacturing processes of nanomaterials. RL agents learn the optimal actions (e.g., parameter settings) to maximize performance outcomes, such as yield or selectivity. Through trial-and-error learning and reward functions, the system improves decision-making over time. RL models can adapt in real-time to variations in raw materials or environmental conditions. This adaptability ensures consistent production quality. It also supports autonomous manufacturing environments.(Zhu et al., 2025)

3.8 Digital Twins for Real-Time Production Adjustment

Digital twins are virtual replicas of physical manufacturing systems that enable real-time simulation, monitoring, and control. By integrating sensor data and AI, digital twins predict system behavior and adjust parameters proactively. In nanomaterial production, this ensures stability and optimal synthesis conditions. Digital twins also identify anomalies or inefficiencies before they escalate. Their predictive capacity enhances scalability, reproducibility, and resource efficiency. They are becoming essential in smart nanofabrication facilities. (Wang et al., 2024)

4. REAL-TIME PERFORMANCE PREDICTION AND ADAPTIVE OPERATION

The application of nanotechnology in water treatment has revolutionized the detection, degradation, and elimination of pollutants. However, the dynamic nature of contaminants and operational variables demands a paradigm shift from static systems to adaptive, predictive technologies. This study shows us how integrating Artificial Intelligence (AI) and real-time monitoring with nano-engineered materials in water treatment systems can make these systems more responsive, efficient, and sustainable. This work highlights key innovations in predictive control of nano-synthesis, AI-enhanced maintenance prediction, membrane failure diagnostics, and intelligent decision-support systems for adaptive operation.

4.1 Prediction and Control of Nano-Synthesis

Precise control over the synthesis of nanomaterials is crucial for achieving desired properties, such as high surface area, catalytic activity, and pollutant selectivity. Conventional synthesis methods often suffer from variability, leading to inconsistent performance. AI-driven predictive models, especially those based on machine learning (ML), can now forecast the physicochemical characteristics of nanomaterials by learning from historical synthesis data and process parameters. For instance, supervised learning algorithms can optimize precursor concentration, temperature, and reaction time to ensure consistent particle size distribution and crystallinity. Real-time feedback mechanisms, supported by AI, can dynamically adjust synthesis parameters, ensuring optimal nanostructure formation and reducing batch-to-batch variability. This predictive approach accelerates innovation while reducing material waste and energy consumption.(Chen & Lv, 2022)

4.2. Integrating AI into Maintenance Prediction

The operational reliability of nanotechnology-based water treatment systems depends on timely maintenance and the longevity of their components. AI algorithms, particularly predictive maintenance models, offer powerful tools to foresee mechanical degradation, chemical fouling, or catalyst deactivation before failure occurs. Sensor networks embedded in nanomaterial reactors can continuously monitor parameters such as flow rate, turbidity, chemical composition, and pressure. AI models trained on historical maintenance logs and sensor data can predict when components, such as nano catalysts or filtration units, require servicing. This not only minimizes downtime but also extends the operational lifespan of expensive nano-enabled systems. Predictive maintenance frameworks also enable condition-based interventions, moving away from routine schedules to a more economical and sustainable predictive regime.(Hairom et al., 2021)

4.3 Integrating AI in Membrane Failure Prediction

Membrane-based nanotechnologies—such as nano-composite membranes or nano-filtration units—are vital in advanced water purification. However, membrane fouling and degradation significantly affect system efficiency. Traditional indicators often detect failure post-occurrence, but AI integration allows preemptive fault detection. By employing deep learning algorithms, large volumes of operational data, including transmembrane pressure, flux rate, and chemical residue, can be analyzed in real-time. These systems identify patterns that signal early-stage fouling, scaling, or mechanical wear. When combined with Internet of Things (IoT) infrastructure, AI-enabled modules can autonomously trigger alerts, recommend cleaning protocols, or reconfigure operational conditions to reduce membrane stress. Such systems reduce water losses, energy consumption, and membrane replacement costs, leading to more efficient and sustainable filtration operations.(Glass et al., 2023)

4.4 Decision-Support Systems for Water Treatment Optimization

The final frontier of intelligent water treatment lies in robust decision-support systems (DSS) that integrate AI, sensor networks, and user interfaces. DSS platforms leverage real-time data from various components of the treatment process, combining them with AI-based predictions to assist operators in making informed decisions. These systems consider multiple variables such as contaminant load, water flow, temperature, energy availability, and treatment goals. Using optimization algorithms, DSS can recommend the best-fit operational strategies, such as adjusting flow rates, switching between treatment modes, or dynamically selecting between nanomaterials for specific contaminants in emergency scenarios, including chemical spills or sudden pollutant surges. DSS can execute rapid simulations, predict outcomes, and suggest mitigation protocols. In remote or resource-limited settings, cloud-connected DSS also enable decentralized control and monitoring, ensuring safe water delivery without constant human supervision(Mehta et al., 2023)

5. METHODOLOGY

This methodology outlines a comprehensive AI-based pipeline for designing advanced water treatment systems that target heavy metals, organics, pathogens, and pharmaceuticals. The framework integrates Internet-of-Things (IoT) sensor networks, deep learning (DL), reinforcement learning (RL), multi-scale simulation tools (such as COMSOL Multiphysics, MATLAB, and EES), and closed-loop optimization. The pipeline proceeds through stages of parameter data collection, AI-driven molecular and membrane design, system-level simulation and optimization, and real-world deployment with adaptive control. Key AI mechanisms include convolutional and recurrent neural networks for data interpretation, deep generative models for materials design, and RL/active learning for process optimization. This novel approach combines molecular-scale design (e.g., designing MOFs, CNTs, and nanocomposites) with real-time plant simulation and feedback control, creating a closed-loop "digital twin" for water treatment.

5.1 AI-Driven Parameter Data Collection

Water quality parameters (pH, temperature, turbidity, electrical conductivity, dissolved oxygen (DO), redox potential, etc.) and pollutant concentrations (heavy metals, pathogens, pharmaceuticals, organics, nutrients) are captured by a distributed IoT sensor network. Each treatment plant or watershed is equipped with multiple sensor types, including pH electrodes, DO probes, turbidity meters, ion-selective electrodes for metals, spectroscopic sensors for organics, and biosensors for pathogens. These sensors (via Zigbee, LoRa, Wi-Fi, cellular, etc.) continuously stream data to a cloud-enabled data platform. As discussed at the beginning of this chapter, IoT-enabled water monitoring can measure a broad spectrum of parameters in real-time, including contaminants such as heavy metals. All sensor data are time-stamped and geotagged, yielding multivariate time series and spectral or image data. Raw data from the sensor network is preprocessed to ensure quality and consistency. Noise filtering, outlier removal, and drift calibration are applied to each sensor stream. Measurements are normalized and fused where necessary, such as combining spectral data (e.g., UV-vis absorption or fluorescence spectra of water samples) with physical parameters (e.g., turbidity) to form feature vectors. Automatic anomaly detection flags spurious readings. The preprocessed dataset contains a high level of contextual metadata, including flow rate, weather conditions, and upstream activities, which are used to capture dynamic influences. These preprocessed data are fed into deep learning models for interpretation and forecasting. Convolutional neural networks (CNNs) are used to analyze image or spectral data – for example, camera images of water turbidity or hyperspectral scans are used for the detection of organic contaminants. Recurrent neural networks (RNNs), such as LSTM or GRU models, process multivariate time-series data to forecast pollutant trends and identify any potential contamination events. A study combining CNN–LSTM models demonstrated the effective prediction of river water quality parameters (TN, TP, TOC) over time. Such models are trained on labeled datasets, using cross-validation and hyperparameter tuning. As the IoT network grows, active learning is employed, and the model also identifies data-deficient regions, such as pollutant levels that are not well represented or novel contamination patterns. It then selectively queries those data points, thereby refining the model. By integrating heterogeneous sensor data, the AI-driven module produces a real-time water quality profile, as well as spatial maps of contaminants and their temporal dynamics. For instance, contours for temperature, pH, DO, and heavy metal concentrations can be generated via geospatial interpolation. These data provide the initial conditions and parameter distributions for subsequent design stages, ensuring that the framework is tailored to the actual water characteristics

5.2 AI-Simulated Design of Advanced Materials and Components

Using the collected water quality data as input, the framework virtually designs and optimizes the treatment materials at the molecular and microstructural level. Implementing generative AI and RL algorithms can propose novel nanomaterials and membrane structures suited to adsorb, degrade, or filter target contaminants. For example, deep generative models like variational autoencoders, generative adversarial networks, and diffusion models are trained on databases of known materials like MOFs, carbon nanotubes (CNTs), graphene oxides, and polymer sorbents to learn the "design space" of nanostructures. We can adapt approaches such as or supramolecular Variational Autoencoders (VAEs) to generate new candidate metal-organic frameworks (MOFs) with pore geometries and functional groups optimized for specific adsorbates (e.g., arsenic, lead, or pharmaceuticals). Similarly, generative models can design polymer or ceramic composite adsorbents by combining monomers/linkers with desired affinities. Reinforcement learning further refines this design: an RL agent iteratively adjusts material parameters (e.g., linker types in MOFs, functionalization levels on CNT surfaces, crosslinking density in nanocomposites) to maximize a simulated reward (e.g., predicted adsorption capacity for a contaminant). Recent work demonstrates that deep reinforcement learning (RL) can perform "inverse materials design" by exploring compositional space to meet multiple objectives (e.g., high uptake and low cost). For instance, a multi-objective RL model identified novel inorganic oxides with target band gaps and low synthesis temperatures. Analogously, our RL module might generate a family of membrane pore size distributions or active surface chemistries. Each AI-generated candidate material is then evaluated in the simulation. We use COMSOL Multiphysics to simulate physicochemical interactions, such as a porous adsorbent model for the diffusion and adsorption of pollutants, or a membrane module to simulate flow-through ultrafiltration or nanofiltration. For adsorption capacity, Molecular Dynamics or Grand Canonical Monte Carlo (GCMC) simulations can be scripted; for example, RASPA software can be integrated via a MATLAB or Python interface to compute the equilibrium adsorption of solutes on candidate MOF structures. In one example, GCMC was used to compute adsorption in AI-designed MOFs, validating that six generated MOFs exceeded 2 mmol/g at 0.1 bar. Where appropriate, COMSOL's porous media and reaction modules simulate fluid flow and multi-species transport through the engineered material. For instance, COMSOL can couple Navier–Stokes flow with surface reaction kinetics to evaluate photocatalytic degradation of organics on a nanocomposite-coated membrane. The simulations provide performance metrics (e.g., adsorption isotherms, breakthrough curves, flux rates, reaction conversion, pressure drop) that are fed back to the AI agent. Surrogate models (such as graph neural networks or regression models trained on simulation outputs) speed up this loop. For example, one can train a crystal graph convolutional neural network (CGCNN) to predict MOF adsorption from structure, greatly accelerating the screening of thousands of candidates. The AI iterates: promising candidates are selected; their designs are mutated (using reinforcement learning) or new samples are generated (using a generative model) and then restimulated. This in silico design loop optimizes the nanomaterial and membrane parameters for maximum contaminant uptake, selectivity, and stability. The AI generative-design and screening workflow yields high-performing adsorbent structures. For illustration, Figure 1 shows six example MOF structures (a–f) generated by a deep generative framework (the GHP-MOF assembly approach). Each MOF exhibits complex 3D connectivity and functional groups tailored to adsorb CO$_2$ (analogous to heavy organic molecules) at high capacity. By analogy, our pipeline produces similarly novel adsorbents or membrane architectures for water pollutants, guided by deep DL simulations and material informatics.

5.3 System-Level Simulation and Optimization

With optimized material components in hand, the next phase simulates the entire water treatment process. We construct a plant model in COMSOL and MATLAB that includes unit operations (adsorption columns, membranes, reactors, bioreactors, UV disinfection, etc.) connected by hydraulic networks. Process parameters (flow rates, contaminant loads, pH adjustments) and thermodynamic data are integrated. COMSOL Multiphysics models fluid flow and contaminant transport, such as Darcy or Navier–Stokes flow in channels, coupled with mass transfer and reaction PDEs. Its Chemical Reaction Engineering Module can simulate biodegradation kinetics or advanced oxidation, while the Porous Media Module handles filtration through packed beds or membranes. In parallel, the Engineering Equation Solver (EES) is used for thermodynamic and equilibrium calculations, such as computing vapor-liquid equilibrium in evaporation pre-treatment or determining the solubility of salts. EES's extensive property library allows quick evaluation of enthalpies, phase equilibria, and chemical equilibria (acid-base, carbonate systems, etc.), ensuring that energy and mass balances are accurate. (As noted in desalination modeling, "EES will be used as it contains the required thermodynamic properties".) The model is driven by AI-processed input data (pollutant concentrations and flow conditions) and by the designed material properties (adsorption isotherms and membrane permeability). For example, a COMSOL simulation of an adsorption column can use the AI-designed isotherm curve as an input, computing breakthrough based on residence time. A membrane reactor model can incorporate pore size and surface charge from the AI design to predict rejection rates. Energy usage (pump power, aeration, thermal loads) is calculated from flow and thermodynamic data. To optimize system performance, we employ active learning and optimization algorithms. A global optimizer (e.g., Bayesian optimization or genetic algorithm) adjusts decision variables —such as adsorbent bed depth, membrane area, reactor volume, and operational setpoints — to maximize objectives (removal efficiency, throughput) while minimizing energy and cost. After each simulation run, an active learner analyzes which parameter regimes yield high uncertainty in performance and then queries additional simulations in those regimes to refine the surrogate models. Active learning is defined as iteratively querying the system at informative inputs until convergence. For example, a Bayesian active learning scheme can tune the pH dosing rate versus adsorbent replacement frequency to meet effluent limits with minimal reagent use. Multi-objective optimization (Pareto front methods) explores trade-offs, such as balancing heavy metal removal against membrane fouling rates. We also integrate reinforcement learning agents at the system level to develop operational policies. A training environment is built by coupling the COMSOL/MATLAB plant model with an RL toolkit (such as MATLAB's Reinforcement Learning Toolbox or a Python RL framework). The agent observes state variables (influent quality, tank levels, energy usage) and takes actions (e.g., adjusting valve positions, pump speeds, chemical dosing rates). Rewards are defined based on overall treatment goals (e.g., high removal, regulatory compliance, low cost). In practice, several RL algorithms are tested, including deep Q-learning (DQN) for discrete control and policy-gradient or actor-critic methods (e.g., TD3, PPO) for continuous control. A recent study evaluated RL control in activated sludge treatment and found that the twin delayed DDPG (TD3) algorithm substantially reduced aeration energy while maintaining effluent quality. In our pipeline, similar agents continually "learn" the best operational strategies via simulation. The learned policy can adaptively steer the plant model under varying loads and disturbances, effectively realizing an optimal control strategy that traditional PID or rule-based controls cannot achieve. Throughout this stage, all data (sensor inputs, material model outputs, process simulations) are centralized. A digital twin database stores the model states and performance metrics. This integrated simulation optimization layer generates an overall design specification, including a process flow diagram with unit sizes, material loadings, and control policies. This data-driven design meets targets on contaminant removal, energy consumption, and cost under real inflow conditions.

5.4 Real-World Integration and Adaptive Deployment

Simulations will enable a guide through the physical implementation of the treatment system and the deployment of AI-based control will be made possible. The AI-recommended materials (e.g., a tailored MOF-based adsorbent or a nano-engineered membrane) are synthesized or fabricated in the laboratory and installed in pilot-scale units. For instance, 3D printing or layer-by-layer coating can produce complex membrane geometries, while chemical synthesis pathways can create the designed nanomaterials. The plant layout from the simulation is constructed (or a modular pilot facility is assembled) with instrumentation that mirrors the digital twin. Real-time feedback loops close the design gap. IoT sensors continuously feed the physical system's operating data (effluent quality, flows, energy use) back into the digital twin. A reinforcement learning agent (trained in simulation) runs on a controller or cloud platform, taking live data as state inputs. The agent adjusts actuators in real time, such as modifying membrane backwash frequency, dosing coagulants, adjusting blower speed, or diverting flows, to maintain optimal performance. This results in a self-optimizing operation: when influent quality or flow conditions change, the RL controller re-optimizes system parameters on the fly. For example, if sensors detect a spike in heavy metal concentration, the controller might automatically increase the adsorbent flow or switch to a secondary polishing unit. If a membrane begins to foul (detected by rising pressure drop), the agent schedules a backwash sooner. Over time, the system learns (via reinforcement feedback) subtle correlations and anticipatory adjustments that human operators may miss. The entire setup thus functions as a digital twin – a virtual replica of the plant running in parallel, driven by AI, to ensure robust performance. This closed-loop deployment enables the system to be highly adaptive. The AI models are periodically retrained as more operational data accrue, improving forecasts and control policies. The framework supports continuous improvement: new material designs or process changes can be iterated by returning to the simulation stage, effectively co-evolving the digital and physical systems.

5.5 AI Models and Control Mechanisms

This framework would leverage some strategic AI algorithms. Convolutional Neural Networks (CNNs) are employed for spatial or image-based data, such as analyzing microscope images of pathogens or spectral fingerprints of organics. Recurrent Neural Networks (RNNs), such as Long Short-Term Memory (LSTM) models, model temporal patterns in time-series sensor data, enabling forecasting and anomaly detection. For materials design, deep generative models (VAEs, GANs, diffusion models) explore the space of feasible nanostructures. We also utilize graph neural networks to predict material properties from structure during screening. On the optimization front, reinforcement learning algorithms play a central role. Both value-based methods and policy-gradient methods are implemented and compared. For example, in wastewater control experiments, TD3 outperformed other methods in minimizing energy usage while meeting treatment objectives. Active learning loops based on Bayesian optimization is also implemented to sample informative simulation points and refine models. Evolutionary algorithms and gradient-based optimizers are used for multi-objective tuning of continuous design variables such as adsorbent dosage vs. flow rate. The AI training process proceeds in stages: first, supervised deep learning (DL) models (CNN, RNN) are trained on historical data using standard loss functions and cross-validation. Next, generative models are trained (often with property predictors in the loop) on libraries of known materials. RL agents are trained in the simulation environment through iterative episodes, using reward functions constructed from performance metrics. Training utilizes GPUs and parallelization (as in the GHP-MOF assembly example, where 120,000 MOFs were assembled in minutes on a supercomputer). Through the MATLAB serves as the integration environment: its toolboxes implement neural nets and RL agents, and its Live Link interface connects to COMSOL. MATLAB also runs custom scripts for data preprocessing, EES calls, and algorithm orchestration. COMSOL serves as the high-fidelity physics engine for simulating units (its flexibility allows coupling with MATLAB/Simulink). EES provides fast thermodynamic property calculations inside the optimization loop.

Figure 2. Step-by-step design of the process

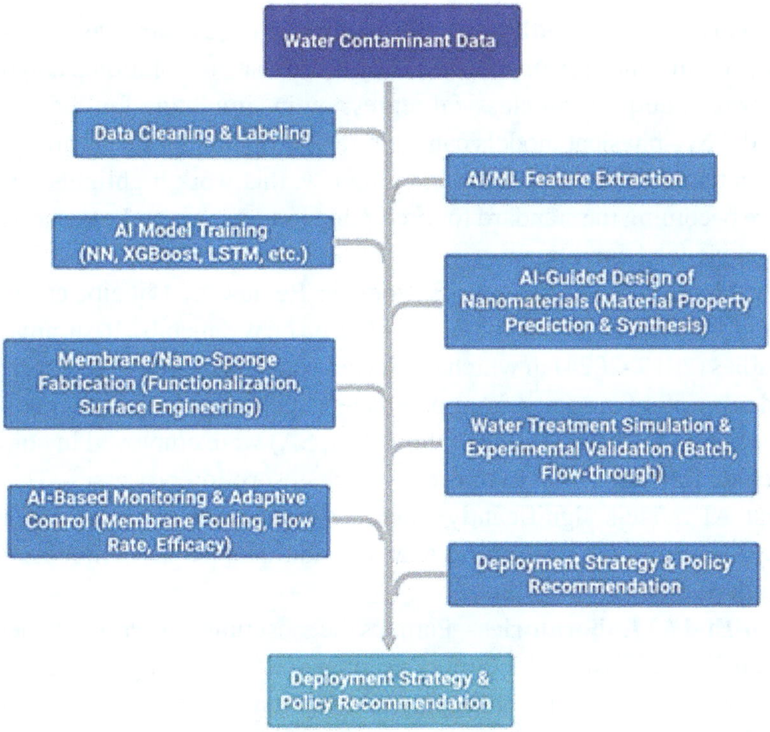

5.6. Case Studies of AI-Guided Design

Several recent studies illustrate AI's concrete impact:

Reverse Osmosis Membranes (RO) – (Gao et al., 2025) compiled metadata on >1,000 RO membranes (structure, chemistry, performance) and used explainable ML to analyze feature importance. Their models revealed which monomers, crosslink densities, and post-treatments have the most significant impact on salt rejection and flux. This meta-analysis suggests new chemistries (e.g., non-polyamide alternatives) and provides guidance for experimentalists on promising modifications.

Membrane Surface Modification – Glass et al. (2024) trained ML on data from ultrafiltration membranes whose surfaces were modified (e-beam, plasma, polymers). The model accurately predicted the permeability and surface charge of newly modified membranes. Crucially, it identified that membrane permeability was most sensitive to factors like polymer additive polarity and treatment temperature. This demonstrates that even small datasets can power useful machine learning (ML) models, and that AI can accelerate membrane development by reducing the number of experimental iterations.

Membrane Fouling Prediction – (Shahouni et al., 2024) surveyed AI models for membrane fouling and flux prediction. They noted that convolutional and recurrent neural networks outperform classical regression in simulating fouling trends, and that hybrid ANN-physical models can capture micropollutant breakthroughs better than conventional isotherms. Although a review, this work highlights a trend: AI models are becoming the standard for predicting long-term membrane performance under dynamic feed conditions.

Heavy Metal Adsorption – A Systematic Review by (Gheibi et al., 2024)) specifically focused on AI in the adsorption of heavy metals. It summarized 49 recent studies (2019–2024) in which neural networks and support vector regressors predicted adsorption capacity from parameters such as pH, contact time, and adsorbent dosage. Metaheuristic algorithms (GA, SA) were employed in conjunction with machine learning (ML) to optimize batch adsorption processes. The authors found that AI models significantly improved prediction accuracy and resource efficiency and suggested integrating AI with traditional isotherm models for better kinetic understanding.

End-to-End AI Laboratories – Perhaps the most futuristic case is (Zheng et al., 2023) Their ChatGPT Research Group" combined large language models, robotics, and ML/BO to self-direct the discovery of MOF and COF syntheses. By having multiple AI "agents" (literature reviewer, chemist, data analyst) work together, they located optimal microwave synthesis conditions for multiple frameworks from millions of options in hours. This approach, although complex, serves as proof of concept for fully AI-driven materials discovery. (Stamatin et al. 2006) It suggests that future water-treatment materials could be found by similar "AI labs" where human oversight is minimal.

5.7 Economic effect and feasibility

Adopting AI-integrated nanotechnologies in water treatment will be largely shaped by dynamic economic contexts and global market pressures. A look at the rising costs of conventional water treatment methods creates a financial push toward smarter, and more cost-efficient alternatives, mostly on industries who are now experiencing tougher economic pressure to comply with very strict environmental regulations. with AI-driven nanotechnologies, industries can comply at with these regulations at reduced operational costs, soaring Pharmaceutical and industrial effluent management costs, are managed when industries move towards adopting predictive AI tools for precision treatment. AI-driven optimization reduces trial-and-error laboratory expenses, thereby cutting research and development costs, and also improve financial feasibility by AI's ability to extend the lifetime of treatment membranes through predictive maintenance. On the other hand, Energy consumption remains a major cost driver in water treatment, and AI-directed nano membranes and sponges promise lower energy demands. (Magni et al 2025). The rising global healthcare costs due to contaminated water encourage governments to subsidize advanced treatment technologies, AI-based material design accelerates innovation cycles, aligning with economic needs for faster product-to-market timelines. water scarcity and droughts impose economic losses that motivate adoption of efficient purification technologies, predictive AI analytics lower operational uncertainties, making capital investment in water treatment more economically secure. It is popularly known that Developing countries face serious economic challenges in maintaining clean water access, water treatment in most developing countries is considered as a luxury because of the cost effect. With the help of AI-nanotech solutions attractive cheaper treatment templates can be achieved. Market competitiveness now favors firms that demonstrate sustainable, tech-driven solutions, boosting investment in AI-nanotech innovations. cost-sharing partnerships between industries, governments, and NGOs increasingly support AI-integrated projects. long-term feasibility is ensured by AI's capacity to optimize system design, reducing recurring costs. AI reduces human labor costs by automating design, monitoring, and optimization processes. Finally, Venture capital and green funds are increasingly directed toward AI-driven sustainability projects. Large corporations are under shareholder pressure to embrace ESG (Environmental, Social, and Governance) goals, driving investment in AI-nano solutions. Global economic trends emphasize resilience, and market differentiation is a strong motivator. Companies adopting AI-nanotech gain competitive branding advantages, providing both economic efficiency and technological robustness. This, in turn, ensures sustained adoption despite fluctuating market pressures. By this, they can meet the expanding global market for water treatment technologies, valued at multi-billion dollars.

6. NOVELTY AND CONTRIBUTIONS

This AI-driven methodology is novel in its holistic integration of molecular design, real-time system simulation, and closed-loop optimization. Unlike traditional approaches that treat materials development and process design separately, our pipeline co-optimizes at all scales. The use of deep generative models to create custom adsorbents and membranes (informed by actual wastewater parameters) is state-of-the-art. Combining this with a digital-twin-based active control system is unprecedented; the framework dynamically adapts the engineered materials and process based on sensor feedback, embodying the actual concept of a digital twin.

Moreover, we emphasize real-time learning: the system does not rely on fixed "if-then" rules but uses RL policies that continuously improve as new data arrive. Active learning in both materials and process spaces further accelerates discovery with minimal experiments. To our knowledge, no existing study has linked generative materials design, COMSOL/MATLAB co-simulation, and reinforcement learning control into a single autonomous design loop. In summary, this methodology presents a novel end-to-end AI framework that automates the design of advanced water treatment systems, spanning sensor data acquisition, AI-driven nanoscale design, and plant-scale operation, to deliver more effective, efficient, and adaptable solutions for clean water.

Table 1. Breakdown of methodology

Module	Purpose	Techniques Used
1. Data Handling	Collect contaminant and material performance data	Experimental data, databases (PubChem, etc.), MATLAB, Python
2. AI/ML Modeling	Predict binding affinity, fouling behavior, and pollutant selectivity.	ML (SVM, RF), DL (LSTM, CNN), AutoML
3. Material Design	AI-guided formulation of nanomaterials	Genetic Algorithms, Inverse Design, DFT calculations
4. Fabrication	Synthesis of optimized materials	Electrospinning, Sol-gel, Hydrothermal
5. Testing & Validation	Treat water in the lab and pilot scale	Real-time sensors, UV-vis, TOC, HPLC
6. Adaptive Optimization	Online learning and re-optimization	Reinforcement Learning, Digital Twin
7. Ethics & Assessment	Evaluating environmental impact	LCA, EIA, ELSI, Green Chemistry
8. Policy & Scale-Up	Recommendation for industrial deployment	Techno-economic Analysis, Road mapping

Figure 3. Schematic diagram for the method and process of AI-motivated water treatment

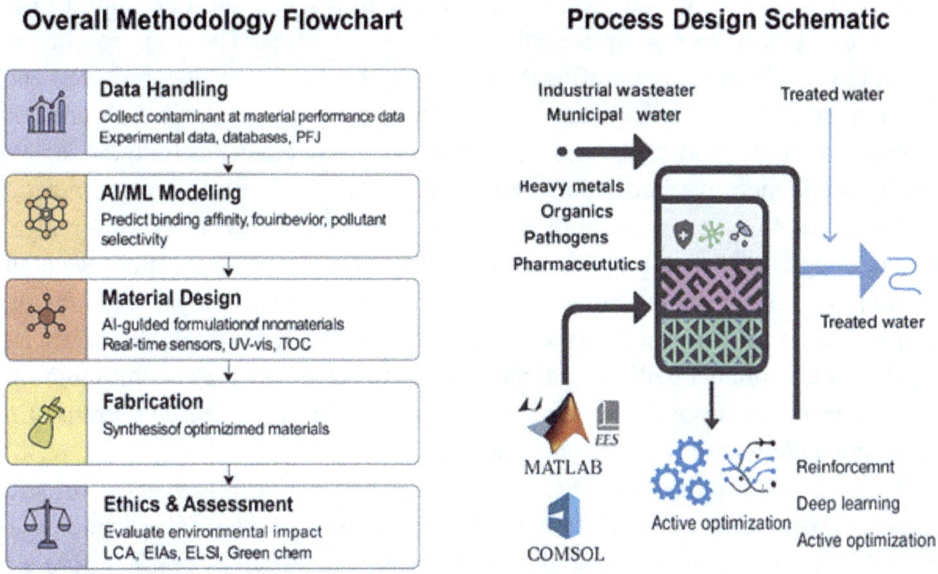

7. CONCLUSION AND RECOMMENDATIONS

Nanotechnology has been revolutionizing materials synthesis and material technologies, significantly increasing the effectiveness and efficiency of the water purification process. Several nanomaterials, including nano-polymers, nanofibers, and nano-biocides, have been synthesized for water treatment purposes and are continually applied in filtration, adsorption, and photocatalysis. From general knowledge, we are aware that nanoparticles and nanomaterials have a high surface area, chemical reactivity, and selectivity; these features enable them to eliminate numerous pollutants from water. The synthesis and application of nanomaterials and membranes for treating contaminated water require precision and monitoring, as several parameters must be considered and constantly controlled. The infusion of AI into this process becomes necessary because AI can carry out these functions with ease. Advances in sensing and monitoring systems utilizing nanotechnology have presented new opportunities for the real-time detection and monitoring of water quality. The integration of AI technology with nanotechnology has the potential to create more innovative and more efficient water treatment systems. The AI-assisted development of nanomaterials for water treatment has made significant advances, including the use of explainable machine learning (ML) models to uncover relationships between membrane structure and performance, deep neural networks predicting adsorbent uptake across chemical families, and the integration of AI laboratory systems to rapidly optimize complex syntheses. These approaches are meant to complement the traditional experimental work rather than replace bench research, and AI helps narrow the search to the most promising materials and conditions. Looking ahead, there will be continued progress in AI techniques, such as graph-based generative models for new nanostructures, multi-objective optimization for balancing flux versus fouling, as well as in their application. Additionally, if we can integrate AI with robotics and high-throughput experimentation, we could revolutionize discovery platforms, where autonomous labs design, manufacture, test, and learn continuously. At the same time, ensuring environmental safety and regulatory compliance will be critical. Interdisciplinary research, combining materials science, environmental engineering, and AI, will be essential to translating these innovations into cleaner water for all.

Environmental and sustainability concerns have become central drivers for bridging AI design with practical fabrication in advanced water treatment nanotechnologies. With AI we can predict eco-friendly design parameters such as optimal pore-size distributions for membranes or sustainable metal–organic framework (MOF) linker ratios. These predictions would then guide laboratory synthesis by informing material selection that minimizes toxic precursors and reduces energy-intensive steps. We can then design fabrication protocols to incorporate green synthesis and more solvents that are reusable, recyclable, and support bio-inspired templating methods. AI-generated pore-size targets are then translated into specific chemical reactions, such as controlled polymerization and other processes. We can also map MOF linker ratios into stoichiometric inputs for hydrothermal or solvothermal synthesis. After these materials have been fabricated, they will undergo experimental testing to determine their adsorption capacity, rejection efficiency, biodegradability, and reusability. The test results are then fed back into the AI model as new training data to refine the accuracy of the prediction. This approach reduces waste by minimizing trial-and-error synthesis and also cuts down water and chemical consumption, addressing sustainability concerns, providing consistency in synthesis and quality, and, lastly, lowering environmental footprints. On the other hand, reinforcement learning further optimizes fabrication by dynamically adjusting synthesis pathways based on real-time lab results, aligning with life cycle assessment (LCA) principles, to ensure environmental compliance at every stage. Connecting predictive AI with experimental fabrication ensures that sustainability goals are met without compromising performance. This represents a transformative model for eco-conscious nanotechnology development, where digital intelligence directly informs green laboratory practice.

Compliance and regulations

Nanomaterials for water treatment must comply with regulatory frameworks such as the U.S. **EPA** TSCA and the EU's REACH, which mandate toxicity and safety evaluations. AI can support compliance by predicting toxicological risks in advance, streamlining approval processes. Similarly, AI-enhanced life-cycle assessment (LCA) methods can estimate environmental impacts from synthesis to disposal, including energy use and carbon footprint. Together, these approaches ensure that AI-designed nanomaterials are not only efficient but also safe, sustainable, and aligned with international standards. The regulatory environment for nanomaterials is guided by frameworks such as the U.S. Environmental Protection Agency's Toxic Substances Control Act (TSCA) and the European Union's REACH (Registration, Evaluation, Authorization and Restriction of Chemicals). These frameworks require extensive data on toxicity, exposure, and environmental safety before new materials can be approved. With the infusion of AI, we can streamline compliance by predicting toxicity and hazard profiles through machine-learning models trained on chemical and toxicological datasets, thereby reducing the need for exhaustive laboratory testing. life-cycle assessment (LCA) provides a systematic way to estimate impacts from synthesis through use and disposal. AI-driven approaches accelerate this process by filling data gaps, analyzing large environmental inventories, and predicting energy use, carbon footprint, and waste outputs. By combining regulatory compliance tools with AI-enhanced LCA, researchers and industry can ensure that novel nanomaterials are not only effective in water treatment but also safe, sustainable, and aligned with global environmental protection standards.

This chapter proposes a novel, AI-motivated methodology for the synthesis, design, simulation, and optimization of nanomaterials, nanomembranes, and nanosponges, targeting the treatment of industrial and municipal wastewater contaminated with heavy metals, organics, pathogens, and pharmaceuticals. By integrating deep learning (DL), reinforcement learning (RL), and active optimization techniques into every phase of the water treatment design pipeline—from molecular simulation to real-world deployment—this approach offers a transformative paradigm for advancing sustainable, intelligent water purification technologies. The methodology begins with a rigorous data collection framework that aggregates physicochemical properties, contaminant profiles, and operational conditions using smart sensors and laboratory instrumentation. These data points form the foundation of feature engineering, where AI models are trained to identify patterns and predict outcomes such as adsorption efficiency, binding affinity, and membrane fouling tendencies. Through deep learning architectures—particularly convolutional neural networks (CNNs) and long short-term memory (LSTM) networks—molecular-level descriptors of materials can be linked to macroscopic treatment performance, enabling the design of functionally tailored nanomaterials for specific contaminant classes. Beyond predictive modeling, the integration of reinforcement learning introduces a dynamic, trial-based learning mechanism that guides iterative simulation loops for material synthesis. The agent-environment interaction paradigm allows the system to learn from simulated and experimental feedback, adapting synthesis parameters to optimize porosity, surface charge, hydrophilicity, and functional group orientation. These optimizations are then modeled and validated using multi-physics software such as COMSOL and MATLAB, facilitating the computational prototyping of nanomaterials under diverse flow and concentration scenarios. A particularly novel element of this methodology is the AI-driven simulation of the water treatment process itself. By feeding optimized nano-material properties into digital twins of treatment systems, real-time performance prediction becomes feasible. This enables dynamic adjustments in residence time, backwash cycles, and membrane replacement schedules. Through the use of active optimization loops, AI models not only monitor key performance indicators (KPIs) but also recommend interventions to enhance efficiency, reduce energy consumption, and minimize waste generation. The final phase involves the transition of these digital insights into practical implementations. Here, the AI system continues to function as a feedback and decision-support mechanism, ensuring the system adapts to real-world variability in contaminant load and operational demands. Ethical and environmental impact considerations are embedded within the optimization logic, ensuring compliance with sustainability frameworks and regulatory standards. Thus, this methodology is not only data-driven and intelligent but also ethically sound and environmentally responsive. Collectively, this approach addresses several limitations inherent in traditional water treatment designs, including low selectivity, membrane fouling, energy inefficiency, and scalability challenges. It also closes the loop between laboratory-scale discovery and field-scale deployment by leveraging AI as both a predictive engine and a design collaborator. The molecular-to-macroscale linkage enabled by this framework represents a major leap forward in realizing precision water treatment systems tailored for 21st-century challenges.

The following recommendations are made:

Adopt AI as a Collaborative Research Partner in Nanomaterial Design

Institutions, researchers, and industry players should treat AI not merely as a computational tool but as a collaborative partner in the research process. Integrating AI from the earliest stages of contaminant identification through to material design will significantly reduce development time and improve design fidelity. Cross-functional teams comprising chemists, environmental scientists, data scientists, and engineers should collaborate to co-develop AI workflows for achieving targeted water treatment goals.

Promote Open Access Environmental Databases for AI Training

A major bottleneck in AI-driven environmental applications is the scarcity of high-quality, open-access datasets. Governments, water agencies, and research labs should invest in standardizing and centralizing water contaminant profiles, treatment results, and nano-material specifications. Such repositories will serve as foundational training data for generalizable AI models that can be adapted to local treatment scenarios.

Utilize Reinforcement Learning for Adaptive Process Optimization

Most current applications of AI in water treatment remain static and predictive. This methodology recommends integrating reinforcement learning as a core driver of adaptive optimization. RL agents can adjust to changing contaminant levels, energy prices, and membrane wear, making the treatment system truly autonomous and self-healing. Pilot testing RL models in digital twin environments is a critical step toward future deployment.

Implement Multi-Scale Modeling Platforms (Molecular to Macro)

The transition from molecular descriptors to pilot-scale system behavior is often fraught with scaling inaccuracies. The use of COMSOL Multiphysics and MATLAB as modeling tools in this methodology bridges this gap. Future efforts should focus on developing standard protocols and plug-ins that enable the seamless integration of AI models trained on molecular simulations with macro-scale hydrodynamic models, thereby facilitating unified simulations.

Develop Smart Membranes with Integrated Sensing Capabilities

Moving forward, it is recommended that nano-membranes and sponges be designed with embedded sensing functionalities, either through optical, electro-chemical, or piezoelectric systems. These smart membranes would feed real-time operational data directly into the AI system, enabling live optimization of pressure, flow, and contaminant removal efficacy. Such smart systems are essential for fully autonomous treatment plants.

Establish Ethical and Environmental AI Frameworks

While AI offers immense potential, its application in environmental engineering must be guided by ethical principles and sustainability goals. Future research should incorporate environmental impact assessments (EIAs), life cycle analyses (LCAs), and ethical, legal, and social implications (ELSI) frameworks into the AI design loop. Doing so will ensure that the technology is both practical and responsible.

Engage Policymakers for Framework Integration and Scalability

For this methodology to transition into national and global water treatment strategies, strong policy alignment is necessary. Governments should be encouraged to fund AI-infrastructure upgrades in municipal water systems and to revise procurement guidelines to include AI-enhanced solutions. Pilot programs demonstrating cost savings and performance improvements can be powerful advocacy tools for such policy shifts.

Expand to Cross-Sectoral Applications Beyond Water Treatment

Although this methodology is primarily designed for water treatment, its core principles are also applicable to other sectors, such as air purification, soil remediation, and industrial process optimization. Researchers and practitioners should explore how the AI-nano-materials paradigm can be extended to other domains of environmental sustainability and public health.

Train the Next Generation of Hybrid Environmental Engineers

Ultimately, academic institutions must begin training a new generation of engineers who are fluent in both environmental science and artificial intelligence. This will require the restructuring of environmental engineering curricula to include courses on machine learning, molecular modeling, and simulation environments like MATLAB and COMSOL. Interdisciplinary graduate programs and dual PhDs will be vital in building this hybrid expertise. The integration of artificial intelligence into the synthesis and simulation of nano-enabled water treatment systems marks the beginning of a new era in environmental engineering. This methodology not only leverages the computational power of AI to solve complex environmental challenges but also redefines the boundaries of design innovation. By uniting molecular-level intelligence with macro-scale process optimization, this framework paves the way for cleaner water, smarter systems, and a more sustainable planet.

REFERENCES

Akhayere, E., & Kavaz, D. (2021). Nano-silica and nano-zeolite synthesized from barley grass straw for effective removal of gasoline from aqueous solution: A comparative study. *Chemical Engineering Communications*, *208*(10), 1419–1435. DOI: 10.1080/00986445.2020.1786373

Akhayere, E., Vaseashta, A., & Kavaz, D. (2020). Novel magnetic nano silica synthesis using barley husk waste for removing petroleum from polluted water for environmental sustainability. *Sustainability (Switzerland)*, *12*(24), 1–16. DOI: 10.3390/su122410646

Anjum, M., Miandad, R., Waqas, M., Gehany, F., & Barakat, M. A. (2019). Remediation of wastewater using various nano-materials. *Arabian Journal of Chemistry*, *12*(8), 4897–4919. DOI: 10.1016/j.arabjc.2016.10.004

Chen, X., & Lv, H. (2022). Intelligent control of nanoparticle synthesis on microfluidic chips with machine learning. *NPG Asia Materials*, *14*(1), 1–20. DOI: 10.1038/s41427-022-00416-1

Gao, W., Wang, G., Li, J., Li, H., Ren, L., Wang, Y., & Kong, L. (2025). Insights into Synthesis and Optimization Features of Reverse Osmosis Membrane Using Machine Learning. *Materials (Basel)*, *18*(4), 1–14. DOI: 10.3390/ma18040840 PMID: 40004363

Gheibi, M., Masoomi, S. R., Magala, M. U., & Amir, M. (2024). Artificial Intelligence (AI) in Adsorption Process of Heavy Metals. *Systematic Reviews*, *2*(2), 57–78.

Glass, S., Schmidt, M., Merten, P., Abdul Latif, A., Fischer, K., Schulze, A., Friederich, P., & Filiz, V. (2023). Design of Modified Polymer Membranes Using Machine Learning. *ACS Applied Materials & Interfaces*. Advance online publication. DOI: 10.1021/acsami.3c18805 PMID: 38600824

Hairom, N. H. H., Soon, C. F., Mohamed, R. M. S. R., Morsin, M., Zainal, N., Nayan, N., Zulkifli, C. Z., & Harun, N. H. (2021). A review of nanotechnological applications to detect and control surface water pollution. *Environmental Technology & Innovation*, *24*, 102032. DOI: 10.1016/j.eti.2021.102032

Iravani, S., & Varma, R. S. (2022). Nanosponges for Water Treatment: Progress and Challenges. *Applied Sciences (Switzerland)*, *12*(9). Advance online publication. DOI: 10.3390/app12094182

Krishnan, A., Sundaram, T., Nagappan, B., & Devarajan, Y., & Bhumika. (2024). Integrating artificial intelligence in nanomembrane systems for advanced water desalination. *Results in Engineering*, *24*(November), 103321. DOI: 10.1016/j.rineng.2024.103321

Magni, M., Jones, E. R., Bierkens, M. F. P., & van Vliet, M. T. H. (2025). Global energy consumption of water treatment technologies. Water Research, 277(December 2024), 123245. DOI: 10.1016/j.watres.2025.123245

Mehta, K. P., Sharma, R., Haldar, S., & Kumar, A. (2023). Advancement in treatment of wastewater with nano technology. *Materials Today: Proceedings*, *80*, 3358–3362. DOI: 10.1016/j.matpr.2021.07.253

Namla, D., Oves, M., Alshaeri, M. A., Al-Maaqar, S. M., Issa, H. N. Y., & Mangse, G. (2025). Nanofiltration as an advanced wastewater treatment technique: A comprehensive review. *Discover Applied Sciences*, *7*(4). Advance online publication. DOI: 10.1007/s42452-025-06576-3

Palansooriya, K. N., Yang, Y., Tsang, Y. F., Sarkar, B., Hou, D., Cao, X., Meers, E., Rinklebe, J., Kim, K. H., & Ok, Y. S. (2020). Occurrence of contaminants in drinking water sources and the potential of biochar for water quality improvement: A review. *Critical Reviews in Environmental Science and Technology*, *50*(6), 549–611. DOI: 10.1080/10643389.2019.1629803

Shahouni, R., Abbasi, M., Dibaj, M., & Akrami, M. (2024). Utilizing Artificial Intelligence to Predict Membrane Behavior in Water Purification and Desalination. *Water (Switzerland)*, *16*(20). Advance online publication. DOI: 10.3390/w16202940

Stamatin, I., Berlic, C., & Vaseashta, A. (2006). On the computer-aided modelling of analyte–receptor interactions for an efficient sensor design. *Thin Solid Films*, *495*(1-2), 312–315.

Vaseashta, A. (2008). Nanoscale Materials, Devices, and Systems for Chem.-Bio Sensors, Photonics, and Energy Generation and Storage. In Vaseashta, A., & Mihailescu, I. N. (Eds.), *Functionalized Nanoscale Materials, Devices and Systems. NATO Science for Peace and Security Series B: Physics and Biophysics*. Springer., DOI: 10.1007/978-1-4020-8903-9_1

Vaseashta, A. (2021). Introduction to Water Safety, Security and Sustainability. In Vaseashta, A., & Maftei, C. (Eds.), *Water Safety, Security and Sustainability. Advanced Sciences and Technologies for Security Applications*. Springer., DOI: 10.1007/978-3-030-76008-3_1

Vaseashta, A. (2022). Nexus of Advanced Technology Platforms for Strengthening Cyber-Defense Capabilities, Vaseashta, DOI: , NATO Science for Peace and Security Series - E: Human and Societal Dynamics. Vol. 155: Practical Applications of Advanced Technologies for Enhancing Security and Defense Capabilities: Perspectives and Challenges for the Western BalkansDOI: 10.3233/NHSDP220003

Vaseashta, A., Duca, G., & Travin, S. (2022). *Handbook of Research on Water Science and Society* (Vols. 1–2). IGI Global publishers., DOI: 10.4018/978-1-7998-7356-3

Vaseashta, A., Gevorgyan, G., Kavaz, D., Ivanov, O., Jawaid, M., & Vasović, D. (2021). Exposome, Biomonitoring, Assessment, and Data Analytics to Quantify Universal Water Quality. In Vaseashta, A., & Maftei, C. (Eds.), *Water Safety, Security and Sustainability. Advanced Sciences and Technologies for Security Applications*. Springer., DOI: 10.1007/978-3-030-76008-3_4

Vishnyakov, A. (2025). Machine Learning in Computational Design and Optimization of Disordered Nanoporous Materials. *Materials (Basel)*, *18*(3). Advance online publication. DOI: 10.3390/ma18030534 PMID: 39942200

Wang, A. J., Li, H., He, Z., Tao, Y., Wang, H., Yang, M., Savic, D., Daigger, G. T., & Ren, N. (2024). Digital Twins for Wastewater Treatment: A Technical Review. *Engineering*, *36*, 21–35. DOI: 10.1016/j.eng.2024.04.012

Wang, Z., Pal, D., Pilechi, A., & Ariya, P. A. (2024). Nanoplastics in Water: Artificial Intelligence-Assisted 4D Physicochemical Characterization and Rapid In Situ Detection. *Environmental Science & Technology*, *58*(20), 8919–8931. DOI: 10.1021/acs.est.3c10408 PMID: 38709668

Yao, Z., Sánchez-Lengeling, B., Bobbitt, N. S., Bucior, B. J., Kumar, S. G. H., Collins, S. P., Burns, T., Woo, T. K., Farha, O. K., Snurr, R. Q., & Aspuru-Guzik, A. (2021). Inverse design of nanoporous crystalline reticular materials with deep generative models. *Nature Machine Intelligence*, *3*(1), 76–86. DOI: 10.1038/s42256-020-00271-1

Yin, H., Xu, M., Luo, Z., Bi, X., Li, J., Zhang, S., & Wang, X. (2024). Machine learning for membrane design and discovery. *Green Energy & Environment*, *9*(1), 54–70. DOI: 10.1016/j.gee.2022.12.001

Zheng, Z., Zhang, O., Nguyen, H. L., Rampal, N., Alawadhi, A. H., Rong, Z., Head-Gordon, T., Borgs, C., Chayes, J. T., & Yaghi, O. M. (2023). ChatGPT Research Group for Optimizing the Crystallinity of MOFs and COFs. *ACS Central Science*, *9*(11), 2161–2170. DOI: 10.1021/acscentsci.3c01087 PMID: 38033801

Zhu, Z., Dong, S., Zhang, H., Parker, W., Yin, R., Bai, X., Yu, Z., Wang, J., Gao, Y., & Ren, H. (2025). Bayesian Optimization-Enhanced Reinforcement learning for Self-adaptive and multi-objective control of wastewater treatment. *Bioresource Technology*, *421*(February), 132210. DOI: 10.1016/j.biortech.2025.132210 PMID: 39933666

Chapter 3

Predictive Modeling for Nanomaterials Using AI-Assisted Design of Experiments

Sai Chaitanya Kishore
https://orcid.org/0000-0002-4524-4231
Srinivasa Ramanujan Institute of Technology, India

T. C. Manjunath
https://orcid.org/0000-0003-2545-9160
Rajarajeswari College of Engineering, India

V. Lavanya
https://orcid.org/0009-0000-0793-5157
CMR University, India

M. R. Arun
https://orcid.org/0000-0002-8291-2956
Vel Tech Rangarajan Dr. Sagunthala R&D Institute of Science and Technology, India

P. Selvakumar
https://orcid.org/0000-0002-3650-4548
Department of Science and Humanities, Nehru Institute of Technology, Coimbatore, India

F. Shabina Fred Rishma
Vel Tech Rangarajan Dr. Sagunthala R&D Institute of Science and Technology, India

ABSTRACT

The convergence of artificial intelligence and nanomaterials science is revolutionizing the discovery, design, and optimization of advanced materials with unprecedented speed and precision. AI-driven methodologies, such as predictive modeling, generative algorithms, and automated design of experiments, enable the rapid identification of novel nanomaterials with tailored properties for diverse applications in energy storage, electronics, healthcare, and environmental sustainability. Traditional materials discovery relies on resource-intensive experimental approaches and iterative

DOI: 10.4018/979-8-3373-2883-6.ch003

trial-and-error methods, often requiring years or even decades to bring new materials from concept to commercial use. This chapter describes how AI accelerates this process by leveraging machine learning, deep learning, and big data analytics to predict material behaviors, optimize synthesis pathways, and identify promising candidates with minimal experimental intervention.

INTRODUCTION TO AI-DRIVEN DISCOVERY OF NANOMATERIALS

Big data, computational modeling, and autonomous experimentation, along with AI-driven discovery methods, are accelerating scientific breakthroughs in medicine, energy, electronics, and environmental sustainability. This paradigm shift is enabling researchers to explore new materials at the atomic and molecular levels, dramatically reducing the time and resources required to develop innovative solutions for global challenges. AI is transforming the discovery of nanomaterials by introducing automation, predictive modeling, and intelligent design strategies that go beyond traditional experimental and theoretical approaches. Key advancements include Machine Learning for Materials Property Prediction (Al-Zoubi et al., 2024): AI algorithms can analyze vast datasets from previous experiments, computational simulations, and quantum mechanical calculations to predict the electronic, mechanical, optical, and thermal properties of novel nanomaterials. Deep Learning for Molecular Structure Optimization: By training deep neural networks on vast chemical and structural datasets, AI can identify patterns and relationships between atomic structures and material properties, leading to the rapid design of optimized nanomaterials. High-Throughput Virtual Screening: AI-driven simulations enable the rapid screening of millions of potential nanomaterial candidates, significantly reducing the time required to identify promising materials. Autonomous Experimental Platforms: AI-powered robotic labs and self-learning AI-driven synthesis systems are capable of conducting autonomous experiments, continuously refining processes without human intervention. AI-powered predictive models are helping scientists design biocompatible nanocarriers for targeted drug delivery, ensuring precision medicine with minimal side effects. AI-driven nanosensors are enabling real-time disease detection by identifying biomarkers at the molecular level, advancing early diagnostics for diseases like cancer and neurodegenerative disorders. As AI models continue to evolve, future breakthroughs in nanomaterials discovery will be driven by the design of Quantum AI-enhanced materials (Apablaza-Campos et al., 2024), which integrate quantum computing with AI to explore high-dimensional material spaces beyond the limitations of classical computing. Autonomous robotic researchers, where AI-powered robotic labs conduct independent experiments, accelerate the synthesis of nanomaterials. Self-optimizing materials, where AI designs adaptive nanomaterials capable of responding dynamically to their environment. The fusion of AI, nanotechnology, and computational modeling is ushering in a new era of intelligent materials science, where conventional trial-and-error methods no longer limit the discovery and development of nanomaterials. AI-driven approaches are unlocking faster, smarter, and more efficient materials design, leading to revolutionary advancements in energy, medicine, electronics, and sustainability (Ashok et al., 2022).

THE CHALLENGE OF NANOMATERIALS DISCOVERY: COMPLEXITY, SCALE, AND SPEED

The discovery of new nanomaterials is one of the most challenging frontiers in materials science, offering enormous potential for breakthroughs in energy, medicine, electronics, and sustainability. However, this field is plagued by fundamental difficulties arising from the complexity of nanoscale interactions, the vast scale of possible material compositions, and the speed at which experiments and simulations must be conducted to meet the growing demands of modern technology. Unlike traditional materials, nanomaterials exhibit size-dependent properties, meaning their behavior is often nonlinear, difficult to predict, and highly sensitive to atomic structure and surface chemistry. These challenges have historically slowed the progress of discovering and deploying next-generation nanomaterials. However, recent advancements in artificial intelligence (AI), quantum computing, and high-throughput experimental methods are beginning to revolutionize this process, enabling the rapid discovery and optimization of novel nanomaterials. Nanomaterials operate at the atomic and molecular scale, where quantum effects, surface interactions, and nanoscale confinement phenomena dominate their properties(Beckett, 2019). This makes their discovery far more complex than that of conventional bulk materials. Some of the primary complexities include nonlinear behavior. At the nanoscale, electronic, optical, mechanical, and chemical properties can change dramatically with slight variations in size, shape, or composition, making it challenging to predict material performance. Quantum Effects: Unlike macroscopic materials, nanomaterials can exhibit quantum confinement, tunneling, and entanglement, requiring advanced computational models and experimental techniques to understand and control their behavior. Another major challenge in nanomaterials discovery is the time-intensive nature of traditional research methods. The typical workflow for discovering a new nanomaterial—synthesizing candidate materials, testing their properties (Beckett, 2023), optimizing their structure, and scaling up production—can take years to decades. Some of the primary speed-related bottlenecks include time-consuming synthesis, as many nanomaterials require complex, multi-step processes, making it challenging to explore multiple candidates efficiently. Additionally, synthesis conditions such as temperature, pressure, reaction time, and precursor concentrations all influence material properties, requiring extensive trial and error.

Slow Characterization Techniques: Traditional characterization methods—such as electron microscopy, X-ray diffraction, and spectroscopy techniques—are time-consuming, requiring hours or days per sample. This limits the rate at which new nanomaterials can be tested. Computational Bottlenecks: Despite advances in computational modeling, quantum chemistry, and molecular simulations, predicting nanomaterial properties with high accuracy remains a time-intensive process, often requiring supercomputers to analyze just a handful of materials at a time. Experimental Validation Lag: Even when a promising nanomaterial is computationally predicted, translating it into a real-world application requires further synthesis, optimization, and testing, which can take years. Recent advancements in artificial intelligence (AI), automation, and high-throughput experimentation are beginning to transform the discovery of nanomaterials by tackling these fundamental challenges head-on. These include AI-Driven Material Design: Machine learning and deep learning models can predict optimal nanomaterials based on prior data, significantly reducing the number of experiments needed. High-Throughput Screening: Automated labs equipped with robotic synthesis platforms and AI-driven analytics can rapidly test thousands of material candidates in a fraction of the time required by traditional methods. Quantum Computing for Material Simulations: Quantum algorithms are expected to revolutionize nanomaterials discovery by solving complex quantum mechanical equations exponentially faster than classical computers. Despite the significant challenges in nanomaterials discovery—complexity, scale, and speed—emerging technologies are offering transformative solutions that will reshape the field of materials science. The integration of AI, quantum computing, high-throughput experimentation, and autonomous robotics is set to revolutionize how we design, test, and apply new nanomaterials(Carlson, 2015). These advances will pave the way for next-generation materials with applications in clean energy, advanced electronics, medicine, and sustainable infrastructure. By overcoming these obstacles, scientists and engineers will unlock the full potential of nanotechnology, ushering in a new era of intelligent materials design that will drive technological progress for decades to come.

THE ROLE OF AI IN MATERIALS SCIENCE: PREDICTIVE MODELING, DESIGN OF EXPERIMENTS, AND MORE

The integration of artificial intelligence (AI) in materials science is revolutionizing the way scientists discover, design, and optimize new materials. Traditionally, the development of materials has been a time-consuming and resource-intensive process, requiring years or even decades of experimental testing and computational simulations. However, AI-driven approaches, including predictive modeling (Caswell & Dörr, 2018), machine learning, automated experimentation, and generative design, are transforming the field by accelerating discoveries, reducing costs, and enhancing precision. By leveraging big data, deep learning, and autonomous systems, AI is enabling breakthroughs in nanomaterials, biomaterials, superconductors (Cools & Diakopoulos, 2023), catalysts, and quantum materials, ultimately leading to next-generation solutions for energy, electronics, medicine, and sustainability. For example, AI-driven models have successfully predicted high-efficiency solar cell materials, superconductors at higher temperatures, and ultra-strong lightweight alloys, all of which were later validated in laboratory settings. AI is also transforming the way experiments are conducted in materials science through automated and intelligent design of experiments (DOE). Instead of relying on traditional, time-consuming trial-and-error methods, AI optimizes experimental workflows by selecting the most informative experiments. AI algorithms determine the minimum number of experiments required to explore a material's full potential, reducing wasted time and resources. Guiding Synthesis Conditions(Cools & Diakopoulos, 2024): AI models analyze reaction parameters (such as temperature, pressure, and precursor composition) to find the optimal synthesis conditions for producing materials with the desired properties. Automating Laboratory Workflows: AI-driven robotic labs can autonomously conduct and adjust experiments in real-time, self-optimizing material fabrication with minimal human intervention. For instance, AI has been used to design and optimize catalysts for clean energy applications, drastically accelerating the discovery of low-cost, high-performance alternatives to platinum-based catalysts. Data Limitations: AI models require large, high-quality datasets, but experimental material data is often scarce or inconsistent. Computational Costs: Training complex AI models for materials and science demands high-performance computing resources, which can be costly and energy intensive. Interpretability (De la Hoz et al., 2023): AI-driven material discoveries must be physically explainable and experimentally verifiable to ensure trust and reliability in scientific research. Integration with Experimental Labs: While AI can design and predict materials, integrating AI seamlessly into laboratory workflows remains a technical challenge. Despite these challenges, AI is fundamentally reshaping the way materials are discovered, designed, and engineered. As AI models improve and datasets grow, we can expect: Quantum AI-assisted materials discovery, leveraging quantum computing for ultra-fast simulations. Fully autonomous AI-driven materials labs, capable of independently designing and testing materials in real-time. AI-powered material databases, where researchers can instantly search for optimal materials for any given application. The fusion of AI and materials science is driving a revolution in predictive modeling, experimental design, and autonomous discovery. AI enables researchers to explore vast material spaces efficiently, uncover new scientific principles, and accelerate the translation of materials from theory to application (Cools & Koliska, 2024). As AI continues to evolve, it will play a crucial role in developing next-generation materials for clean energy, advanced electronics, sustainable manufacturing, and biomedical innovations. The future of materials science is not just about discovery; it is about intelligent, data-driven, and AI-powered breakthroughs that will transform industries and improve human life.

PREDICTIVE MODELING OF MATERIAL PROPERTIES: MACHINE LEARNING AND DEEP LEARNING TECHNIQUES

The ability to accurately predict material properties is crucial for accelerating discoveries in materials science, nanotechnology, and engineering. Traditionally, researchers have relied on trial-and-error experimentation and physics-based simulations to study materials, which can be slow, resource-intensive, and computationally expensive. However, the emergence of machine learning (ML) and deep learning (DL) techniques is revolutionizing predictive modeling, enabling researchers to rapidly screen materials, optimize properties, and discover novel compounds with unprecedented speed and accuracy(Díaz-Campo & Chaparro-Domínguez, 2020). These AI-driven methods extract patterns from vast datasets, generalize complex relationships, and make highly accurate predictions, significantly reducing the time and cost associated with materials research. By leveraging ML and DL, researchers can predict mechanical, electronic, thermal, and optical properties of materials based on atomic structure, chemical composition, and processing conditions. These techniques are being applied to high-throughput material screening, inverse materials design, and accelerated discovery of superconductors, semiconductors, catalysts, and biomaterials. This article explores how machine learning and deep learning are transforming predictive modeling in materials science, discussing key methodologies, applications, and challenges. Researchers have utilized supervised ML models to predict the band gaps of semiconductors, a critical property in solar cells and optoelectronics. By training ML models on datasets of known semiconductors and band gaps, scientists can rapidly identify new materials with ideal electronic properties without conducting labor-intensive experiments. Unsupervised learning methods help identify hidden patterns in materials data without requiring labeled outputs. These techniques help in grouping materials with similar properties(Díaz Noci et al., 2024), discovering novel material classes, and detecting anomalies in experimental datasets.

Applications of Predictive Modeling in Materials Science: From Nanomaterials to Bulk Materials

These advanced techniques allow for accurate forecasting of material behavior, reducing the reliance on traditional trial-and-error experimentation and costly physical testing. From nanomaterials with atomic-level precision to bulk materials used in large-scale manufacturing, predictive modeling plays a crucial role in diverse industries, including electronics, energy storage, aerospace, healthcare, and structural engineering (Fırat, 2019). This article explores the applications of predictive modeling across different material classes, highlighting how AI-driven methodologies are accelerating innovation in nanomaterials, polymers, biomaterials, composites, and metals. Nanomaterials, defined by their nanoscale dimensions (1–100 nm), exhibit unique optical, mechanical, electrical, and catalytic properties that differ significantly from their bulk counterparts. (Forja-Pena et al., 2024) Predictive modeling is essential for understanding and tailoring these properties, as quantum effects, surface interactions, and size-dependent phenomena govern nanoscale behavior. AI-driven graph neural networks (GNNs) have been used to predict the stability of hypothetical 2D materials, leading to the discovery of new ultrathin semiconductors with high carrier mobility for next-generation transistors. Predictive modeling is redefining materials discovery and engineering, enabling faster, smarter, and more cost-effective research. From nanomaterials to bulk materials, AI-driven simulations and ML techniques are accelerating the design of next-generation materials for electronics, energy, aerospace, healthcare, and structural applications. As AI continues to evolve, predictive modeling will become an essential tool for developing innovative, high-performance, and sustainable materials, shaping the future of technology and industry.

CHALLENGES AND LIMITATIONS OF PREDICTIVE MODELING IN MATERIALS SCIENCE: DATA QUALITY, MODEL INTERPRETABILITY, AND MORE

Predictive modeling has revolutionized materials science, enabling researchers to accelerate material discovery, optimize properties, and reduce experimental costs. However, despite its transformative potential, predictive modeling faces significant challenges and limitations that hinder its full integration into materials research. These challenges include data quality and availability, model interpretability, computational costs, and experimental validation constraints. Addressing these limitations is critical to ensuring that predictive models produce reliable, scalable, and practically useful results for industries such as electronics, energy, aerospace, healthcare, and manufacturing. This article explores the key challenges in predictive modeling for materials science, focusing on issues such as data scarcity(García-Orosa et al., 2023), bias, model explainability, transferability, and integration with experimental techniques. By understanding these limitations, researchers can develop more robust, transparent, and generalizable AI-driven approaches to materials design and discovery. These approaches suffer from data scarcity, inconsistency, and bias, limiting the effectiveness of machine learning (ML) and deep learning (DL) models. Limited Experimental Data: Unlike fields like image recognition (which have millions of labeled examples), materials science datasets are often small and fragmented. Many materials have never been synthesized or tested, leading to data gaps. Noisy and Inconsistent Measurements: Material properties can vary due to differences in synthesis conditions, measurement techniques, and environmental factors. This introduces noise and reduces the reproducibility of predictive models. Many datasets focus on well-known materials, leading to a lack of diversity in training data. As a result, models struggle to predict properties for novel or underexplored materials. (Helberger & Diakopoulos, 2023) Deep learning models trained on biased materials datasets may overestimate band gaps for novel semiconductors because the training data primarily consists of oxide-based materials, neglecting organic and hybrid materials. Combining machine learning with first-principles physics simulations (e.g., Density Functional Theory - DFT) to ensure predictions follow known scientific principles. Feature Importance Analysis involves identifying which atomic, electronic, or structural features most contribute to predicted properties. Few-Shot and Zero-Shot Learning: Training models to make predictions for new materials without requiring a large dataset (Hermann, 2022). Domain Adaptation Techniques: Using transfer learning to adapt models trained on one material type to another. Physics-Informed Machine Learning: Incorporating physical laws and constraints to improve model robustness. Training Deep Learning Models: Requires high-performance GPUs or specialized AI hardware, which may not be available to all researchers. High-Fidelity Quantum Simulations: Techniques like DFT and molecular dynamics can take weeks to months for complex materials. Scalability Issues: Many predictive models are optimized for small datasets but struggle to handle large-scale industrial applications. Even when predictive models suggest new materials with promising properties, these materials must be synthesized and tested experimentally—a major bottleneck in materials science. Despite significant challenges, predictive modeling remains a powerful tool for materials discovery.

Addressing data quality, model interpretability, generalizability, computational costs, and experimental validation will be key to unlocking AI's full potential in materials science. As quantum computing, automated laboratories, and physics-informed AI continue to evolve, the future promises more accurate, scalable, and trustworthy predictive models, accelerating breakthroughs in nanomaterials, quantum materials, energy storage, and advanced manufacturing.

DESIGN OF EXPERIMENTS (DOE) FOR MATERIALS SCIENCE: TRADITIONAL METHODS AND AI-DRIVEN APPROACHES

The Design of Experiments (DoE) is a fundamental methodology in materials science that systematically explores how different factors influence material properties. Traditionally, DoE has been used to optimize material formulations, processing conditions, and performance characteristics through structured experimental planning (Ioscote et al., 2024). However, with the rise of artificial intelligence (AI) and machine learning (ML), DoE is undergoing a transformation, where AI-driven approaches are significantly enhancing efficiency, reducing costs, and accelerating materials discovery. This article explores both traditional and AI-enhanced DoE methodologies, highlighting their advantages, limitations, and applications in nanomaterials, polymers, alloys, composites, and energy materials. By integrating statistical techniques, computational simulations, and machine learning, researchers can optimize experiments, minimize resource consumption, and predict material behaviors with unprecedented accuracy. High Experimental Costs require significant material, time, and labor resources. Limited Scope includes struggles with complex materials having nonlinear interactions. Being time-consuming, it cannot efficiently explore large parameter spaces. AI-driven DoE enhances traditional methods by using machine learning, deep learning, and Bayesian optimization to select the most informative experiments and reduce unnecessary trials intelligently. Faster Material Discovery: Identifies promising materials in weeks instead of years. Reduced Costs: Minimizes the number of required experiments while maintaining accuracy. Explores Complex Interactions: Captures nonlinear and high-dimensional relationships that traditional methods miss. Data Dependency: Requires high-quality data to train models effectively. Computational Costs: Some AI methods demand significant processing power (Israel & Amer, 2022). Model Interpretability: Predictions may lack clear explanations. The combination of traditional statistical methods and AI-driven approaches is revolutionizing DoE in materials science. While traditional DoE provides a structured foundation, AI enhances it by accelerating discovery, reducing costs, and optimizing experimental workflows. As AI, robotics, and computational modeling continue to advance, DoE will play an even greater role in unlocking next-generation materials for energy, electronics, aerospace, and healthcare applications.

AI-DRIVEN DOE FOR NANOMATERIALS DISCOVERY: OPTIMIZING EXPERIMENTAL CONDITIONS AND REDUCING TRIAL-AND-ERROR

The discovery and optimization of nanomaterials play a crucial role in advancing technologies in electronics, energy storage, catalysis, and biomedical applications. However, the traditional approach to nanomaterials research relies heavily on trial-and-error experimentation, which is time-consuming, resource-intensive, and inefficient. The complexity of nanoscale systems, where properties are influenced by atomic structure, synthesis conditions, and environmental factors, makes experimental design even more challenging. To overcome these limitations, researchers are increasingly turning to AI-driven Design of Experiments (DoE), leveraging machine learning (ML), deep learning (DL), and Bayesian optimization to guide experimental workflows intelligently. AI-driven DoE enhances the discovery of high-performance nanomaterials by systematically identifying optimal synthesis conditions, minimizing the number of required experiments, and revealing nonlinear relationships that traditional methods often overlook (JournalismAI, 2024). This article explores how AI-driven DoE transforms nanomaterials research, reducing trial-and-error, accelerating discovery, and improving reproducibility across various domains, including quantum dots, 2D materials, nanocatalysts, and nanocomposites. Multidimensional Parameter Space: Nanomaterial synthesis involves multiple variables, including precursor composition, temperature, pressure, reaction time, pH, and surfactant concentration. Traditional DoE requires a large number of experiments to explore all possible interactions. Nonlinear and Stochastic Behaviors: Nanomaterial properties are often influenced by complex and unpredictable interactions at the atomic level, making traditional linear models inadequate. Resource Constraints: Synthesizing and characterizing nanomaterials can be expensive and time-consuming, requiring high-precision instrumentation (e.g., TEM, XPS, AFM). (Jones, 2023)Traditional DoE methods struggle to efficiently navigate this complex experimental space, often requiring hundreds to thousands of experiments to find optimal conditions. AI-driven approaches offer a transformative solution by predicting promising experimental conditions with fewer trials. AI-driven DoE integrates machine learning, deep learning, and Bayesian optimization to select experiments that maximize information gain while minimizing trial-and-error intelligently. By leveraging AI, researchers can explore vast experimental spaces efficiently and accurately, leading to faster nanomaterial discoveries. Bayesian Optimization: Iteratively selects the most informative experiments, prioritizing promising conditions while discarding less relevant ones. Active Learning: AI models dynamically adjust experimental strategies based on real-time feedback, refining predictions with each new experiment. Neural Networks & Deep Learning: Capture nonlinear relationships between synthesis parameters and material properties, improving prediction accuracy. Reinforcement Learning (Krausová & Moravec, 2022) (RL): Models experimental design as an AI-driven Design of Experiments (DoE), which is revolutionizing nanomaterials discovery by optimizing experimental conditions, reducing trial-and-error, and accelerating innovation. By integrating machine learning, Bayesian optimization, reinforcement learning, and deep learning, researchers can rapidly explore vast experimental spaces, reduce costs, and enhance reproducibility. As AI continues to evolve, the integration of autonomous laboratories, quantum computing, and real-time adaptive learning will further enhance nanomaterials research, unlocking new materials for energy, medicine, electronics, and beyond. The future of nanotechnology and materials science is being shaped by AI-driven experimentation, leading to breakthroughs at the atomic scale with unprecedented speed and precision.

CASE STUDY: AI-DRIVEN DOE FOR THE DISCOVERY OF NEW NANOCATALYSTS

The discovery of high-performance nanocatalysts is critical for advancing clean energy technologies, environmental remediation, and industrial chemical processes. Traditional catalyst development relies on time-consuming trial-and-error synthesis and empirical optimization, making it expensive and inefficient. However, AI-driven Design of Experiments (DoE) is transforming this process by enabling data-driven catalyst discovery, significantly reducing experimental workloads while improving performance and scalability. This case study explores how AI-driven DoE has been successfully applied to nanocatalyst discovery, specifically focusing on the development of platinum-free hydrogen evolution reaction (HER) catalysts for green hydrogen production(Lao & You, 2024). By integrating machine learning (ML), Bayesian optimization, and high-throughput experimentation, researchers have accelerated the identification of cost-effective, high-efficiency catalytic materials, demonstrating the power of AI-driven DoE in revolutionizing nanomaterials research. This case study highlights how AI-driven Design of Experiments (DoE) is transforming nanocatalyst discovery, offering a faster, cost-effective, and highly efficient approach to materials research. By leveraging machine learning, Bayesian optimization, and active learning, researchers have significantly reduced the trial-and-error process, leading to the discovery of platinum-free HER catalysts with remarkable efficiency. Key Takeaways: AI-driven DoE accelerates nanomaterials discovery, reducing experimental trials by up to 80%. Machine learning models enable predictive synthesis, improving efficiency and cost-effectiveness. The approach has broad applications, from clean energy to sustainable chemical production. The future lies in autonomous AI-driven laboratories, capable of self-optimizing nanomaterial synthesis (López-García & Vizoso, 2021). As AI-driven experimentation continues to evolve, the field of nanocatalysis is poised for unprecedented breakthroughs, paving the way for next-generation energy solutions and sustainable chemistry.

AI-ENABLED DISCOVERY OF NEW NANOMATERIALS FOR ENERGY STORAGE APPLICATIONS: A CASE STUDY ANALYSIS

The increasing demand for efficient and sustainable energy storage solutions has driven significant advancements in nanomaterial research. Traditional experimental approaches for discovering new materials are often time-consuming and costly, necessitating the adoption of artificial intelligence (AI) techniques to accelerate the discovery process. AI-enabled materials discovery integrates machine learning (ML), high-throughput simulations, and data-driven approaches to predict and design novel nanomaterials with enhanced properties for energy storage applications. This case study explores how AI is transforming the field of nanomaterials for energy storage, focusing on its methodologies, successes, and challenges. AI is revolutionizing nanomaterials discovery by leveraging vast datasets, computational models, and predictive analytics to identify promising materials. Machine learning algorithms analyze large databases of known materials and their properties, enabling researchers to predict new materials with optimal performance characteristics. One notable approach is the use of high-throughput computational screening, where AI models rapidly evaluate thousands of potential nanomaterials to identify candidates with desirable energy storage properties(Murcia-Verdú(Pavlik, 2023) & Ufarte-Ruiz, 2019). For instance, AI can predict the electrochemical stability, ion conductivity, and mechanical properties of battery materials, thereby reducing the reliance on expensive and time-consuming laboratory experiments. Additionally, AI-driven molecular dynamics simulations enable researchers to study the behavior of nanomaterials at the atomic level, providing insights into their structural and electronic properties. AI-enabled discovery of nanomaterials has led to breakthroughs in various energy storage technologies, including lithium-ion batteries (LIBs), supercapacitors, solid-state batteries, and hydrogen storage AI techniques have accelerated the discovery of high-performance cathode and anode materials for LIBs. By analyzing vast datasets of electrode materials, AI models have identified novel lithium-rich oxides, silicon-based anodes, and solid electrolytes with improved energy density and cycle life.

For example, Google's DeepMind and researchers from the University of Cambridge utilized AI to predict promising (Peña-Fernández et al., 2023) electrolyte. AI-driven material discovery has enhanced the development of nanomaterials for supercapacitors, which require high surface area and excellent electrical conductivity. AI models predict optimal nanoporous carbon structures, transition metal oxides, and MXenes that exhibit superior charge storage capabilities. By analyzing structural-property relationships, AI enables the design of nanomaterials with enhanced energy and power densities. The discovery of solid-state electrolytes is a crucial challenge in battery research, as they offer improved safety and higher energy density compared to conventional liquid electrolytes. AI has been instrumental in identifying novel ceramic and polymer-based solid electrolytes with high ionic conductivity. Through data-driven approaches, researchers have successfully discovered new lithium superionic conductors that enable efficient charge transport in solid-state batteries AI facilitates the design of nanomaterials for hydrogen storage by predicting metal-organic frameworks (MOFs) and complex hydrides with optimal adsorption and desorption properties. By leveraging AI models, researchers can rapidly identify materials with high hydrogen storage capacity and stability, accelerating the development of next-generation hydrogen storage technologies. Another success story is the collaboration between Toyota Research Institute (TRI) and MIT, where AI-driven simulations identified new solid-state electrolyte materials with superior conductivity. By integrating AI with robotic experimentation, researchers reduced the time required for material synthesis and testing, leading to the rapid discovery of high-performance battery components.

Moreover, IBM's AI-driven platform, called "Deep Search," has been employed to analyze scientific literature and extract insights for battery material development. By processing vast amounts of research papers and patents, AI identifies emerging trends and promising materials that may have been overlooked through conventional research methods. Despite its promising potential, AI-enabled nanomaterial discovery faces several challenges. One of the primary limitations is the quality and availability of data. Another challenge is the interpretability of AI predictions. While AI algorithms can predict material properties with high accuracy, understanding the underlying physical and chemical mechanisms remains difficult. Researchers often struggle to translate AI-generated predictions into practical experimental designs due to the complexity of nanomaterial interactions. Additionally, the integration of AI with experimental workflows requires significant computational resources and expertise (Pérez-Seijo et al., 2020). Developing reliable AI models necessitates collaboration between materials scientists, data scientists, and computational experts. Bridging the gap between AI predictions and real-world material synthesis remains a critical challenge in accelerating the adoption of AI in nanomaterials research. The future of AI-enabled nanomaterial discovery is promising, with advancements in deep learning, quantum computing, and automation expected to accelerate material innovation further. Emerging AI techniques, such as self-learning algorithms and AI-guided robotic experimentation, will enhance the efficiency and accuracy of material discovery. The integration of AI with cloud-based databases and automated synthesis platforms will enable real-time optimization of nanomaterial properties, revolutionizing energy storage technologies.

AI-DRIVEN DESIGN OF NEW NANOCOMPOSITES FOR AEROSPACE APPLICATIONS: A CASE STUDY DISCUSSION

Nanocomposites—materials composed of nanometer-scale reinforcements embedded in a matrix—have emerged as game-changing solutions for aerospace applications. However, the discovery and optimization of new nanocomposites through traditional experimental methods are often costly and time-intensive. Artificial intelligence (AI) has revolutionized material design by enabling the rapid discovery, optimization, and characterization of high-performance nanocomposites tailored for aerospace applications. This case study explores the role of AI in the development of nanocomposites, highlighting key methodologies (Salaverría, 2019), applications, breakthroughs, and challenges in the field. AI is transforming the process of designing nanocomposites by integrating machine learning (ML), computational modeling, and big data analytics. Traditional materials discovery relies on trial-and-error experimentation and physical simulations, which are resource-intensive. AI accelerates this process by learning from existing datasets, predicting material behavior, and guiding the synthesis of novel nanocomposites with optimal properties. One of the most powerful applications of AI in this domain is high-throughput computational screening. AI can rapidly evaluate thousands of nanomaterial combinations and predict their mechanical strength, fatigue resistance, and weight-to-performance ratios. This approach significantly reduces the time and cost required to develop aerospace-grade nanocomposites compared to conventional experimentation. AI-enabled design of nanocomposites has led to significant advancements in aerospace materials, enhancing performance in structural components, thermal protection systems, and multifunctional applications. Several real-world case studies demonstrate the successful application of AI in nanocomposite design for aerospace applications. One notable example is Airbus's AI-driven materials research initiative, where machine learning models have been employed to optimize carbon fiber-reinforced composites for aircraft wings. By analyzing data from thousands of material samples, AI predicted the optimal fiber-matrix interactions, leading to stronger and lighter aerospace components. Another breakthrough involves NASA's collaboration with AI researchers to develop next-generation heat-resistant nanocomposites for spacecraft. Using AI-guided simulations, NASA identified new ceramic nanocomposite formulations with enhanced thermal and mechanical performance, significantly reducing material testing time. (Shi & Sun, 2024)MIT and Lockheed Martin have also leveraged AI to design graphene-based composites for aerospace applications. AI models successfully predicted the ideal graphene dispersion and alignment within polymer matrices, resulting in nanocomposites with exceptional mechanical properties and weight savings of up to 30%.

AI models require large, high-quality datasets to make accurate predictions. However, data on advanced nanocomposites is often limited, inconsistent, or proprietary, restricting AI's effectiveness. The need for standardized and openly available materials databases is critical for further advancements. The future of AI-driven nanocomposite design for aerospace applications is promising, with ongoing advancements in deep learning, quantum computing, and automated experimentation(Johri, S., et al. 2025). Self-learning AI models, combined with robotic synthesis and high-throughput experimentation, will further accelerate material discovery. The integration of AI with digital twin simulations and real-time monitoring systems will enhance predictive maintenance and improve aerospace component reliability. In conclusion, AI has emerged as a powerful tool for the discovery and optimization of nanocomposites tailored for aerospace applications. By leveraging machine learning, computational simulations, and data analytics, researchers can rapidly design materials with superior strength, thermal stability, and multifunctional properties. While challenges such as data availability, model interpretability, and experimental validation persist, the continued evolution of AI-driven materials science will revolutionize aerospace engineering. The collaboration between AI researchers, material scientists, and aerospace engineers will be instrumental in developing next-generation nanocomposites that enhance aircraft performance, safety, and sustainability. AI-driven materials discovery represents a paradigm shift in aerospace innovation, paving the way for lighter, stronger, and more resilient aerospace structures in the future.

AI-ENABLED DISCOVERY OF NEW NANOMATERIALS FOR BIOMEDICAL APPLICATIONS: AN IN-DEPTH CASE STUDY ANALYSIS

The integration of artificial intelligence (AI) in nanomaterials research has revolutionized biomedical applications by accelerating the discovery, optimization, and functionalization of nanomaterials for diagnostics, drug delivery, tissue engineering, and biosensing. Traditional methods of developing nanomaterials for biomedical applications rely on extensive experimental screening and trial-and-error approaches, which are often costly and time-consuming. AI, particularly machine learning (ML) and deep learning (DL), offers a data-driven approach to predict and design novel nanomaterials with tailored properties, enhancing their biocompatibility, stability, and therapeutic efficiency. This case study explores how AI-driven nanomaterial discovery is transforming biomedical applications, discussing key methodologies, applications, real-world breakthroughs, and challenges in the field. A groundbreaking AI-driven biosensor was developed for COVID-19 detection, utilizing AI-optimized gold nanoparticles functionalized with viral RNA probes. This AI-designed biosensor exhibited high specificity and rapid detection capability, significantly improving point-of-care diagnostics. The future of AI-enabled nanomaterial discovery in biomedical applications is promising, with advancements in quantum computing, self-learning(Soler, 2011), AI models, and automated high-throughput experimentation. AI-integrated robotic laboratories will further accelerate material synthesis and testing, reducing the gap between computational predictions and real-world biomedical applications. In conclusion, AI has revolutionized the discovery of nanomaterials for biomedical applications by enhancing efficiency, reducing development time, and enabling the design of highly functional materials. From targeted drug delivery and biosensing to imaging and regenerative medicine, AI-driven nanomaterials hold immense potential to transform healthcare. While challenges remain in data quality, validation, and regulatory frameworks, continued research and interdisciplinary collaboration will drive the next generation of AI-enabled biomedical nanotechnology, shaping the future of precision medicine and advanced healthcare solutions.

FUTURE DIRECTIONS: EMERGING TRENDS AND TECHNOLOGIES IN AI-DRIVEN MATERIALS DISCOVERY

Traditional methods of material discovery rely on labor-intensive experimentation and complex computational modeling, which can take years to yield viable materials. AI-driven approaches, including machine learning (ML), deep learning (DL), and reinforcement learning (RL), have significantly accelerated this process by predicting material properties, optimizing synthesis conditions, and guiding experimental design(Ufarte Ruiz & Manfredi Sánchez, 2019). As AI continues to evolve, new trends and technologies are emerging, reshaping the landscape of materials discovery(Shende, A. D., et al. 2025). This article explores the future directions of AI-driven materials discovery, highlighting emerging AI algorithms, novel data sources, automated experimentation, and breakthrough applications in various scientific fields. AI-driven materials discovery is entering a new era, with emerging trends in generative AI, autonomous laboratories, and multi-modal data integration shaping the future of the field. As AI models become more sophisticated, they will continue to accelerate the discovery of advanced materials for energy, electronics, healthcare, and environmental sustainability. The combination of AI with automated experimentation and real-time data collection promises a future where new materials can be designed, synthesized, and optimized in a matter of days rather than years. With continued advancements in AI algorithms, data infrastructure, and experimental automation, the next decade will see unprecedented breakthroughs in materials science, driving innovation across multiple industries and unlocking new technological frontiers.

THE POWER OF AI-DRIVEN DISCOVERY IN MATERIALS SCIENCE: KEY FINDINGS AND FUTURE IMPLICATIONS

AI-driven approaches, particularly machine learning (ML), deep learning (DL), and reinforcement learning (RL), are transforming this process by analyzing vast datasets, predicting material properties, and guiding experimental synthesis (Ufarte-Ruiz et al., 2023). This article summarizes the key findings of AI-driven materials discovery, explores its transformative impact across multiple scientific domains, and discusses the future implications for industry, sustainability, and technological advancement. By shifting from traditional trial-and-error methods to AI-guided discovery, researchers can accelerate the innovation cycle in materials science, enabling breakthroughs in fields such as quantum computing, energy storage, and nanomedicine. This AI-driven approach enhances efficiency, reduces material waste, and enables the rapid deployment of novel materials for commercial and industrial applications. By revolutionizing these key domains, AI is driving advancements that will reshape industries and improve global technological capabilities. These advancements signal a future where materials discovery is fully automated, allowing for rapid prototyping and commercialization of new materials. AI is making materials discovery more accessible by reducing reliance on costly experimental facilities and computational resources (Debnath, P., et al. 2025). Cloud-based AI platforms and open-access materials databases enable researchers worldwide to participate in AI-driven discovery, fostering global collaboration. AI is playing a critical role in the design of sustainable materials that minimize environmental impact. Future AI-driven research will focus on developing materials that balance innovation with ethical responsibility to ensure AI-driven materials benefit society as a whole. AI-driven discovery is transforming materials science, enabling faster, smarter, and more sustainable innovation across energy, electronics, healthcare, and environmental science. The integration of AI with automated experimentation, robotics, and quantum computing is paving the way for an era of accelerated materials development, where new materials can be designed and tested in weeks rather than years. Looking ahead, AI will not only enhance existing research methodologies but also unlock entirely new classes of materials that were previously unimaginable. As AI-powered self-learning laboratories and collaborative global databases continue to evolve, the materials of the future will be more efficient, sustainable, and tailored to human needs than ever before. With continued advancements, interdisciplinary collaboration, and ethical oversight, AI-driven materials discovery will play a central role in shaping the next generation of scientific and technological breakthroughs, ushering in a new era of innovation and industrial transformation.

RECOMMENDATIONS FOR PRACTITIONERS: IMPLEMENTING AI-DRIVEN MATERIALS DISCOVERY IN YOUR RESEARCH OR ORGANIZATION

Artificial intelligence (AI) is rapidly transforming materials science by enabling researchers and organizations to discover, design, and optimize materials with unprecedented speed and efficiency. The successful implementation of AI-driven materials discovery requires a strategic approach, integrating advanced AI algorithms, high-quality data, computational tools, and experimental validation. Whether you are a researcher in academia, a materials scientist in industry, or part of an organization looking to leverage AI for innovation, adopting best practices is essential for maximizing the benefits of AI-driven materials discovery. This article provides practical recommendations for successfully implementing AI in materials research, covering key aspects such as data management (Van Dalen, 2024), AI model selection, computational infrastructure, collaboration strategies, and experimental integration. AI-driven materials discovery is revolutionizing the way new materials are designed, synthesized, and optimized. By adopting best practices in data management, AI model selection, computational infrastructure, interdisciplinary collaboration, and ethical AI implementation, researchers and organizations can maximize the benefits of AI-driven materials innovation (Aragani, V. M., et al. 2025). As AI tools continue to advance, the future of materials science will see faster breakthroughs, automated laboratories, and highly efficient materials tailored for emerging technologies. Organizations that embrace AI-driven discovery today will lead the next generation of innovation in energy, electronics, healthcare, and sustainable materials. By strategically integrating AI into materials research workflows, practitioners can unlock new possibilities, accelerate discovery, and shape the future of materials science in the coming decades.

EPILOGUE: THE POTENTIAL OF AI-DRIVEN MATERIALS DISCOVERY TO TRANSFORM INDUSTRIES AND SOCIETIES

AI accelerates the identification and optimization of novel materials with unprecedented precision and efficiency. Traditionally, materials discovery has relied on labor-intensive trial-and-error experiments and incremental advancements, often taking decades to bring new materials from conception to application. However,(Vázquez-Herrero et al., 2020) AI-powered techniques—such as generative models, automated high-throughput screening, and predictive simulations—have revolutionized this process by dramatically reducing discovery timelines and costs while improving the accuracy of material performance predictions. One of the most transformative applications of AI-driven materials discovery lies in the development of sustainable and environmentally friendly materials. Climate change and resource depletion present pressing global challenges, necessitating innovative solutions to reduce carbon footprints, enhance energy efficiency, and minimize waste (Ramakrishnan T. et al., 2022). AI enables the rapid identification of novel catalysts for carbon capture, next-generation battery materials for efficient energy storage, and biodegradable polymers to replace conventional plastics. By accelerating the search for high-performance and sustainable alternatives, AI contributes directly to the global push toward a circular economy and net-zero emissions. The healthcare industry also stands to benefit significantly from AI-driven materials discovery, particularly in pharmaceuticals, biomaterials, and medical devices. AI algorithms can screen vast chemical spaces to identify new drug candidates with optimal efficacy and safety profiles, expediting the drug development pipeline and reducing reliance on costly experimental methods. Furthermore, AI-assisted design of biomaterials has led to breakthroughs in tissue engineering, regenerative medicine, and drug delivery systems. For instance, AI models can predict how different biomaterials interact with human cells, leading to the development of customized implants and prosthetics that enhance patient outcomes. This fusion of AI and materials science has the potential to revolutionize personalized medicine, enabling treatments tailored to individual genetic and physiological characteristics. In the field of manufacturing and industrial applications, AI is redefining how materials are designed, processed, and deployed. From aerospace to consumer electronics, AI-powered predictive modeling allows for the optimization of materials with enhanced strength, durability, and lightweight properties. AI-driven design tools enable engineers to create advanced composites and smart materials that respond to external stimuli, opening new frontiers in self-healing materials, adaptive surfaces, and nanotechnology (Sreenivasulu A., et al. 2022). Additionally, AI streamlines supply chain management by predicting material availability, optimizing production processes, and reducing waste, leading to more sustainable and cost-effective manufacturing practices. Beyond industry-specific applications, AI-driven materials discovery holds profound societal implications.

The ability to rapidly develop and deploy new materials has the potential to bridge technological gaps between developed and developing nations, fostering greater global equity in access to essential resources. For instance, AI can help identify low-cost, locally sourced materials for infrastructure, energy production, and clean water solutions, improving the quality of life in underserved regions. Furthermore, the democratization of AI tools in materials science allows researchers and startups worldwide to participate in innovation, reducing barriers to entry and fostering a more inclusive technological ecosystem. However, the transformative power of AI in materials discovery also comes with ethical and regulatory considerations that must not be overlooked(Mahadev Madgule et al. 2023). The rapid pace of AI-driven innovation raises concerns about data privacy, intellectual property rights, and the potential monopolization of advanced materials by a few powerful entities. Ensuring open-access databases, collaborative research frameworks, and equitable distribution of AI-generated discoveries is crucial in preventing technological disparities. Moreover, the responsible development and deployment of AI require transparency, interpretability, and adherence to ethical guidelines to mitigate risks associated with biased algorithms, unintended consequences, and environmental impact (Ventura-Pocino, 2021). As AI continues to evolve, interdisciplinary collaboration between material scientists, computer scientists, policymakers, and industry leaders will be essential in shaping a future where AI-driven materials discovery serves the greater good. By fostering responsible AI development, prioritizing sustainability, and promoting equitable access to technological advancements, societies can harness the full potential of AI to create a world that is more resilient, efficient, and inclusive. Traditional constraints no longer bind the future of materials science—AI has unlocked a new era of accelerated discovery, where the following groundbreaking material may be just an algorithm away.

CONCLUSION

The integration of artificial intelligence (AI) into nanomaterials discovery marks a transformative era in materials science, redefining the speed, efficiency, and precision with which novel materials are designed, synthesized, and optimized. AI-driven methodologies—such as predictive modeling, machine learning-assisted design of experiments, and autonomous materials screening—have significantly accelerated the identification of nanomaterials with tailored properties, revolutionizing industries ranging from energy and electronics to healthcare and environmental sustainability. Data-driven, algorithmically guided approaches are replacing the traditional paradigm of trial-and-error experimentation. These new methods allow for rapid exploration of vast chemical and structural spaces, minimizing resource-intensive testing while maximizing the potential for groundbreaking discoveries. has emerged as a cornerstone of AI-driven materials research, enabling scientists to anticipate the performance of nanomaterials before they are physically synthesized. By leveraging extensive databases, quantum mechanics simulations, and machine learning algorithms, AI can accurately predict key material attributes such as electronic structure, mechanical strength, thermal conductivity, and catalytic activity. These predictive capabilities not only accelerate the discovery process but also enhance the design of high-performance materials for specific applications, from next-generation batteries and semiconductors to nanocatalysts and biomaterials. The synergy between AI and physics-based models—such as density functional theory (DFT) and molecular dynamics simulations—has further refined the accuracy of predictions, bridging the gap between theoretical modeling and experimental realization. represents another fundamental shift in materials discovery, optimizing how nanomaterials are synthesized and characterized. Unlike traditional experimental approaches that rely on exhaustive parameter variations and human intuition, AI-driven design of experiments employs adaptive learning, Bayesian optimization, and reinforcement learning to identify the most informative experimental conditions. These techniques allow for the real-time refinement of hypotheses, dramatically reducing the number of required experiments while ensuring the most efficient exploration of material compositions and processing parameters. The integration of autonomous robotic laboratories, capable of conducting closed-loop experiments with minimal human intervention, further amplifies the speed and reproducibility of materials research, paving the way for unprecedented advancements in nanomaterial synthesis and application. across diverse technological domains. In the energy sector, AI has facilitated the identification of novel battery materials with higher energy densities, improved cycle life, and enhanced safety, expediting the transition toward more efficient energy storage solutions. Similarly, AI-assisted discovery of advanced nanocatalysts has led to breakthroughs in sustainable hydrogen production, carbon capture, and fuel cell technologies, accelerating progress toward decarbonization. In electronics, AI-guided identification of two-dimensional materials has revolutionized transistor design, optoelectronics, and flexible electronics, enabling the development of ultra-fast and energy-efficient devices. The biomedical field has also benefited from AI-driven materials innovation, with machine learning accelerating the design of nanocarriers for targeted drug delivery, biocompatible scaffolds for tissue engineering, and responsive biomaterials for regenerative medicine. These real-world case studies illustrate AI's potential to drive cutting-edge material solutions across industries, reshaping the landscape of scientific discovery and technological innovation. While AI-driven nanomaterials discovery offers immense opportunities, it also presents challenges that must be carefully addressed to ensure responsible and equitable advancement. Ethical considerations, data transparency, and model interpretability are crucial in maintaining trust and reliability in AI-assisted research. The development of open-access materials databases, collaborative research frameworks, and standardized validation protocols will be essential in preventing biases and ensuring the reproducibility of AI-generated discoveries. Furthermore, interdisciplinary collaboration between material scientists, data scientists, engineers, and policymakers will be vital in establishing guidelines for the ethical and sustainable application of AI in materials science. Looking ahead, the future of AI-driven nanomaterials discovery holds boundless potential for innovation, as advancements in machine learning algorithms, quantum computing, and automated experimentation continue to push the boundaries of what is scientifically possible. By harnessing the power of AI to accelerate materials research, industries can unlock revolutionary solutions to some of the most pressing global challenges, from climate change and renewable energy to next-generation healthcare and smart manufacturing. The convergence of AI and nanomaterials science represents not just an evolution, but a revolution—one that is poised to redefine the way materials are discovered, developed, and deployed in the 21st century and beyond.

REFERENCES

Al-Zoubi, O., Ahmad, N., & Hamid, N. A. (2024). Artificial intelligence in newsrooms: Ethical challenges facing journalists. *Studies in Media and Communication*, *12*, 6587. DOI: 10.11114/smc.v12i1.6587

Apablaza-Campos, A., Wilches Tinjacá, J. A., & Salaverría, R. (2024). Generative artificial intelligence for journalistic content in Ibero-America: Perceptions. Challenges and Reg. Proj. *BiD*, *52*, 1–8. DOI: 10.1344/bid2024.52.06

Aragani, V. M., Anumolu, V. R., & Selvakumar, P. (2025). Democratization in the Age of Algorithms: Navigating Opportunities and Challenges. In Wongmahesak, K., Wekke, I., Seftyono, C., & Nurdin, N. (Eds.), *Democracy and Democratization in the Age of AI* (pp. 39–56). IGI Global Scientific Publishing., DOI: 10.4018/979-8-3693-8749-8.ch003

Ashok, M., Madan, R., Joha, A., & Sivarajah, U. (2022). Ethical framework for artificial intelligence and digital technologies. *International Journal of Information Management*, *62*, 102433. DOI: 10.1016/j.ijinfomgt.2021.102433

Beckett, C. (2019). New powers, new responsibilities. A global survey of journalism and artificial intelligence. The Journalism AI report. Available at: https://drive.google.com/file/d/1utmAMCmd4rfJHrUfLLfSJ-clpFTjyef1/view (Accessed July 14, 2024).

Beckett, C. (2023). Generating change. A global survey of what news organizations are doing with AI. The journalism AI report. Available at: https://static1.squarespace.com/static/64d60527c01ae7106f2646e9/t/656e400a1c23e22da0681e46/1701724190867/Generating+Change+_+The+Journalism+AI+report+_+English.pdf (Accessed July 14, 2024).

Carlson, M. (2015). The robotic reporter. *Journalism*, *3*, 416–431. DOI: 10.1080/21670811.2014.976412

Caswell, D., & Dörr, K. (2018). Automated journalism 2.0: Event-driven narratives. *Journalism Practice*, *12*, 477–496. DOI: 10.1080/17512786.2017.1320773

Cools, H., & Diakopoulos, N. (2023). Towards guidelines for guidelines on the use of generative AI in newsrooms. Medium. Available at: https://generative-ai-newsroom.com/towards-guidelines-for-guidelines-on-the-use-of-generative-ai-in-

Cools, H., & Diakopoulos, N. (2024). Uses of generative AI in the newsroom: Mapping journalists ' perceptions of perils and possibilities. *J. Pract.*, *27*, 1–19. DOI: 10.1080/17512786.2024.2394558

Cools, H., & Koliska, M. (2024). News automation and algorithmic transparency in the newsroom: the case of the Washington Post. Journal. Stud. 25, 662–680. doi: DOI: 10.1080/1461670X.2024.2326636

De la Hoz, K., Coelho, F., & Prodigioso, V. (2023). IA para periodistas. Available at: https://www.prodigiosovolcan.com/sismogramas/ia-periodistas/

Debnath, P., Deshmukh, S. M., Laroia, M., Selvakumar, P., Manjunath, T. C., & Bhattacharya, S. (2025). Challenges Faced by Countries in Renewable Energy Adoption. In Özsungur, F., Chaychi Semsari, M., & Küçük Bayraktar, H. (Eds.), *Geopolitical Landscapes of Renewable Energy and Urban Growth* (pp. 295–320). IGI Global Scientific Publishing., DOI: 10.4018/979-8-3693-8814-3.ch010

Díaz-Campo, J., & Chaparro-Domínguez, M. A. (2020). Periodismo computacional y ética. *Revista ICONO, 14*(18), 10–32. DOI: 10.7195/ri14.v18i1.1488

Díaz Noci, J., Peña-Fernández, S., Meso-Ayerdi, K., & Larrondo-Ureta, A. (2024). The influence of AI in the media workforce: How companies use an Array of legal remedies. *Trípodos., 55*, 33–54. DOI: 10.51698/tripodos.2024.55.03

Fırat, F. (2019). "Robot journalism", in the International Encyclopedia of Journalism Studies. Int. Encycl. J. Stu. 14, 1–5. doi: DOI: 10.1002/9781118841570.iejs0243

Forja-Pena, T., García-Orosa, B., & López-García, X. (2024). The ethical revolution: Challenges and reflections in the face of the integration of artificial intelligence in digital journalism. *Communicatio Socialis, 27*, 237–254. DOI: 10.15581/003.37.3.237-254

García-Orosa, B., Canavilhas, J., & Vázquez-Herrero, J. (2023). Algorithms and communication: A systematized literature review. *Comunicar, 31*, 9–21. DOI: 10.3916/C74-2023-01

Helberger, N., & Diakopoulos, N. (2023). ChatGPT and the AI act. *International Policy Review, 12*, 1682. DOI: 10.14763/2023.1.1682

Hermann, E. (2022). Artificial intelligence and mass personalization of communication content—An ethical and literacy perspective. *New Media & Society, 24*, 1258–1277. DOI: 10.1177/14614448211022702

Ioscote, F., Gonçalves, A., & Quadros, C. (2024). Artificial intelligence in journalism: A ten-year retrospective of scientific articles (2014-2023). *J. Media, 5*, 873–891. DOI: 10.3390/journalmedia5030056

Israel, M. J., & Amer, A. (2022). Rethinking data infrastructure and its ethical implications in the face of automated digital content. *AI and Ethics*, *3*, 427–439. DOI: 10.1007/s43681-022-00169-1

Johri, S., Selvakumar, P., Satyanarayana, P., Das, A., Gupta, S., & Bhattacharya, S. (2025). Urbanization and Energy: The Nexus of Global Challenges. In Özsungur, F., Chaychi Semsari, M., & Küçük Bayraktar, H. (Eds.), *Geopolitical Landscapes of Renewable Energy and Urban Growth* (pp. 453–478). IGI Global Scientific Publishing., DOI: 10.4018/979-8-3693-8814-3.ch016

Jones, B. (2023). *Generative AI and journalism. A rapid risk-based review*. Edinburgh Research Explorer.

Journalism, A. I. (2024). AI journalism starter pack. Available at: https://www.skeyesmedia.org/documents/bo_filemanager/AI-journalism-Starter-Pack-_-A-guide-by-JournalismAI.pdf (Accessed July 14, 2024).

Krausová, A., & Moravec, V. (2022). Disappearing authorship: Ethical protection of AI-generated news from the perspective of copyright and other Laws. *JIPITEC*, *13*, 132.

Lao, Y., & You, Y. (2024). *Unraveling generative AI in BBC News: application, impact, literacy, and governance*. TGPPP. [Epub ahead of print], DOI: 10.1108/TG-01-2024-0022

López-García, X., & Vizoso, A. (2021). Periodismo de alta tecnología: Signo de los tiempos digitales del tercer milenio. *EPI*, *30*, 1. DOI: 10.3145/epi.2021.may.01

Murcia-Verdú, F. J., & Ufarte-Ruiz, M. J. (2019). Mapa de riesgos del periodismo hi-tech 18, 47–55. doi: DOI: 10.31009/hipertext.net.2019.i18.05

Pavlik, J. (2023). Collaborating with ChatGPT: Considering the implications of generative artificial intelligence for journalism and media education. *J. Mass Commun. Educ.*, *78*, 84–93. DOI: 10.1177/10776958221149577

Peña-Fernández, S., Meso-Ayerdi, K., Larrondo-Ureta, A., & Díaz-Noci, J. (2023). Without journalists, there is no journalism: The social dimension of generative artificial intelligence in the media. *El Profesional de la Información*, *32*, 27. DOI: 10.3145/epi.2023.mar.27

Pérez-Seijo, S., Gutiérrez-Caneda, B., & López-García, X. (2020). Periodismo digital y alta tecnología: De la consolidación a los renovados desafíos. *Index Comun.*, *10*, 129–151. DOI: 10.33732/ixc/10/03Period

Ramakrishnan, T., Mohan Gift, M. D., Chitradevi, S., Jegan, R., & Subha Hency Jose, P. Nagaraja, Rajneesh Sharma H.N., Selvakumar P., Sintayehu Mekuria Hailegiorgis, (2022). "Study of Numerous Resins Used in Polymer Matrix Composite Materials", Advances in Materials Science and Engineering, vol. 2022, Article ID 1088926, 8 pages, 2022. https://doi.org/DOI: 10.1155/2022/1088926

Salaverría, R. (2019). Digital journalism: 25 years of research. *EPI*, *28*, 1. DOI: 10.3145/epi.2019.ene.01

Shende, A. D., Tekade, S. A., Deshmukh, A. A., Tembhurkar, S. P., & Selvakumar, P. (2025). Water Quality Monitoring and Control With AI. In Ruiz-Vanoye, J., & Díaz-Parra, O. (Eds.), *Smart Water Technology for Sustainable Management in Modern Cities* (pp. 51–80). IGI Global Scientific Publishing., DOI: 10.4018/979-8-3693-8074-1.ch003

Shi, Y., & Sun, L. (2024). How generative AI is transforming journalism: Development. *Appl. Ethics J. Medi.*, *5*, 582–594. DOI: 10.3390/journalmedia5020039

Soler, P., (2011). "La investigación cualitativa. Un enfoque integrador," in La investigación en comunicación; métodos y técnicas en la era digital, coord. L. Vilches Manterola, 189–236.

Sreenivasulu, A., Manimozhi, T., Satheesh Kumar, P., Selvakumar, P., & Singh, P. (2022). "The Model of Smart Sensing Device For Sensitive Nanoclusters Modification in Sensing Properties," 2022 2nd International Conference on Advanced Computing and Innovative Technologies in Engineering (ICACITE), Greater Noida, India, 2022, pp. 1043–1047, doi: DOI: 10.1109/ICACITE53722.2022.9823846

Ufarte Ruiz, M. J., & Manfredi Sánchez, J. L. (2019). Algorithms and bots applied to journalism. The case of Narrativa Inteligencia artificial: Structure, production, and informative quality. *Doxa Comun.*, *2019*, 213–233. DOI: 10.31921/doxacom.n29a11

Ufarte-Ruiz, M. J., Murcia-Verdú, F. J. y Túñez-López, J. M. (2023). Use of artificial intelligence in synthetic media: first newsrooms without journalists. Prof. Inf. 32:2. Doi: .. 2023. mar.03DOI: 10.3145/epi

Van Dalen, A. (2024). Revisiting the algorithms behind the headlines. How journalists respond to the professional competition of Generative AI. *J. Prac.*, *14*, 1–18. DOI: 10.1080/17512786.2024.2389209

Vázquez-Herrero, J., Negreira-Rey, M. C., & López-García, X. (2020). Let's dance the news! How the news media are adapting to the logic of TikTok. *Journalism*, *23*, 1717–1735. DOI: 10.1177/1464884920969092

Ventura-Pocino, P. (2021). *Algorithms in the newsrooms: Challenges and recommendations for artificial intelligence with the ethical values of journalism.* Catalan Fundació Consell de la Información de Catalunya/Press Council.

Viner, K., & Bateson, A. (2024). The Guardian's approach to generative AI | | The Guardian. The Guardian. Available at: https://www.theguardian.com/help/insideguardian/2023/jun/16/the-guardians-approach-to-generative-ai (Accessed October 2, 2024).

Chapter 4
Next Generation Sustainable Biocomposites Through Machine Learning:
Performance Optimization and Life Cycle Evaluation

Safad Ismail
https://orcid.org/0000-0001-5048-1920
Muthoot Institute of Technology and Science, India

Shibu K. R.
https://orcid.org/0000-0001-6556-1173
Saintgtis College of Engineering, India

F. Ajesh
TKM Institute of Technology, India

G. Sreekumar
College of Applied Science Perissery, IHRD, India

ABSTRACT

In the pursuit of environmentally responsible innovation, sustainable bio-composites have emerged as vital materials in construction, automotive, and aerospace. However, traditional methods for material optimization, defect detection, and Life Cycle Assessment are often labor-intensive and experimentally challenging. This chapter explores the role of advanced Machine Learning (ML) models in revolutionizing

DOI: 10.4018/979-8-3373-2883-6.ch004

the design, evaluation, and regulation of sustainable bio-composites. Emphasis is placed on Graph Neural Networks, Vision Transformers, and Physics-Informed Neural Networks. GNNs model fiber-matrix interactions to predict performance and biodegradability, while ViTs enable real-time defect detection. PINNs integrate physics-based and data-driven learning to simulate material behavior. The chapter also addresses ethical, legal, and regulatory issues in AI integration—such as data privacy in ML-based LCA tools, intellectual property, and environmental compliance—offering a roadmap for responsible AI in sustainable bio-composites.

INTRODUCTION

The global shift towards sustainable materials has been driven by increasing concerns over the depletion of fossil fuels, the rise in plastic waste, and the growing carbon footprint of products. Plastics, which are largely made from petrochemicals derived from fossil fuels, have been widely used due to their low cost and durability (Singh et al., 2022). However, these same qualities make plastics a major environmental issue. In 2019, the global market for plastics was valued at USD 568.9 billion, and it is expected to grow at a rate of 3.2% annually in the coming years. Despite their widespread use, plastics don't decompose easily, and they accumulate in landfills and oceans once discarded. Environmental experts warn that by 2050, the amount of plastic waste in landfills could reach 12 million tons (Dey et al., 2024). Currently, around 400 million tons of plastic waste are generated every year, with packaging contributing to nearly half of this total. Of that, about 150 million tons end up in the ocean, creating a significant threat to marine life. Plastics are difficult to break down due to their chemical structure. Moreover, the addition of stabilizers extend their life, but make them resistant to degradation. This paradox—where the very properties that make plastics useful also contribute to their environmental harm—highlights the urgency of finding more sustainable alternatives.

One promising solution is the increasing use of bio-composites, particularly in industries like construction, automotive, and aerospace. Bio-composites, made from renewable resources, offer the potential to reduce environmental impact, improve biodegradability, and support circular economy practices (Carvalho et al., 2024). However, creating high-performance bio-composites presents challenges, including variations in material properties, differences in life cycle impacts, and the need to comply with strict regulatory standards. Traditional methods of designing bio-composites, conducting LCA, and detecting defects have relied on manual testing and fixed models (Ramesh et al., 2024). While effective, these approaches are time-consuming, resource-intensive, and often fail to adapt to the changing demands of sustainability. Additionally, they tend to overlook critical issues such as ethical concerns, legal compliance, and the need for transparency in emerging technologies like AI. These gaps reflect a disconnect between technological advances and the responsible development of materials that are both innovative and sustainable. Recent advancements in ML offer promising avenues to address these challenges. ML techniques have been employed to enhance various stages of LCA, including life cycle inventory generation, impact assessment, and interpretation. For instance, a comprehensive review by (Romeiko et al. 2024) analyzed 40 studies combining ML and LCA methods, revealing that ML approaches improved prediction accuracy, pattern discovery, and computational efficiency in LCAs. However, the study also emphasized the need for continuous data collection, robust model selection, and uncertainty evaluation to support more reliable ML applications in LCA .

Furthermore, integrating AI into LCA processes has shown potential in identifying significant contributors to environmental impacts and quantifying uncertainties. For example, Lie et al. 2020) utilized ML to pinpoint key factors influencing life cycle impacts in agricultural practices, demonstrating the value of AI in enhancing the interpretation phase of LCA .

Despite these advancements, there remains a notable lack of comprehensive frameworks that synergize advanced AI models with sustainability assessment, regulatory compliance, and ethical governance, tailored explicitly to bio-composite materials. Existing studies often focus on isolated tasks, such as property prediction or LCA estimation, without addressing the multidisciplinary complexities of compliance, ethics, and real-world validation. This chapter aims to bridge the identified gaps by proposing a unified, AI-driven framework that optimizes bio-composite performance, automates sustainability assessments, enhances regulatory compliance, and incorporates ethical and legal dimensions. By using advanced ML models, including GNNs, ViTs, and PINNs, the chapter presents a holistic methodology to accelerate sustainable material innovation while ensuring transparency, responsibility, and regulatory alignment. Through detailed case studies, simulations, and experimental validations, this work contributes to establishing a responsible, AI-enhanced pipeline for future bio-composite technologies, aligning technical innovation with societal and environmental expectations.

Figure 1. Biodegradable Packaging and Utensils.

Fundamentals of Bio-Composites and Sustainability

Bio-composites are materials made from a combination of natural fibers or biodegradable fillers and a polymer matrix. Unlike traditional composites, which use synthetic fibers like glass or carbon, bio- composites use renewable materials such as jute, flax, hemp, coir, bamboo, or agricultural waste (see Figure 1). The polymer matrix in bio-composites can also come from renewable sources, such as Poly Lactic Acid (PLA), Poly Hydroxy Alkanoates (PHA), or starch-based plastics (Naser et al., 2021). These materials are lightweight, biodegradable, and have a smaller environmental impact compared to conventional materials. They also provide good strength for various applications. The growing focus on sustainability and circular economy has led to increased use of bio-composites in many sectors. In contrast to the petroleum-based plastics, bio-composites are designed to be eco-friendly throughout their life cycle. LCA methods are used to measure the environmental impact of a product throughout its life, from material extraction and production to use and disposal. For bio-composites, LCA helps to make informed decisions to choose sustainable materials and improve overall systems. LCA studies show that bio-composites have a lower environmental impact than traditional plastics (Al-

Maharma et al., 2022). They use less energy, produce fewer emissions, and create less waste that ends up in landfills. To ensure their performance, safety, and environmental compliance, the development and use of bio-composites are guided by several international standards and regulatory frameworks. The ISO 14040 series provides comprehensive guidelines for evaluating environmental impacts through LCA and is widely adopted in sustainable manufacturing practices (Khanna et al., 2022). The American Society for Testing and Materials (ASTM) has established specific standards—such as ASTM D5338—for assessing the strength, thermal resistance, and compostability of bio-based materials (Nizamuddin et al., 2024). In parallel, European Union regulations, including the Packaging and Packaging Waste Directive, the EU Green Deal, the REACH Regulation, and the Ecodesign Directive, collectively promote the adoption of sustainable materials while ensuring they meet strict safety and environmental standards (Barkhausen et al., 2022).

Figure 2. Integrating AI, Sustainability, and Compliance for a Resilient Future

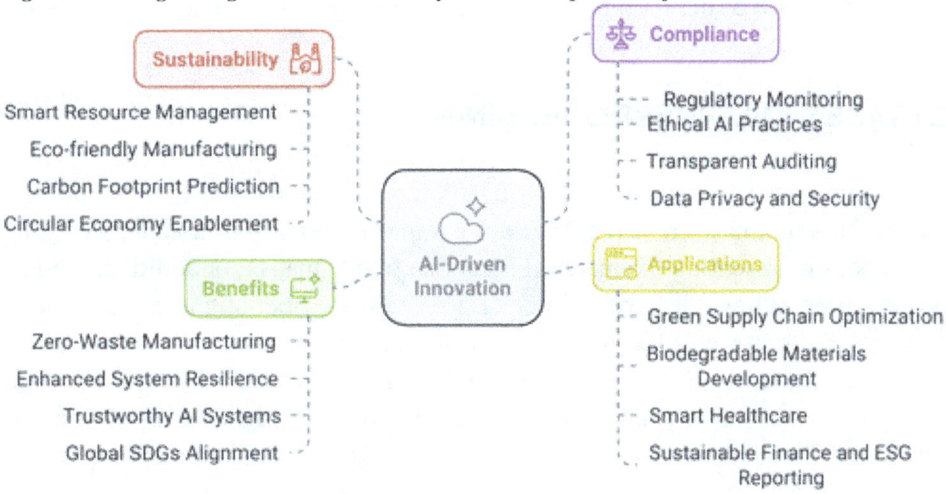

Machine Learning (ML) Techniques for Bio-Composite Innovation

Material Science explores the relationship between the structure, properties, and performance of materials. With the growing power of computers, ML has become a useful tool in this field. It helps the scientists to make sense of large amounts of data, spot patterns, and make predictions about how materials will perform (Ismail et al., 2024). As a result. researchers can find and develop new materials faster and more efficiently, without relying much on time-consuming trial-and-error methods. Based on this researchers have started to use more focused ML methods. This helps them to better understand how materials behave and to meet industry needs as illustrated in Figure 2. In the following subsection, we explore how advanced ML techniques can be applied in materials science with proven applications.

Graph Neural Networks (GNNs) for Fiber-Matrix Interaction and Performance Prediction

GNNs are a type of DL technique developed for making predictions based on graph-structured data. These networks can be used directly on graphs. It enables tasks like predicting information at the node, edge, or entire graph level (Zhu et al., 2021). In the context of bio-composite materials, GNNs can model the intricate relationships between fibers, matrices, and other components at the micro- scopic level (Ismail et al., 2025). This helps to predict their mechanical behavior, durability, and failure points. By understanding these details, GNNs improves the design and performance of bio- composites, making them more sustainable and effective.

DATASET AND PREPROCESSING

We used the data from the Cambridge Polymer Composites Dataset (Cencer et al., 2022) as summarized in Table 1. Each composite material is modeled as a graph with nodes representing the material parts like fibers, matrix, or additives, and the edges represent how these parts interact, for example, fiber alignment or bonding strength. To construct these graphs, microstructure images were first segmented to isolate fiber and matrix regions. Spatial relationships between these regions were then modeled using Voronoi tessellation and region-growing algorithms. Key features were then extracted at both the node and edge levels. These include shape, orientation, and material properties, while edge features capture local bonding strength and stiffness variations. All features were normalized to a [0,1] range using min-max scaling. Additionally, graph sizes were standardized through techniques such as sub-sampling and aggregation.

Table 1. Experimental Data for Polymer Composites

Composite ID	Fiber Type	Matrix Type	Tensile Strength (MPa)	Elastic Mod- ulus (GPa)	Degradation Rate (% mass loss over 30 days)
C-001	Glass Fiber	Epoxy	450	30	1.2
C-002	Carbon Fiber	Polyester	600	40	0.5
C-003	Natural Fiber	PLA	120	5.2	7.8
C-004	Basalt Fiber	Polypropylene	380	20	2.5
C-005	Hemp Fiber	Bio-Epoxy	150	6.0	6.3
C-006	Flax Fiber	PBS	140	5.5	8.0

RESULTS AND DISCUSSION

Figure 3 illustrates the step-by-step process of converting a composite microstructure into a graph: (a) Original biocomposite microstructure, b) Segmented fiber–matrix regions, c) Voronoi-based graph overlay capturing spatial relationships. Each node in the graph corresponds to a distinct fiber region. From these regions, quantitative node-level features were extracted and normalized, as shown in Table 2.

Figure 3. a) Original biocomposite microstructure, b) segmented fiber–matrix regions, c) Voronoi graph capturing spatial relationships for feature extraction

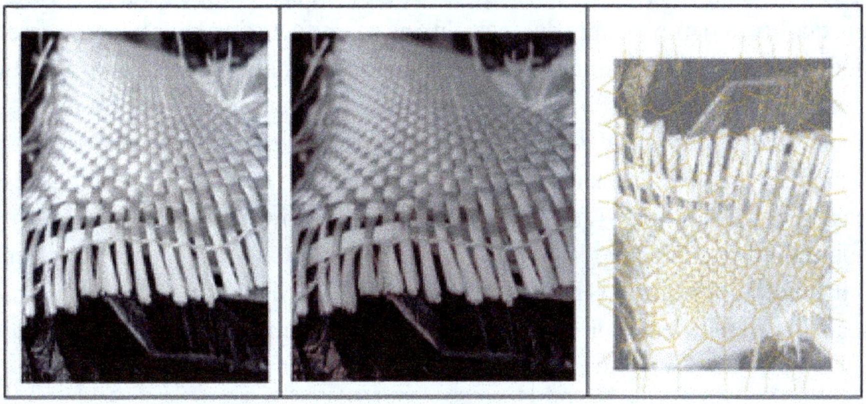

a b c

Table 2. Normalized Node Features Extracted from Microstructure Image

Node ID	Shape (Eccentricity)	Orientation	Material Property (Area)
1	0.72	0.51	0.65
2	0.60	0.45	0.53
3	0.84	0.63	0.78
4	0.69	0.59	0.61
5	0.55	0.41	0.48
.	.	.	.
N	0.77	0.58	0.70

In Figure 4, subplot (a), the histogram represents the distribution of tensile strength across various composite samples, reflecting how different fiber and matrix combinations influence mechanical performance. Subplot (b) displays a scatter plot relating elastic modulus to tensile strength, where the trend suggests that materials with greater stiffness tend to exhibit higher strength. Subplot (c) features box plots that compare the degradation rate and fiber orientation, indicating variability in biodegradation behavior and structural alignment among the samples. Subplot (d) shows a basic network diagram outlining the fiber-matrix connections for each composite, emphasizing differences in bonding strength, which play a crucial role in determining overall material behavior.

Vision Transformers (ViTs) for Real-Time Defect Detection and Quality Control of Bio-Composites

In recent years, the demand for high performance and eco friendly materials has driven significant advancements in the production of bio-composites. Ensuring consistent quality and minimizing de- fects during manufacturing have become essential to meet both industrial standards and sustainability goals. The usage of traditional machine vision systems such as using Convolutional Neural Networks (CNNs), have demonstrated potential in detecting surface flaws. However, they often struggle with the complex textures and subtle patterns inherent in bio-composite materials. To address these limitations, ViTs have emerged as a powerful alternative. By using self-attention mechanisms, ViTs excel at cap- turing long-range dependencies and global contextual features (Hatamizadeh et al., 2023). This makes them well-suited for real-time defect detection and quality assessment in material science applications.

Figure 4. Visual analysis of composite properties: (a) Tensile Strength Distribution, (b) Elastic Mod- ulus vs Tensile Strength, (c) Degradation Rate and Fiber Orien- tation, (d) Fiber–Matrix Bonding Net- work.

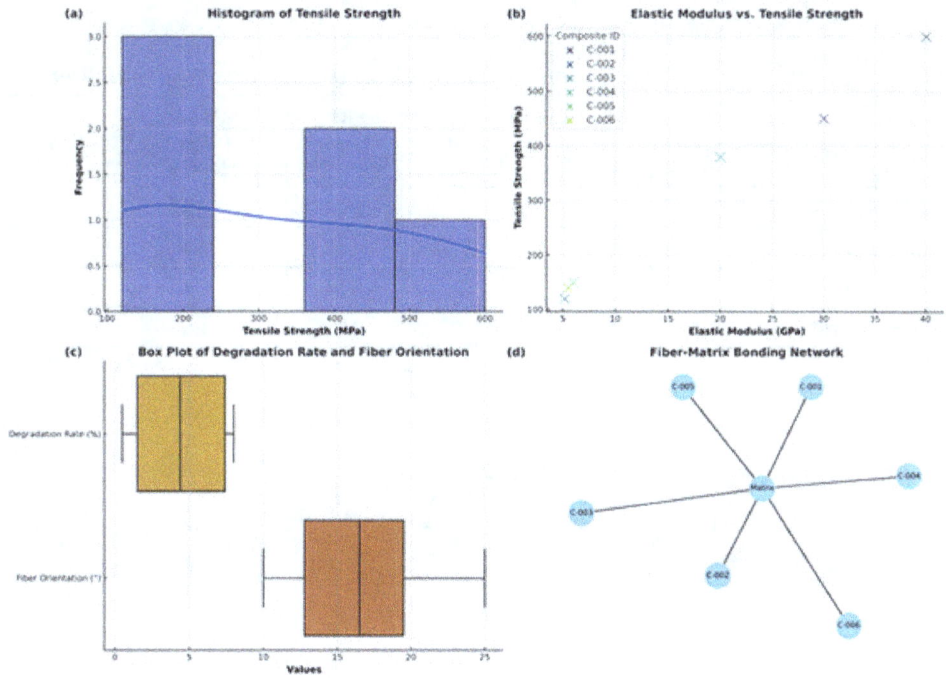

Dataset

This study is based on a hybrid dataset designed to capture a wide range of surface defects in composite materials. The first component of the dataset is the widely recognized Composite Material Surface Defects (CMSD) dataset, which includes more than 5,000 annotated images of synthetic composite surfaces exhibiting various types of defects (Liu et al., 2024). Along with this a custom dataset was developed in a controlled laboratory environment. This was done because, most publicly available datasets like the CMSD are based on synthetic composites (e.g., carbon fiber). The custom dataset fo- cus on bio-composites made from jute-PLA and flax-PLA sheets. High-resolution imaging equipment was used to capture detailed surface features during the production process. Each image was carefully annotated using the LabelMe tool to identify and classify common defects such as fiber pull-out, void formations, micro-cracks, and resin-rich zones. This combination provides a robust foundation for evaluating defect detection models in both synthetic and bio-based composite materials. The Data Preprocessing Workflow is shown in Figure 5. The Figure 6 visualize the patch-splitting and embedding process applied to a carbon bio-composite surface image.

Table 3. Model Performance Comparison for Composite Property Prediction

Model	MAE (Tensile Strength)	R^2 (Tensile Strength)	MAE (Elastic Modulus)	R^2 (Elastic Modulus)
Linear Regression (LR)	38.2	0.72	4.3	0.68
Random Forest (RF)	25.6	0.85	2.9	0.81
Feedforward NN (FNN)	21.4	0.88	2.5	0.84
Graph Neural Network (GNN)	**14.7**	**0.93**	**1.6**	**0.91**

Figure 5. Data Preprocessing Workflow

ViT Model Setup and Training

The ViT-B/16 model from the Hugging Face Transformers library was fine-tuned using PyTorch on an NVIDIA A100 GPU cluster. The model was trained using a stratified split with 70% training, 15% validation, and 15% testing (see Table 4 for configuration details).

Table 4. ViT Model Configuration for Bio-Composite Defect Detection

Configuration	Details
Architecture	12 Transformer blocks, 768 hidden dimensions, 12 heads
Optimizer	AdamW with weight decay (0.01)
Learning Rate Scheduler	Cosine Annealing
Loss Function	Cross-entropy (for classification), Dice Loss (for segmentation)
Epochs	30
Batch Size	32

Figure 6. Patch-splitting and embedding of a carbon bio-composite surface image: a) Original image, b) Patch-split view, c) Embedded representation

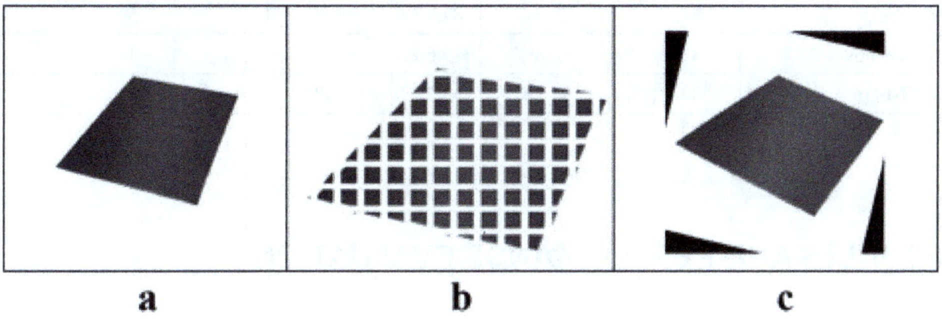

Figure 7. ROC Curves of ResNet50, EfficientNet-B3, and ViT-B/16

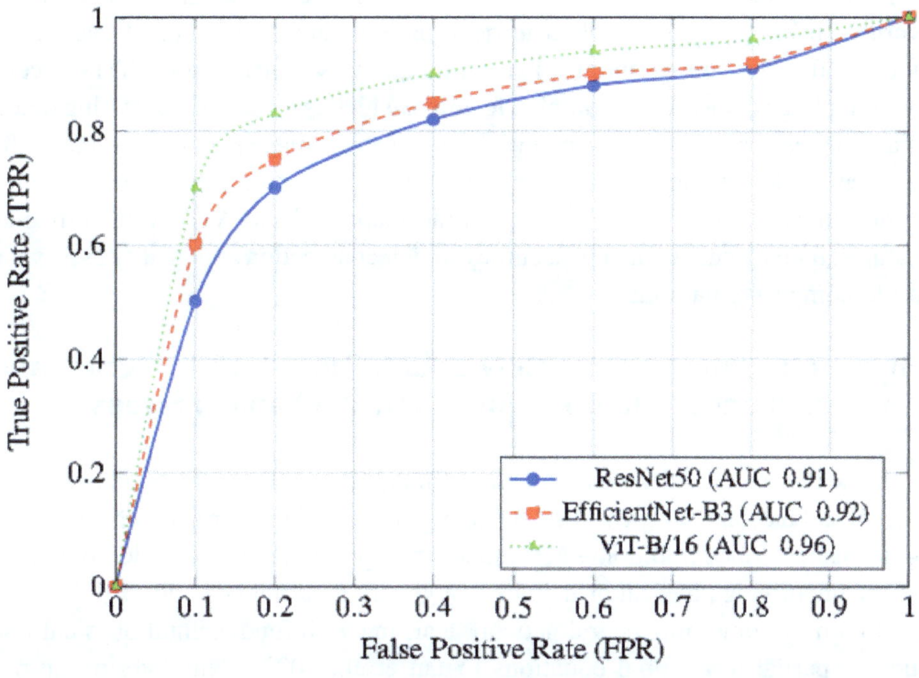

The ROC curve in Fig 7 compares the classification performance of three models. ViT-B/16 achieves the highest AUC (0.96), indicating the best performance, followed by EfficientNet-B3 (0.92) and ResNet50 (0.91).

Table 5. Performance Metrics of Each Model

Model	Accuracy	Precision	Recall	F1-Score	Inference Time (ms)
ResNet50	89.4%	88.7%	87.9%	88.3%	52
EfficientNet-B3	91.1%	90.8%	89.6%	90.2%	45
ViT-B/16	**94.6%**	**94.1%**	**93.5%**	**93.8%**	**34**

RESULTS AND PERFORMANCE EVALUATION

The ViT model achieved superior results compared to CNN-based baselines (ResNet50 and EfficientNet-B3- B3), as summarized in Table 5. Unlike CNNs, which mainly focus on extracting local features through convolutional operations, ViT utilizes self-attention mechanisms to capture long-range dependencies and global context across the entire material surface. This ability to model global relationships is crucial in bio-composites, where intricate patterns and spatial interactions between fibers and resins are essential for accurate analysis. Furthermore, ViTs excel in handling large datasets and complex features, which is critical for detecting defects or anomalies in bio-composite materials. By treating image patches as tokens, ViTs can better address the intricate surface textures and subtle variations present in composite structures. The self-attention mechanism allows ViTs to prioritize key features, thereby enhancing the accuracy of defect detection and quality assessment in bio-composite materials.

Physics-Informed Neural Networks (PINNs) for Bio-Composite Simulation under Environmental and Mechanical Stress

It is essential to understand how biodegradable composites respond to environmental and mechanical stress for developing sustainable materials with predictable performance. Since traditional ML models rely purely on data and do not account for the underlying physical principles that govern material behavior. PINNs offers a solution by combining neural network learning with fundamental physical laws, such as partial differential equations (Yuan et al., 2022). This hybrid approach enables more accurate and realistic simulations, especially when data is limited or expensive to obtain, and ensures that predictions remain grounded in the physics of material degradation and stress distribution.

Dataset

We use an open-source experimental dataset based on PLA–Hemp fiber bio-composite sheets (Gioffre et al., 2021). The materials were exposed to varying environmental conditions, including temperatures between 25–80°C, humidity levels from 30–95% RH, and UV exposure ranging from 8 to 48 hours. Mechanical properties were evaluated through uniaxial tensile tests conducted under both dry and hu- mid conditions at multiple degradation intervals (0, 5, 10, 20, and 30 days). The dataset is summarized in Table 6.

Table 6. Summary of Used Dataset

Category	Details
Material	PLA–Hemp fiber bio-composite sheets
Environmental Conditions	Temperature: 25–80°C Relative Humidity: 30–95% RH UV Exposure: 8–48 hours
Mechanical Testing	Uniaxial tensile tests under dry and humid conditions Test intervals: 0, 5, 10, 20, and 30 days
Recorded Features	Time (days) Temperature (°C) Relative Humidity (%) UV Exposure Time (hrs) Tensile Strength (MPa) Elongation at Break (%) Young's Modulus (GPa)

Governing Equations Embedded in PINN Loss

The PINN integrates domain-specific governing equations into its loss function.

Mechanical Stress (Equilibrium Equation):

$$\nabla \cdot \sigma + f = 0 \quad (1)$$

Constitutive Relation:

$$\sigma = E(t, T, H) \cdot \varepsilon \quad (2)$$

where $E(t, T, H)$ is the environment-dependent Young's modulus, modeled as a degradation function of time (t), temperature (T), and humidity (H).

Environmental Diffusion (Fick's Law):

$$\partial C/\partial t = D\nabla^2 C \qquad (3)$$

where C is the concentration and D is the diffusion coefficient.

Neural Network Architecture

The PINN consists of:

- 6 hidden layers with 64 neurons in each layer.
- Sinusoidal activation functions to improve the learning of PDE-based features.

The overall loss function combines multiple components:

$$L = \lambda 1 \cdot L\text{PDE} + \lambda 2 \cdot L\text{data} + \lambda 3 \cdot L\text{boundary} \qquad (4)$$

Here, λ_1, λ_2, and λ_3 are empirically tuned coefficients for balancing the loss terms.

Optimization

The training is carried out in two stages:

- **Stage 1:** Adam optimizer is used with an initial learning rate of 0.001 for 10,000 epochs.
- **Stage 2:** The L-BFGS optimizer is applied for fine convergence.

Table 7. Comparison of MAE between PINN and vanilla DNN

Model	MAE (MPa)
PINN	0.3
Vanilla DNN	0.7

Figure 8. PINN Generalization at Unobserved Conditions vs. Experimental Data

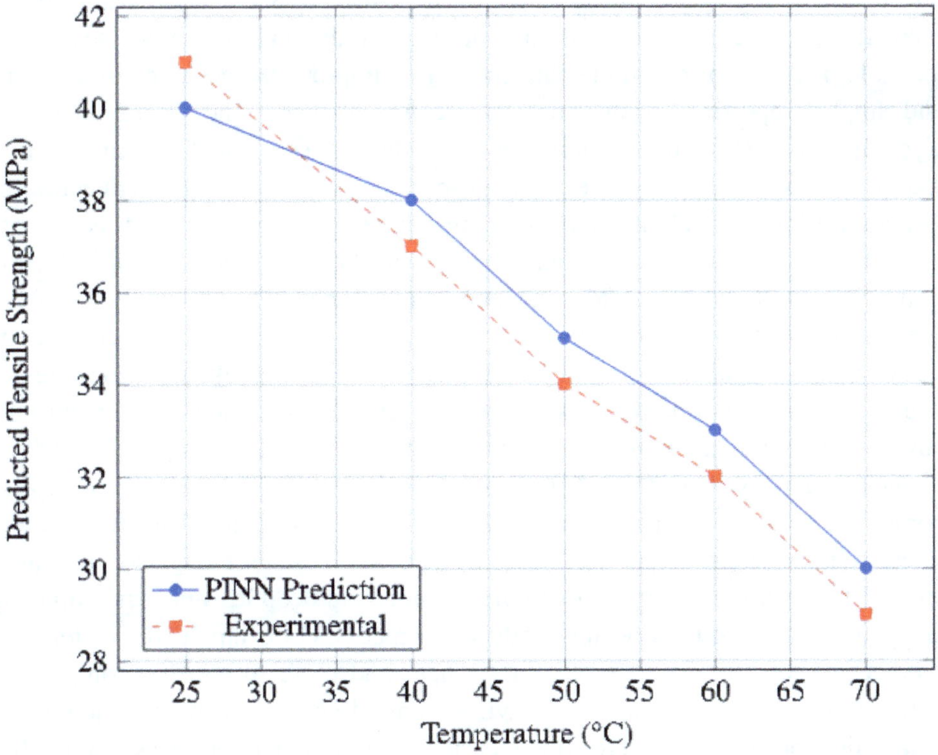

The usage of PINNs demonstrates strong potential for accurately modeling biodegradable composites under varying environmental and mechanical stresses. Figure 8 shows that the predicted tensile strength closely aligns with experimental data across increasing temperatures, confirming the model's ability to generalize under unobserved conditions. By embedding physical laws into the learning process, PINNs not only enhance predictive accuracy but also offer insights into material performance over time, crucial for sustainable material design. Future work aims to extend this capability to more complex behaviors, such as anisotropy, and integrate uncertainty modeling through Bayesian PINNs.

AI-Enhanced Life Cycle Assessment (AI-LCA)

LCA is an essential approach that helps businesses to evaluate the environmental impact of their products, services, or processes at every stage—from the extraction of raw materials to final disposal. Although LCA provides a thorough evaluation, it can be slow and labor-intensive. The process typically depends on historical data and manual data collection, which may result in assessments that are outdated or not specific to regional contexts. In contrast, AI-driven LCA uses ML algorithms, predictive analytics, and automation to optimize traditional LCA processes. These technologies help to streamline data collection, improve the accuracy of analysis, and simplify reporting. In particular, these can be applied to the generation of Life Cycle Inventory (LCI) data, where automated data collection and pattern detection reduce manual work and make the data more consistent. The data collection of bio composite materials involves tracking complex, multi-stage processes, from raw material cultivation (e.g., plant fibers or bio-based resins) to manufacturing, transportation, and disposal. ML can automate the process by collecting data from sensors, production systems, and supply chain records. This allows continuous tracking of material flows, energy use, and waste. Models like decision trees or neural networks can find patterns in the data, giving a clear view of the material's environmental impact without manual data entry. A notable example comes from Cymbio company that uses ML to monitor and optimize the production of bio-composite materials (Nagaraja et al., 2024). They collect data from sensors and integrate it into their production line, tracking material flows, energy consumption, and waste generation. By using neural networks and deep learning, the company analyzes this data to improve the production process. For example, the system can predict the best mix of materials for stronger and more sustainable composites, helping to reduce waste and lower environmental impact. This process automates data collection and analysis, making it easier to monitor and improve sustainability without manual input.

Predictive Impact Assessment Using ML

ML algorithms can predict missing or future LCI values by analyzing correlations between production variables and environmental outputs. The regression models could predict the carbon footprint of a bio-composite material based on inputs such as fiber type, resin composition, and manufacturing energy. These predictive models help to overcome data gaps and provide insights into how different production choices (e.g., material substitution, transportation methods) may affect the environmental performance of bio-composites in real-time. Table 8 explains a dataset for the prediction of the carbon footprint of a bio-composite material using a Support Vector Regression (SVR) model.

Table 8. Bio-composite Dataset

Fiber Type	Resin Composition (%)	Manufacturing Energy (kWh)	Transportation Method	Carbon Footprint (gCO2e)
Natural	30	50	Road	120
Synthetic	40	60	Sea	160
Natural	50	40	Air	200
Synthetic	60	70	Road	180
Natural	35	45	Sea	130
Synthetic	45	65	Air	210

Figure 9. SVR-predicted vs actual carbon footprint values

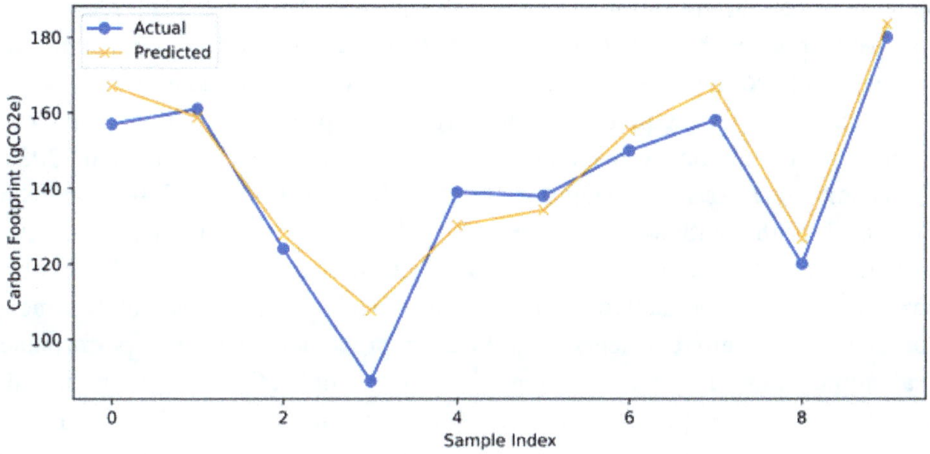

As evident from Table 9, SVR model demonstrated strong predictive performance, achieving a low MAE of 7.12 and an R^2 score of 0.880, indicating it explains 88% of the variance in carbon footprint

Table 9. Performance Metrics of SVR Model

Metric	Value
Mean Absolute Error (MAE)	7.12
R^2 Score	0.880

based on key input features. As shown in Figure 9, the predicted values closely follow the actual carbon footprint trends, confirming the model's accuracy and generalization across varying composite parameters.

CASE STUDY: ML MODELS FOR BIODEGRADABLE COMPOSITE EVALUATION

In recent years, the global shift towards environmental sustainability has led to significant advancements in the development of biodegradable composite materials. As seen earlier, these materials offer an eco-friendly alternative to traditional polymers, especially in applications such as packaging, construction, and textiles. Researchers and industries have increasingly adopted ML techniques to streamline the evaluation and optimization of these composites. This case study presents recent developments from 2020 to 2025, showing how ML has transformed the way biodegradable composites are evaluated in both research and industry.

Researchers at Chongqing University have advanced the prediction of compressive strength in geopolymer composites using hybrid ML models (Zhang et al., 2024). They combined Harris Hawks Optimization with Random Forest (HHO-RF) and Sine Cosine Algorithm with Random Forest (SCA- RF) to predict material performance. The SCA-RF model achieved a coefficient of determination (R^2) of 0.987, indicating a very high level of predictive reliability. This model's ability to accurately forecast compressive strength represents a significant improvement in material performance evaluation. These advancements have important implications for industries like construction, where optimized material design can reduce costs and environmental impact. Furthermore, this research also opens the door for exploring other optimization methods and using larger datasets, which can help to create stronger and more environmentally friendly materials for future construction projects. A study at the National Institute of Technology Rourkela focused on predicting erosion wear in ramie-epoxy composites reinforced with sponge iron slag (Mahapatra et al., 2023). The researchers used Gradient Boosting Machine (GBM) models to accurately predict erosion rates. The study found that the filler content, amount of sponge iron slag, significantly impacted wear resistance. The higher the slag content, the more resistant the composite was to erosion. This research shows how ML models like GBM, can optimize composite materials for better performance, especially in industries like automotive and aerospace, where durability is crucial. It also helps to make material testing more efficient and cost-effective. Researchers have also applied Artificial Neural Networks (ANN) to optimize the erosion performance of polyester composites filled with biochars derived from cashew nutshells, sugarcane, and PET waste. ANN models were employed to predict and optimize the erosion resistance of these composites under various conditions (Thimmaiah et al., 2023). The study revealed that the addition of biochar, particularly from cashew nutshells, enhanced the erosion resistance of the polyester composites. Among the biochar variants, the cashew nutshell biochar-filled composites performed the best, showing superior resistance to erosion, especially at a 60° impact angle, which is a common condition for assessing material durability. The ANN model helped to identify the optimal biochar content and distribution within the polyester matrix, contributing to the composite's enhanced performance. These results suggest that biochar can improve the durability of composites while promoting waste reuse. The use of ANN in this context highlights the potential of ML techniques to optimize material properties efficiently. It reduces the need for extensive trial-and-error testing. This approach is valuable for industries looking to develop high-performance, eco-friendly materials, such as automotive and construction, where erosion resistance is a critical factor. Researchers at Maryland University integrate robotics, ML, and molecular simulations to accelerate the discovery of all-natural plastic substitutes. This innovative approach helped to speed up the development of biodegradable nano composites for packaging applications. By using robotics for automated testing, the team could efficiently experiment with different material formulations. The ML algorithms played a critical role in analyzing large datasets to identify promising candidates for biodegradability and material performance, while molecular simulations provided insights into the atomic-level behavior of the composites. The results showed that the biodegradable nano composites developed through this process exhibited competitive mechanical properties and were capable of decomposing naturally without leaving harmful residues. This achievement significantly reduced the development time for biodegradable materials, offering a promising alternative to traditional plastics in packaging. The research highlights the potential of combining these advanced techniques to create sustainable, high-performance materials, which could lead to a reduction in plastic waste and environmental pollution in the packaging industry.

Industry Applications

Several companies are using ML to improve biodegradable composites, advancing both environmental sustainability and material performance. Spinnova is a Finnish company that has developed groundbreaking technology to turn wood pulp into textile fibers without using harmful chemicals. Founded in 2014,it has created a sustainable fiber production process (Photopoulos et al., 2023). This process eliminates the toxic solvents traditionally used in producing textile fibers like viscose and polyester. This innovation helps to reduce the textile industry's environmental impact by providing a biodegradable alternative to traditional fibers. The process starts with sustainably sourced wood pulp, which is turned into a fiber-like material. The fibers are then spun into yarn for textile production. The fibers produced are biodegradable and sustainable, meeting the growing demand for eco-friendly alternatives in the fashion industry. Their fiber production is shown in Figure 10. The company has opened a commercial-scale production facility with Suzano, the world's largest pulp producer. This facility is called Woodspin, produces 1,000 tonnes of sustainable fiber each year, with plans to increase production. It has strategic partnerships with lead- ing global fashion brands, including Adidas, H&M, and Marimekko. These collaborations show that Spinnova's technology can scale and offer a more sustainable future for the fashion industry. Despite this, the company faces challenges like sourcing raw materials, relying on virgin wood pulp from Suzano, which raises concerns about deforestation. Scaling its technology to meet global demand while maintaining sustainability is also complex, requiring significant investments. Additionally, the company faces growing competition from other eco-friendly fiber producers. As demand for sustainable textiles grows, Spinnova's fiber production process offers a promising model for the future of fashion. Its focus on reducing carbon emissions, conserving water, and eliminating microplastics. Similarly, a UK-based company, Polymateria has developed an innovative Bio transformation technology that enables plastics to biodegrade in real world conditions within a year, without leaving behind harmful microplastics. Their Lyfecycle product line offers a new way to tackle plastic pollution. It helps plastic to break down faster when exposed to sunlight, moisture, and heat, turning it into safe, biodegradable materials. This technology is already being used in areas like fashion—for things like synthetic fabrics and garment tags—and in farming, where it replaces plastic sheets that can harm the soil. Despite its potential, Polymateria faces challenges related to market adoption, cost competitiveness, and regulatory frameworks. However, the company's work supports the concept of a circular economy and could make a significant difference in reducing plastic waste.

Figure 10. Spinnova's Eco-Friendly Fiber Production Process

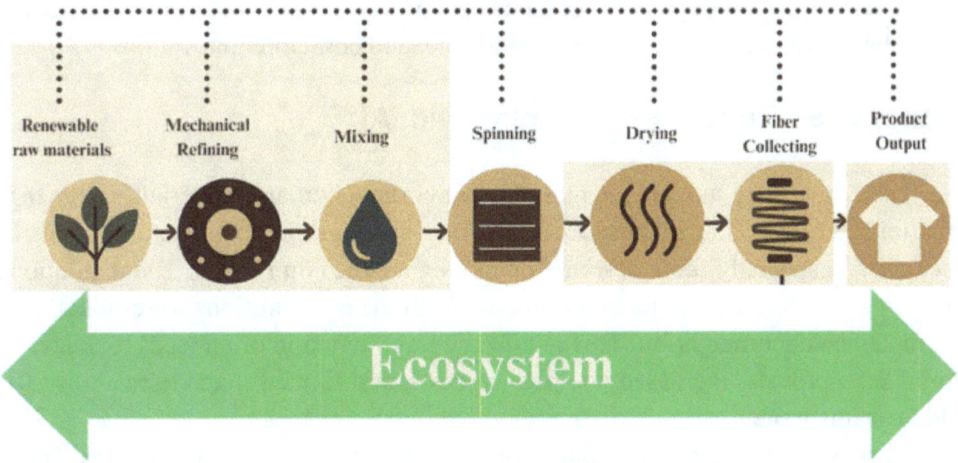

AMSilk GmbH is a German biotechnology company that has developed Biosteel® Fiber. It is a biodegradable and high-performance material inspired by spider silk. Biosteel® is made through fermentation process using engineered bacteria. It is strong, stretchy, and soft, and it breaks down naturally in both soil and ocean water. This new fiber is already being used in different fields (Bargel et al., 2024). In fashion, AMSilk worked with Adidas to make the Future craft Bio fabric shoe, which has an upper made completely from Biosteel® and is designed to break down after use. In the car industry, Mercedes-Benz used Biosteel® inside its VISION EQXX concept car, including parts like door handles, showing a move toward eco-friendly and vegan materials (Mendanha et al., 2024). In medicine and cosmetics, Biosteel® is used in coatings for medical tools and in skincare products because it is safe for the body and helps to fight germs.

The above case studies highlights how innovative technologies, including ML and bioengineering, are transforming biodegradable composite production. While each faces unique challenges, their so- lutions reflect a growing commitment to environmental sustainability, reduced resource consumption, and the advancement of a circular economy. Their work shows a responsible future for industries like fashion, farming, and cars.

Regulatory Compliance and Ethical Considerations

As ML technologies become more integrated into the design, optimization, and quality control of biodegradable composites, several regulatory and ethical concerns must be addressed to ensure safe, responsible, and equitable innovation.

Automated Regulatory Mapping with AI

The complexity and fragmentation of environmental and materials safety regulations across countries pose a significant challenge for manufacturers and researchers, especially those working on biodegradable composites. The regulations may vary widely across jurisdictions, with differing definitions, thresholds, and compliance documentation. This regulatory fragmentation often leads to delays in product approvals, increased compliance costs, and limited market scalability. To address this issue

AI-powered regulatory mapping tools have emerged as powerful solutions. These system use NLP techniques and ML tools to automatically extract, interpret, and align legal standards with product specifications. They can analyze vast amount of legal texts to identify compliance requirements for biodegradable composites, streamlining certification pathways and accelerating market entry. For instance, compliance.ai provides an AI-driven platform that helps financial institutions and other regulated industries monitor and interpret regulatory changes (Tillu et al., 2025). In the context of environmental compliance, companies like LogicGate and Ascent RegTech are developing intelligent compliance tools that can be adapted to analyze environmental product safety standards. Similarly RegGenome, a spinout from the University of Cambridge, is using AI to structure and map global regulatory content. This can be used to specific domains such as biodegradable materials. In the field of manufacturing, Systems Applications and Products in Data Processing (SAP) has integrated AI-based compliance checks into its SAP Responsible Design and Production tool. This helps the companies to track Extended Producer Responsibility (EPR) and plastic taxes across jurisdictions, en- suring that material and packaging choices meet evolving regulations. This model could easily extend to biodegradable composites as part of circular economy strategies. These AI driven platforms not only speed up the compliance process but also assist in streamlining certification pathways, reducing human error, and accelerating the go-to-market timeline for eco-friendly innovations. For example, if a biodegradable composite material needs to meet EU REACH regulations and U.S. EPA standards, an AI system can automatically flag conflicting requirements and recommend product modifications or highlight alternative certification routes.

Figure 11. AI Process for Regulatory Compliance of Biodegradable Composites

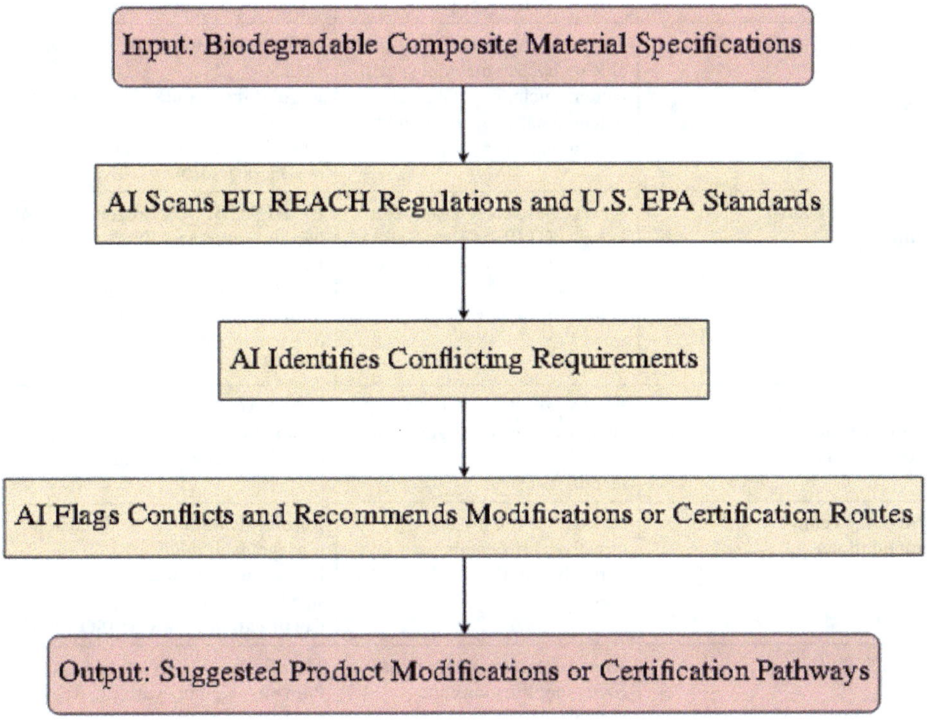

Table 10. AI-Driven Regulatory Compliance Process for Biodegradable Composites

Step	Description	Example Output
Step 1: Input	The input consists of biodegradable composite material specifications, including composition, potential hazards, and environmental impact.	**Material Composition:** PLA with flax fiber **Regulatory Standards:** EU REACH, U.S. EPA Hazardous Content: Minor use of non-biodegradable additives
Step 2: AI Scannin	AI system scans relevant regulations (EU REACH, U.S. EPA, etc.) to extract key regulatory re- quirements applicable to the ma- terial.	**EU REACH:** No more than 0.1% of harmful substances like lead or cadmium **U.S. EPA:** The product must be tested for biodegradability with 90% degradation in 6 months.
Step 3: Identifying Conflicting Requirements	AI analyzes the material specifications against the regulations to identify conflicts.	**EU REACH:** No conflict. **U.S. EPA:** Conflict detected due to non-biodegradable additives.
Step 4: Recommendations	AI suggests modifications or alternative certification pathways to resolve conflicts.	**Recommendation 1:** Replace non-biodegradable additives. **Recommendation 2:** Perform EPA-standard testing. **Certification Pathway:** ISO 14001.
Step 5: Output	AI generates a report summarizing modifications and compli- ance suggestions.	**Modifications**: Replace additives. **Pathway:** Comply with EU REACH and EPA.

However, this AI system comes with challenges. They require constant updating, legal validation, and contextual understanding to avoid misinterpretation and ensure relevance across jurisdictions. De- spite the challenges, AI is turning compliance into a strategic advantage for innovators in sustainable materials and green technologies.

Legal Challenges: Material Safety, Intellectual Property, and Certification

The legal rules for AI-generated biodegradable materials are complex, particularly in the realms of product safety, Intellectual Property (IP), and certification protocols. Legal systems globally mandate that new materials, especially those intended for commercial or public use, meet strict safety and environmental standards. However, existing regulatory regimes—such as the EU's REACH Regulation (EC No 1907/2006) or the U.S. Toxic Substances Control Act (TSCA)—were not designed with AI-generated materials in mind. These materials, often developed through black-box algorithms, pose challenges in terms of traceability and explainability of their chemical and physical properties (Rayasam et al., 2022). A pertinent example is the C-673/18 Organizationen Greenpeace Norden vs European Chemicals Agency, where the court emphasized precautionary principles in chemical safety, underscoring that novel materials must be thoroughly assessed for their long-term ecological impacts. This has implications for biodegradable materials whose degradation products might be unpredictable without extensive testing. One of the biggest legal issues with AI-generated inventions is who owns them. Current intellectual property laws—like the U.S. Patent Act and the TRIPS Agreement—require that an inventor must be human. This was confirmed in cases like Thaler v. Comptroller-General of Patents (UK, 2021) and Thaler v. Vidal (U.S., 2022), where courts ruled that AI systems cannot be named as inventors under current laws. This creates uncertainty for AI-designed biodegradable materials, especially when there is little or no human involvement. If these inventions can't be patented, companies might choose to keep them as trade secrets, which reduces transparency. Or, they may risk losing control over their innovations. This could discourage open research and prevent the sharing of environmentally important discoveries.

Traditional standards-setting bodies, such as ISO and ASTM International, primarily rely on conventional Research and Development (RD) methods to determine material properties. These standards often fail to account for the outcomes generated by AI models, which are increasingly influencing product design, especially in the realm of biodegradable products and other environmentally sensitive innovations. Currently, AI-generated content and outcomes are not always recognized within existing certification frameworks. This disconnect can lead to certification delays or even refusals, significantly slowing the commercialization of AI-enhanced biodegradable products. For instance, the lack of protocols to verify the properties of AI-influenced materials may hinder the deployment of such innovations, limiting their market potential. To address this, legal reforms are necessary. Initiatives like the European AI Act (Draft 2021) and U.S. proposals on trustworthy AI aim to establish clear guidelines for the use of AI in product design. These efforts are focused on risk-based classifications and ensure that the influence of AI systems on product outcomes is well-documented. However, these legal frameworks are still in development, and achieving global harmonization will likely take years.

Ethical AI in Sustainability: Transparency, Bias, and Accountability

As AI becomes increasingly integrated into sustainability initiatives—particularly in the prediction of material properties, life cycle assessment (LCA), and the design of biodegradable materials—ethical considerations must be prioritized. The responsible use of AI in these domains ensures that its deployment does not inadvertently harm vulnerable communities or contribute to environmentally damaging practices.

Sustainable AI must be grounded in transparency, fairness, and accountability to support equitable and effective environmental solutions. In the context of AI-powered material design, companies may use machine learning to predict biodegradability or performance characteristics of novel materials. However, if the decision-making processes behind these models are opaque or lack interpretability, stakeholders such as regulatory agencies and environmental certification bodies cannot verify whether AI predictions align with real-world outcomes. This undermines the credibility and reliability of sustainability claims. A relevant example is the U.S. Supreme Court case Monsanto vs. Geertson Seed Farms (2010), which underscored the need for transparency in evaluating the environmental impact of genetically modified crops. This principle equally applies to AI-driven innovations, where understanding and clearly explaining the risks and performance of new technologies is essential. To foster trust and compliance, Explainable AI (XAI) frameworks such as SHAP (SHapley Additive exPlanations) and LIME (Local Interpretable Model-agnostic Explanations) are crucial. These tools make AI decisions more understandable to humans, enabling regulators and stakeholders to assess and validate the environmental implications of AI-assisted material innovations. Moreover, regulations like the General Data Protection Regulation (GDPR) in the EU mandate the right to explanation for automated decisions, reinforcing the importance of transparency in AI systems. Bias in AI models also poses significant ethical challenges. Since AI systems learn from data, biased or incomplete training data can skew results, leading to solutions that work well in some contexts but fail in others.

For example, if a model is trained using composting data primarily from North America and Western Europe, it may suggest materials optimized for those regions. However, these recommendations might underperform in countries across South Asia or Sub-Saharan Africa, where microbial environments, climate conditions, and waste management systems differ significantly. Such mismatches can result in ineffective degradation, environmental harm, and missed opportunities for sustainability in regions that need it most. Ensuring that training data is diverse, inclusive, and geographically representative is therefore vital to creating globally viable eco-materials. Accountability is another critical component in the ethical deployment of AI in sustainability. If an AI-designed bio-composite material—certified under standards such as ISO 17088—fails to decompose as expected and leads to issues like microplastic pollution or system inefficiencies in municipal composting, who is held responsible? The AI developers, the manufacturers implementing the AI outputs, or the certification bodies that approved the product? This dilemma mirrors the European Court of Justice's 2017 ruling on autonomous vehicle liability, where the court explored the attribution of responsibility between software developers, manufacturers, and end users. A similar approach is needed in sustainable material science to assign accountability when AI-driven outcomes result in environmental harm. To address these concerns, several steps are essential: independent third-party audits of AI models used in material innovation; implementation of explainability requirements within AI certification workflows; and empowerment of regulatory bodies like the European Chemicals Agency (ECHA) or the U.S. Environmental Pro- tection Agency (EPA) to oversee AI integration in environmental technologies. Aligning emerging AI governance frameworks—such as the European AI Act—with product certification laws will help establish clear legal and ethical standards for AI-generated sustainable materials. In conclusion, in- tegrating ethical principles of transparency, bias mitigation, and accountability into the AI lifecycle is essential for ensuring that biodegradable innovations are not only scientifically effective but also socially equitable and environmentally safe across diverse global contexts.

CASE STUDIES AND EXPERIMENTAL VALIDATION

To ensure that AI-driven material innovation translates into practical, scalable, and environmentally sound solutions, real-world case studies and experimental validation are essential. This section explores how various industries—including automotive, aerospace, and construction—have adopted AI-enabled sustainable materials.

Automotive Industry: Lightweight Biocomposite Components

The automotive industry is a leading sector in adopting AI for sustainable and lightweight material development. AI helps to optimize biocomposite materials, which are both strong and environmentally friendly, by improving their performance, durability, and biodegradability.

- Ford Motor Company: Ford uses AI to develop lightweight and sustainable biocomposites, specifically for car interiors as showin in Figure 12. In collaboration with Hollandia Plastic Products (HPP) and BASF, Ford uses AI-driven simulations to optimize the properties of mate- rials like wheat straw-reinforced plastics (Rahman et al., 2023). These materials reduce vehicle weight and improve fuel efficiency. Example: By employing machine learning (ML) models to predict the mechanical behavior of different fiber-resin combinations, Ford is able to vali- date these models through physical stress testing and ensure compliance with standards such as ISO 527 and ASTM D638. This process significantly shortens product development timelines, making the production process more efficient and sustainable.

Figure 12. Ford's use of recycled and bio-based materials in automotive manufacturing

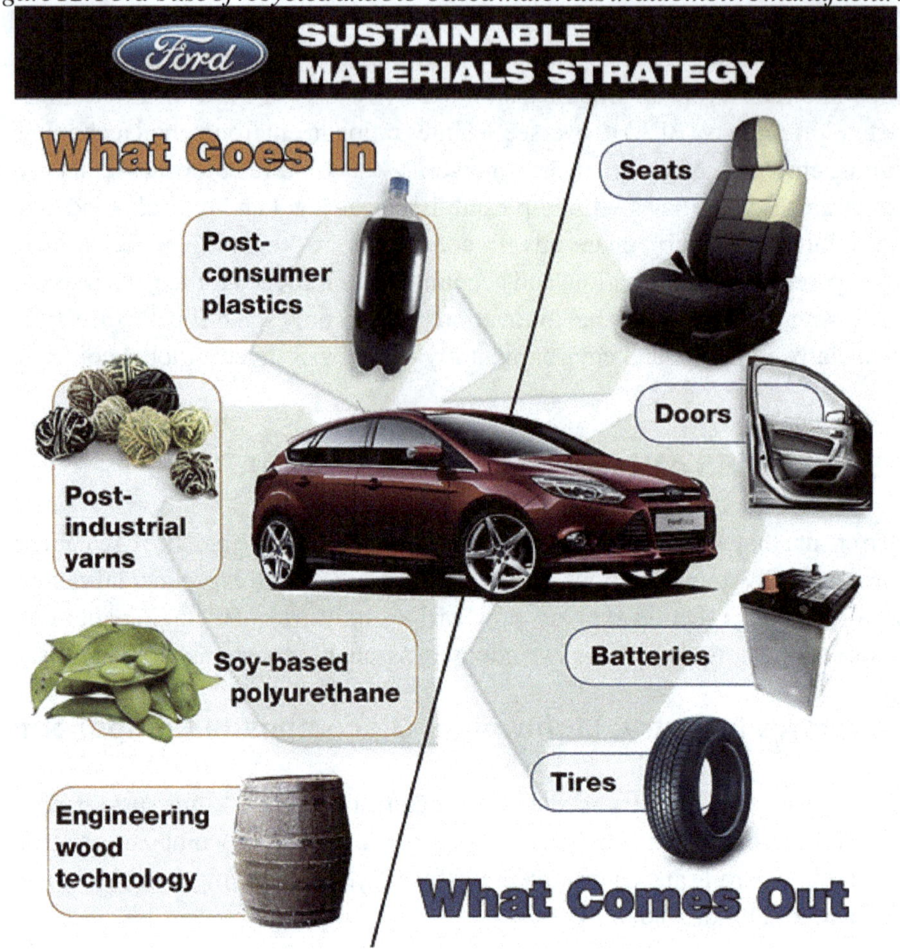

- BMW Group: BMW uses AI in the development of biodegradable and lightweight biocom- posites. In collaboration with SGL Carbon and Fraunhofer UMSICHT, BMW uses machine learning to optimize plant-based fibers such as flax and hemp for car parts (Carvalho et al., 2024). These fibers replace traditional plastics, thus reducing CO2 emissions and improving sustainability. Example: AI models help BMW to predict the mechanical properties of these biocomposites, ensuring they meet the durability standards for automotive applications.
- Volkswagen Group: Volkswagen employs AI models in its 'Sustainable Materials' initiative to optimize the use of biocomposites in car parts. AI helps to identify the best plant-based materials, like hemp and kenaf, for re-

placing conventional plastics in car components (Oh et al., 2024). Example: AI is used not only to predict the biodegradability of these materials but also to ensure that the parts meet stringent performance and environmental standards, ensuring sustainability in vehicle production.

- Toyota Motor Corporation: Toyota uses AI to predict the durability and environmental perfor- mance of biocomposite materials made from agricultural byproducts such as rice husks and kenaf (Saha et al., 2024). These materials are being explored for lightweight, eco-friendly parts for their vehicles. Example: Toyota employs machine learning to simulate the long-term degradation of these materials, ensuring that they will perform well in various environmental conditions while maintaining a lower environmental impact.
- General Motors (GM): GM has adopted AI tools to explore the use of biocomposites such as soy-based foam and natural fiber-reinforced materials in automotive interiors (De et al., 2023). Example: AI simulations are used to predict how these biocomposites will behave in the long term, assessing their strength, durability, and biodegradability. This allows GM to create more sustainable, cost-effective, and eco-friendly components for their vehicles.

Aerospace Industry: Sustainable Materials for Aircraft

The aerospace industry is also leveraging AI to create sustainable, lightweight materials. Aerospace companies are exploring biocomposites to reduce fuel consumption, increase efficiency, and minimize environmental impact.

- Airbus: Airbus uses AI to develop new lightweight composite materials that are both strong and eco-friendly. AI is applied in the design and optimization of biocomposites for aircraft structures, reducing overall weight and improving fuel efficiency (Mazlan et al., 2024). Example: Airbus uses AI simulations to optimize the integration of natural fibers in composite materials, which are then tested for strength and durability. These AI-driven designs have helped Airbus reduce the weight of certain aircraft components, resulting in lower fuel consumption. As shown in Figure 13, Airbus utilizes Sustainable Aviation Fuel (SAF) and biocomposites to promote sustainable aviation.

Figure 13. Airbus's use of SAF and biocomposites for sustainable aviation.

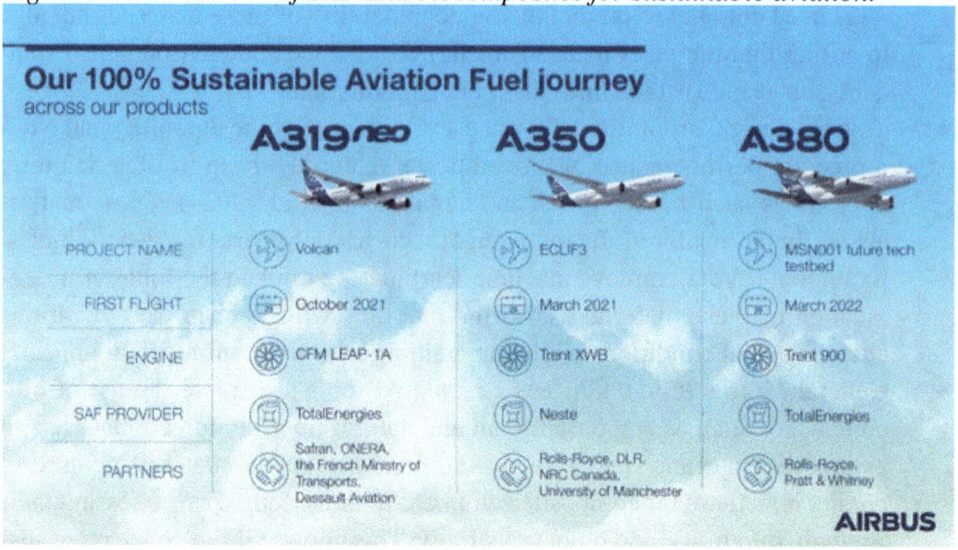

- Boeing: Boeing applies AI in the development of biocomposites for airplane interiors, such as seats, panels, and other components. Example: Boeing has developed bio-based composite ma- terials for seating, which have been optimized using machine learning to improve strength and reduce weight. AI models also help ensure that these materials meet regulatory standards and perform well over the lifespan of the aircraft. As shown in Figure 14, biocomposite materials play a significant role in modern aircraft structure and design.

Figure 14. Bio-composite materials in modern aircraft structure and design

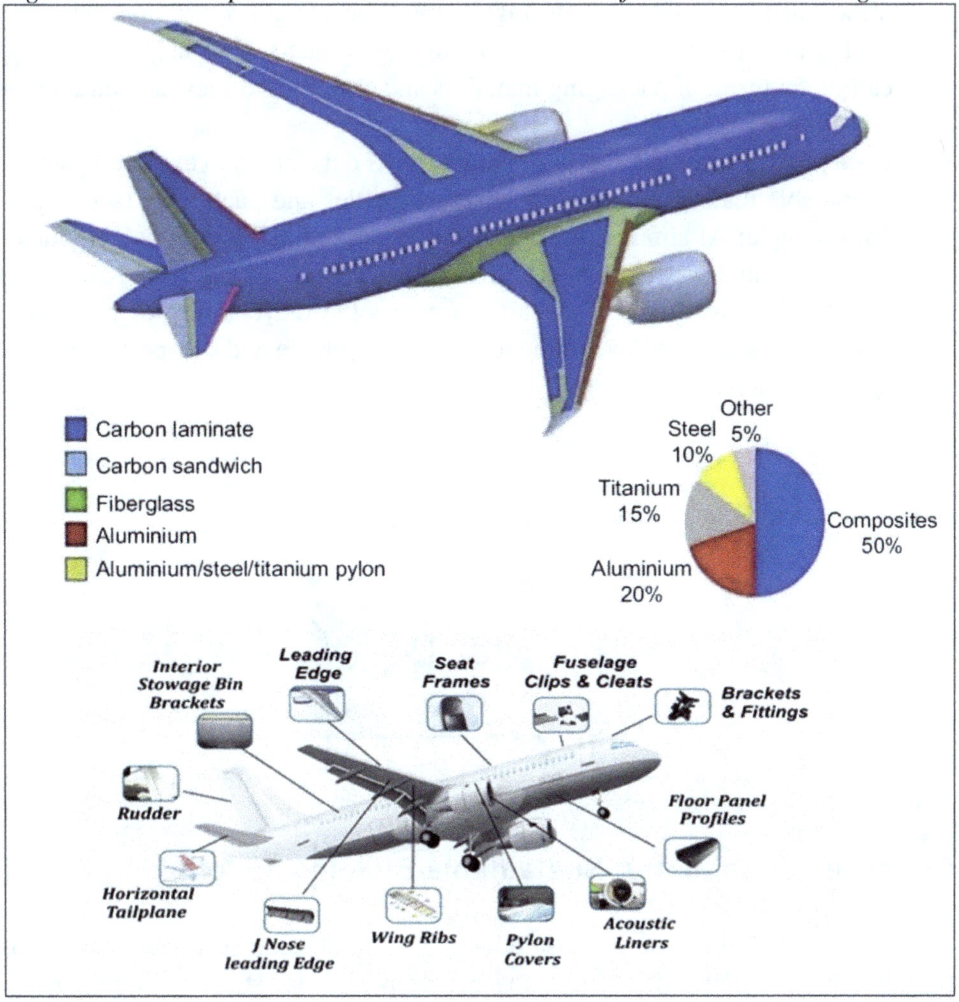

Packaging Industry: Sustainable Packaging Solutions

The packaging industry is increasingly adopting AI to design biodegradable and recyclable materials for sustainable packaging solutions. This helps reduce waste and promotes circular economy practices.

- Tetra Pak: Tetra Pak uses AI to improve the production of sustainable packaging materials that use biodegradable plastics and natural fibers. AI is employed to predict the properties of various bioplastics and natural fiber combinations. Example: Tetra Pak uses AI-driven simulations to predict how these materials will behave during the manufacturing process, improving the quality and sustainability of their packaging products. This helps reduce the carbon footprint of packaging materials and ensures that they are suitable for recycling.

- Coca-Cola: Coca-Cola has partnered with AI developers to explore new biodegradable mate- rials for their beverage bottles and packaging (see Figure 15. Example: AI simulations help identify the optimal biodegradable materials that balance performance, environmental impact, and cost (Kim et al., 2022). AI tools also help Coca-Cola assess the recyclability of packaging materials, aiding the development of fully recyclable and compostable packaging solutions.

Figure 15. Coca-Cola's biocomposite packaging

Construction Industry: Sustainable Building Materials

The construction industry is also adopting AI to develop sustainable and eco-friendly building mate- rials, including biocomposites for structural components as shown in Figure 16.

- LafargeHolcim: LafargeHolcim uses AI to develop sustainable concrete mixes that incorporate bio-based materials to reduce carbon emissions. Example: AI models predict the performance of bio-concretes made from agricultural byproducts, which are then tested in real-world condi- tions for durability and environmental impact.
- Skanska: Skanska is exploring biocomposites for building materials, using AI to optimize their design and performance. Example: AI is used to predict the strength and environmental impact of materials made from agricultural waste, such as straw, and to identify the best combinations of these materials for specific building applications.

Figure 16. Usage of biobased materials to lower carbon emissions

The integration of AI into the development of lightweight and sustainable bio-composites is trans- forming multiple industries, including automotive, aerospace, packaging, and construction. From reducing vehicle weight to creating eco-friendly building materials and sustainable packaging, AI models help companies optimize material properties, predict performance, and ensure environmental sustainability. As these industries continue to innovate, AI will play an increasingly important role in creating materials that are not only efficient but also contribute to a more sustainable future.

CHALLENGES, LIMITATIONS, AND FUTURE DIRECTIONS

While ML has shown great potential in revolutionizing bio-composite research, several challenges need to be addressed for its widespread adoption and effective implementation. These challenges include data-related limitations, model interpretability, and integration with traditional computational methods. Additionally, ensuring sustainability in bio-composite development remains a key concern. This section discusses these obstacles and proposes future research directions to improve the reliability, accuracy, and applicability of ML models in this field. Fig 17 illustrates the major challenges of data processing in ML.

Figure 17. ML Challenges in Sustainable Materials

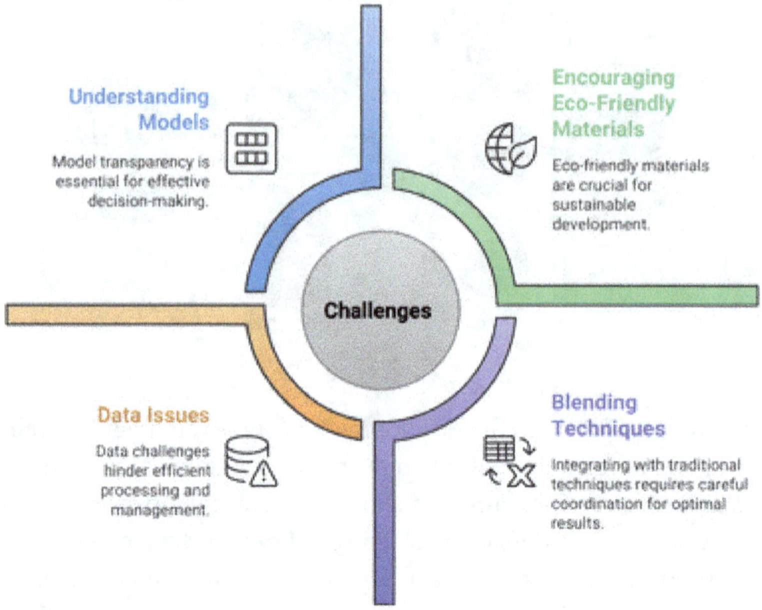

- Data-Related Challenges: The limited availability of quality data is a major hurdle in applying ML to biocomposites. Since, biocomposites are made from a wide variety of natural fibers and biodegradable poly- mers, each with unique characteristics. This diversity makes it hard to gather consistent data across different studies. Additionally, many biocomposites are developed for specific applica- tions, and the data often remains unpublished or proprietary. Without enough data, ML models may become overfitted — meaning they work well on the training data but fail to perform ac- curately on new or unseen samples.
- Issues with Data Consistency and Format: Another key issue is the absence of standardized data formats in biocomposite research. Dif- ferences in testing methods, measurement units, and data reporting practices can lead to inconsistent datasets. These variations make it difficult to combine or compare data from different sources. Furthermore, experimental data can sometimes include errors or noise, reducing the reliability of ML predictions. For ML to be effectively used in bio-composite development, it is essential to create clean, consistent, and standardized datasets across the field.
- Balancing Accuracy and Interpretability: In the field of bio-composites, it is not enough for ML models to just make accurate predictions — scientists also

need to understand why a prediction was made. Many advanced ML models, like deep learning networks, can produce very accurate results, but they often work like 'black boxes', meaning their internal workings difficult to understand. This lack of transparency can be a problem when researchers need to link predictions to real-world material behavior. For exam- ple, knowing that a certain bio-composite has high strength is useful, but understanding which features (like fiber content or matrix type) contribute to that strength is even more valuable. Therefore, it is important to find a balance — using models that are accurate, but also explain- able enough to provide insights into the physical and mechanical behavior of bio-composites.

Despite the challenges, the future of ML in bio-composite research looks promising. The rising need for eco-friendly, high-performance materials, combined with rapid progress in computational techniques and material science, is pushing the field forward. As tools for data analysis and integration become more refined, we are entering a new era where ML helps to drive smarter, more sustainable innovations in bio-composites.

Future Prospects: Towards Self-Evolving Sustainable Materials

Recent developments in automated materials discovery platforms—such as the Materials Acceler- ation Platforms (MAPs) by institutions like the Toyota Research Institute and Harvard's Materials Intelligence Research are transforming how materials are designed. These platforms integrate ML al- gorithms with robotics and high-throughput experiments to autonomously explore vast material spaces and optimize composites for targeted performance traits (e.g., tensile strength, biodegradability, ther- mal resistance). In parallel, cutting-edge technologies like Reinforcement Learning (RL) and gener- ative design algorithms are beginning to shape the next generation of materials (Kumar et al., 2024). At MIT's CSAIL lab, for instance, researchers have created materials that can adapt or reconfigure themselves in response to environmental conditions. These systems rely on predictive ML models and real-time operational feedback, allowing them to improve continuously without human interven- tion (Kantaros et al., 2024). Another key innovation lies in multi-objective optimization frameworks that evaluate multiple factors like performance, cost, and environmental footprint—at once. These models are particularly valuable for the development of bio-composites, where balancing function with sustainability is crucial. A notable example is a 2024 study published in Nature Communications, which introduced an AI-powered platform capable of designing bio-resins that offer both high recyclability and low embodied energy. Such advances mark a pivotal step toward materials that evolve intelligently and align with long-term environmental goals (El-Rayes et al., 2024). Startups and col- laborative research groups—including the Ellen MacArthur Foundation, Circularize, and Reselo are advancing sustainable innovation by integrating blockchain and AI into material tracking systems. These technologies facilitate end-to-end visibility across the material lifecycle. Through LCA data, ML algorithms can suggest modifications in product design to encourage ease of disassembly, reuse, and natural decomposition, thereby supporting closed-loop manufacturing. Simultaneously, current research is exploring the incorporation of carbon-negative materials into bio-composites. Materials such as hemp fibers and mycelium-based structures are gaining attention not only for their biodegrad- ability but also for their ability to store atmospheric carbon. ML can be employed to model the carbon sequestration potential of these materials across their life span, helping in carbon accounting and compliance with environmental regulations. Further, researchers at institutions like ETH Zurich and Lawrence Berkeley National Lab have begun integrating ML into Carbon Capture and Utilization (CCU) pipelines, using bio-composite materials as carbon sinks. These efforts involve the use of bio- composite materials as functional carbon sinks (Saxena et al., 2024). Advanced materials equipped with sensors and adaptive ML systems are being designed to continuously monitor and respond to parameters like carbon adsorption rates, structural longevity, and ecological degradation. Such innovations are helping to create better bio-composite materials that can protect the environment and helps to fight climate change.

CONCLUSION

This chapter introduced the ASB-ML framework, showing how advanced AI models like GNNs, Vision Transformers, and Physics-Informed Neural Networks enhance bio-composite design, analysis, and sustainability. Integrated with life cycle analysis, these models improve prediction, defect detection, and data-driven decision-making. Beyond technical progress, ASB-ML supports industry, society, and environmental goals by enabling efficient, low-waste, and traceable materials. It also underscores the need for responsible AI—ensuring transparency, ethical data use, and regulatory compliance. While challenges such as data limitations and standardization persist, ASB-ML offers a collaborative path forward. It bridges technology and policy to advance sustainable innovation and realize a circular economy.

REFERENCES

Al-Maharma, A. Y., Patil, S. P., & Markert, B. (2022). Environmental impact analysis of plant fibers and their composites relative to their synthetic counterparts based on life cycle assessment approach. In *Advances in bio-based fiber* (pp. 741–781). Woodhead Publishing.

Bargel, H., & Scheibel, T. (2024). A bio-engineering approach to generate bioinspired (spider) silk protein-based materials. at-Automatisierungstechnik, 72(7), 657-665.

Barkhausen, R., Durand, A., & Fick, K. (2022). Review and analysis of ecodesign directive implementing measures: Product regulations shifting from energy efficiency towards a circular economy. *Sustainability*, *14*(16), 10318.

Carvalho, D., Ferreira, N., Franc¸a, B., Marques, R., Silva, M., Silva, S., ... & Oliveira, C. (2024). Advancing sustainability in the automotive industry: Bioprepregs and fully bio-based composites. Composites Part C: Open Access, 100459.

Carvalho, D., Ferreira, N., Franc¸a, B., Marques, R., Silva, M., Silva, S., ... & Oliveira, C. (2024). Advancing sustainability in the automotive industry: Bioprepregs and fully bio-based composites. Composites Part C: Open Access, 100459.

Cencer, M. M., Moore, J. S., & Assary, R. S. (2022). Machine learning for polymeric materials: An introduction. *Polymer International*, *71*(5), 537–542.

De, S., James, B., Ji, J., Wasti, S., Zhang, S., Kore, S., & Zhao, X. (2023). Biomass-derived composites for various applications. In *Advances in Bioenergy* (Vol. 8, pp. 145–196). Elsevie.

Dey, S., Veerendra, G. T. N., Babu, P. A., Manoj, A. P., & Nagarjuna, K. (2024). Degradation of plastics waste and its effects on biological ecosystems: A scientific analysis and comprehensive review. *Biomedical Materials & Devices*, *2*(1), 70–112.

El-Rayes, N., Chang, A., & Shi, J. (2024). Blockchain Technology Toward a Circular Economy: A Future Paradigm. In *Circular Economy and Sustainable Development: A Necessary Nexus for a Sustainable Future* (pp. 335–351). Springer International Publishing.

Hatamizadeh, A., Yin, H., Heinrich, G., Kautz, J., & Molchanov, P. (2023, July). Global context vision transformers. In *International Conference on Machine Learning* (pp. 12633-12646). PMLR.

Ismail, S., & Vasudev, H. (2024, September). EIROB: An Emotionally Intelligent ROBot for Empathetic Support. In *2024 IEEE International Conference on Signal Processing, Informatics, Communication and Energy Systems (SPICES)* (pp. 1-6). IEEE.

Ismail, S., & Vasudev, H. (2025). Artificial Intelligence for Personalized Electric Vehicle Energy Management. In Emerging Multisector Applications of AI and IoT (pp. 221-260). IGI Global Scientific Publishing.

Ismail, S., & Vasudev, H. (2025). Artificial Intelligence for Personalized Electric Vehicle Energy Management. In *Emerging Multisector Applications of AI and IoT* (pp. 221–260). IGI Global Scientific Publishing.

Ismail, S., Vasudev, H., & Cheriyan, J. (2025). Enhancing Seed Quality and Predictive Accuracy in Indian Agriculture: Case Study at RARS, Kottayam, Kerala. In *Advances in Seed Quality Evaluation and Improvement* (pp. 491–528). Springer.

Kantaros, A., Ganetsos, T., & Petrescu, F. I. T. (2024). Transforming object design and creation: Biomaterials and contemporary manufacturing leading the way. *Biomimetics*, *9*(1), 48. PMID: 38248622

Khanna, N., Wadhwa, J., Pitroda, A., Shah, P., Schoop, J., & Sarıkaya, M. (2022). Life cycle assessment of environmentally friendly initiatives for sustainable machining: A short review of current knowledge and a case study. *Sustainable Materials and Technologies*, *32*, e00413.

Kim, T., & Lee, S. D. (2022). Designing for green and grey: Insights from single-use plastic water bottles. *International Journal of Environmental Research and Public Health*, *19*(3), 1423. PMID: 35162446

Kumar Singh, S., Rai, R., Pradip Khawale, R., Patel, D., Bielecki, D., Nguyen, R., & Zhang, Z. (2024). Deep learning in computational design synthesis: A comprehensive review. *Journal of Computing and Information Science in Engineering*, *24*(4), 040801.

Liu, M., Li, H., Zhou, H., Zhang, H., & Huang, G. (2024). Development of machine learning methods for mechanical problems associated with fibre composite materials: A review. *Composites Communications*, •••, 101988.

Mahapatra, S. K., & Satapathy, A. (2023). Parametric analysis of erosion wear of sponge iron slag-filled ramie–epoxy composites using Taguchi and preference selection index methods. *Proceedings of the Institution of Mechanical Engineers. Part E, Journal of Process Mechanical Engineering*, •••, 09544089231193927.

Mazlan, N., Shah, A. U. M., Hua, T. C., Najmi, I., & Ramli, N. (2024). Biocomposites as Aircraft Materials. Sustainable Materials and Manufacturing Techniques in Aviation, 27-53.

Mendanha, D. (2024). Driving sustainability in the automotive industry: bio-coated materials and modern strategies. Academia Materials Science, 1(2).

Nagaraja, S., Anand, P. B., & Ammarullah, M. I. (2024). Synergistic advances in natural fibre composites: A comprehensive review of the eco-friendly bio-composite development, its charac- terization and diverse applications. *RSC Advances*, *14*(25), 17594–17611. PMID: 38828274

Naser, A. Z., Deiab, I., Defersha, F., & Yang, S. (2021). Expanding poly (lactic acid) (PLA) and polyhydroxyalkanoates (PHAs) applications: A review on modifications and effects. *Polymers*, *13*(23), 4271. PMID: 34883773

Nizamuddin, S., Baloch, A. J., Chen, C., Arif, M., & Mubarak, N. M. (2024). Bio-based plas- tics, biodegradable plastics, and compostable plastics: Biodegradation mechanism, biodegrad- ability standards and environmental stratagem. *International Biodeterioration & Biodegradation*, *195*, 105887.

Oh, E., Godoy Zu´niga, M. M., Nguyen, T. B., Kim, B. H., Trung Tien, T., & Suhr, J. (2024). Sus- tainable green composite materials in the next-generation mobility industry: review and prospec- tive.

Photopoulos, J. (2023). Spinning sustainability: Meet the physicists turning wood into clothes. *Physics World*, *36*(11), 37.

Rahman, M. Z., Rahman, M., Mahbub, T., Ashiquzzaman, M., Sagadevan, S., & Hoque, M. E. (2023). Advanced biopolymers for automobile and aviation engineering applications. *Journal of Polymer Research*, *30*(3), 106.

Ramesh, M., Deepa, C., Kumar, L. R., Sanjay, M. R., & Siengchin, S. (2022). Life-cycle and environmental impact assessments on processing of plant fibres and its bio-composites: A critical review. *Journal of Industrial Textiles*, •••, 5518S–5542S.

Rayasam, S. D., Koman, P. D., Axelrad, D. A., Woodruff, T. J., & Chartres, N. (2022). Toxic Substances Control Act (TSCA) implementation: How the amended law has failed to protect vulnerable populations from toxic chemicals in the United States. *Environmental Science & Technology*, *56*(17), 11969–11982. PMID: 35980084

Romeiko, X. X., Zhang, X., Pang, Y., Gao, F., Xu, M., Lin, S., & Babbitt, C. (2024). A review of machine learning applications in life cycle assessment studies. *The Science of the Total Environment*, *912*, 168969. PMID: 38036122

Saha, A., Pugno, N. M., & Kundanati, L. (2025). Agricultural Waste-based Composites for Electric Vehicle Interiors: A Step Towards Sustainability Using Waste to Value Approach.

Saxena, A., Prakash Gupta, J., Tiwary, J. K., Kumar, A., Sharma, S., Pandey, G., & Raghav Chaturvedi, K. (2024). Innovative Pathways in Carbon Capture: Advancements and Strategic Approaches for Effective Carbon Capture, Utilization, and Storage. *Sustainability*, *16*(22), 10132.

Singh, N., Ogunseitan, O. A., Wong, M. H., & Tang, Y. (2022). Sustainable materials alternative to petrochemical plastics pollution: A review analysis. *Sustainable Horizons*, *2*, 100016.

Thimmaiah, S. H., Narayanappa, K., Thyavihalli Girijappa, Y., Gulihonenahali Rajakumara, A., Hemath, M., Thiagamani, S. M. K., & Verma, A. (2023). An artificial neural network and Taguchi prediction on wear characteristics of Kenaf–Kevlar fabric reinforced hybrid polyester composites. *Polymer Composites*, *44*(1), 261–273.

Tillu, R., Muthusubramanian, M., & Periyasamy, V. (2023). From data to compliance: the role of ai/ml in optimizing regulatory reporting processes. Journal of Knowledge Learning and Science Technology ISSN: 2959-6386 (online), 2(3), 381-391.

Yuan, L., Ni, Y. Q., Deng, X. Y., & Hao, S. (2022). A-PINN: Auxiliary physics informed neural networks for forward and inverse problems of nonlinear integro-differential equations. *Journal of Computational Physics*, *462*, 111260.

Zhang, P., Mao, Y., Yuan, W., Zheng, J., Hu, S., & Wang, K. (2024). A critical review on modeling and prediction on properties of fresh and hardened geopolymer composites. *Journal of Building Engineering*, •••, 109184.

Zhu, Y., Xu, W., Zhang, J., Liu, Q., Wu, S., & Wang, L. (2021). Deep graph structure learning for robust representations: A survey. arXiv preprint arXiv:2103.03036, 14, 1-1.

Chapter 5
Nanomaterials Discovery Using AI-Based Technology Accelerators:
Innovation Through Predictive Modeling and Intelligent Experimentation

Ramadevi Vitthal Salunkhe
https://orcid.org/0009-0005-0247-0115
G.H. Raisoni College of Engineering, Nagpur, India

Tusharkumar Rajaram Sathe
Tatyasaheb Kore Institute of Engineering and Technology, India

Rushikesh Moreshwarrao Shete
https://orcid.org/0009-0008-1793-7584
St. Vincent Pallotti College of Engineering and Technology, Nagpur, India

Pooja Wagh
https://orcid.org/0009-0001-7389-539X
St. Vincent Pallotti College of Engineering and Technology, Nagpur, India

Roshan Kolte
KDK College of Engineering, Nagpur, India

Renuka Raut
https://orcid.org/0009-0001-9917-9498
St. Vincent Pallotti College of Engineering and Technology, Nagpur, India

ABSTRACT

Integration of artificial intelligence into nanomaterials research is revolutionizing the discovery, design, and optimization of advanced materials. This chapter presents a comprehensive framework for AI-driven discovery of nanomaterials, encompassing data acquisition, predictive modeling, intelligent experimentation, and iterative

DOI: 10.4018/979-8-3373-2883-6.ch005

feedback loops. It explores sector-specific applications across energy, environment, biomedical, manufacturing, and agriculture, while addressing technical, integration, and ethical challenges that influence large-scale adoption. Emerging tools such as quantum computing, federated learning, and digital twins are discussed in the context of accelerating innovation. The proposed roadmap highlights the crucial roles of infrastructure, policy, and education in establishing a sustainable and ethically informed ecosystem for AI-enabled nanomaterials research. By bridging interdisciplinary gaps, the chapter offers strategic insights for researchers, policymakers, and industry stakeholders aiming to harness AI for transformative advancements in nanotechnology.

1. INTRODUCTION

The rapid advancement of materials science in recent decades has been marked by the emergence of nanotechnology as a transformative force. Nanomaterials—engineered at dimensions between 1 and 100 nanometres—possess unique and often unexpected properties that differ significantly from their bulk counterparts, enabling innovations across multiple sectors. Their integration into industrial applications, consumer products, and high-tech systems underscores their strategic importance in addressing global challenges such as sustainable energy, healthcare innovation, environmental remediation, and advanced manufacturing. Nevertheless, despite these opportunities, the journey from conceptualizing a novel nanomaterial to its large-scale deployment remains a complex and resource-intensive process. This reality has catalyzed a growing interest in leveraging computational and intelligent systems—particularly Artificial Intelligence (AI)—to accelerate discovery, reduce costs, and enhance the precision of experimental outcomes (Chen et al., 2021; Jablonka et al., 2020).

1.1 Context: Nanomaterials in Modern Science and Industry

Nanomaterials are materials with structural features in the range of 1 to 100 nanometres, a scale at which unique physical, chemical, optical, and electronic phenomena emerge. These properties arise due to effects such as quantum confinement, large surface area-to-volume ratios, and atomic-level structural control. As a result, nanomaterials have become indispensable in various domains, ranging from electronics and energy to medicine, aerospace, environmental engineering, and advanced manufacturing.

In electronics, carbon-based nanomaterials such as graphene and carbon nanotubes have demonstrated exceptional electrical conductivity, mechanical strength, and flexibility, enabling the development of high-speed transistors, flexible displays, and next-generation sensors. In energy applications, nanostructured materials are used in high-capacity lithium-ion batteries, supercapacitors, and photovoltaic devices, improving charge storage, energy conversion efficiency, and durability. The biomedical field leverages nanoparticles for targeted drug delivery (Chou et al., 2025), imaging, and diagnostics, offering new approaches for personalized medicine and minimally invasive therapies (Hou et al., 2021; Mirkin, 2025). Similarly, nanomaterials such as titanium dioxide and zinc oxide are employed in environmental remediation, catalyzing pollutant degradation and enabling efficient water purification.

From a global economic perspective, the nanomaterials market has grown rapidly, with projections suggesting it could exceed USD 20B in the coming years, driven by increased adoption across industries and ongoing technological innovation. However, despite these advances, the development cycle for novel nanomaterials remains slow and resource-intensive, hampered by the complexity of nanoscale phenomena and the limitations of conventional research methodologies.

1.2 Limitations of Traditional Discovery Approaches

Historically, nanomaterial discovery has relied heavily on empirical experimentation and incremental improvements. Materials vary in composition, morphology, and synthesis conditions in laboratory settings, and their resulting properties are analyzed through characterization techniques such as scanning electron microscopy (SEM), X-ray diffraction (XRD), and spectroscopy. While this approach has produced significant breakthroughs, it suffers from several critical limitations:

1. **High Dimensionality of the Design Space**

The possible combinations of elements, structures, and processing parameters create an astronomical number of potential nanomaterial candidates. Exploring even a fraction of this design space experimentally is infeasible. For example, synthesizing and testing every possible binary or ternary nanoparticle combination would require decades of work and significant financial resources.

2. **Time-Intensive and Costly Experiments**

Each synthesis and characterization cycle can take days or weeks, depending on the complexity of the material and the number of variables involved. Scaling up these cycles to explore an ample parameter space leads to prohibitive time and cost demands.

3. **Limited Reproducibility**

Slight variations in experimental conditions—such as temperature fluctuations, precursor purity, or equipment calibration—can significantly affect material properties. Ensuring reproducibility across laboratories remains a significant challenge.

4. **Fragmented Data Ecosystem**

Nanomaterials research generates vast quantities of data, but these are often siloed in individual laboratories, publications, or proprietary databases. The lack of standardized formats and accessible repositories hinders large-scale comparative analysis.

5. **Human Intuition Bias**

Traditional discovery relies on the expertise and intuition of researchers to select promising candidates for synthesis and testing. While invaluable, this human-driven selection process may overlook unconventional but potentially high-performing material configurations.

These challenges are compounded by the increasing demand for materials with tailored, multi-functional properties, such as nanocomposites that are simultaneously lightweight, conductive, and chemically stable under extreme conditions. The inability of conventional methods to efficiently navigate such complex requirements has prompted the exploration of new computational and automated strategies (Singh et al., 2022; Radulescu et al., 2023).

1.3 The Role of AI in Accelerating Design and Experimentation

Artificial Intelligence (AI) is redefining how nanomaterials are conceptualized, designed, and validated. By leveraging computational models, machine learning (ML) algorithms, and automated experimentation platforms, AI enables researchers to explore and optimize vast design spaces far more efficiently than traditional trial-and-error methods.

Key contributions of AI in nanomaterial discovery include:

1. Predictive Modeling of Properties

ML models trained on experimental and computational datasets can predict physical, chemical, and mechanical properties of new nanomaterials before they are synthesized. For example, convolutional neural networks (CNNs) can analyze structural images to forecast electronic band gaps, while regression models can estimate thermal conductivity or tensile strength (Li et al., 2022).

2. Inverse Design Approaches

Instead of starting with a candidate material and testing its properties, AI can begin with a target property profile—such as high electrical conductivity and thermal stability—and work backward to suggest optimal compositions and structures. Generative models, including variational autoencoders (VAEs) and generative adversarial networks (GANs), are increasingly used for this purpose.

3. Accelerated High-Throughput Screening

AI can guide high-throughput experimental setups to focus on the most promising regions of the design space, significantly reducing the number of physical experiments needed. This is particularly impactful in combinatorial chemistry for nano catalysts or nanoparticle libraries.

4. Integration with Autonomous Laboratories

Robotic synthesis platforms integrated with AI-driven decision-making can conduct experiments 24/7, adjusting parameters in real time based on model predictions and experimental outcomes (MacLeod et al., 2022; Volk et al., 2023; Szymanski et al., 2023). This creates a closed-loop system where AI continuously learns from new data to refine its predictions.

5. Data Mining and Knowledge Extraction

Natural language processing (NLP) tools can scan vast corpora of scientific literature and patents to extract relevant synthesis protocols, property measurements, and trends. This enables the identification of underexplored material classes or synthesis methods that have shown promise in related contexts.

6. Enhanced Reproducibility and Standardization

AI systems can enforce standardized data acquisition protocols and automatically correct for measurement noise, improving reproducibility across different research groups and facilities (Molnar, 2023; Zhong et al., 2022).

Impact on Research Timelines:

Studies have shown that AI-driven nanomaterial discovery can reduce the timeline from initial concept to validated prototype from several years to mere months (Karniadakis et al., 2021). This acceleration not only increases the pace of innovation but also opens opportunities for rapid response to emerging global challenges—such as the development of new filtration membranes during a public health crisis or high-capacity energy storage materials for renewable energy integration.

Figure 1. AI-Driven Framework for Nanomaterials Discovery

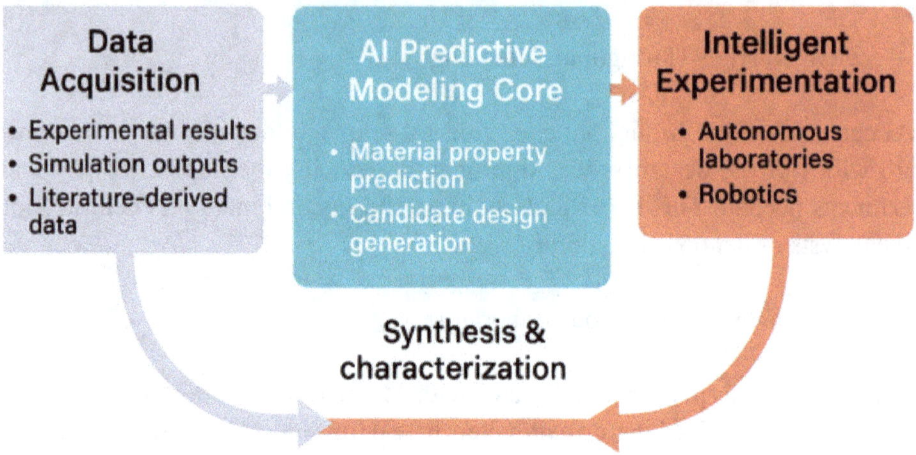

The framework is depicted as a cyclical process. Data Acquisition modules on the left include experimental results, simulation outputs, and literature-derived data. These feed into the AI Predictive Modeling Core, which applies machine learning algorithms to forecast material properties and generate candidate designs. On the right, Intelligent Experimentation modules, powered by autonomous laboratories and robotics, synthesize and characterize the candidates. Results are fed back into the AI models in a continuous learning loop, enabling iterative refinement of both predictions and materials performance.

2. FUNDAMENTALS OF NANOMATERIALS AND AI TECHNOLOGIES

The convergence of nanotechnology and artificial intelligence (AI) has the potential to revolutionize materials science by enabling faster, more precise, and cost-effective discovery processes. This section outlines the key types of nanomaterials and the AI technologies that are transforming their design, optimization, and deployment (Mrdjenovich et al., 2021; Tran et al., 2022).

2.1 Nanomaterials: Types, Properties, and Applications

Nanomaterials are classified based on their dimensionality, composition, and structure:

- **Zero-Dimensional (0D) Nanomaterials:** Nanoparticles, quantum dots.

Properties: Strong quantum confinement, high surface energy.
Applications: Targeted drug delivery, bioimaging, and quantum computing components.

- **One-Dimensional (1D) Nanomaterials:** Nanowires, nanotubes.

Properties: High aspect ratio, directional conductivity, and mechanical flexibility.
Applications: Flexible electronics, high-strength composites, nano sensors.

- **Two-Dimensional (2D) Nanomaterials:** Graphene, MoS_2, h-BN.

Properties: Atomic-scale thickness, tunable band gaps, high carrier mobility.
Applications: Energy storage devices, catalysis, transparent conductive films.

- **Three-Dimensional (3D) Nanomaterials:** Nanocomposites, nanoporous frameworks.

Properties: Large surface area, customizable porosity.
Applications: Gas storage, heterogeneous catalysis, structural reinforcement.

Unique Advantages:

Nanomaterials exhibit high reactivity, tunable electronic/optical properties, and lightweight mechanical strength, making them indispensable in electronics, renewable energy, biomedical technologies, environmental remediation, and aerospace.

2.2 Artificial Intelligence in Materials Science

Artificial Intelligence provides computational strategies to navigate the high-dimensional design space of nanomaterials. By integrating experimental, simulated, and literature-derived datasets, AI models can predict properties, optimize synthesis pathways, and propose novel material designs.

Machine Learning (ML)

- Algorithms: Random Forests (RF), Support Vector Machines (SVM), Gradient Boosting Machines (GBM).
- Applications: Predicting bandgap, thermal conductivity, and catalytic activity.
- Example: Random Forest models trained on Density Functional Theory (DFT) datasets predicted perovskite bandgaps with high accuracy.

Deep Learning (DL)

- Algorithms: Convolutional Neural Networks (CNNs) for SEM/TEM image analysis; Graph Neural Networks (GNNs) and Crystal Graph Convolutional Neural Networks (CGCNNs) for structure–property prediction.
- Applications: Predicting formation energies, crystal stability, defect tolerance.
- Example: CGCNN achieved a mean absolute error (MAE) of 0.04 eV in bandgap prediction. Crystal Graph Convolutional Neural Networks (CGCNNs) have achieved a mean absolute error (MAE) of 0.04 eV in bandgap prediction (Beyerle et al., 2023).

Reinforcement Learning (RL)

- Algorithms: Deep Q-Networks (DQN), Policy Gradient, Actor–Critic methods.
- Applications: Optimizing experimental conditions in autonomous labs (e.g., chemical vapor deposition).
- Example: RL agents improved carbon nanotube (CNT) yields by >30% through adaptive process control.

Natural Language Processing (NLP)

- Algorithms: Transformer-based models (BERT, SciBERT, ChemBERTa).
- Applications: Mining synthesis protocols, extracting material-property trends from literature.

- Example: NLP pipelines identified optimal MOF candidates for CO_2 capture by scanning >500,000 papers.

Generative Models

- Algorithms: Variational Autoencoders (VAEs), Generative Adversarial Networks (GANs).
- Applications: Inverse design of materials with targeted properties.
- Example: VAEs proposed stable perovskite structures for photovoltaics, while GANs generated novel nanoparticle morphologies.

2.3 Comparative Analysis: AI-Driven vs. Traditional Approaches

The following table summarizes the differences between traditional experimental workflows and AI-augmented discovery pipelines in nanomaterials research.

Table 1. Comparison of Traditional and AI-Driven Nanomaterials Discovery

Parameter	Traditional Approach	AI-Driven Approach
Design Space Exploration	Limited by manual selection and human intuition	Automated, data-driven search of vast parameter spaces
Speed	Weeks to months per material candidate	Hours to days for predictions and short synthesis cycles
Data Utilization	Experimental results only	Integrates experimental, simulated, and literature data
Cost Efficiency	High due to extensive lab work	Lower through targeted, high-probability experiments
Reproducibility	Dependent on lab conditions and human expertise	Standardized protocols via automation and AI monitoring
Innovation Potential	Incremental improvements	Discovery of unconventional, high-performance materials

2.4 Mapping AI Methods to Nanomaterial Challenges

Table 2.2 maps core AI methods to their corresponding algorithms and typical applications in nanomaterials research. It illustrates how approaches such as machine learning, deep learning, reinforcement learning, natural language processing, and generative models are applied to challenges, including property prediction, literature mining, synthesis optimization, and inverse material design.

Table 2. AI Methods and Applications in Nanomaterials

AI Method	Algorithms	Typical Applications
Machine Learning (ML)	RF, SVM, GBM	Bandgap prediction, mechanical strength
Deep Learning (DL)	CNN, GNN, CGCNN	Image-based analysis, property forecasting
Reinforcement Learning (RL)	DQN, Policy Gradient	Autonomous synthesis optimization
Natural Language Processing (NLP)	BERT, SciBERT	Literature mining, data extraction
Generative Models	VAE, GAN	Inverse design of nanostructures

3. AI-DRIVEN PREDICTIVE MODELING IN NANOMATERIALS RESEARCH

Artificial Intelligence (AI)-driven predictive modeling is reshaping nanomaterials research by enabling accurate forecasting of material properties, accelerating design cycles, and reducing reliance on exhaustive experimental trials. By integrating data from experiments, simulations, and literature, predictive models capture complex structure–property relationships and guide targeted synthesis.

3.1 Foundations of Predictive Modeling

Predictive modeling involves constructing computational functions that map input descriptors (composition, structure, synthesis conditions) to output properties (bandgap, conductivity, catalytic activity). In nanomaterials research, these models are trained on diverse datasets:

- Experimental data: XRD, SEM, TEM, Raman spectroscopy.
- Simulated data: Density Function Theory (DFT), Molecular Dynamics (MD), Monte Carlo simulations.
- High-throughput screening: Combinatorial testing of material variants.
- Literature mining: Extracting datasets from publications and patents via NLP tools.

This integration allows researchers to identify correlations between atomic-scale descriptors and macroscopic performance.

3.2 Data Sources and Preprocessing

High-quality data is critical for robust predictions. Common preprocessing steps include:

- Dimensionality reduction: Principal Component Analysis (PCA), t-SNE, or autoencoders to reduce descriptor space.
- Feature engineering: Creating descriptors such as bond length, electronegativity, and atomic radii.
- Normalization and cleaning: Ensuring reproducibility across datasets with standardized metadata.

3.3 Machine Learning Approaches

Before diving into individual algorithms, it is important to note that machine learning approaches in nanomaterials research rely heavily on the choice of descriptors and representation schemes. Descriptors may include physicochemical features such as electronegativity, ionic radii, lattice parameters, or topological fingerprints derived from crystal graphs. The performance of predictive models is often constrained not by the algorithm itself but by how well these descriptors capture the underlying physics and chemistry of the system. For instance, handcrafted descriptors are effective for regression-based models, whereas deep learning frameworks such as CNNs and GNNs can learn hierarchical features directly from raw structural or imaging data. Consequently, selecting the appropriate representation—whether tabular descriptors, molecular graphs, or microstructural images—dictates which machine learning technique will be most effective for property prediction.

3.3.1 Regression Models

Regression models predict continuous variables such as thermal conductivity, elastic modulus, or bandgap energy.

Equation for Linear Regression:

$$y = \beta_0 + \beta_1 x_1 + \beta_2 x_2 + \cdots + \beta_n x_n + \epsilon$$

where y is the target property, x_i are descriptors βi $beta_i$ are learned coefficients and ϵ is the error.

- Support Vector Regression (SVR): Captures non-linear property relationships.
- Random Forest Regression: Handles noisy datasets with good interpretability.
- Gradient Boosting: Achieves high accuracy in predicting mechanical and thermal properties.

3.3.2 Classification Models

Classification models categorize materials into discrete classes, e.g., "photocatalytic" vs. "non-photocatalytic."

- Decision Trees and Random Forests: Widely applied for photocatalyst screening.
- Neural Networks: Classify crystal phases based on XRD spectra.
- Example: A Random Forest classifier achieved >90% accuracy in predicting photocatalytic activity of TiO_2 variants.

3.3.3 Deep Learning Models

Deep learning is particularly suited for high-dimensional, image- and graph-based datasets.

- CNNs: Extract features from microscopy images, predicting bandgaps and defect distributions (Zhang et al., 2022; Dubrovsky et al., 2024).
- Graph Neural Networks (GNNs): Represent atoms as nodes and bonds as edges. Crystal Graph Convolutional Neural Networks (CGCNNs) have achieved a mean absolute error (MAE) of 0.04 eV in bandgap prediction.
- Transformers: Emerging models for sequential/graph embeddings of materials data.

3.3.4 Transfer Learning

Transfer learning allows models trained on large general datasets (e.g., Materials Project) to be adapted to smaller, domain-specific datasets.

- Example: Transferred knowledge from >100,000 crystal structures to accurately predict properties of metal-organic frameworks (MOFs).

3.4 Inverse Design and Generative Modeling

Instead of predicting properties for given structures, **inverse design** starts with desired properties and proposes candidate structures.

- **Bayesian Optimization:** Guides candidate selection by balancing exploration and exploitation.

$x^* = arg\ max_x\ ;\ \mu(x) + \kappa\sigma(x)$

where $\mu(x)$ is the predicted mean property, $\sigma(x)$ is uncertainty and $\kappa\sigma(x)$ controls trade-off.

- **Generative Models:**
    - o **Variational Autoencoders (VAEs):** Map material descriptors into latent space, generating novel candidates with smooth property variations.
    - o **Generative Adversarial Networks (GANs):** Produce synthetic nanostructures resembling high-performing materials.
- **Example:** VAEs were used to design perovskite solar absorbers with optimized stability and efficiency.

3.5 Integration with Autonomous Experimental Platforms

Predictive models achieve their maximum impact when integrated with autonomous or semi-autonomous experimental systems, creating a closed-loop discovery framework. In such systems, AI does not just generate predictions but also actively guides laboratory experiments in real time.

- **Step 1** – Candidate Generation: The predictive model identifies promising nanomaterial compositions or synthesis conditions based on structure–property correlations.
- **Step 2** – Automated Synthesis: Robotic platforms, such as programmable reactors and automated pipetting systems, synthesize these candidates with high precision and repeatability.
- **Step 3** – High-Throughput Characterization: Automated tools such as X-ray diffraction (XRD), Raman spectroscopy, or scanning electron microscopy (SEM) rapidly assess the physical and chemical properties of synthesized materials.
- **Step 4** – Feedback to Model: The characterization results are automatically digitized and fed back into the AI model to refine predictions.

This continuous learning cycle accelerates discovery, minimizes human error, and ensures reproducibility.

Example:

- Reinforcement Learning (RL)-based controllers have been applied to chemical vapor deposition (CVD) systems for carbon nanotube synthesis. The AI autonomously adjusted parameters such as temperature and gas flow rates, improving yield by 35% compared to human-guided processes (Li et al., 2024).

- Bayesian Optimization has been used in robotic laboratories to tune nanoparticle synthesis conditions, significantly reducing the number of required experimental runs.

3.6 Case Studies in Predictive Modeling for Nanomaterials

Real-world case studies demonstrate the practical effectiveness of AI-driven predictive modeling:

- **Case Study 1** – TiO_2 Photocatalysts: Random Forest algorithms were trained on datasets of TiO_2 nanoparticles with different dopants and synthesis routes. The model predicted photocatalytic efficiency with >90% accuracy, guiding researchers toward high-yield formulations.
- **Case Study 2** – Bandgap Prediction in 2D Materials: A CNN was trained on microscopy images of layered 2D materials. The model extracted structural cues such as layer thickness and defects to predict electronic bandgaps, outperforming conventional regression models.
- **Case Study 3** – Carbon Nanotube Growth: RL agents in a CVD setup autonomously adjusted experimental parameters, leading to a 35% yield increase and 40% process time reduction.
- **Case Study 4** – MOFs for Gas Adsorption: Graph Neural Networks (GNNs) analyzed structural features of Metal–Organic Frameworks (MOFs) to predict CO_2 capture efficiency. This enabled the rapid identification of record-breaking MOFs validated within weeks.

Significance: These studies highlight that predictive modeling is not a theoretical exercise but a practical enabler of faster, cheaper, and more reliable materials discovery (Yang et al., 2024; Chakroborty et al., 2025; Song et al., 2025).

3.7 Challenges in AI-Driven Predictive Modeling

Despite promising results, predictive modeling in nanomaterials faces several limitations:

1. Data Scarcity
    - Unlike fields like image recognition, nanoscience datasets are often small and heterogeneous.

- High-quality labeled data for rare materials (e.g., quantum dots or novel nanocomposites) is limited.
- Small datasets increase risks of overfitting and poor generalization.

2. Model Interpretability
    - Many deep learning models act as black boxes, providing predictions without scientific explanations.
    - In materials science, researchers require mechanistic insights (e.g., *why* a bandgap narrows) rather than just outputs.
    - Explainable AI (XAI) tools such as SHAP, LIME, and Grad-CAM are being adapted to materials datasets to improve transparency (Davydiuk et al., 2025).
3. Generalization Across Domains
    - Models trained on one material class (e.g., oxides) often fail to transfer knowledge to another (e.g., metal–organic frameworks).
    - Transfer learning offers a partial solution but requires careful domain adaptation.
4. Computational Requirements
    - Training deep learning models using DFT- or MD-generated data is computationally expensive.
    - Access to high-performance computing (HPC) clusters or GPU-based architectures is often limited to well-funded labs, creating disparities.

3.8 The Predictive Modeling Workflow

The predictive modeling workflow in nanomaterials research is a cyclical, self-improving process (Figure 3.1).

Step 1 – Data Acquisition

- Sources: Experimental data (XRD, SEM, spectroscopy), simulated data (DFT, MD), high-throughput screening, and NLP-extracted literature datasets.
- Data is curated and annotated with standardized metadata to ensure reproducibility.

Step 2 – Data Preprocessing

- Feature engineering: Handcrafted descriptors (atomic radius, electronegativity, bond lengths).
- Dimensionality reduction: PCA, autoencoders, or t-SNE to reduce high-dimensional spaces.

- Data cleaning: Normalization and outlier removal.

Step 3 – Model Training and Validation

- ML models (RF, SVM, GBM) and DL models (CNNs, GNNs, Transformers) are trained on datasets.
- Validation strategies: Cross-validation, bootstrapping, or transfer learning for small datasets.

Step 4 – Prediction and Inverse Design

- Models predict material properties for unseen candidates.
- Bayesian Optimization, VAEs, and GANs propose novel candidates with desired performance profiles.

Step 5 – Experimental Validation

- Autonomous or semi-automated laboratories synthesize AI-suggested materials.
- Characterization tools provide feedback to update the predictive model.

Figure 2 illustrates the typical workflow for AI-driven predictive modeling in nanomaterials research. The process begins with **data acquisition**, followed by **data preprocessing**, **model training**, and **validation**. Predictions are then used to guide **experimental synthesis**, and feedback from experiments continuously refines the AI models.

Figure 2. Workflow for AI-Driven Predictive Modeling in Nanomaterials

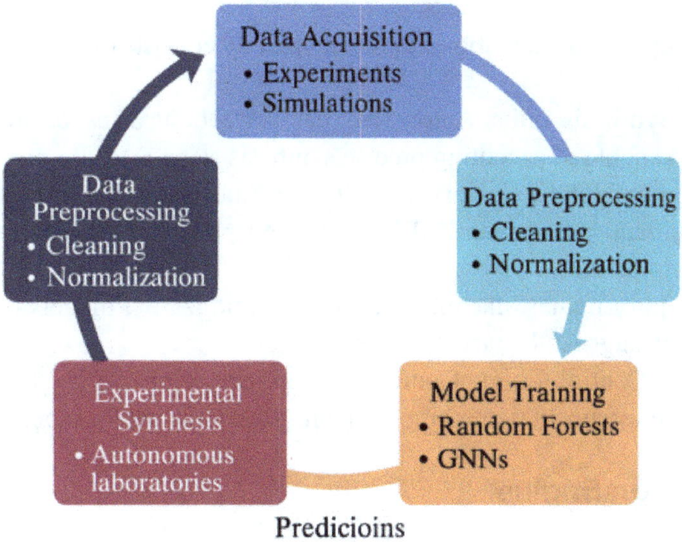

The workflow diagram depicts a circular process. Starting at the top, Data Acquisition gathers information from experiments, simulations, and literature mining. Data Preprocessing cleans, normalizes, and structures this data for machine learning use. Model Training applies algorithms such as Random Forests, CNNs, or GNNs to learn structure–property relationships. Model Validation ensures predictive accuracy using separate test datasets. Predictions inform Experimental Synthesis, which is carried out in autonomous or semi-autonomous laboratories. Results are fed back into the system, forming a continuous improvement loop that enhances both prediction quality and material performance.

4. INTELLIGENT EXPERIMENTATION PLATFORMS

AI-driven predictive modeling becomes most impactful when combined with intelligent experimentation platforms. These systems integrate machine learning (ML), robotics, and automation to create closed-loop laboratories that can autonomously design, execute, and refine experiments. By minimizing human intervention and leveraging high-throughput experimentation, such platforms accelerate the discovery of nanomaterials with desired properties.

4.1 Components of Intelligent Experimentation

An intelligent experimentation platform typically consists of:

- Robotic Synthesis Units: Automated reactors, robotic arms, or droplet-based systems capable of executing precise synthesis steps.
- High-Throughput Characterization Tools: Automated instruments such as XRD, Raman spectroscopy, SEM/TEM, and AFM, integrated with data acquisition pipelines.
- AI Decision Engine: Machine learning or optimization models that analyze results and suggest the next set of experiments.
- Feedback Control Loop: Continuous refinement of the predictive model based on experimental outcomes, enabling adaptive exploration.

4.2 AI Decision Engine

The decision engine acts as the "brain" of the platform, guiding experimentation through advanced optimization techniques:

- **Reinforcement Learning (RL):**
  - RL agents explore experimental conditions by balancing exploitation (using known high-performing conditions) and exploration (testing new ones).
  - Example: RL was applied to optimize chemical vapor deposition (CVD) conditions for carbon nanotube growth, achieving a 35% yield improvement.
- **Bayesian Optimization (BO):**
  - BO models experimental outputs with probabilistic surrogates, efficiently identifying optimal synthesis conditions with limited trials.
  - Example: BO identified optimal nanoparticle synthesis parameters with 70% fewer experimental runs compared to grid search.
- **Evolutionary Algorithms (EA):**
  - Algorithms such as Genetic Algorithms (GA) and Particle Swarm Optimization (PSO) simulate natural selection to explore large design spaces.
  - Example: GA was used to evolve nanoparticle alloy compositions, leading to catalysts with superior CO_2 reduction efficiency.
- **Multi-Objective Optimization:**
  - Many nanomaterial problems involve trade-offs (e.g., maximizing strength while minimizing weight).

- o Algorithms such as NSGA-II (Non-dominated Sorting Genetic Algorithm II) generate Pareto-optimal fronts for balanced solutions.

4.3 Integration with Predictive Modeling

Predictive modeling and intelligent experimentation are synergistic:

1. Predictive models suggest candidate materials or synthesis conditions.
2. Intelligent platforms test these candidates in the laboratory.
3. Results are fed back into the predictive model, refining its accuracy.
4. The loop continues until optimal materials are discovered.

This closed-loop framework accelerates discovery cycles from months/years to days/weeks. Hybrid quantum–classical methods, such as parameterized quantum circuits with quantum kernels, have been proposed to enhance machine learning performance in material discovery (Chang, 2022)

4.4 Comparative Analysis of Optimization Methods

Table 4.1 compares leading optimization techniques applied within intelligent experimentation platforms for nanomaterials discovery. It outlines their specific use cases, strengths, and limitations, highlighting how reinforcement learning, Bayesian optimization, genetic algorithms, particle swarm optimization, and NSGA-II contribute to adaptive design, efficient parameter control, and multi-objective optimization in nanoscience.

Table 4. Optimization Techniques in Intelligent Experimentation Platforms

Method	Use Case in Nanomaterials	Strengths	Limitations
Reinforcement Learning (RL)	CNT synthesis via CVD	Adaptive, learns from feedback	Needs large training data
Bayesian Optimization (BO)	Nanoparticle size control	Efficient with small datasets	Slower in very high dimensions
Genetic Algorithms (GA)	Alloy nanoparticle discovery	Good for global search	Computationally costly
Particle Swarm Optimization (PSO)	MOF structure optimization	Simple, effective	May get trapped in local minima
NSGA-II	Multi-objective design	Generates Pareto fronts	Complex to implement

4.5 Case Studies of Intelligent Platforms

- Autonomous Nano catalyst Discovery:
  - o A robotic laboratory combined Bayesian optimization with automated synthesis to discover platinum-based nano catalysts with 20% higher efficiency than existing ones.
- Reinforcement Learning for CNT Growth:
  - o RL agents-controlled gas flow and temperature in CVD systems, improving CNT yield and structural uniformity.
- Genetic Algorithms for MOFs:
  - o GA-based optimization identified MOF structures with record-high CO_2 adsorption capacities, validated experimentally within weeks.
- Particle Swarm Optimization for Photocatalysts:
  - o PSO optimized the doping levels in TiO_2 photocatalysts, resulting in a 25% increase in visible-light activity compared to conventional approaches (Musa et al., 2025).

4.6 Architecture of an Intelligent Experimentation Platform

While intelligent experimentation platforms represent a paradigm shift in nanomaterials research, several challenges still limit their widespread deployment. It summarizes these key barriers and highlights directions for future progress.

1. Scalability and Cost:
   - o Robotic laboratories and autonomous synthesis systems are expensive to build and maintain, restricting their adoption to elite research canters.
   - o Future solutions may involve modular, low-cost robotic platforms and cloud-based sharing of experimental protocols.
2. Integration of Multimodal Systems:
   - o Synchronizing robotic synthesis units, high-throughput characterization tools, and AI decision engines remains technically complex.
   - o Research is moving toward standardized APIs and interoperable platforms that allow seamless communication between different instruments.
3. Data Standardization and Sharing:
   - o Experimental data are often stored in non-standard formats, making it difficult to reproduce and benchmark results across laboratories.
   - o Initiatives such as the Materials Genome Initiative and Open Catalyst Project are working toward universal data standards and federated learning approaches.
4. Model Interpretability:

- o Black-box deep learning models make it difficult for researchers to understand why certain conditions are suggested.
- o The integration of Explainable AI (XAI) techniques such as SHAP and Grad-CAM into experimental platforms will improve trust and scientific insight.

5. Balance Between Autonomy and Human Oversight:
   - o Fully autonomous platforms risk generating results without adequate expert interpretation.
   - o The future lies in human-in-the-loop frameworks, where AI handles repetitive tasks while scientists focus on interpretation and theory-building.

Figure 3. Architecture of an Intelligent Experimentation Platform

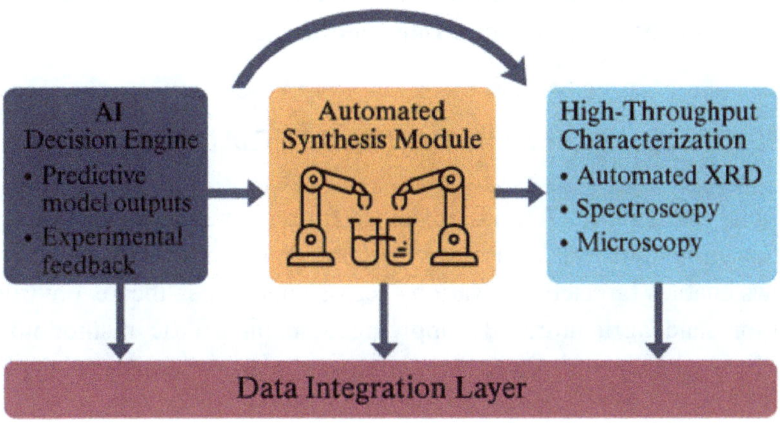

Figure 3 shows an architecture of an Intelligent Experimentation Platform for Nanomaterials Research. The schematic diagram presents a modular architecture. On the left, AI Decision Engine processes predictive model outputs and experimental feedback to determine synthesis parameters. This connects to the Automated Synthesis Module in the center, represented by robotic arms and programmable reactors. The High-Throughput Characterization Module on the right analyzes the synthesized materials using automated XRD, spectroscopy, and microscopy. A Data Integration Layer below connects all components, feeding results back to the AI Decision Engine in a closed-loop cycle. Arrows indicate continuous data flow, ensuring iterative optimization of material properties.

4.7 Future Directions in Intelligent Experimentation Platforms

The evolution of IEPs in nanomaterials research will likely include:

- **Integration with Digital Twins**: Real-time virtual simulations of experimental setups to predict outcomes before physical execution.
- **Cloud-Connected Laboratories**: Global collaboration through remotely accessible autonomous labs.
- **Quantum-Enhanced Optimization**: Using quantum algorithms to navigate ultra-high-dimensional synthesis parameter spaces.
- **Sustainability-Driven AI Models**: Incorporating environmental impact assessments into experimental decision-making.

As these platforms mature, they will serve as critical enablers of rapid, sustainable and globally collaborative nanomaterials innovation.

5. APPLICATIONS OF AI IN NANOMATERIAL DISCOVERY

Artificial Intelligence (AI) has advanced from theoretical promise to practical application in nanomaterials discovery. Its integration into design and synthesis workflows enables targeted innovations across energy, healthcare, environmental remediation, and agriculture. By applying domain-specific methodologies, AI bridges the gap between predictive modeling and real-world material deployment.

5.1 Energy and Environmental Applications

Energy storage and conversion technologies demand nanomaterials with tailored properties such as high conductivity, catalytic activity, and stability.

- **Battery Materials:**
  - o Graph Neural Networks (GNNs) trained on Materials Project datasets predict ion mobility and stability in solid electrolytes.
  - o Bayesian Optimization is applied to optimize synthesis parameters for lithium-ion and sodium-ion batteries.
  - o Case Example: A GNN model identified promising lithium superionic conductors with experimental validation More recent work demonstrates that advanced graph neural network architectures can further enhance scalability and predictive accuracy in energy materials discovery (Khakpour et al., 2025).

- **Catalysis and Fuel Cells:**
  - o CNNs analyze microscopy images to detect active sites in catalysts.
  - o Reinforcement Learning (RL) is used to optimize catalyst deposition conditions in fuel cells.
  - o Example: RL-guided platinum nanoparticle synthesis improved oxygen reduction reaction (ORR) activity by 30%.
- **Environmental Remediation:**
  - o Support Vector Machines (SVMs) predict adsorption efficiency of nanomaterials for heavy metal removal from water.
  - o Machine learning regression models are applied to model photocatalytic degradation rates of organic pollutants using TiO_2 nanostructures.

5.2 Biomedical Applications

Nanomaterials play a critical role in drug delivery, bioimaging, and diagnostics, where safety, selectivity, and biocompatibility are paramount.

- **Drug Delivery:**
  - o Deep Learning (DL) models predict nanoparticle–protein interactions, enabling the design of carriers with controlled release properties.
  - o Bayesian networks optimize nanoparticle surface modifications for targeted drug delivery.
- **Bioimaging:**
  - o CNNs process fluorescence microscopy and MRI images to evaluate nanoparticle uptake in cells and tissues.
  - o Example: A CNN trained on fluorescence data achieved >90% accuracy in distinguishing healthy vs. cancerous tissue with nanoparticle contrast agents.
- **Diagnostics:**
  - o Natural Language Processing (NLP) extracts biomedical nanomaterial data from literature to identify biomarkers.
  - o AI-powered biosensors use ML classifiers to enhance detection of pathogens at the nanoscale.

5.3 Agricultural and Food Applications

AI-driven nanomaterials are transforming sustainable agriculture and food systems.

- **Smart Fertilizers:**

- Regression models predict release kinetics of nano-encapsulated fertilizers based on polymer coatings.
- Optimization algorithms balance nutrient release with soil pH and crop type.

- **Food Safety:**
  - SVM and Random Forest classifiers detect pesticide residues using nanomaterial-based biosensors.
  - Case Example: A Random Forest model improved the detection sensitivity of nanoparticle-enabled food sensors by 25% compared to traditional spectroscopy.
- **Crop Monitoring:**
  - DL models process hyperspectral images enhanced with nanoparticle tags for real-time stress detection in plants.
  - Example: Nanoparticle-based imaging coupled with CNN analysis identified drought stress in maize with 95% accuracy.

5.4 Case Study Highlights

- **Energy**: GNN-predicted lithium conductors validated experimentally for solid-state batteries.
- **Biomedical:** CNNs classified nanoparticle-enhanced tissue images with >90% accuracy.
- **Environmental:** SVMs predicted pollutant adsorption on nanomaterials with high reliability.
- **Agriculture:** Random Forest-based biosensors enhanced pesticide detection by 25%.

5.5 Challenges and Outlook

- Safety and Regulation: Biomedical and agricultural applications require strict toxicity testing.
- Scalability: High-throughput synthesis is still costly for large-scale production.
- Data Scarcity: Domain-specific datasets (e.g., agricultural nanomaterials) are limited.
- Future Prospects: Integration of Explainable AI (XAI) and federated learning can enhance trust and enable collaborative, global datasets.

5.6 Summary of AI-Driven Nanomaterials Applications

Table 5.1 provides a consolidated overview of AI-driven nanomaterials applications across key sectors. It summarizes the role of AI, representative nanomaterial types, and the resulting benefits in energy, environment, biomedical, manufacturing, and agriculture domains. This tabular synthesis highlights how AI enhances material discovery and deployment by improving efficiency, sustainability, and performance across diverse industries.

Table 5. AI-Driven Nanomaterials Applications Across Sectors

Sector	AI Role	Nanomaterial Types	Key Benefits
Energy	Predict material stability, optimize catalysts	Metal oxides, nanocarbons, quantum dots	Higher efficiency, longer lifespan
Environment	Screen adsorbents, optimize photocatalysts	MOFs, TiO_2 nanoparticles	High pollutant removal, selective adsorption
Biomedical	Design targeted delivery systems	Gold nanoparticles, liposomes	Improved targeting, reduced side effects
Manufacturing	Predict conductivity, optimize coatings	Graphene composites, CNTs	Better performance, lightweight
Agriculture	Formulate nano-fertilizers, nano-pesticides	Metal oxide nanoparticles, polymer nanocarriers	Enhanced yield, reduced waste

5.7 Cross-Sector AI-Driven Nanomaterials Applications

Figure 4. Cross-Sector Applications of AI-Driven Nanomaterials

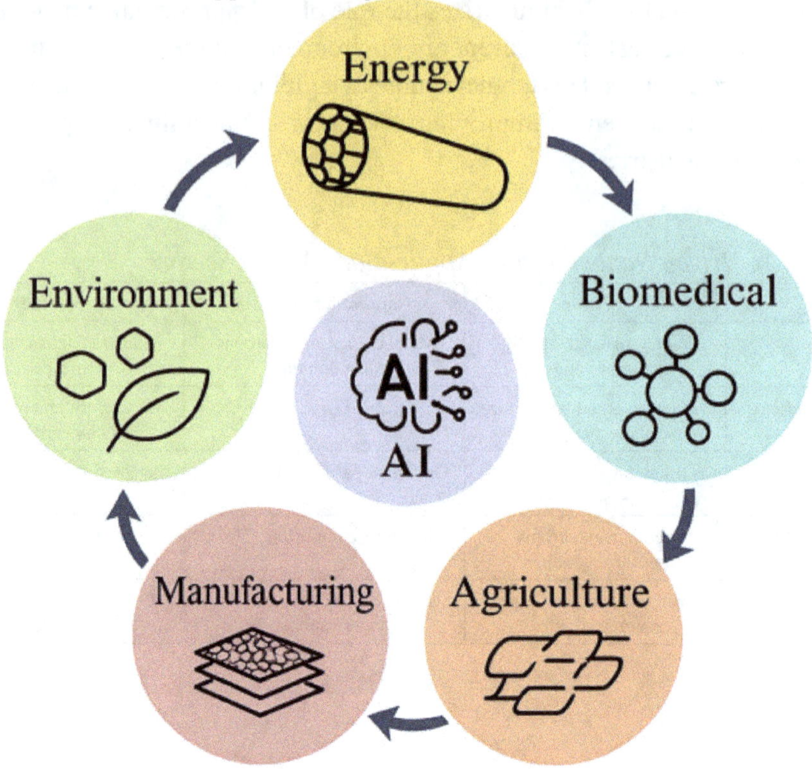

The infographic presents five interconnected circles, each representing a sector: Energy, Environment, Biomedical, Manufacturing, and Agriculture. Arrows between circles highlight cross-domain synergies, such as the use of photocatalysts in both environmental and energy applications. The centre features an AI core, symbolizing the predictive modeling and intelligent experimentation that power all sectors. Icons within each sector circle depict representative nanomaterial forms, such as nanotubes, nanoparticles, and nanosheets.

5.8 Emerging Trends in Applications

- **Multi-Functional Nanomaterials:** AI is increasingly used to design materials that can serve multiple roles, e.g., energy storage and pollutant removal.
- **Personalized Nanomedicine:** Predictive models tailor nanomaterial-based treatments to individual patient profiles.

- **Circular Economy in Nanotechnology:** AI helps design recyclable and environmentally safe nanomaterials, reducing waste.
- **Quantum Computing Integration:** Future AI models for nanomaterials discovery may leverage quantum computing for ultra-high-dimensional optimization. Digital twins offer a powerful framework for simulating and optimizing material behaviour in real time, bridging computational predictions with experimental validation (Kalidindi, 2022).

These applications demonstrate that AI is not merely an accelerant for nanomaterials research but a fundamental enabler of innovations that span the entire industrial and societal spectrum.

6. CHALLENGES IN AI-DRIVEN NANOMATERIALS DISCOVERY

Although AI offers transformative opportunities for nanomaterial design, predictive modeling, and synthesis, its integration into practical workflows is constrained by several challenges. These challenges are not only technical but also data-driven, infrastructural, and ethical in nature.

6.1 Data Scarcity and Quality Issues

High-quality, labelled datasets are the foundation of AI models, yet nanomaterials research suffers from:

- Small Sample Sizes: Experimental data on novel nanostructures (e.g., quantum dots, MOFs) are limited compared to fields like image recognition.
- Heterogeneity: Data sources include experiments, simulations, and literature mining, which often vary in quality and format.
- Noisy and Incomplete Data: Experimental variability and missing metadata reduce model robustness.

Example: Training CNNs on microscopy images of nanostructures is challenging due to image artifacts and limited annotated datasets.

6.2 Model Interpretability

Many AI models, particularly deep learning networks, function as black boxes, making it difficult to extract scientific insight:

- **Problem:** A GNN may predict a bandgap value but cannot easily explain *which atomic interactions* influenced the outcome.
- **Solutions:**
  - o Explainable AI (XAI) tools like SHAP, LIME, and Grad-CAM are being adapted for materials science.
  - o Physics-informed ML integrates fundamental laws into AI models, improving interpretability (Karniadakis et al., 2021; Molnar, 2023; Zhong et al., 2022).

6.3 Transferability and Generalization

- Models trained on one material class (e.g., oxides) often fail when applied to others (e.g., perovskites, polymers).
- Transfer learning offers partial solutions but requires careful domain adaptation.
- Example: Models trained on DFT datasets often underperform when predicting real-world experimental outcomes due to domain shift (Chen et al., 2021; Jablonka et al., 2020).

6.4 Computational Resource Demands

- Training deep learning models on high-dimensional datasets (e.g., DFT simulations of 100,000+ compounds) requires GPU clusters or supercomputers.
- Smaller labs face barriers to entry due to infrastructure costs.
- Cloud-based and federated learning frameworks are emerging as scalable alternatives (Huang & Barnard, 2022; Rieke et al., 2020).

6.5 Integration with Experimental Workflows

The widespread adoption of AI-driven nanomaterials discovery is not solely a technical issue of combining algorithms with laboratory systems—it is also hindered by systemic challenges that span data, modeling, integration, and ethics. Figure 6.1 summarizes these barriers to large-scale adoption.

1. Data Challenges
   - o **Scarcity:** High-quality, labelled datasets for nanomaterials remain limited, especially for emerging classes such as quantum dots and metal–organic frameworks (MOFs).
   - o **Quality:** Experimental datasets often contain noise, inconsistencies, or incomplete metadata, which reduce model reliability.

- o **Bias:** Training data biased toward well-studied materials (e.g., oxides) results in models that underperform for novel or less-represented material classes.

2. Model Challenges
   - o **Interpretability:** Black-box models such as deep neural networks provide predictions without mechanistic insights, limiting trust and adoption in scientific workflows.
   - o **Transferability:** Models trained on simulated data (e.g., DFT calculations) often fail to generalize to real-world experimental conditions.
   - o **Computation:** Training large-scale models on high-dimensional datasets demands significant computational resources, limiting access to smaller laboratories.
3. Integration Challenges
   - o **Automation:** Robotic synthesis and characterization platforms remain costly and technically challenging to integrate into existing laboratories (MacLeod et al., 2022; Volk et al., 2023; Szymanski et al., 2023).
   - o **Compatibility:** AI frameworks must be interoperable with diverse laboratory hardware and software systems.
   - o **Collaboration:** Effective integration requires cross-disciplinary collaboration between computer scientists, chemists, materials scientists, and engineers.
4. Ethical and Social Challenges
   - o **Privacy:** Nanomaterials research data often involve sensitive intellectual property, raising concerns over secure data sharing.
   - o **Environment:** The environmental impact of large-scale nanoparticle production and disposal is poorly understood.
   - o **Accountability:** Defining responsibility in AI-driven discoveries (e.g., errors in autonomous synthesis leading to hazardous materials) remains an unresolved issue.

Figure 5. Challenges in AI-Driven Nanomaterials Discovery

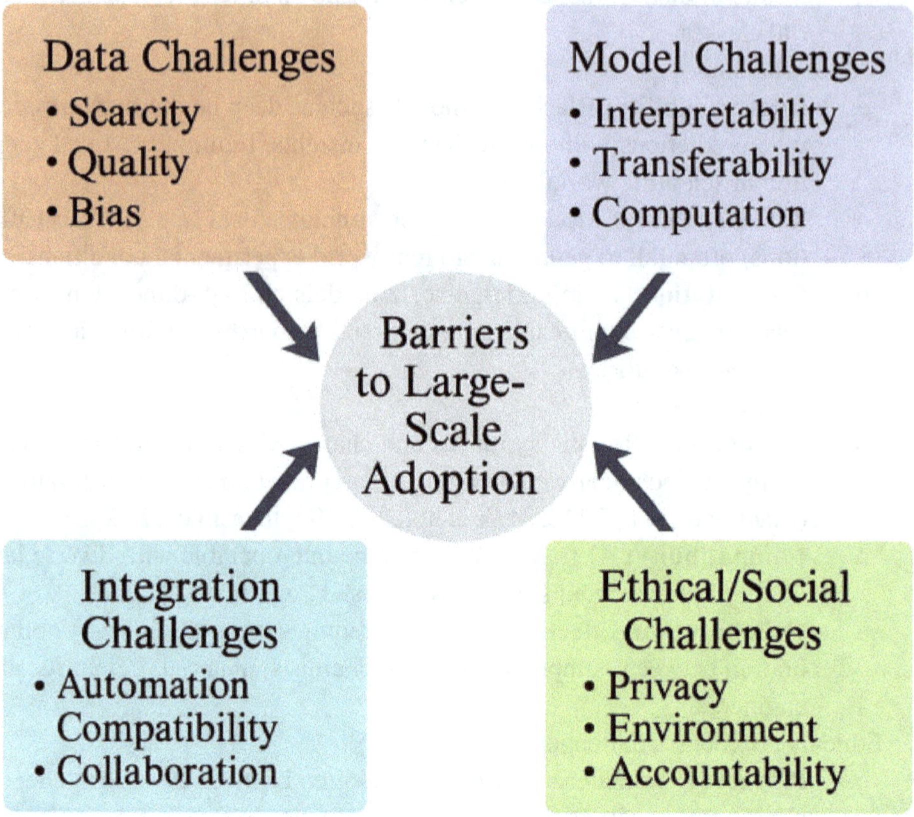

Figure 5 shows the framework of Challenges in AI-Driven Nanomaterials Discovery. The diagram is a four-quadrant model. The top-left quadrant represents Data Challenges (scarcity, quality, bias), the top-right represents Model Challenges (interpretability, transferability, computation), the bottom-left shows Integration Challenges (automation compatibility, collaboration), and the bottom-right depicts Ethical/Social Challenges (privacy, environment, accountability). Arrows converge toward a central circle labelled "Barriers to Large-Scale Adoption," illustrating how each category interconnects to influence adoption speed and trust in AI-driven nanomaterials discovery.

6.6 Ethical and Regulatory Concerns

- Biomedical applications: Nanoparticles designed by AI must undergo strict toxicity testing before approval.

- Agricultural nanomaterials: Require environmental impact assessments to avoid unintended ecosystem effects.
- Data Governance: Ownership, privacy, and security of shared nanomaterials datasets remain unresolved (Floridi, 2021; OECD, 2021).

6.7 Summary of Challenges

Table 6.1 summarizes the key challenges associated with AI-driven nanomaterials discovery. It outlines the major obstacles, provides concise descriptions, and highlights emerging solutions such as federated learning, explainable AI, and hybrid quantum-classical approaches. This overview underscores the need for integrated technical, ethical, and policy frameworks to ensure responsible and scalable adoption of AI in nanomaterials research.

Table 6. Challenges in AI-Driven Nanomaterials Discovery

Challenge	Description	Emerging Solutions
Data Scarcity	Limited, noisy, heterogeneous datasets	Data curation, federated learning
Interpretability	Black-box nature of DL/GNN models	XAI, physics-informed ML
Transferability	Poor generalization across material classes	Transfer learning, domain adaptation
Computational Cost	High GPU/cluster demand	Cloud AI, hybrid quantum-classical ML
Workflow Integration	Difficulty linking AI with labs	Autonomous labs, modular robotics
Ethics & Regulation	Toxicity, environmental risks	Policy frameworks, safety testing

7. FUTURE DIRECTIONS AND EMERGING TRENDS

The integration of Artificial Intelligence (AI) into nanomaterials research is still in its formative stages. While current applications demonstrate the power of predictive modeling, intelligent experimentation, and data-driven design, future research will expand these capabilities into more autonomous, interpretable, and globally connected frameworks. Several promising directions are outlined below.

7.1 Explainable and Trustworthy AI (XAI)

- **Challenge:** Current deep learning and graph neural network models often function as black boxes, limiting adoption in materials science where mechanistic understanding is crucial.
- **Emerging Trend:** Development of explainable AI (XAI) tools such as SHAP (SHapley Additive Explanations), LIME (Local Interpretable Model-agnostic Explanations), and Grad-CAM for microscopy data.
- **Impact:** These tools will allow researchers to identify the structural or chemical features most responsible for predicted properties, thereby improving trust, interpretability, and scientific insight.

7.2 Federated and Collaborative Learning

- Challenge: Data scarcity and privacy restrictions limit model training.
- Emerging Trend: Federated Learning (FL) enables multiple laboratories to collaboratively train AI models without sharing raw datasets. Instead, only model parameters are exchanged.
- Impact: This approach improves dataset diversity and model generalization while protecting intellectual property and sensitive biomedical/agricultural data.

7.3 Quantum Machine Learning (QML)

- **Challenge:** High-dimensional design spaces in nanomaterials exceed the computational limits of classical algorithms.
- **Emerging Trend:** Quantum Machine Learning (QML) leverages quantum computing to accelerate search in exponentially large design spaces.
- **Impact:** Hybrid quantum–classical algorithms show promise for solving problems such as bandgap prediction, molecular orbital calculations, and optimization of catalytic surfaces.

7.4 Autonomous Self-Driving Laboratories

- Challenge: Most current intelligent experimentation platforms require human-in-the-loop monitoring.
- Emerging Trend: Self-driving laboratories (SDLs) integrate AI, robotics, and high-throughput characterization into fully closed-loop systems.
- Impact: These platforms can autonomously generate hypotheses, design experiments, and validate results in real time. Early demonstrations have re-

duced experimental cycles from months to days (Tom et al., 2024; Tobias et al., 2025).

7.5 Multi-Objective and Sustainable Design

- **Challenge:** Many nanomaterials require balancing trade-offs, e.g., maximizing catalytic activity while minimizing toxicity.
- **Emerging Trend:** Multi-objective optimization algorithms (e.g., NSGA-II, Bayesian multi-objective optimization) are increasingly applied to discover Pareto-optimal materials.
- **Impact:** This allows the discovery of sustainable nanomaterials that are not only high-performing but also environmentally safe and economically viable.

7.6 Integration of AI with Digital Twins

- Emerging Trend: The concept of a digital twin—a virtual replica of a material or experimental system—offers a powerful framework for testing hypotheses in silico before real-world experiments.
- Impact: By coupling AI with digital twins, researchers can simulate thousands of design variations, drastically reducing experimental costs and failures.

7.7 Policy, Infrastructure, and Education

- **Policy:** Clear governance structures are needed for AI-driven materials discovery, especially in biomedical and environmental contexts.
- **Infrastructure:** Shared data repositories, cloud platforms, and standardized APIs will improve interoperability.
- **Education:** Training the next generation of researchers requires interdisciplinary curricula combining materials science, computer science, and ethics.

7.8 Outlook

The future of AI-driven nanomaterials discovery lies in scalable, interpretable, and collaborative frameworks. By combining XAI, federated learning, quantum-enhanced models, and autonomous laboratories, researchers can accelerate innovation while ensuring safety and sustainability. The convergence of AI with policy, infrastructure, and education will determine how quickly these technologies move from niche demonstrations to mainstream adoption in global nanoscience (Asghari & Naderi, 2024).

7.9 Global Nanomaterials Knowledge Graphs

Knowledge graphs link **experimental data, computational predictions, literature insights, and patents** into interconnected networks. AI can mine these for patterns and novel correlations.

- **Example:** A knowledge graph could reveal that certain dopants used in electronic materials also enhance catalytic properties, prompting cross-disciplinary innovations.
- **Research Opportunity:** Creating **real-time updating** global nanomaterials knowledge graphs that pull from open literature and industry contributions.
- **Impact:** Faster identification of **research gaps** and **emerging trends**.

7.10 Standardization and Governance Frameworks

As AI-assisted nanomaterials discovery becomes widespread, standardized **data protocols, ethical guidelines, and regulatory frameworks** will be essential.

- **Example:** International standards for nanoparticle characterization metadata would make datasets more interoperable across AI systems.
- **Research Opportunity:** Multi-stakeholder collaborations between scientists, policymakers, and industry leaders to develop **responsible AI use frameworks**.
- **Impact:** Enhanced transparency, safety, and public trust in AI-powered nanotechnology.

Table 7.1 outlines detailed future research directions in AI-driven nanomaterials discovery. It highlights emerging areas such as quantum–AI integration, autonomous laboratories, sustainability-driven design, and federated learning, alongside advances in multi-scale modeling, personalized nanomedicine, and digital twins. By mapping each research focus to concrete examples and anticipated impacts, the table provides a forward-looking framework for innovation, standardization, and responsible governance in the field.

Table 7. Detailed Future Research Directions

Future Direction	Research Focus	Example	Expected Impact
Quantum–AI Integration	Hybrid quantum–classical workflows for nanoscale simulations	Quantum-dot bandgap tuning via QML	Ultra-fast, precise material discovery
Autonomous Laboratories	AI–robot closed-loop discovery cycles	Catalyst optimization in self-driving labs	Reduces discovery cycles from years to weeks
AI-Driven Sustainability	Life-cycle assessment (LCA)-integrated optimization	Biodegradable nanocarriers for drug delivery	Eco-safe and sustainable innovations
Federated Learning (FL)	Secure collaborative model training across global labs	Drug-delivery nanoparticle datasets	Expanded, privacy-preserving datasets
Multi-Scale Modeling	Cross-scale AI integration (atomic → device)	AI models for Li-ion battery electrode design	Holistic design optimization
Personalized Nanomedicine	AI-guided patient-specific nanocarrier design	mRNA lipid nanoparticle formulations	Precision and personalized therapy
Digital Twins	Virtual replicas of experimental materials and workflows	Photocatalyst lifespan modeling	Reduced testing cost and faster iteration
Explainable AI (XAI)	Feature attribution and interpretable models	Catalyst design with SHAP-based insights	Greater trust and scientific interpretability
Knowledge Graphs	Linked nanomaterials databases and ontologies	Dopant–catalysis relationship discovery	Accelerated innovation via connected data
Governance Frameworks	Policy, ethics, and standardization	Metadata protocols for nanomaterials datasets	Safe, transparent, and ethical adoption

Figure 6. Roadmap for AI-Driven Nanomaterials Research

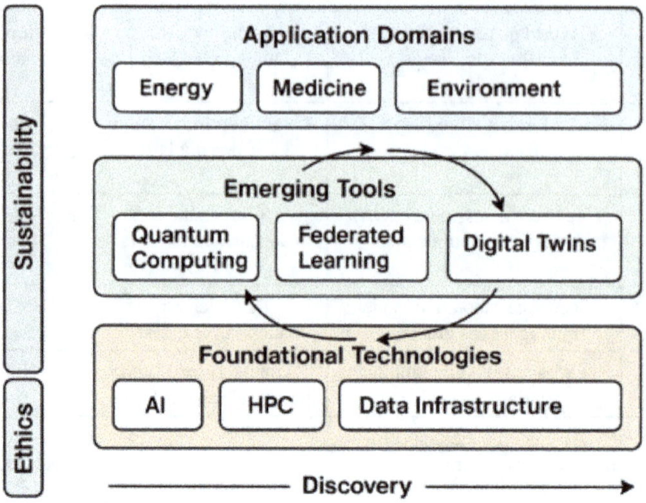

Figure 6 illustrates A layered roadmap with three tiers — Foundational Technologies (AI, HPC, data infrastructure) at the base, Emerging Tools (quantum computing, federated learning, digital twins) in the middle, and Application Domains (energy, medicine, environment, manufacturing) at the top. Vertical pillars labelled "Sustainability" and "Ethics" span all tiers, indicating their cross-cutting importance. Arrows represent iterative feedback loops between discovery, validation, and deployment stages.

8. POLICY, INFRASTRUCTURE, AND EDUCATIONAL ROADMAP

The successful integration of Artificial Intelligence (AI) into nanomaterials discovery extends beyond algorithms and laboratory automation. It requires supportive policy frameworks, robust digital and experimental infrastructure, and a new generation of interdisciplinary researchers. This section outlines a roadmap to ensure sustainable, ethical, and impactful adoption of AI in nanoscience.

8.1 Policy and Governance Frameworks

AI-driven nanomaterials research intersects with issues of data governance, intellectual property, and safety regulations.

- **Standardization:** Global organizations such as OECD (2021) and ISO are developing metadata and interoperability standards to ensure reproducibility in AI-driven materials research.
- **Ethics and Safety:** Policies must address toxicity testing, environmental impact of large-scale nanoparticle production, and AI transparency in biomedical applications.
- **Data Sharing:** Secure frameworks for federated data sharing can balance intellectual property protection with collaborative innovation.
- **Accountability:** Clear guidelines are needed to assign responsibility when autonomous laboratories make material design decisions.

8.2 Infrastructure for AI-Driven Nanomaterials Discovery

In addition to well-curated repositories, the effectiveness of digital infrastructure depends on how efficiently researchers can access, manage, and analyze large-scale nanomaterials datasets. As experimental and computational outputs grow, data pipelines must incorporate automated preprocessing, metadata tagging, and quality control mechanisms. Such infrastructure ensures that raw outputs from high-throughput synthesis or simulations are converted into machine-readable formats suitable for AI training. Without these standardized pipelines, even the best repositories remain underutilized, and predictive models suffer from inconsistent inputs.

8.2.1 Digital Infrastructure

- **Data Repositories:** Open-access platforms such as the Materials Project, Open Catalyst Project, and emerging nanomaterials databases must be expanded with richer metadata.
- **High-Performance Computing (HPC):** Access to GPU clusters and cloud platforms is essential for deep learning, generative modeling, and quantum–classical workflows.
- **Federated Platforms:** Cloud-based federated learning can allow multiple labs to collaborate without direct data exchange.

8.2.2 Experimental Infrastructure

- **Autonomous Laboratories:** Modular, AI-integrated robotic systems for synthesis and high-throughput characterization.
- **Interoperability:** Standardized APIs to ensure compatibility between AI models and laboratory equipment.

- **Digital Twins:** Infrastructure to support virtual replicas of experiments for pre-validation of designs.

8.3 Educational Roadmap

To harness AI in nanomaterials, researchers must be equipped with interdisciplinary training.

- **Curriculum Integration:** Universities should develop joint programs combining materials science, computer science, and data science.
- **Skill Development:** Training in machine learning (ML), deep learning (DL), reinforcement learning (RL), and explainable AI (XAI) tailored to nanoscience datasets.
- **Ethics and Policy Training:** Understanding of data governance, sustainability, and ethical AI deployment.
- **Industry–Academia Collaboration:** Partnerships to create internships, joint research centres, and translational labs that bridge theoretical AI with real-world nanomaterials applications.

8.4 Roadmap Summary

Table 8.1 presents a roadmap for AI-driven nanomaterials discovery, structured across four key dimensions: policy and governance, digital infrastructure, experimental infrastructure, and education and training. It highlights actionable steps, example initiatives, and the anticipated impacts needed to ensure transparent, scalable, and future-ready adoption of AI in nanomaterials research.

Table 8. Roadmap for AI-Driven Nanomaterials Discovery

Dimension	Key Actions	Example Initiatives	Expected Impact
Policy & Governance	Standardization, safety testing, accountability	OECD AI principles, ISO metadata standards	Transparent and safe adoption
Digital Infrastructure	Expand databases, HPC/cloud access, federated learning	Materials Project, Open Catalyst Project	Scalable, collaborative innovation
Experimental Infrastructure	Build autonomous labs, interoperability standards	Modular robotic labs, digital twins	Faster discovery cycles
Education & Training	Interdisciplinary curriculum, ethics integration	AI+Nanotech graduate programs	Next-gen skilled workforce

Figure 7. Ecosystem for AI-Driven Nanomaterials Discovery

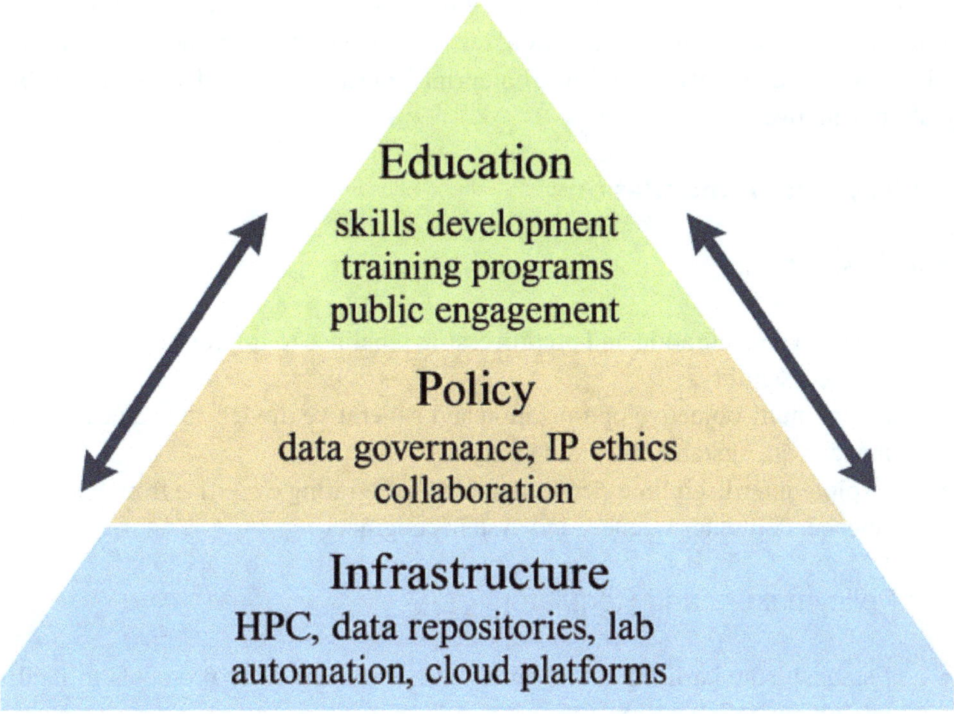

Figure 7, A three-tiered pyramid diagram. The base tier is labeled Infrastructure (HPC, data repositories, lab automation, cloud platforms), the middle tier is Policy (data governance, IP, ethics, collaboration), and the top tier is Education (skills development, training programs, public engagement). Arrows connect the tiers, emphasizing that policy enables infrastructure, and infrastructure supports education, all working together to enable sustainable AI-powered nanomaterials innovation.

9. CONCLUSION AND RECOMMENDATIONS

The convergence of Artificial Intelligence (AI) and nanomaterials science represents a paradigm shift in materials discovery. By leveraging predictive modeling, intelligent experimentation, and data-driven design, AI enables rapid identification of promising candidates, optimization of synthesis parameters, and reduction of costly trial-and-error cycles. Across domains such as energy storage, catalysis, environmental remediation, and biomedical applications, AI has already demonstrated measurable improvements in efficiency, accuracy, and scalability.

However, realizing the full potential of AI in nanoscience requires addressing challenges related to data scarcity, model interpretability, experimental integration, and ethical governance. Recent advances in explainable AI (XAI), federated learning (FL), and quantum machine learning (QML) indicate promising pathways to overcome these barriers. In addition, the emergence of autonomous self-driving laboratories and digital twins offers a blueprint for faster, safer, and more sustainable materials innovation.

9.1 Key Recommendations

For Researchers

- Invest in data curation and preprocessing pipelines to improve dataset quality and accessibility.
- Adopt multi-objective optimization and generative models to balance performance with sustainability and safety.
- Explore interdisciplinary collaborations, integrating expertise from materials science, computer science, and systems engineering.

For Policymakers and Regulators

- Establish standardized data formats and interoperability protocols to facilitate global collaboration.
- Implement regulatory frameworks for toxicity, environmental impact, and ethical AI use in nanomaterials research.
- Support open-science initiatives and incentivize federated learning platforms that balance innovation with intellectual property protection.

For Educators and Institutions

- Design interdisciplinary curricula combining AI, nanotechnology, and ethics to train a next-generation workforce.
- Promote industry–academia partnerships that provide real-world AI applications in materials discovery.
- Encourage the development of open-access educational resources and training datasets for researchers worldwide.

9.2 Author's Contribution

The present chapter goes beyond reviewing existing literature by organizing and connecting insights across artificial intelligence methods and nanomaterials applications. The main contribution lies in the way these diverse strands have been brought together into a coherent framework that highlights the relationships between predictive modeling, intelligent experimentation, and autonomous laboratory systems.

Several points may be emphasized as the original contribution of this work:

- The chapter structures the discussion of machine learning, deep learning, reinforcement learning, and generative approaches in direct relation to practical nanomaterial challenges. This mapping provides a clear reference for researchers who seek to align AI tools with domain-specific problems.
- Comparative perspectives are developed through the use of case examples and summary tables, covering biomedical, energy, agricultural, and environmental domains. These serve as a consolidated view that is not commonly available in single studies.
- The narrative links technical methods with broader research infrastructures, particularly the concept of self-driving laboratories, which is positioned here as part of a continuous discovery cycle rather than as isolated technological advances.
- Finally, the chapter outlines a set of forward-looking directions—such as explainable AI, federated learning, quantum approaches, and digital twins—that may guide the next stage of interdisciplinary research in this area.

Taken together, these elements represent the author's contribution in framing AI-driven nanomaterials discovery not only as a field of technical progress but also as a structured, strategic research agenda.

9.3 Outlook

Looking ahead, the integration of AI into nanomaterials research will evolve from isolated case studies into mainstream scientific workflows. As computational power scales, datasets expand, and policy frameworks mature, AI will increasingly function as a co-pilot in scientific discovery, complementing human expertise with predictive insights, optimization strategies, and automated experimentation. The future lies in a collaborative ecosystem where AI-powered laboratories, federated data networks, and global governance frameworks work together to accelerate innovation while ensuring ethical, sustainable, and socially responsible outcomes. If these challenges are addressed, AI-driven nanomaterials discovery has the potential to shorten development timelines from decades to years, unlocking breakthroughs that can transform energy systems, healthcare, agriculture, and beyond.

REFERENCES

Asghari, J., & Naderi, M. (2024). Artificial Intelligence in Nanomaterials Studies: A Statistical Overview. *Nanoscale and Advanced Materials*, *1*(2), 73–86. DOI: 10.22034/nsam.2024.04.01

Beyerle, E., ... (2023). Recent advances in describing and driving crystal nucleation using machine learning and artificial intelligence. *arXiv preprint*, arXiv:2304.13815. https://arxiv.org/abs/2304.13815

Chakroborty, S., Nath, N., & Sahoo, S.. (2025). A review of emerging trends in nanomaterial-driven AI for biomedical applications. *Nanoscale Advances*, *7*, 3619–3630. DOI: 10.1039/D5NA00032G PMID: 40370571

Chang, D. T. (2022). Parameterized quantum circuits with quantum kernels for machine learning: A hybrid quantum–classical approach. *arXiv preprint*, arXiv:2209.14449. https://arxiv.org/abs/2209.14449

Chen, W.. (2021). Machine learning for multi-scale modeling of battery materials. *Joule*, *5*(8), 1934–1953. DOI: 10.1016/j.joule.2021.07.007

Chou, W. C. (2025). Machine learning and artificial intelligence in nanomedicine: Connecting nanoparticle design with toxicology and translational applications. *Wiley Interdisciplinary Reviews. Nanomedicine and Nanobiotechnology*, *17*(3), e70027. DOI: 10.1002/wnan.70027 PMID: 40813104

Davydiuk, N., Krieg, E., & Gaitzsch, J.. (2025). The rising danger of AI-generated images in nanomaterials science and what we can do about it. *Nature Nanotechnology*, *20*, 567–569. DOI: 10.1038/s41565-025-02009-9 PMID: 40954352

Dubrovsky, I., Dmitrenko, A., Dmitrenko, A., Serov, N., & Vinogradov, V. (2024). Unveiling the potential of AI for nanomaterial morphology prediction. *arXiv preprint*, arXiv:2406.02591. https://arxiv.org/abs/2406.02591

Floridi, L. (Ed.). (2021). *Ethics, governance, and policies in artificial intelligence*. Springer., DOI: 10.1007/978-3-030-81907-1

Hou, X., Zaks, T., Langer, R., & Dong, Y. (2021). Lipid nanoparticles for mRNA delivery. *Nature Reviews. Materials*, *6*, 1078–1094. DOI: 10.1038/s41578-021-00358-0 PMID: 34394960

Huang, W., & Barnard, A. S. (2022). Federated data processing and learning for collaboration in the physical sciences. *Machine Learning: Science and Technology*, *3*(4), 045023. DOI: 10.1088/2632-2153/aca87c

Jablonka, K. M., Ongari, D., Moosavi, S. M., & Smit, B. (2020). Big-data science in porous materials: Materials genomics and machine learning. *Chemical Reviews*, *120*(16), 8066–8129. DOI: 10.1021/acs.chemrev.0c00004 PMID: 32520531

Kalidindi, S. R. (2022). Digital twins for materials. *Frontiers in Materials*. Advance online publication. DOI: 10.3389/fmats.2022.818535

Karniadakis, G. E.. (2021). Physics-informed machine learning. *Nature Reviews. Physics*, *3*, 422–440. DOI: 10.1038/s42254-021-00314-5

Khakpour, A., Florescu, L., Tilley, R., Jiang, H., Swaminathan Iyer, K., & Gustavo Carneiro. (2025). *arXiv preprint*, arXiv:2503.13798. https://arxiv.org/abs/2503.13798

Li, X.. (2022). Deep learning for microscopy image analysis of nanostructures. *ACS Nano*, *16*(4), 5678–5690. DOI: 10.1021/acsnano.1c11025

Li, Y., & Wang, S. Lv advancement of precision medicine, Z., Wang, Z., Zhao, Y., Xie, Y., Xu, Y., Qian, L., Yang, Y., Zhao, Z., & Zhang, J. (2024). Transforming the synthesis of carbon nanotubes with machine learning models and automation (CARCO). *arXiv preprint*, arXiv:2404.01006. https://arxiv.org/abs/2404.01006

MacLeod, B. P., Parlane, F. G. L., & Rupnow, C. C.. (2022). A self-driving laboratory advances the Pareto front for material properties. *Nature Communications*, *13*, 995. DOI: 10.1038/s41467-022-28580-6 PMID: 35194074

Mirkin, C. A. (2025). Blueprints for better drugs: The structural revolution in nanoscale therapeutics. *ACS Nano*, Article ASAP. https://doi.org/DOI: 10.1021/acsnano.5c06380

Molnar, C. (2023). *Interpretable machine learning: A guide for making black box models explainable* (2nd ed.). Springer., https://christophmolnar.com/books/interpretable-machine-learning

Mrdjenovich, D., Hyde, C., & Lo, W. S.. (2021). Propnet: A knowledge graph for materials science. *Patterns (New York, N.Y.)*, *2*(9), 100325. DOI: 10.1016/j.patter.2021.100325 PMID: 34278363

Musa, S. S., Ibrahim, A. M., Alhassan, M. Y., Musa, A. H., Jibo, A. G., Auwal, A. R., Okesanya, O. J., Othman, Z. K., Abubakar, M. S., Ahmed, M. M., Barroso, C. J. V., Sium, A. F., Garcia, M. B., Flores, J. B., Maikifi, A. S., Kouwenhoven, M. B. N., & Lucero-Prisno, D. E. (2025). Nanotechnology and machine learning: A promising confluence for the advancement of precision medicine. *Intelligence-Based Medicine*, *12*, 100267. DOI: 10.1016/j.ibmed.2025.100267

OECD. (2021). *Artificial intelligence in science: Challenges, opportunities, and the future of research.* OECD Publishing. https://www.oecd.org/en/publications/artificial-intelligence-in-science_a8d820bd-en.html

Radulescu, D.-M., Surdu, V.-A., Ficai, A., Ficai, D., Grumezescu, A.-M., & Andronescu, E. (2023). Green synthesis of metal and metal oxide nanoparticles: Principles and biomedical applications. *International Journal of Molecular Sciences*, *24*(20), 15397. DOI: 10.3390/ijms242015397 PMID: 37895077

Rieke, N., Hancox, J., Li, W., Milletarì, F., Roth, H., Albarqouni, S., Bakas, S., Galtier, M. N., Landman, B. A., Maier-Hein, K., Ourselin, S., Sheller, M., Summers, R. M., Trask, A., Xu, D., Baust, M., & Cardoso, M. J. (2020). The future of digital health with federated learning. *npj. Digital Medicine*, *3*, 119. DOI: 10.1038/s41746-020-00323-1 PMID: 33015372

Singh, R.. (2022). Sustainability in nanomaterials through AI-driven life cycle assessment. *Journal of Cleaner Production*, *356*, 131885. DOI: 10.1016/j.jclepro.2022.131885

Song, X., Gao, X., & Wang, H.. (2025). Artificial intelligence for nanomedicine. *Science China. Chemistry*. Advance online publication. DOI: 10.1007/s11426-025-2942-5

Szymanski, N. J.. (2023). An autonomous laboratory for the accelerated synthesis of inorganic materials. *Nature*, *614*, 413–420. DOI: 10.1038/s41586-023-06734-w PMID: 38030721

Tobias, A. V., Yip, J., Rouhanipour, A., & Ziolek, F. (2025). Autonomous 'self-driving' laboratories: A review of technology, challenges, and future directions. *Royal Society Open Science*, *12*, 250646. DOI: 10.1098/rsos.250646 PMID: 40852582

Tom, G., Steiner, A., Kwon, S., & Rahman, T. (2024). Self-Driving Laboratories for Chemistry and Materials Science. *Chemical Reviews*, *124*(20), 17245–17369. DOI: 10.1021/acs.chemrev.4c00055 PMID: 39137296

Tran, R., Lan, J., Shuaibi, M., Wood, B. M., Shoghi, N., Sriram, A., Therrien, F., ... (2022). The Open Catalyst 2022 (OC22) dataset and challenges for oxide electrocatalysts. *arXiv preprint*, arXiv:2206.08917. https://arxiv.org/abs/2206.08917

Volk, A. A., Epps, R. W., Yonemoto, D. T., Masters, B. S., Castellano, F. N., Reyes, K. G., Abolhasani, M., et al. (2023). AlphaFlow: Autonomous discovery and optimization of multi-step chemistry using a self-driven fluidic lab guided by reinforcement learning. *Nature Communications, 14*, Article. https://doi.org/DOI: 10.1038/s41467-023-37139-y

Yang, L., Wang, H., Leng, D., Fang, S., Yang, Y., & Du, Y. (2024). Machine learning applications in nanomaterials: Recent advances and future perspectives. *Chemical Engineering Journal*, *500*, 156687. DOI: 10.1016/j.cej.2024.156687

Zhang, Y., ... (2022). Interdisciplinary discovery of nanomaterials based on convolutional neural networks. *arXiv preprint*, arXiv:2212.02805. https://arxiv.org/abs/2212.02805

Zhong, X., et al. (2022). Explainable machine learning in materials science. *npj Computational Materials*. https://doi.org/DOI: 10.1038/s41524-022-00884-7

Chapter 6
Integrating AI and Nanotechnology Into STEM Education:
Curriculum Development, Case Studies, and Future Trends

Bogdan Dobrica
3NanoSAE Technology Transfer Center, University of Bucharest, Romania

Ioan Stamatin
3NanoSAE Technology Transfer Center, University of Bucharest, Romania

Catalin Berlic
Faculty of Physics, University of Bucharest, Romania

Adriana Balan
https://orcid.org/0000-0003-3060-3761
Faculty of Physics, University of Bucharest, Romania

Ashok Vaseashta
https://orcid.org/0000-0002-5649-0067
International Clean Water Institute, USA

ABSTRACT

We explore the integration of AI and nanotechnology in STEM education, highlighting their potential to transform pedagogical approaches and prepare students for future technological progress. The chapter underscores the importance of embedding AI and nanotechnology into STEM curricula by reviewing recent scientific literature. It explores how AI can enhance teaching practices, assessment methods, and ad-

DOI: 10.4018/979-8-3373-2883-6.ch006

ministrative tasks, thereby offering personalized learning experiences that boost student engagement and performance. Furthermore, the chapter highlights how nanotechnology can bridge the gap between theoretical knowledge and practical application, providing students with hands-on experience using advanced materials and techniques. An analysis is provided of the curriculum development frameworks, effective integration strategies, and the challenges and solutions related to implementing these technologies in education. A sizable section is dedicated to a case study at the University of Bucharest, Faculty of Physics, where AI and nanotechnology have been successfully incorporated into STEM programs, demonstrating their impact on students' skillsets.

1. INTRODUCTION

The incorporation of Artificial Intelligence (AI) and nanotechnology into STEM education has been shaped by a series of pivotal milestones spanning the last six decades. Initial explorations of AI in educational contexts can be traced to the late 1960s, when early Intelligent Tutoring Systems (ITS) such as SCHOLAR demonstrated the potential of computer-based instruction to adapt to learners' needs (Carbonell, 1970). During the 1980s and 1990s, rule-based expert systems and emerging machine learning techniques began to influence instructional software design, laying essential groundwork for contemporary AI-driven learning analytics and adaptive educational environments (Woolf, 2010). The rapid expansion of data-driven methods and deep learning during the 2010s further accelerated this trend, enabling the deployment of adaptive learning platforms, predictive analytics, and intelligent feedback systems that support personalized instruction and real-time assessment of student progress (Luckin et al., 2016; Holmes et al., 2019).

Developments in nanotechnology followed a parallel trajectory of progressive integration into education. The conceptual foundations were articulated as early as 1960 in Richard Feynman's seminal lecture "There's Plenty of Room at the Bottom" (Feynman, 1960), while the invention of scanning tunneling microscopy in 1981 provided the first means of visualizing and manipulating matter at the atomic scale (Binnig & Rohrer, 1986). The launch of the U.S. National Nanotechnology Initiative (NNI) in 2000 marked a decisive moment, stimulating the inclusion of nanoscale science and engineering content in university curricula worldwide (Roco, 2003). By the following decade, international programs such as NanoHUB and the NSF's Nanotechnology Undergraduate Education (NUE) initiative had introduced virtual laboratories, simulation-based modules, and interdisciplinary teaching resources that brought nanoscience concepts into mainstream undergraduate STEM education (Sweeney et al., 2006; Murday et al., 2019). These successive advances have established the foundation for the present convergence of AI and nanotechnology as complementary pillars within STEM curricula.

The integration of AI and nanotechnology into STEM (Science, Technology, Engineering, and Mathematics) education appears critical for enhancing educational methodologies and preparing students for the challenges of future technology. Indeed, recent advancements have influenced how research and innovation are conducted in education across a broad spectrum of sciences, driven by parallel developments in AI and nanotechnology (Wang & Lester, 2023). AI opens new possibilities for data analysis, predictive modeling, and intelligent tutoring while permeating academia and industry. Conversely, nanotechnology permits the atomic manipulation of matter to open novel materials, devices, and even therapeutic strategies (Roco, 2011). These two rapidly evolving fields now call for educational curricula that can keep up with novel ideas, experimental techniques, and conceptual advancements. As a result, teachers have started to give way to new ways of teaching- that is, adaptive learning platforms, virtual simulation environments, and collaborative, project-based workshops involving nanoscale experimentation (Sacha & Varona, 2013).

AI technologies have demonstrated incredible potential in transforming STEM education by enhancing instructional practices, assessment strategies, and administrative processes (Almasri, 2024). AI-powered tools, such as intelligent tutoring systems, automated assessment platforms, and learning analytics, provide personalized learning experiences, thereby improving student engagement and performance (Xu & Ouyang, 2022; Sarı et al., 2020). For instance, AI can analyze student data to identify learning gaps and tailor educational content to individual needs, fostering a more effective and adaptive learning environment (Chiu & Li, 2023).

Nanotechnology offers unique opportunities to bridge the gap between theoretical knowledge and practical application in STEM education. By incorporating nanomaterials and nanotechnology concepts into the curriculum, students gain hands-on experience with advanced materials and techniques that are crucial for modern scientific and engineering practices (Chiu & Li, 2023). This practical approach not only enhances students' understanding of complex scientific principles but also prepares them for careers in high-tech industries where nanotechnology plays a critical role (Chiu & Li, 2023).

The integration of AI and nanotechnology fosters interdisciplinary learning, encouraging students to apply concepts from multiple STEM fields to solve real-world problems. In this sense, real-world interdisciplinary learning is promoting innovation and critical thinking, since students learn how to use AI algorithms for data extraction and manipulation, and nanotechnology for designing and manipulating materials (Xu & Ouyang, 2022; Sarı et al., 2020; Chiu & Li, 2023). Such types of cross-disciplinary deployable skills are essential for solving today's challenges in science and technology, making students versatile and competitive for the job market (Xu & Ouyang, 2022).

By staying abreast of current developments in AI and nanotechnology, their seamless integration into the STEM educational landscape will ensure that future generations of learners are equipped to navigate and contribute to emerging technological advances. While automating certain high-complexity tasks, AI now appears to unlock the potential of nanotechnology to create an irreversible revolution in materials science, underscoring the importance of students being introduced to both technologies early in their academic careers (Almasri, 2024; Xu & Ouyang, 2022). This is crucial for developing a workforce capable of driving innovation and technological leadership on a global scale (Chiu & Li, 2023).

This chapter covers the impact of AI and nanotechnology in the field of STEM education and aims to highlight how these technologies can bring a change in the modernization of educational practices and prepare students for future technological challenges. Synthesizing findings from recent scientific articles, this chapter clarifies the importance of integrating AI and nanotechnology into STEM curriculum. For instance, AI technologies could potentially enhance teaching methods, assessment techniques, and even administrative processes, achieving personalized learning experiences that engage and improve student performance. With the above-stated, curriculum development frameworks and research-based strategies for the highly effective integration of such technologies into education, along with the challenges and solutions related to these novel technologies in education, are analyzed. A significant part of the chapter deals with a case study within the University of Bucharest, Faculty of Physics, where AI and nanotechnology integration materialized in STEM programs. The results of this case study provide real curriculum development examples, hands-on projects, and their measurable impact on students' abilities and career preparedness. Thus, this chapter will guide educators and policymakers enthusiastic about influencing STEM education by analyzing the practical application of effective and ineffective practices in integrating AI and nanotechnology into STEM education.

2. CURRICULUM DEVELOPMENT

2.1. Rationale for Integration

In this section, we discuss how teachers can merge AI-assisted teaching tools with hands-on work in nanotechnology, pointing to the main pedagogical challenges and benefits associated with this type of interdisciplinary instruction. One of the significant issues is the content and its complexity, as students need to grasp fundamental quantum phenomena, advanced material characterization, and safety/ethical considerations while also interfacing with AI applications like machine learning (ML) models or intelligent tutoring systems (Kay, 2008; Koedinger et al., 2012). Adaptive learning solutions are preferred by an ever-increasing number of instructors who personalize instruction by using rich virtual labs and simulations to provide cost-effective and scalable experiments. AI is, in turn, used to augment live tutoring by providing timely and detailed feedback. All these approaches fit well into the creative, discovery-oriented hands-on nanotechnology workshops, an essential bridge between extended theoretical conversations and actual lab experiences (Bothun et al., 2019; Jiao & Barakat, 2011).

While the integration of AI and nanotechnology into the STEM curriculum prepares students for impending technological advancements, it is increasingly vital for the educational environment to equip these learners with the skills and knowledge to navigate and influence this evolving landscape (Zhai & Krajcik, 2024; Ningtyas et al., 2024). AI has demonstrated substantial potential in transforming STEM education by enhancing instructional practices, assessment strategies, and administrative processes. AI-powered tools, such as intelligent tutoring systems, automated assessment platforms, and learning analytics, provide personalized learning experiences, thereby improving student engagement and performance (Xu & Ouyang, 2022). By analyzing student data to identify learning gaps and tailor educational content to individual needs, AI fosters a more effective and adaptive learning environment (Almasri, 2024). By incorporating nanomaterials and nanotechnology concepts into the curriculum, students gain hands-on experience with advanced materials and techniques that are crucial for modern scientific and engineering practices (Zhai & Krajcik, 2024; Ningtyas et al., 2024). This practical approach enhances students' understanding of complex scientific principles and prepares them for high-tech industries where nanotechnology plays a critical role. The integration of AI and nanotechnology fosters interdisciplinary learning, encouraging students to apply concepts from multiple STEM fields to solve real-world problems. This inter-disciplinary culture, in turn, nourishes the innovation environment and critical thinking, thereby positioning students in a versatile position to challenge competitors within the job market (Christopoulos et al., 2020).

Using AI-oriented software tools in STEM education at the university level has many advantages. AI can analyze student data and create educational content according to individual needs, presenting personalized learning experiences that enhance student engagement and performance. Interactive AI technologies, like simulation and virtual labs, will create hands-on experiences to make learning lively and bring accessibility. AI-assisted tools give immediate feedback on student performance, allowing students to identify their mistakes and improve the learning process. The analytical ability of AI becomes a tool for maintaining an appropriate assessment approach to student progress while giving real-time insights to the educator regarding learning trends and areas needing intervention. AI can also adjust itself to students' pace and learning style, thus rendering a more efficient and adaptive learning ecosystem. AI-oriented platforms can maximize the resource utilization of educational inputs and make quality STEM education reachable across a wider pool of potential students. By incorporating AI into STEM education, students are better prepared to handle future technological challenges and acquire skills that will be essential in a technologically advanced world. AI can sift through large volumes of data, identifying the patterns and insights that will inform different instructional strategies. Such strategies can help educators tailor teaching materials to different student demographics based on feedback from AI analyses. AI tools will encourage students to engage in solving problems within a learning environment that emphasizes interaction and cooperation. Through AI-driven educational platforms, students will have access to resources, teachers, and experts that may not be available within their geographic confines, thereby raising the overall quality of education.

Thus, there is a need to design courses that integrate AI into nanotechnologies, focusing on developing skills that align with job market needs. A flowchart showing three interlinked gears, each representing the main steps of the course design process, is illustrated in Figure 1 The steps are defining course objectives, identifying key topics, and developing a curriculum. The course objectives include specifying learning outcomes, emphasizing interdisciplinary skills, defining measurable goals, consulting with stakeholders, and considering the ethical ramifications the course might have. Next comes deciding what key topics should be included in the course. Curriculum development involves preparing a course plan, hands-on projects, and developing assessment strategies. Further on, the flowchart leads to reviewing and revising contents, implementing courses, and evaluating effectiveness. The flowchart visually represents a structured approach to developing content in a STEM course that integrates AI, highlighting the essential elements and their interrelations.

Moreover, developing and nurturing soft skills in STEM education becomes crucial as interdisciplinary approaches are integrated with artificial intelligence and nanotechnologies. Communication, teamwork, problem-solving, and emotional intelligence are essential for both personal and professional development. These soft skills improve relationships, help overcome obstacles, and create an innovative and motivating learning environment. Most importantly, the outcome is better-prepared, more employable professionals.

Figure 1. A structured approach to developing content for a STEM course integrating AI, highlighting the key components and their interconnections.

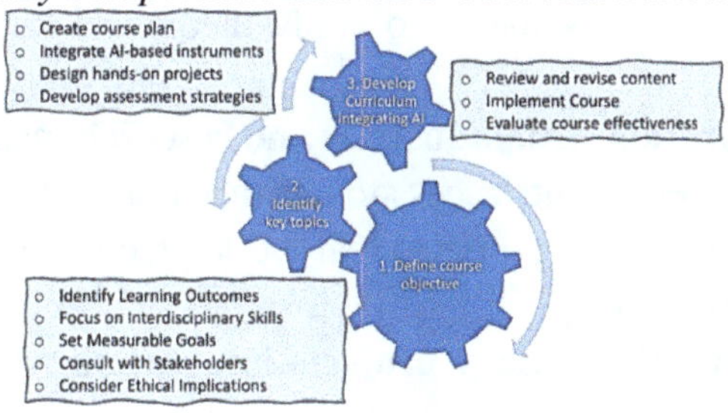

2.2. Frameworks and Strategies

The meaningful frameworks and pathways for integrating AI and nanotech into the existing STEM curriculum should include several key elements. Firstly, it is necessary to define the future technologies that will guide curriculum development and the competencies required for future careers. In this case, the curriculum could incorporate the most relevant and potentially influential technologies by aligning with industry experts to identify worthwhile technologies. Secondly, these AI-enabled tools can help teachers in personalizing the learning experience while improving their instructional practices. Intelligent tutoring and automated assessment platforms can also be integrated into learning analytics with curriculum plans to stimulate student learning and hence, enhance educational outcomes (Zawacki-Richter et al., 2019).

To ensure that AI and nanotechnology are meaningfully integrated into STEM education, the following strategies can guide the development and implementation of effective frameworks:

- **Define future technologies and competencies in collaboration with industry experts:** Align curriculum design with emerging technological trends by consulting industry professionals to identify which AI-driven and nanotechnology-based tools are most relevant for future STEM careers.
- **Leverage AI-enabled tools to enhance teaching and personalize learning:** Integrate intelligent tutoring systems, automated assessment platforms, and learning analytics to support teachers in tailoring instruction and improving educational outcomes (Dorouka et al., 2021).
- **Utilize online AI platforms and digital resources:** Incorporate tools such as those listed in Table 1 to enrich STEM learning; these platforms provide access to AI-based content, simulations, and analytics that can be embedded into classroom instruction.

- **Promote hands-on learning experiences in nanotechnology:** Include laboratory-based activities, experiments with nanomaterials, and design challenges that allow students to apply theoretical knowledge to practical contexts.
- **Foster collaboration with industries and research institutions:** Build partnerships that give students access to advanced project materials, techniques, and mentorship, exposing them to real-world practices in nanomanufacturing and materials science.
- **Encourage interdisciplinary projects integrating AI and nanotechnology:** Design project-based learning tasks that require students to apply knowledge from multiple STEM domains, using AI algorithms to analyze data from nanotechnology experiments and optimize processes (Cai et al., 2024).
- **Incorporate interactive virtual environments for nanoscience education:** Use platforms such as NanoHub and the PhET Interactive Simulations from the University of Colorado to let students explore nanoscale phenomena virtually, providing collaborative and self-guided learning opportunities (Klick et al., 2017; Marchi et al., 2022).

The following diagram illustrates how these strategies work together to guide the structured integration of AI and nanotechnology into STEM education.

Figure 2. Proposed Framework for Integrating AI and Nanotechnology in STEM Education.

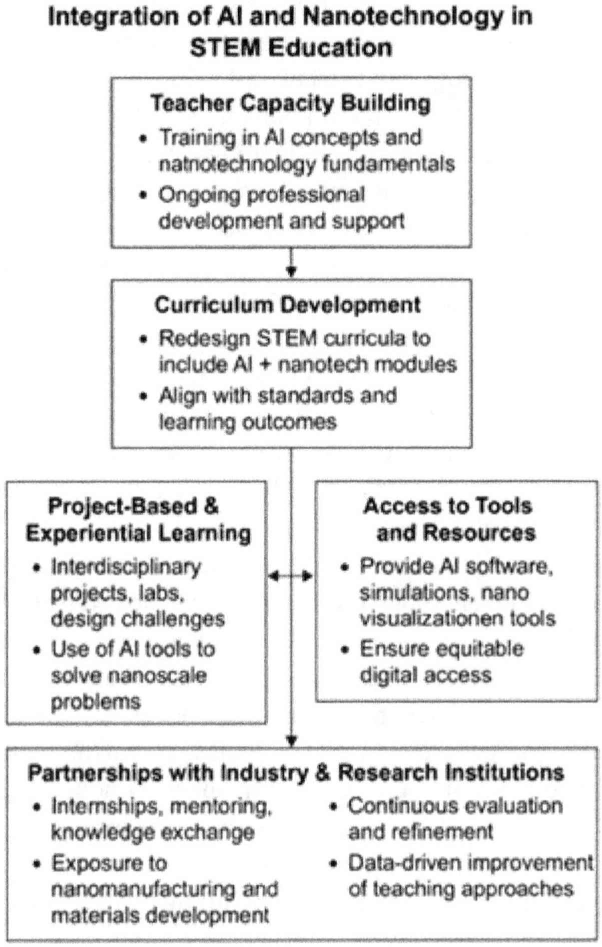

In Table 1, an overview of various online platforms that effectively utilize AI to enhance STEM subject learning is provided. It also includes a short explanation of their features and functionalities, the access link to the platform, and accessibility conditions: open access, subscription-based, or a combination of free and premium. The table can be useful for educators, students, and institutions integrating AI-based tools into their STEM curriculum.

It is increasingly necessary to include hands-on learning experiences in teaching nanotechnology. Students should be made to experiment and engage in projects involving nanomaterials and nanotechnology concepts, applying their theoretical knowledge to practical purposes. This can be accomplished through collaborations with industries and research institutions, giving students access to advanced project materials and techniques. Interdisciplinary projects, where students are required to apply knowledge from multiple STEM domains, should also be encouraged. These interdisciplinary projects provide a great avenue to integrate AI algorithms with data generated from nanotechnology experiments, allowing for the analysis of results and optimization of processes. This collaboration and innovation will prepare students for the complex challenges of tomorrow (Dorouka et al., 2021).

Interactive virtual environments for nanoscience education allow students to interact with phenomena that are inaccessible through conventional laboratory practices due to their microscopic scale, and thereby advance education significantly (Dorouka et al., 2021; Cai et al., 2024; Tawiah et al., 2024). The integration of AI and nanotechnology looks forward to a tremendous possibility to revolutionize many domains, from advanced materials, medicine, to energy and environmental science (Sacha & Varona, 2013). Established platforms like NanoHub and the PhET Interactive Simulations from the University of Colorado offer virtual experiments across various nanoscience concepts, providing room for collaborative experiments as well as self-guided explorations.

NanoHub is an interactive online environment created and managed by the Network for Computational Nanotechnology (NCN), primarily hosted by Purdue University. It offers a broad selection of simulation tools, online lectures, and learning materials designed to provide equitable access to high-level nanotechnology resources for students, educators, and researchers worldwide (Klick et al., 2017). By integrating real-world research projects into virtual modules, NanoHub enables learners to explore nanoscience concepts in a hands-on manner without needing specialized hardware or physical lab space. Many of these modules incorporate live data from experiments and advanced simulation engines, further enriching the learning experience and fostering greater engagement. By allowing users to post questions or feedback, share and remix simulations, and access a vast library of research articles and instructional materials, the platform also promotes collaboration and helps to create a positive community around nanotechnology education.

PhET Interactive Simulations, developed at the University of Colorado Boulder, bridging conceptual gaps in physics and nanotechnology, provides open-access, research-based simulations that cover a broad range of topics in science, technology, engineering, and mathematics (Marchi et al., 2022). These simulations employ intuitive, game-like interfaces enabling users to adjust variables, manipulate virtual equipment, and observe real-time results. PhET is especially useful in both formal and informal learning environments because it provides a wealth of support resources, such as lesson plans, teacher guides, and student prompts. PhET supports students in developing strong conceptual frameworks by emphasizing inquiry-driven design and interactive components. This encourages students to try new things, fail safely, and eventually gain a deeper understanding of difficult ideas. Nanoscience-specific modules promote increased student engagement and comprehension of subjects like quantum effects and nanoscale structures by enabling students to visualize phenomena at the molecular or atomic level.

Table 1. Online platforms for AI-assisted learning in STEM subjects.

AI-based platforms	Brief description	Access Link	Access conditions
Labster	Virtual labs powered by AI, interactive and immersive STEM learning experiences	https://www.labster.com/	Some open educational resources (OER) through partnerships, but the whole platform generally requires a subscription or institutional access.
Cognii	Personalized tutoring and assessment in STEM subjects	https://www.cognii.com/	Requires a subscription or institutional access.
NanoHub,	Simulation tools, online lectures, and learning materials	https://nanohub.org/	Open access
PhET Interactive Simulations	Research-based simulations that cover a broad range of topics in science, technology, engineering, and mathematics	https://phet.colorado.edu/	Open access
Amrita Virtual Labs	Online lab simulations for various STEM domains	https://vlab.amrita.edu/	Open access
QuantumWise,	Intuitive graphical interface, multi-scale modeling	https://www.european-mrs.com/quantum-wise	Offers trial license for academic or material research purposes
ALEKS	An adaptive learning platform that uses AI to assess student knowledge and tailor instruction accordingly	https://www.aleks.com/	Requires a subscription or institutional access.
Knewton	Adaptive learning technology that personalizes educational content based on individual student needs	https://www.knewton.com/	Requires a subscription or institutional access.
Carnegie Learning	AI-driven math tutoring systems that offer personalized learning experiences	https://www.carnegielearning.com/	Requires a subscription or institutional access.
Smart Sparrow	An adaptive learning platform that uses AI to create personalized learning experiences	https://www.smartsparrow.com/	Requires a subscription or institutional access.
Querium	AI-driven tutoring for STEM subjects, focusing on personalized learning and real-time feedback	Https://www.querium.com/	Requires a subscription or institutional access.
Squirrel AI Learning	Personalized tutoring and adaptive learning experiences	Https://squirrelai.com/	Requires a subscription or institutional access.
Thinkster Math	An AI-driven math tutoring platform that offers personalized learning plans and real-time feedback	Https://hellothinkster.com/	Requires a subscription or institutional access.
Century Tech	Personalized learning experiences and real-time analytics	Https://www.century.tech/	Requires a subscription or institutional access.
Edmentum	AI-driven personalized learning solutions for STEM education	Https://www.edmentum.com/	Requires a subscription or institutional access.
Knowre	An AI-powered math learning platform that offers personalized learning paths and real-time feedback	Https://www.knowre.com/	Requires a subscription or institutional access.

continued on following page

Table 1. Continued

AI-based platforms	Brief description	Access Link	Access conditions
Brainly	Provides personalized homework help and tutoring in STEM subjects	Https://www.brainly.com/	Offers some free access, but premium features require a subscription.
Quizlet	Personalized learning experiences and adaptive learning paths	Https://quizlet.com/	Offers some free access, but premium features require a subscription.
Khan Academy	Personalized learning experiences and adaptive learning paths	Https://www.khanacademy.org/	Khan Academy is open access and free to use.
Coursera	Offers AI-driven personalized learning experiences and adaptive learning paths in various STEM subjects	Https://www.coursera.org/	Coursera offers some free courses, but many courses and features require a subscription.

Labster offers immersive virtual lab simulations across various STEM fields, including specialized nanotechnology modules. These simulations guide learners through realistic scenarios, such as examining nanoscale materials with electron microscopy or synthesizing quantum dots, enhancing conceptual understanding and practical problem-solving within a safe, cost-effective environment. Users receive real-time feedback on their experimental decisions and have access to supplementary theoretical resources, supporting blended learning approaches.

Amrita Virtual Labs: The Virtual Labs project by Amrita Vishwa Vidyapeetham in India, developed under the support of the Ministry of Education, provides online lab simulations for diverse STEM domains, including nanotechnology. Students can explore virtual experiments on nanoscale material characterization, nanosensors, and nano-biotechnology. The platform also integrates instructional aids, such as step-by-step experimental guidelines and interactive quizzes.

QuantumWise, now part of Synopsys, provides a commercial simulation environment known as Virtual NanoLab. This platform is geared towards advanced users conducting research-level projects in nanotechnology and materials science. Virtual NanoLab allows students and researchers to design atomic-scale models, run electronic-structure calculations, and analyze complex nanosystems through an intuitive graphical interface. The software supports multi-scale modeling, bridging quantum-level phenomena with larger-scale properties of materials. Such environments also aid in bridging the conceptual gap between theoretical knowledge and practical application by simulating realistic experimental conditions (Klimeck et al., 2008; Perkins et al., 2010).

2.3. Challenges and Solutions

Developing curricula for AI and integrating nanotechnology into STEM education presents a variety of challenges that typically coexist. The primary obstacle involves the knowledge a teacher must have about these technologies. Therefore, comprehensive training and professional development programs are mandatory to prepare teachers to deliver AI and nanotechnology competently. Training can happen through workshops, online classes, and working with people in the industry (Zawacki-Richter et al., 2019).

Cost and availability of sophisticated materials and technologies also form a serious barrier. However, Educational institutions can use some common components and resources, such as less expensive microcontrollers and open-source software. They can share resources with industry and research institutions to access more advanced materials and techniques, thus reducing barriers to use.

The curriculum must integrate the ethical and social implications of AI and nanotechnology. Educators should scaffold discussions about ethical, social, and environmental influences of this technology in anticipation of producing responsible and sustainable behavior in students (Klimeck et al., 2008). This can be achieved through case studies, debates, and interdisciplinary projects exploring the broader implications of technological advancements. By addressing these challenges and implementing effective frameworks and strategies, educators can successfully integrate AI and nanotechnology into STEM curricula, preparing students for future technological advancements and fostering a more innovative and interdisciplinary approach to education.

Integrating AI into STEM education can play a pivotal role in channeling students' interest toward careers in nanotechnology manufacturing and materials science. Exposure to AI-based tools for data analysis, materials simulation, and process optimization allows students to engage with authentic problem-solving scenarios that mirror real industrial practices (Perkins et al., 2010; Nixon et al., 2024). Such experiences not only build computational and analytical competencies but also increase students' awareness of the diverse roles AI plays in nanomaterials design, quality assurance, and automated fabrication. Embedding interdisciplinary project work that integrates AI algorithms with nanoscale experimentation can inspire students to envision future career trajectories in advanced manufacturing sectors, thereby strengthening the talent pipeline for the emerging AI–nanotechnology workforce.

The accelerating development of advanced materials and intelligent manufacturing systems has further heightened the need for graduates who can combine AI-based analytical skills with foundational knowledge of nanoscale engineering. Sectors such as semiconductor fabrication, energy storage, and biomedical device manufacturing increasingly rely on AI to optimize design parameters, automate quality control, and shorten discovery cycles in nanomaterials research (Khan & Joya, 2021; Kalra et al., 2020). Yet, many STEM programs offer only limited exposure to the interdisciplinary skills required for these fields, leaving a gap between academic preparation and industrial expectations. Embedding AI modules—covering areas such as machine learning, predictive modeling, and algorithmic design—within STEM curricula can foster the computational thinking essential for addressing complex challenges in nanomanufacturing contexts.

At the same time, meaningful application of nanotechnology in manufacturing depends on the ability to interpret large experimental datasets and link them to real-time production environments. Including targeted nanoscience content on materials characterization, fabrication techniques, and ethical considerations—delivered alongside AI-based design and analysis tools—can help bridge this gap (Lee et al., 2015). Such integration would enable students to contribute to emerging domains such as sustainable nanomaterials, autonomous fabrication lines, and cyber-physical systems aligned with Industry 4.0 and the transition toward Industry 5.0. By explicitly connecting educational practice to these technological trends, STEM programs can prepare students to drive future innovations in materials, devices, and manufacturing.

Building on this connection, introducing AI into STEM education at earlier stages can serve as a catalyst for developing students' interest and competencies in fields such as nanotechnology manufacturing and advanced materials. Early exposure to AI-driven modeling, automation, and data analysis tools not only equips students with transferable technical skills but also enables them to envision practical applications of their knowledge in real-world production environments. By highlighting how AI supports precision, scalability, and innovation in nanomanufacturing workflows, educators can help students see clear academic and professional pathways. This targeted preparation can channel their curiosity and focus on future careers at the intersection of AI, nanotechnology, and advanced materials development.

3. CASE STUDY: SUCCESSFULLY INTEGRATED AI AND NANOTECHNOLOGY INTO STEM PROGRAMS AT THE UNIVERSITY OF BUCHAREST, FACULTY OF PHYSICS

3.1. A Model of Modern Education: Transforming Learning through Practice

In a world where traditional education faces major challenges, Pro-Youth Association has proposed an innovative program at the University of Bucharest that aims to demonstrate that learning can be connected to current economic and technological realities. The course, designed as a practical alternative to the usual theoretical model, focused on developing applied skills and preparing participants for the challenges of the job market. The program originated from Pro-Youth's desire to offer a concrete solution to one of the most significant deficiencies of the Romanian educational system: the disconnect between theory and practice. Inspired by the growing need for well-prepared professionals, this course aimed to provide young people with an environment where they could learn through direct application and be guided by industry experts.

During the course, participants had the opportunity to develop technical skills through projects that simulated real-world situations. For example, one of the most appreciated modules was dedicated to creating a do-it-yourself (DIY) personal assistant using modern technologies such as Raspberry Pi and Python libraries. The course also covered topics such as machine learning, using Google Colab, and dataset analysis, all through an interactive and collaborative methodology. Additionally, interaction with industry professionals offered unique perspectives on the job market. Participants learned the skills employers seek, how to prepare for interviews, and how to navigate the corporate environment.

3.2 Course Objectives

Starting from the idea of developing a learning community and transmitting knowledge in the field of IoT and AI, Pro-Youth created a mini-course support available for free and distributed online. A unique aspect of the course support is its autonomous structure, which allows participants in the IoT and AI Course to teach what they have learned. Thus, graduates of the IoT and AI Course become vectors of knowledge, extending the impact of this program through the communities they are part of. This support was built around three key objectives, aimed at maximizing educational impact:

- Learning from Zero to a Functional Machine Learning Model: Participants are guided step by step in building a model to predict incident solar radiation. This process familiarizes them with essential machine learning concepts and teaches them how to apply these concepts in a concrete project with practical applicability.
- Using Accessible Hardware: To eliminate barriers related to costs and availability, the course uses accessible components such as the Raspberry Pi Pico W microcontroller, breadboards, sensors, and small solar panels. Thus, the course is a viable solution even in the context of the global semiconductor crisis.
- Building and Interacting with an IoT System: From connecting hardware to a WiFi network to collecting and analyzing data, participants learn how to develop a complete IoT system that can be integrated into industrial solutions or personal projects.

3.3 Course Support Structure

The material is structured to allow progressive learning, covering all aspects necessary for the proposed project. Among the main chapters are:

- Introduction to Python: This chapter includes fundamental concepts such as using Google Colab, data types, control structures, functions, and popular modules (NumPy, pandas, matplotlib), preparing participants for the practical part of the course.
- Raspberry Pi Pico W – Hardware for Everyone: Participants discover how to configure and program this accessible microcontroller, using it for practical applications such as monitoring electric current and voltage or charging NiMH batteries with a solar panel.
- Building a Machine Learning Model: The guide explains in detail how to collect and preprocess data, build the theoretical model of incident solar radiation, and validate results using real datasets.
- Impact of MPNetv2 in Retrieval Augmented Search: Accessibility and Performance: One of the essential steps in the evolution of machine learning and Artificial Intelligence technologies is the creation of solutions that allow rapid processing and analysis of information from a wide range of sources. In this context, AiDocker represents Pro-Youth's effort to make machine learning models available to the public, packaged in Docker containers, accessible through a simple REST interface. These are designed to facilitate the rapid and efficient integration of machine learning models into various applications, without requiring advanced technical knowledge from users.
- MPNetv2: Innovation in Retrieval Augmented Search: One of the most interesting additions in 2024 to this project is MPNetv2, a model used for Retrieval Augmented Search (RAS). RAS is an advanced technique that combines traditional searches with machine learning to provide more precise and personalized answers to users. For example, when a user asks a question in a search engine, RAS helps find the most relevant documents from a database, using machine learning to improve the response selection process. MPNetv2 is an improved version of MPNet, a pre-trained deep learning model that uses neural networks to understand better the meaning of words and their context in a text. In the case of MPNetv2, the focus is on retrieval – the process of searching and extracting relevant information from a large amount of data. An essential aspect of MPNetv2 is that it uses Redis.

3.4 Measurable Impact

The innovative IoT and AI programs brought remarkable results, reflected in the success of participants and the creation of an active community:

- Participants: Of the 18 initially enrolled, seven were prepared for interviews, and 6 received job offers from companies such as Short Pixel and XPath Global.
- Internships: 7 students were accepted into the program, including at Short Pixel.
- Active Community: 8 participants continue to collaborate through Slack and participate in meet-up events.
- Duration and Structure: The course lasted 11 weeks, with weekly sessions lasting 6 hours and 30 minutes. Two interviews (technical and team-fit) evaluated the course.

Table 2. Investment and Resources

Category	Justification	Total amount (Euro)
Infrastructure	Equipment: 10 fully equipped workstations, necessary consumables, 3D printer, Kubernetes mini-cluster with three nodes for deep learning	26,000 €
IT Services	Cloud (AWS) and project management (Confluence, Jira)	8,000 €
Meals	Meals for participants	1,000 €
Administrative	Administrative expenses (accounting and legal consultancy)	2,900 €
Total		**37,900 €**

These results highlight a practical model connected to job market requirements, through which Pro-Youth has surpassed the educational system's classic paradigms.

Figure 3. Activities in the IoT and AI course at the Faculty of Physics, University of Bucharest

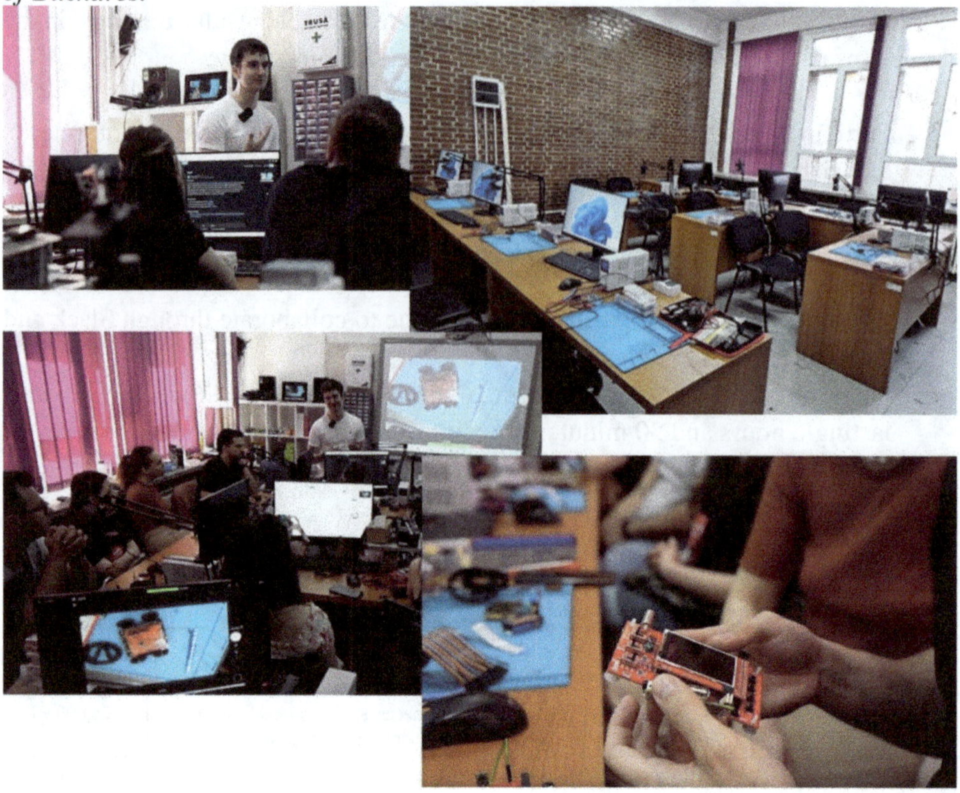

3.5 Participant Feedback

Table 3. Participants highlighted several elements that made the course unique and valuable

Practical approach	the only course that encouraged real development, not just passing an exam
Hands-on activities	running code, detailed explanation of each step, IoT projects, and programming were highly appreciated
Collaborative experiences	the hackathon, teamwork, and simulated interviews were considered beneficial and engaging
Relaxed atmosphere	pizza, jokes, and social moments contributed to a pleasant learning environment
Innovative technological projects	extracting weather data, configuring a DIY Home Assistant, and using technologies like Flask, Docker, and Picovoice were major points of interest

Table 4. Participants listed many new concepts learned

Machine learning	using Google Colab, understanding CNN networks, ML workflow, and dataset interpretation
Programming	advanced Python concepts, classes, embeddings, tokens, and using libraries like e-speak
IoT	configuring and using a Raspberry Pi, programming a personal assistant, and interacting with sensors.
Management and industry	lessons about working in corporations, the interview process, and entrepreneurship concepts

Some points for improvement were suggested:

- Time management for module completion.
- Better access to resources like the internet or timely pizza delivery.
- Clear structure for internships and subsequent activities.
- General sentiment of melancholy towards the end of the course, reflecting attachment to colleagues and the environment.

Representative Comments

- "The only course that encourages development, not just passing an exam."
- "I configured my own Home Assistant and worked with IoT sensors."
- "The coolest 'prof' I've ever had."
- "I learned many new programming concepts and practical industry perspectives."
- "We discussed every line of code and understood what it does."
- "I've always wanted a DIY personal assistant – the project was super engaging."
- "Discussions about interviews and industry stories were very educational and interesting."
- "Internet was a challenge sometimes, especially during practical moments."
- "Pizza was sometimes delayed."
- "We had to wait 12 hours for model training – it would have been good to have faster alternatives."

Feedback received demonstrates the positive impact of this program. Participants praised the practical approach of the course: "The only course that encourages development, not just passing an exam." Others appreciated the applied projects: "I've always wanted a DIY personal assistant – the project was super engaging."

However, the course was not without challenges. Logistics posed some difficulties, such as issues related to internet access or the long time required for model training. Even so, these aspects were overcome through a collective commitment to learning and collaboration.

3.6 Connecting with the Job Market

One of the most important distinctions between this course and traditional programs was the close link with employment market conditions. Through workshops, mock interview sessions, and inspirational industry success stories, the course bridged education and employer requirements and gave participants a competitive edge. "We worked through every line of code and knew what it did," said one of the participants, which reflects the worth of this learning model.

3.7 Concluding remarks for the case study

This course is an example of innovation in education. By introducing emerging technologies, interacting with industry partners, and focusing on practical competencies, it demonstrated that education could be reinvented to develop young people capable of solving problems in the future. It is a success story that shows that change is possible where vision, resources, and the passion to make a difference converge.

4. FUTURE TRENDS IN INTERDISCIPLINARY EDUCATION

Innovation in technology and the growing need for innovative answers to global problems are driving changes in the field of interdisciplinary education. Artificial intelligence and nanotechnology are two new areas that will shape interdisciplinary STEM education in the future. A more integrated and collaborative learning style is emerging due to these technologies transforming teaching methods. AI technologies are transforming STEM education through the development of new tools and approaches that enhance learning outcomes and experiences. Current research recognizes several unfolding trends. Personalized learning is becoming increasingly popular as AI-powered systems analyze student data to tailor education content to fit individual needs, thereby increasing engagement and performance. Intelligent tutoring systems and adaptive learning platforms offer students personalized feedback and guidance, making their learning experience more enriched (Xu & Ouyang, 2022; Kalidindi & De Graef, 2015; Fonash, 2001).

Another important trend is automated marking and feedback. AI technology helps students learn more effectively by marking their work and providing immediate feedback to recognize their errors. This enables educators to concentrate less on administrative tasks and more on one-on-one instruction (Chen et al., 2020).

Learning analytics is another area where AI is making a substantial impact. By identifying patterns and trends in student performance, AI enables data-informed decision-making, leading to improved teaching methods and learning outcomes (Cai et al., 2024).

AI makes it easier for students to work on cross-disciplinary projects that require them to apply knowledge from multiple STEM fields. For instance, to examine outcomes and improve procedures, AI algorithms could be combined with experimental data from nanotechnology (Hwang & Tu, 2021). The inclusion of AI in the curriculum requires teachers to be acquainted with such technologies. Recent research emphasizes the need for comprehensive AI materials and professional development opportunities to support teachers in this transition (Zawacki-Richter et al., 2019).

Nanotechnology is another area that is full of potential to reshape STEM education. The addition of nanotechnology to the course materials allows students to apply experiments and projects practically, bridging the gap between theory and reality. It serves to stimulate scientific inquiry and critical thinking (Tawiah et al., 2024). Nanotechnology, by definition, involves several STEM disciplines, promoting interactions among students and instructors across disciplines. This kind of interdisciplinarity is required to solve complex scientific and technological problems (Park et al., 2023).

Industrial partnerships can provide students with hands-on experience and understanding of the latest advancements in nanotechnology. These partnerships help connect education with the requirements of industry, preparing students for their prospective careers (Dorouka et al., 2021). With rapid advancement in nanotechnology, addressing its social, environmental, and ethical implications has become essential. Educators are incorporating these discourses into their syllabus to inculcate an understanding of responsibility and sustainability.

Gazing into the future, certain things can be anticipated about interdisciplinary STEM education. The integration of AI and nanotechnology in STEM education will become progressively seamless and ubiquitous, enhancing learning and better preparing students for the coming technological world. The ability to transfer knowledge between disciplines will become increasingly important, as educational institutions focus on developing these abilities to promote innovation and critical thinking. Advances in technology will facilitate greater collaboration and networking among students, teachers, and industry professionals, making the learning landscape more dynamic and interconnected. With continued growth in technology, there will be an increased emphasis on the need for a lifelong learning cycle. Institutions will have to provide ongoing support and training to get people up to date with the latest developments.

Policymakers and educators must take into account the following important factors in order to successfully integrate AI and nanotechnology into interdisciplinary STEM education:

- Investing in teacher training is imperative to providing teachers with comprehensive resources and support when integrating these technologies into their pedagogy.
- Creating industry partnerships can connect educational programs to industry demands and give students valuable experience.
- The curriculum must incorporate discussions about the ethical, social, and environmental ramifications of AI and nanotechnology to promote moral behavior.
- Promoting interdisciplinary cooperation by allowing educators and students to work together across disciplines strengthens a more cohesive approach to STEM education.

By embracing these challenges, teachers and legislators can guarantee that students are adequately prepared to navigate and shape the future technological environment.

5. CONCLUSIONS AND PERSPECTIVES

In the context of STEM education, this chapter focuses on the complex interplay between nanotechnology and AI. It has drawn significant attention to their exciting and revolutionary potential to change how education is practiced. The discussion began with a clear justification for integrating these cutting-edge technologies into STEM curricula, emphasizing their potential to enhance teaching strategies and creative evaluation methods. Additionally, it has been shown that AI-supported technologies, such as refined learning analytics and intelligent tutoring systems, provide customized learning experiences that improve students' academic performance and engagement.

The chapter offered effective models and approaches for incorporating AI and nanotechnology into current STEM curricula, highlighting the necessity of teamwork with industry partners and of hands-on training. The chapter also introduced typical pitfalls in curriculum design, such as teacher training needs and the expense of cutting-edge materials, and offered solutions to tackle these challenges. An extensive section of the chapter was dedicated to a comprehensive case study at the University of Bucharest, developed in cooperation with the Pro-Youth Association. This educational institution has effectively achieved an extremely impressive and ground-breaking incorporation of AI and cutting-edge nanotechnology. All of this originated in the context of STEM programs, whose goal is to advance science, technology, engineering, and math education. This comprehensive case study presented various real-world examples and hands-on projects to optimize learning experiences. Furthermore, it demonstrated the significant influence of these group efforts in enhancing students' competencies and preparing them for professional careers in various fields.

In the future, the combination of AI and nanotechnology in STEM education will remain instrumental in equipping students for technological innovations. As these technologies continue to develop, their integration into teaching practices will become increasingly seamless and prevalent, improving the learning process. The ability to transfer learning from outside disciplines will become increasingly important, with education systems prioritizing the development of interdisciplinary skills to promote innovation and critical thinking.

The ongoing advancements in technology will facilitate more networking opportunities and collaboration between learners, educators, and business professionals. This will go a long way in creating a more dynamic, interactive, and interconnected learning process, which is vital nowadays. As technology continues to evolve rapidly, the necessity for lifelong learning will become increasingly imperative, and learning institutions will need to recognize this and provide ongoing training and support mechanisms. This will be crucial for allowing people to stay tuned and abreast of the current developments and advances in various fields, including safety, security, and resilience (Holmes et al., 2019; Curreli & Rakich, 2020). To effectively incorporate AI and nanotechnology into interdisciplinary STEM education, policymakers and educators must invest in teacher training, build partnerships with industry, promote ethical and responsible practice, and support interdisciplinary collaboration. Educators can promote innovation and uphold technological leadership globally by implementing these strategies and ensuring that students are sufficiently equipped to succeed and contribute to future technological advances.

REFERENCES

Almasri, F. (2024). Exploring the impact of artificial intelligence in teaching and learning of science: A systematic review of empirical research. *Research in Science Education, 54*(5), 977–997. DOI: 10.1007/s11165-024-10176-3

Binnig, G., & Rohrer, H. (1986). Scanning tunneling microscopy. *IBM Journal of Research and Development, 30*(4), 355–369. DOI: 10.1147/rd.304.0355

Bothun, G. D., Oyanedel-Craver, V., & Park, K. (2019). Using "nano tools" as the basis for a hands-on experiential course in nanotechnology. In *Nanotechnology in undergraduate education* (pp. 291–308). Elsevier. https://doi.org/DOI: 10.1016/B978-0-12-814835-8.00011-X

Cai, C., Zhu, G., & Ma, M. (2024). A systemic review of AI for interdisciplinary learning: Application contexts, roles, and influences. *Education and Information Technologies*. Advance online publication. DOI: 10.1007/s10639-024-13193-x

Carbonell, J. R. (1970). AI in CAI: An artificial-intelligence approach to computer-assisted instruction. *IEEE Transactions on Man-Machine Systems, 11*(4), 190–202. DOI: 10.1109/TMMS.1970.299939

Chen, L., Chen, P., & Lin, Z. (2020). Artificial intelligence in education: A review. *IEEE Access : Practical Innovations, Open Solutions, 8*, 75264–75278. DOI: 10.1109/ACCESS.2020.2988510

Chiu, T. K. F., & Li, Y. (2023). How can emerging technologies impact STEM education? *Journal for STEM Education Research, 6*(3), 375–384. DOI: 10.1007/s41979-023-00113-w

Christopoulos, A., Pellas, N., & Laakso, M.-J. (2020). A learning analytics theoretical framework for STEM education virtual reality applications. *Education Sciences, 10*(11), 317. Advance online publication. DOI: 10.3390/educsci10110317

Curreli, M., & Rakich, S. S. (2020). An integrated approach to teaching and learning nanotechnology: The Omni Nano model. *Journal of Research in Innovative Teaching & Learning, 13*(1), 141–145. DOI: 10.1108/JRIT-02-2020-0012

Dorouka, P., Papadakis, S., & Kalogiannakis, M. (2021). The contribution of the health crisis to young children's nano-literacy through STEAM education. *Hellenic Journal of STEM Education, 2*(1), 1–7. DOI: 10.51724/hjstemed.v2i1.18

Feynman, R. P. (1960). There's plenty of room at the bottom. *Engineering and Science, 23*(5), 22–36.

Fonash, S. J. (2001). Education and training of the nanotechnology workforce. *Journal of Nanoparticle Research, 3*(1), 79–82. DOI: 10.1023/A:1011408311390

Holmes, W., Bialik, M., & Fadel, C. (2019). *Artificial intelligence in education: Promises and implications for teaching and learning*. Center for Curriculum Redesign. https://curriculumredesign.org/wp-content/uploads/AI-in-Education.pdf

Hwang, G.-J., & Tu, Y.-F. (2021). Roles and research trends of artificial intelligence in mathematics education: A bibliometric mapping analysis and systematic review. *Mathematics, 9*(6), 584. Advance online publication. DOI: 10.3390/math9060584

Jiao, L. H., & Barakat, N. (2011). Incorporation of hands-on activities in learning nanomaterials. In *ASME 2011 International Mechanical Engineering Congress and Exposition* (pp. 109–115). American Society of Mechanical Engineers. https://doi.org/DOI: 10.1115/IMECE2011-62598

Kalidindi, S. R., & De Graef, M. (2015). Materials data science: Current status and future outlook. *Annual Review of Materials Research, 45*(1), 171–193. DOI: 10.1146/annurev-matsci-070214-020844

Kalra, R., Patel, D., & Vaya, D. (2020). Role of artificial intelligence in nanotechnology: A review. *Journal of Computational Science, 46*, 101189. Advance online publication. DOI: 10.1016/j.jocs.2020.101189

Kay, J. (2008). Lifelong learner modeling for lifelong personalized pervasive learning. *IEEE Transactions on Learning Technologies, 1*(4), 215–228. DOI: 10.1109/TLT.2009.9

Khan, F., & Joya, Y. (2021). Artificial intelligence applications in nanotechnology: Opportunities and challenges. *Nano Today, 37*, 101071. Advance online publication. DOI: 10.1016/j.nantod.2020.101071

Klick, A., Kim, S., & Kim, M.-W. (2017). Using virtual reality (VR) for education in nanoscience. *Microscopy and Microanalysis, 23*(S1), 2302–2303. DOI: 10.1017/S143192761701217X

Klimeck, G., McLennan, M., Brophy, S. P., Adams, G. B.III, & Lundstrom, M. S. (2008). nanoHUB.org: Advancing education and research in nanotechnology. *Computing in Science & Engineering, 10*(5), 17–23. DOI: 10.1109/MCSE.2008.120

Koedinger, K. R., Corbett, A. T., & Perfetti, C. (2012). The knowledge-learning-instruction framework: Bridging the science-practice chasm. *Cognitive Science, 36*(5), 757–798. DOI: 10.1111/j.1551-6709.2012.01245.x PMID: 22486653

Lee, J., Bagheri, B., & Kao, H. A. (2015). A cyber-physical systems architecture for Industry 4.0-based manufacturing systems. *Manufacturing Letters, 3*, 18–23. DOI: 10.1016/j.mfglet.2014.12.001

Luckin, R., Holmes, W., Griffiths, M., & Forcier, L. B. (2016). *Intelligence unleashed: An argument for AI in education*. Pearson. Holmes, W., Bialik, M., & Fadel, C. (2019). *Artificial intelligence in education: Promises and implications for teaching and learning*. Center for Curriculum Redesign.

Marchi, F., Friot, J., Fousse, L., Bain, N., Carrez, J., Bolcato, V., & Castagné, N. (2022). "Learning by doing" en nanosciences et nanotechnologies au moyen des réalités virtuelles et de l'interaction multisensorielle avec systèmes haptiques au CIME Nanotech de Grenoble. *J3eA, 21*, Article 1009. https://doi.org/DOI: 10.1051/j3ea/20221009

Murday, J. S., Jones, S. S., & Stanley, G. M. (2019). Nanoscale science and engineering education: Issues, trends, and future directions. *Journal of Nano Education, 11*(1), 1–12. DOI: 10.1166/jne.2019.1048

Ningtyas, P. K., Widarti, H. R., Parlan, P., Rahayu, S., & Dasna, I. W. (2024). Enhancing students' abilities and skills through science learning integrated STEM: A systematic literature review. *International Journal of Education in Mathematics, Science, and Technology, 12*(5), 1161–1181. DOI: 10.46328/ijemst.4292

Nixon, N., Lin, Y., & Snow, L. (2024). Catalyzing equity in STEM teams: Harnessing generative AI for inclusion and diversity. *arXiv*. https://doi.org//arXiv.2402.00037DOI: 10.48550

Park, J., Teo, T. W., Teo, A., Chang, J., Huang, J. S., & Koo, S. (2023). Integrating artificial intelligence into science lessons: Teachers' experiences and views. *International Journal of STEM Education, 10*(1), 61. Advance online publication. DOI: 10.1186/s40594-023-00454-3

Perkins, K. K., Adams, W., Dubson, M., Finkelstein, N. D., & Wieman, C. E. (2010). PhET: Interactive simulations for teaching and learning physics. *The Physics Teacher, 48*(4), 225–230. DOI: 10.1119/1.3392045

Roco, M. C. (2003). Converging science and technology at the nanoscale: Opportunities for education and training. *Nature Biotechnology, 21*(10), 1247–1249. DOI: 10.1038/nbt1003-1247 PMID: 14520410

Roco, M. C. (2011). The long view of nanotechnology development: The National Nanotechnology Initiative at 10 years. *Journal of Nanoparticle Research, 13*(2), 427–445. DOI: 10.1007/s11051-010-0192-z

Sacha, G. M., & Varona, P. (2013). Artificial intelligence in nanotechnology. *Nanotechnology*, *24*(45), 452002. Advance online publication. DOI: 10.1088/0957-4484/24/45/452002 PMID: 24121558

Sarı, U., Duygu, E., Şen, Ö. F., & Kırındı, T. (2020). The effect of STEM education on scientific process skills and STEM awareness in simulation-based inquiry learning environment. *Journal of Turkish Science Education*, *17*(3), 387–405. DOI: 10.36681/tused.2020.34

Sweeney, A. E., Vaidyanathan, P., & Seal, S. (2006). Undergraduate research and education in nanotechnology: Development, implementation and evaluation of a Research Experiences for Undergraduates program in nanomaterials processing and characterization. *International Journal of Engineering Education*, *22*(1), 157–170.

Tawiah, B., Seidu, R. K., Fobiri, G. K., & Chichi, C. A. (2024). Nanotechnology and education: Preparing the next generation of future leaders. In George, S. C., & Tawiah, B. (Eds.), *Nanotechnology in societal development* (pp. 141–161). Springer., DOI: 10.1007/978-981-97-6184-5_7

Wang, N., & Lester, J. C. (2023). K-12 education in the age of AI: A call to action for K-12 AI literacy. *International Journal of Artificial Intelligence in Education*, *33*(2), 228–260. DOI: 10.1007/s40593-023-00358-x

Woolf, B. P. (2010). *Building intelligent interactive tutors: Student-centered strategies for revolutionizing e-learning*. Morgan Kaufmann.

Xu, W., & Ouyang, F. (2022). The application of AI technologies in STEM education: A systematic review from 2011 to 2021. *International Journal of STEM Education*, *9*(1), 59. Advance online publication. DOI: 10.1186/s40594-022-00377-5

Zawacki-Richter, O., Marín, V. I., Bond, M., & Gouverneur, F. (2019). Systematic review of research on artificial intelligence applications in higher education – where are the educators? *International Journal of Educational Technology in Higher Education*, *16*(1), 39. Advance online publication. DOI: 10.1186/s41239-019-0171-0

Zhai, X., & Krajcik, J. (2024). Artificial intelligence-based STEM education. In Zhai, X., & Krajcik, J. (Eds.), *Uses of artificial intelligence in STEM education* (pp. 1–24). Oxford University Press., DOI: 10.1093/oso/9780198882077.003.0001

Chapter 7
Educational Tools to Integrate AI and Nanotechnology in Future-Oriented Studies

Hemant Kumar Upadhyay
https://orcid.org/0000-0002-2102-1084
Poornima University, India

Rocky Kumar
https://orcid.org/0009-0002-6995-2037
Poornima University, India

Himanshi Sharma
https://orcid.org/0009-0007-8458-6836
Poornima University, India

Tushar Ghorpade
D.Y. Patil University, India

ABSTRACT

Artificial Intelligence (AI) and Nanotechnology are two of the transforming subjects in science, innovations, and industry, where there are far-reaching consequences in education. Given that such technologies are developing faster than ever before, it is crucial to incorporate them into the learning curriculum to produce future-oriented students. The chapter provides a thorough analysis of key concepts in AI and Nanotechnology, highlighting them and making them easily approachable to maintain conceptual clarity for both students and educators. It examines the potential of digital solutions to increase engagement, develop an understanding of

DOI: 10.4018/979-8-3373-2883-6.ch007

concepts, and bridge the gap between knowing and doing, or between theoretical and practical knowledge, using interactive applications, virtual simulations, and adaptive learning platforms. The chapter presents a perceptive look at the issues confronting teachers and schools in teaching advanced technology-based syllabi, such as a lack of money, uneven distribution of technological facilities, and the need to train instructors. Moreover, it also provides viable recommendations to policymakers, curriculum designers, and educators that they can use to promote equity, inclusivity, and successful learning. Using theoretical background and practice guidelines, the chapter provides a blueprint on how AI and Nanotechnology in education can be effectively utilized to foster innovation, critical thinking, and digital literacy in learners.

INTRODUCTION

The use of Artificial Intelligence (AI) and Nanotechnology is rapidly revolutionizing several sectors. There is an increasing need in these areas to develop new and effective educational tools to deliver fundamental knowledge, promote an interdisciplinary mindset, and equip learners with the ability to work in real-life contexts. (Chichi, n.d.). Traditional pedagogical approaches do not usually succeed in teaching advanced information in a way that is easy to understand and interesting. Cutting-edge education methodologies, such as simulations, virtual laboratories, interactive tutorials, augmented/virtual reality, coding, and Artificial Intelligence-driven tutoring solutions, are revolutionizing how students engage with these complex topics. Educational applications are becoming cross-disciplinary, helping students understand the interface between AI and nanotechnology by providing educational tools that bridge the gap between the two domains. (Ünlütabak et al., 2023). Such tools are also becoming very important to informal and lifelong learning settings, including those in online courses, interactive MOOCs, and learning platforms. This chapter reviews the available types of educational applications that can be used in AI and nanotechnology, evaluates their impact on learning outcomes, identifies best practices in their implementation, and explores future directions of innovation in education within these emerging fields. The aim is to provide educators, policymakers, and learners with an understanding of how education can keep pace with the seismic changes brought about by AI and nano, ensuring the next generation is well-prepared to play their role and lead in these game changers. (Díaz-Marcos & Mendoza Gonzálvez, 2025a).

These developments also bring with them the possibility of a global sense of urgency: education systems must rapidly change to prepare students not only to master these technologies but also to use them to innovate responsibly. AI and

Nanotechnology, as the drivers of the Fourth Industrial Revolution, are no longer luxuries but prerequisites for 21st-century competitiveness in a knowledge-based economy that requires high-level education. Such an education needs to be more interdisciplinary and adaptive, allowing students to apply advanced technical expertise in real, ethical, and socially engaged situations. (Husseiny, n.d.).

BACKGROUND

Nanotechnology and Artificial Intelligence (AI) have evolved as new formations of technological revolution, restructuring industries, education, and everyday life over the past few decades. The mix of AI involved in crafting systems that are motivated by human intelligence and Nanotechnology involved working on a very minute level, boasts of immense potential in introducing inventions in many areas (Hassan et al., 2023a).

The development of new pedagogical solutions and tools, including virtual labs, machine learning, and augmented reality (AR), is the force behind this change and is transforming how students learning AI and nanotechnology are being educated. In addition to this, adaptive learning systems offer students individual learning paths through which they can learn at their own pace according to their potential. This wave of innovation nonetheless brings along specific difficulties in the sense that there must be acceptable access to all students, and the teachers must be sufficiently equipped, in addition to finding appropriate ways of evaluating the progress of the students (Ofozoba et al., 2025).

In anticipation of such problems, schools have already established interdisciplinary study courses that incorporate the use of AI and Nanotechnology. This is meant to fill knowledge gaps and enable students to deal with the technologically oriented future. Obstacles, however, are still in the way, with the main ones involving the scalability, teacher readiness, and the difficulty of developing new evaluation mechanisms beyond memorization (Díaz-López et al., 2025).

Nanotechnology and Artificial Intelligence (AI) are the new forces that have transformed industry, education, and personal life over the past several decades. AI aims to develop systems based on human intelligence, while nanotechnology operates at the atomic and molecular levels to create new functions. All these sectors have enormous potential in the health care sector, manufacturing, education, and optimal development. (Zafar & Rana, 2025).

The invention of virtual laboratory units, flexible learning technology, and hyper-realistic environments, including the possibility of adding augmented reality (AR), is transforming the process of learning and teaching complex scientific ideas to students. Such innovations in learning enable customized learning trajectories that may enable students to learn at their own speed and at their own capacity. Even with these changes, there are issues of equitable access, training teachers in the skills they need, and the creation of new kinds of assessments that are not based on memorization (Dorouka & Kalogiannakis, 2024a).

Students are being prepared for a technologically oriented future by introducing interdisciplinary courses that include AI and nanotechnology. Nevertheless, the problems of scaling, regulatory compliance, cost-efficiency, and morality remain a drag. This chapter describes how convergent technologies can be used to create inclusive, active, and compelling learning experiences, and what the opportunities and challenges facing teachers, students, and policymakers are (Ghattas & Carver, 2012a).

LITERATURE REVIEW

The AI-nanotechnology nexus in education is an increasingly popular research topic, but there is still little evidence-based literature available. The current studies focus on the importance of immersive and digital applications in marketing the outcomes of the learning process. Moreover, meta-analyses of virtual and remotely operated lab utilization state that conceptual learning, retention, and equity scores are improved, making them objective and superior to physical laboratories (Wu et al., 2020; Potkonjak et al., 2022). The AR/VR research on STEM education also shows impressive results in retrieval skills (spatial) and interaction between the learners, but has inevitable trade-offs like cost, availability, and motion sickness (Radianti et al., 2020; Hamilton et al., 2021).

Adaptive learning using AI has proven helpful in personalizing learning as well. In controlled experiments, adaptive platforms were also found to have moderate and high-size ability-modulated difficulty and feedback effects on student performance (OECD, 2021; Khosrowavi et al., 2022). Nonetheless, the issue of algorithm bias, along with their fairness and transparency, introduces the necessity to implement humans as the leaders of the evaluation systems (Holmes et al., 2021).

Research in nanotechnology education has focused on awareness-building and conceptual learning through project-based learning and simulations. Despite this optimism, some studies would somehow not take into consideration the regulatory and safety factors. The biosafety policies in institutions also address the use of personal protective equipment (PPE), nanomaterials, and waste management as part of learning laboratories (NIOSH, 2020; OECD, 2023). However, a small number of research studies in education introduce such safety measures into the curriculum.

Another area that is not developed is the alignment of the workforce. The European Skills Agenda (EU, 2020) and the U.S. National Nanotechnology Initiative roadmaps (NNI, 2021) highlight how AI and nanotechnology education skills can be mapped to competencies in manufacturing, robotics, and device prototyping. Regardless of these policy directions, there is very little empirical research where workforce outcomes are correlated with classroom interventions.

1.1 Overview of AI and Nanotechnology in Modern Education

Artificial Intelligence (AI) and nanotechnology are two of the most advanced technologies in the world, making a significant presence in various domains such as business, healthcare, the environment, and education. These technologies are defining the future of education since they bridge the gap between knowledge of various disciplines and could create some significant transformations (Dorouka & Kalogiannakis, 2024b).

AI is explicitly taking on an increasing role in the education sector. It has not only become a subject to study but also a resource to improve the study process for students. Since AI is a model that replicates human-like thinking and learning capabilities, it contributes to elevating learning experiences and text learning processes, making them more specific and efficient for any student. The applications of AI in education include adaptive educational systems, auto-grading systems, and smart-tutoring systems, among other uses (Russell & Norvig, 2020).

Technologically speaking, Nanotechnology concerns the manipulation of matter on the atomic and molecular level, most usually between 1 and 100 nanometers (Sharma et al., 2024). A myriad of scientific domains, including physics, chemistry, biology, and materials science, are incorporated with nanotechnology. It educates the students to think interdisciplinarily and exposes them to new scientific ideas in the classroom. Early introduction to such things leads to a creative and research-oriented mindset, coupled with the development of curiosity. Therefore, more schools have been incorporating nanotechnology science into their curricula, with most opting to use interactive resources like lab sets, simulations, and visual models to highlight the intricacies of nanotechnology.

AI and nanotechnology provide access to inquiry-based, hands-on, and highly interactive learning (Roco, 2023a). Nevertheless, as they entail their technicalities, educators must incorporate the existing learning pedagogies into the learning experiences. Therefore, as these technologies continue to dominate the STEM education sector and cross-curricular programs, there is a demand for education that can be expanded within sight. (Sweeney & Seal (2018)

1.2. Importance of Educational Tools in Emerging Technologies

Introducing advanced scientific disciplines like AI and Nanotechnology into the mainstream education system would require the use of special learning tools. Such tools as virtual labs and simulations, machine learning environments, and nano kits promote the concept of experiential learning and understanding of complex concepts. These technological aids have proved successful in enhancing learning, improving student participation, critical thinking, and facilitating the real-life expression of theoretical knowledge (Chatterjee et al., 2021).

Table 1. Educational Tools Usage Ratio (AI vs. Nanotechnology)

Tool Type	AI Usage	Nanotech Usage	Ratio (AI: Nanotech)	References
Virtual Labs	1	1	1:1	Chatterjee et al., 2021; Klimeck et al., 2008
Gamified Platforms	1	0	1:0	Lee, 2019
3D Modeling Tools	0.5*	1	0.5:1	Sweeney & Seal, 2018; Humphrey et al., 1996
AI Programming Tools	1	0	1:0	Ghosh et al., 2020
Microscopy Simulators	0	1	0:1	Humphrey et al., 1996

Learning tools are essential for providing access to new technologies. They offer online and face-to-face study environments that favor different types of learning and reduce the reliance on lecture-led methods. An example is the Goggles AI technologies, such as TensorFlow Playground and Teachable Machine. The former allows users to play with neural networks using only a browser, with no prior programming skills required. Similarly, online nanotechnology facilities help learners observe nanoparticle interactions, which cannot be observed with the naked eye (Oliveira et al., 2019).

Moreover, technology will allow the acquisition of 21st-century skills (problem-solving, creativity, and digital literacy). It helps to build active interactions and collaboration by adopting gamified modules, AR, and project-based learning designs. Such methods not only increase motivation but also open up learning to underrepresented individuals in the STEM educational setting (Lee, 2019).

As an illustration, in the learning environment of South Korean classrooms, AI-powered discussion boards enable learners to discuss and explore solutions related to the applications of nanotechnology, aligning theoretical knowledge with social and ethical consequences. (Spector, 2013). Similarly, at the sharp end of the resource divide are affordable VR headsets in rural Nigeria to simulate worlds at the nanoscale. The involvement of these tools accelerates the pace of understanding and makes advanced science accessible, enabling students from diverse backgrounds to participate in next-generation industries effectively. The learning tools are also evolving to make learning multilingual and culturally competent, making them mandatory for global access. (Mendoza et al., 2024).

Table 2. Comparison of Educational Tools in AI vs Nanotechnology

Tool Type	Used in AI Education	Used in Nanotech Education	Purpose
Virtual Labs	Yes	Yes	Simulated experimentation and demonstrations
Gamified Platforms	Yes	No	Interactive and engaging learning
3D Modeling Tools	Rarely	Yes	Structure and nanoscale visualization
AI Programming Tools	Yes	No	Hands-on coding experience
Microscopy Simulators	No	Yes	Understanding atomic-level properties

Figure 1. AI & Nanotechnology Integration (Use-Case Diagram)

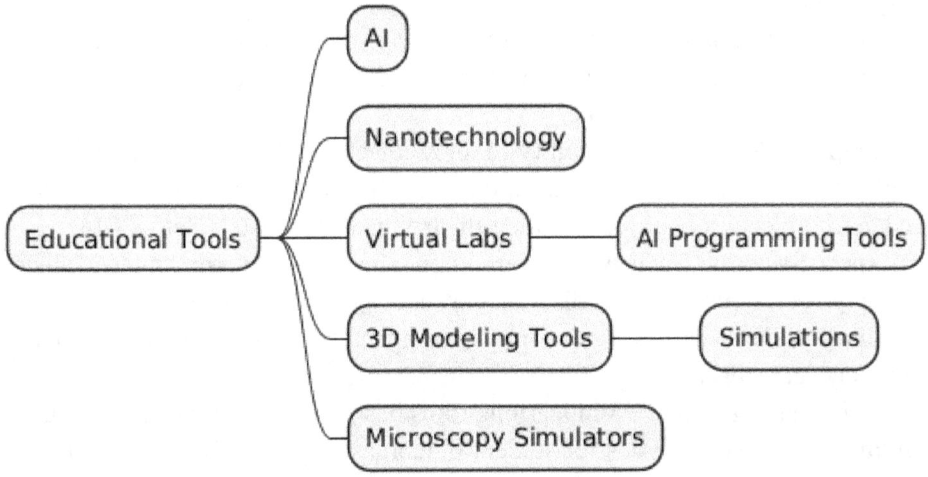

2. FOUNDATIONS OF AI AND NANOTECHNOLOGY

Before learning how Artificial Intelligence (AI) and Nanotechnology are used in education, one must know some of the main principles behind these technologies. Both of them rely on various scientific ideas, yet both are directed at innovation and problem-solving in complex fields. The main aspects that constitute the two disciplines are explained in this section and can serve as the theoretical basis for integrating them into the education process. (Roco, 2023b).

2.1 Key Concepts in Artificial Intelligence

Artificial Intelligence (AI) is an interdisciplinary field that draws inspiration from many different directions, including computer science, psychology, mathematics, and neuroscience. The central aim of AI is to develop systems or machines capable of performing tasks usually performed by human mental abilities. Among such activities, there is learning based on the past, making decisions and choices, problem solving, pattern recognition, reading and understanding of verbal and written words, and communication (Russell Norvig, 2020).

AI is based on the following subfields:

Machine Learning (ML): Another sector of AI where the system learns from data without a particular programming. Contemporary AI applications depend on ML algorithms, such as supervised learning, unsupervised learning, or reinforcement learning.

Natural Language Processing (NLP): This domain deals with computer-human language interaction. It makes it possible for applications such as chatbots, translators, and voice assistants.

Computer Vision: Is the ability to allow machines to understand and extract insights from pictures and videos captured by the camera from the real world, as needed for object recognition and the self-driving car industry(Neto, 2014).

Expert Systems: Such AI software intends to mimic human expert decision-making power, typically in medical diagnoses, engineering, or financial fields.

Robotics: Deals with designing intelligent agents that can be applied to problems in the physical world.

The increasing role of AI in learning tools has given rise to intelligent tutoring systems, predictive analytics for student outcomes, and adaptive content delivery platforms (Baker & Smith, 2019). These developments underscore the importance of learning AI basics as part of a contemporary education system.

2.2 Core Principles of Nanotechnology

Nanotechnology involves matter control and manipulation at the nanometer scale (1–100 nm). Materials with such dimensions possess distinct physical, chemical, and biological properties that are very unique and different from those of their bulk analog. Such nanoscale manipulation creates innovation in medicine, electronics, materials science, and environmental engineering (Roco & Bainbridge, 2005).

Key principles of nanotechnology include:

Quantum Effects: Material properties at the nanoscale are significantly dominated by quantum mechanics, including electrical conductivity, optical properties, and magnetic properties.

Surface Area-to-Volume Ratio: Nanoparticles have a substantial surface area-to-volume ratio and, therefore, are highly reactive and interactive with other substances.

Self-assembly: It is the process by which molecules autonomously arrange themselves into a logical structure according to specific chemical criteria(Sanders, 2018).

Nanoscale Fabrication: Techniques like molecular beam epitaxy and electron beam lithography enable very accurate material fabrication at the atomic or molecular level.

Education in nanotechnology emphasizes learning across disciplines, blending physics, chemistry, biology, and engineering principles. Computer-aided microscopy software, kits, and interactive models allow viewing nanoscale events in visually comprehensible ways (Sweeney & Seal, 2018). Teaching nanotechnology to students at the introductory level encourages instructors to enhance scientific curiosity and interest in innovation(Kulkarni, 2015).

3. EDUCATIONAL NEEDS AND CHALLENGES

There are many complex challenges as well as exciting new opportunities when integrating AI and nanotechnology into traditional education. Even if these fields open up new possibilities for scientific advancement, they also require innovative teaching methods, sufficient teacher preparation, and equitable access to state-of-the-art equipment for all students. This section presents key barriers to the successful teaching of these new fields. (He, 2025).

3.1 Pedagogical Challenges in Teaching AI and Nanotechnology

Teaching AI and Nanotechnology demands a departure from conventional pedagogy. Both these fields are highly interdisciplinary and draw impact from mathematics, computer science, engineering, physics, and life sciences, which differs from the conventional subject matter that progresses gradually and has more established disciplinary origins. Thus, instructors are challenged to craft structured curricula that remain up to date and pedagogically appropriate (Touretzky et al., 2019). One of the most significant challenges is the shortage of trained instructors with both subject-matter expertise and pedagogical proficiency. A large number of faculty members do not know how to make complex or technical information understandable to a heterogeneous student body, especially at the secondary or undergraduate levels. Further, such topics are significantly computational and conceptual in nature, rendering them hard to impart using lecturing methods alone. In the case of AI, students must understand algorithms, programming languages, and logic, whereas in Nanotechnology, they need atomic-scale chemistry and physics, which occasionally require sophisticated lab facilities or simulation software (Srikantaiah & Muralidhar, 2020).

Curriculum fragmentation is yet another obstacle. AI and Nanotechnology are taught as standalone electives or advanced courses, isolating students from their interdisciplinary interfaces. Without cross-module integration and practical applications, student engagement might decline.

Testing practices must also be reformed. Traditional tests are not sufficient to assess competence in project-based, hands-on fields. Instead, teachers must create authentic measures like simulations, coding exercises, or design-based learning portfolios.

Brazilian and South African educators are piloting “flipped classroom” methodologies to bring greater interactivity into nanotech and AI instruction in instances of tight lab space. The methods permit students to investigate simulations or demonstrations of AI from home before small-group discussions on them in the classroom. Organized training and curriculum integration must be provided to implement such methods, which most schools lack. Furthermore, the time constraints in rigid curricula typically do not allow teachers to incorporate interdisciplinary instruction or design-thinking-oriented learning, which are what make these subjects truly effective. (He, 2025).

Table 3. Pedagogical Challenges in Teaching AI and Nanotechnology

Challenge	AI	Nanotechnology
Lack of trained faculty	✓	✓
High equipment cost		✓
Curriculum fragmentation	✓	✓
Computational difficulty	✓	
Need for visual tools	✓	✓

3.2 Access and Equity in Technology Education

Access to quality education in AI and Nanotechnology is not uniformly distributed across geographies, socio-economic strata, and institutions. A vast digital divide between urban and rural schools, and between affluent and under-resourced institutions, particularly in developing countries (UNESCO, 2022), can impede the democratization of knowledge and exacerbate underlying educational inequalities.

Infrastructure constraints—such as the absence of high-speed internet, advanced computing infrastructure, and state-of-the-art laboratories—may prohibit schools from implementing AI tools or nanotechnology kits. Further, open-source platforms and learning materials are typically accessible only in dominant world languages, precluding students in linguistically diverse areas.

Gender inequalities also exist in STEM, with girls and women being underrepresented in AI and nanotechnology programs. Such inequalities tend to be perpetuated by stereotypes, societal expectations, and a lack of role models (UN Women, 2021).

To address access challenges, policymakers and educators must adopt **inclusive strategies**:

- Foster open-access digital platforms and multilingual resources
- Offer training programs for underrepresented students and teachers
- Form alliances with NGOs and tech firms to provide educational toolkits to marginalized communities
- Integrate diversity and ethics into the curriculum to make it relevant and inclusive

4. DIGITAL AND INTERACTIVE TOOLS FOR AI EDUCATION

Technology-based learning technologies have been pivotal in making Artificial Intelligence (AI) learning interactive and accessible. Not only do the technologies lower the entry point for students, but they also help teachers communicate complex concepts in terms of concrete, visual, and interactive experiences.(Alam & Mohanty, 2023). A comprehensive digital learning environment for AI includes online tutorials, simulation rooms, AI construction tools, game-based learning interfaces, and conversational agents like chatbots. (Nur Fitria, 2021).

Figure 2.

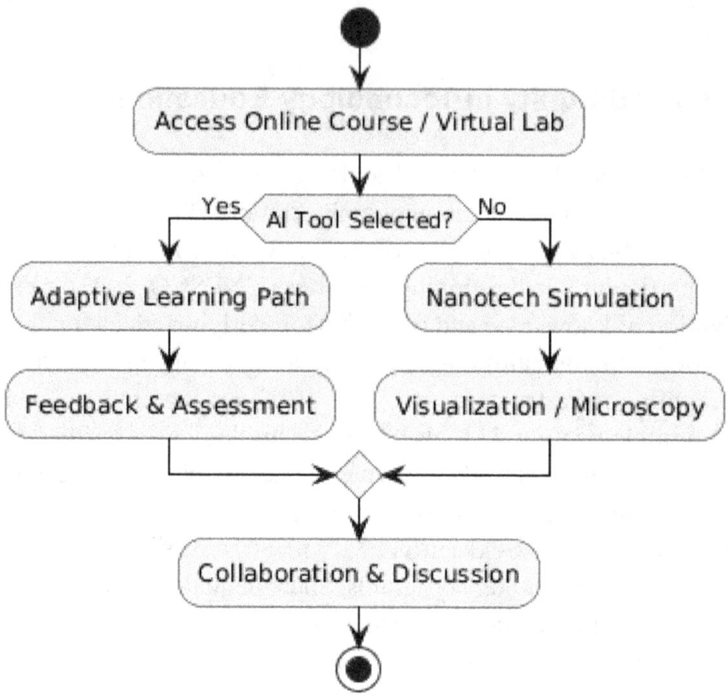

Figure 3. Flow of Digital Learning (Activity Diagram)

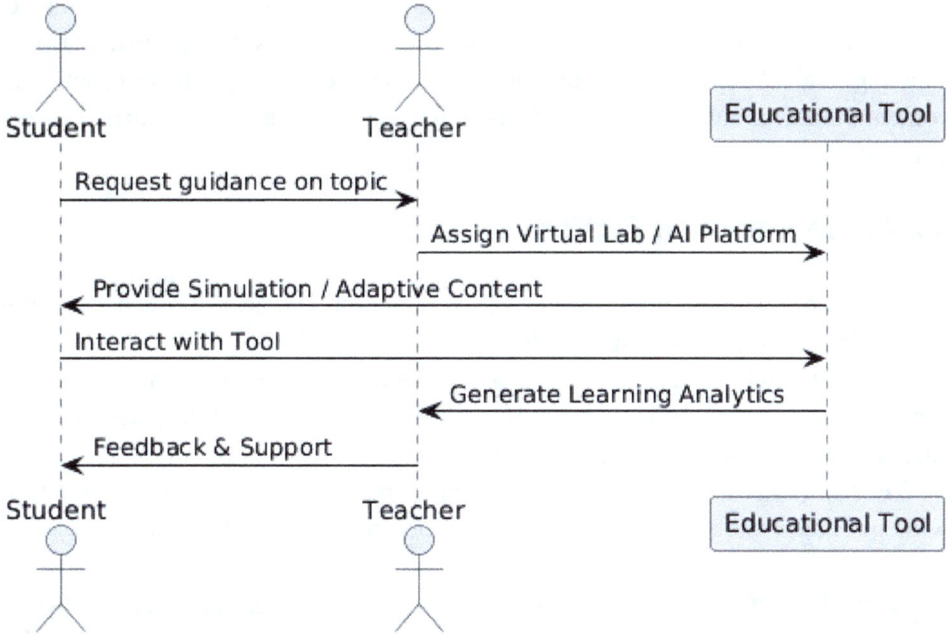

4.1 Online Courses, MOOCs, Simulation Tools

Large-scale Open Online Courses (MOOCs) and niche simulation packages have been a significant force for democratizing learning in AI. Sites such as Coursera, edX, and Udacity provide systematically designed, mostly university-branded courses on machine learning, deep learning, and data science. These are supplemented with video lectures, online quizzes, and capstone projects that enable users to implement abstract concepts in real-world applications (Jordan, 2015).

Simulation platforms facilitate the visualization and manipulation of AI algorithms without the need for advanced programming knowledge. Platforms such as TensorFlow Playground enable drag-and-drop exploration of neural networks, which makes it easier for students to grasp the effects of hyperparameters, layers, and activation functions on model behavior. (Alam & Mohanty, 2023).

These tools facilitate experiential learning, enabling students to move from abstract theory to concrete understanding. Moreover, they typically offer community discussion forums and peer review, enhancing collaborative learning and critical thinking. (Imamguluyev et al., 2024).

MOOCs are particularly powerful because they can enable peer-to-peer learning as well as global collaboration. Students from other continents can collaborate on

AI-based design challenges or nano-simulation lab work, gaining both technical skills and cultural insights. Also, higher-end MOOCs nowadays have artificial intelligence tutors who adapt pace and depth of content according to student participation—showing how AI can enhance education on AI. The resulting platform is not just a content platform, but a social environment for learning that self-optimizes(Yaseen et al., 2025).

4.2 AI Programming Platforms

For older students, coding websites offer collaborative spaces in which to develop and deploy AI solutions. A few of the best websites include:

Google Collab: Cloud-based Jupiter notebook where students can code and run Python programming code, especially with the use of machine learning libraries such as TensorFlow, PyTorch, and Scikit-learn.

Microsoft Azure ML Studio: A drag-and-drop tool for creating AI workflows without extensive coding, perfect for educators to introduce AI without programming overhead.

IBM Watson Studio: Allows students to develop, train, and deploy models based on a collection of AI capabilities, such as AutoAI and visual data exploration capabilities(Ganesan, n.d.).

They both offer unrestricted academic use and rich documentation, making them ideal for classroom instruction and self-study. They support a project-oriented learning mode, where students can work on actual problems like image classification, sentiment analysis, or recommendation systems (Ghosh et al., 2020).

They also facilitate the development of students' soft skills, such as teamwork, problem-solving, and debugging skills. For example, real-time code-sharing features in Google Collab foster collaborative work and joint learning, even from various countries. Instructors can monitor code construction, drop inline comments, and provide instant feedback—a necessary element to learn AI logic patterns. The integration of cloud-based technologies also reduces hardware limitations, allowing students to engage in high-level machine learning practice with just a typical internet connection.(Luckin & Cukurova, 2019).

4.3 Gamified Learning Platforms and Chatbots

To make AI learning more engaging for students, many educational tools use gamification—adding elements like rewards, challenges, and storytelling. Platforms such as AI Dungeon and Codable (designed for children) use interactive puzzles and story-based learning to teach the fundamentals of AI logic and algorithms in a fun and accessible way (Neumann et al., 2023).

Such bots, built into learning management systems (LMS), respond to students' queries, give hints on assignments, and provide individualized feedback. For instance, chatbots such as Jill Watson, which were created at Georgia Tech, performed well as virtual TAs for online classes by responding to thousands of student queries (Goel & Polepeddi, 2016).

By providing students with dynamic, tailored, and real-time learning environments, gamification and intelligent learning tools significantly enhance student learning. By preventing mental overload and adjusting the presentation of material to the learners' pace and progress, these techniques help students remember and comprehend AI principles more effectively. Such an adaptive method not only maintains students motivated but also facilitates greater learning and long-term retention (Hidayatulloh et al., 2021).

5. LAB SAFETY AND REGULATORY CHECKLIST FOR NANOTECHNOLOGY EDUCATION

Nanotechnology is an emerging field being implemented in educational institutions, but it must be done in strict compliance with safety, ethical, and regulatory protocols. Although virtual and simulation-based laboratories minimize most hazards, some activities in nanotechnology require dealing with chemicals, nanomaterials, and some specialized equipment. To provide a safe and legal learning environment, teachers and schools should establish a coherent safety and regulation system. The checklist below offers feasible advice on the design and implementation of nanotechnology educational laboratories. (Ahn et al., 2016):

5.1 Personal Protective Equipment (PPE)

- Wearing lab coats, safety goggles and gloves when performing all practical experiments.
- Respirators/special masks when working with nanomaterial powders or aerosols.
- Well-identified areas where PPE may be worn and removed.

5.2 Chemical and Nanomaterial Handling

- Only institution-approved nanomaterials have known safety data sheets (SDS).

- Label all chemicals and nanomaterials with the symbols of hazard.
- Store nanomaterials in airtight packages under desired conditions of temperature and moisture(Omari Shekaftik et al., 2022).

5.3 Waste Disposal and Environmental Safety

- Separate nanomaterial waste from general chemical waste.
- Dispose of according to institutional and local Environmental Health and Safety (EHS) requirements.
- Disposing of a document in a lab logbook to comply with the audit(Ghosh & Kumar, 2024a).

5.4 Supervision and Access Control

- Keep a ratio of student-instructor of 10:1 in wet labs.
- Only direct supervision of the trained students should be restricted to high-risk activities.
- Safety orientation and assessment prior to participation in the lab.

5.5 Institutional and Regulatory Approvals

- Obtain explicit permission for nanomaterials use by the Institutional Biosafety Committee (IBC) or its equivalent.
- Get a risk assessment done by the Environmental Health and Safety (EHS) office of the institution.
- Adhere to national requirements regarding chemical handling, waste disposal, and workplace safety(Ghosh & Kumar, 2024b).

5.6 Emergency Preparedness

- Keep a supply of spill kits, eyewash stations, and fire extinguishers in all laboratories.
- Educate trainers on the standard emergency responses, such as spill containment and first aid.

    Post emergency contacts and procedures in obvious places.(Vega-Baudrit et al., 2023).

6. EDUCATIONAL TOOLS FOR NANOTECHNOLOGY

Nanotechnology education is a special case because of the abstract scale, interdisciplinary material, and costly infrastructure needs. To fill these knowledge chasms, teachers are looking to digital and interactive resources that deepen understanding, support experimentation, and model real-world uses. The following section emphasizes major types of resources, virtual labs, microscopy simulation software, and 3D nanomaterial modeling packages, and discusses their pedagogical significance.

6.1 Virtual Labs and Simulation Software

Virtual laboratories are now imperative in providing realistic learning experiences without infrastructure. The virtual platforms provide learners with a chance to perform simulated nanoscale experiments and promote conceptual clarity using dynamic visualization.

Nano Hub is one of the most remarkable platforms of nanotechnology education since it gives web access to interactive simulations, scientific calculators, and systematic nanotechnology courses (Klimeck et al., 2008). With a user base of over 1.4 million people around the globe, Nano Hub facilitates most of the nano-scale simulations, including particle-wave duality, quantum tunneling, and transistor action (Lundstrom & Klimeck, 2009).

An issue of nanotechnology education described here is the importance of virtual labs that provide high-quality, cost-effective, and scalable laboratory experiences. Physical nanofabrication is too expensive and advanced, making virtual tools an alternative and easier option for resource-limited institutions.

Among the most popular ones, Nano Hub is an open-source platform providing over 500 learning modules and simulation tools in nanoscience and nanotechnology. It enables students to explore quantum dots, carbon nanotubes, and nanoelectronics without requiring that they write application programs or learn how to set up and control hardware (Klimeck et al., 2008). The other one is PhET interactive simulations that consist of graphic modules explaining the basic concepts of physics and chemistry employed in nanoscience(Kilani et al., 2018).

With such systems, students can vary variables, visualize at the nanoscale, and perform simulations in real time--facilitating inquiry-based and exploratory learning. Further, they facilitate access at any time of the day and are therefore suitable for the blended and distance learning model.

Virtual labs are being used to enable secondary schools in third-world countries like Kenya and Bangladesh to access experiments that would not be accessible due to the high cost of infrastructure. The partnerships with global initiatives like the Nano Hub or the Virtual Labs project of India have ensured that under-resourced institutions can deliver quality STEM education without incurring the financial burden of expensive machinery. This is what it requires to demonstrate the power of virtual simulation in making the educational playing field more global and preparing students with skills that meet international standards in the sciences.(Khan, 2020a).

6.2 Microscopy and Visualization Tools

Nanoscale principles may be difficult to learn because they are based on structures too tiny to be seen with regular microscopes. To help students learn the details, small educational tools in the form of simulators of Atomic Force Microscopes (AFM) and Scanning Electron Microscopes (SEM) are used. These electronic tools allow students to learn the surface aspects and performance of materials in a nanometer scratch, providing a more interactive and clearer perspective.(Khan, 2020b).

An example of such is in Open SEM and AFM Simulator, which has a realistic interface that simulates actual operations of microscopy. Such simulators teach students how to interpret the data produced by nanoscale imaging, make magnification settings, and identify atomic structures or surface defects.

VMD (Visual Molecular Dynamics) is another advanced platform that is widely used to visualize the dynamics and structure of biological nanostructures. It supports molecular modeling, trajectory analysis, and visualization of nanoscale systems, making it suitable for inclusion in nanobiotechnology courses (Humphrey et al., 1996).

These technologies enable making concepts such as molecular interactions, surface topology, and crystal structures much more tangible and feasible. They also provide practical experience in working with scientific programs- this is a handy set of skills in both research and industry.(Mandrikas et al., 2020a).

6.3 3D Modeling and Nanomaterial Simulators

Three-dimensional modeling software enables the students to construct, revise, and simulate the nanostructures in a computerized setting. Software like Avogadro, CheSketch, and Crystal Viewer in Nano hub software enable students to create atomic and molecular structures, analyze geometries, and predict chemical interactions.

These tools enable multi-scale modeling, allowing the switch between microscale interactions at the molecular level and bulk material properties. As an example, students may create a simulation of graphene sheets' stress response or allow nanoparticles to assemble themselves into a configuration of choice.(Adams et al., 2004).

Further, there are advanced quantum mechanics and molecular dynamics computation modules in Materials Studio and the LAMMPS (Large-scale Atomic/Molecular Massively Parallel Simulator). Although they are primarily used in research settings, educational-simplified versions of this software are more widely available for teaching.

These tools encourage spatial thinking, hypothesis-driven experimentation, and teamwork in their design and construction within school settings. Students gain a better understanding of the structure-property relations at the nanoscale, which is the most significant aspect of materials science and nanotechnology.(Moeck et al., n.d.).

7. INTEGRATED LEARNING ENVIRONMENTS

The disciplines of Artificial Intelligence (AI) and Nanotechnology are very intricate and intertwined, which cannot always be explored in their entirety within a typical educational setting. It requires contemporary learning environments that go beyond traditional teaching methods to facilitate the teaching of such subjects. Such environments are known as Integrated Learning Environments (ILEs), where the subjects overlap and where learners use virtual tools and immersive technologies. (Ghattas & Carver, 2012b).

Such types of environments aid in developing a more interactive and collaborative learning environment. In this section, the argument is that the overlap of subjects with flexible pedagogy and the supply of new forms of technology, such as augmented reality (AR) and virtual reality (VR), can positively impact learning AI and Nanotechnology. The value of creating courses, which can integrate ideas across disciplines and offer a more interactive and practical approach to learning, is also discussed in this section.(Curreli & Rakich, 2020).

Figure 4. Component diagram of an Integrated Learning Environment (ILE) combining AI-driven and nanotechnology-based educational tools.

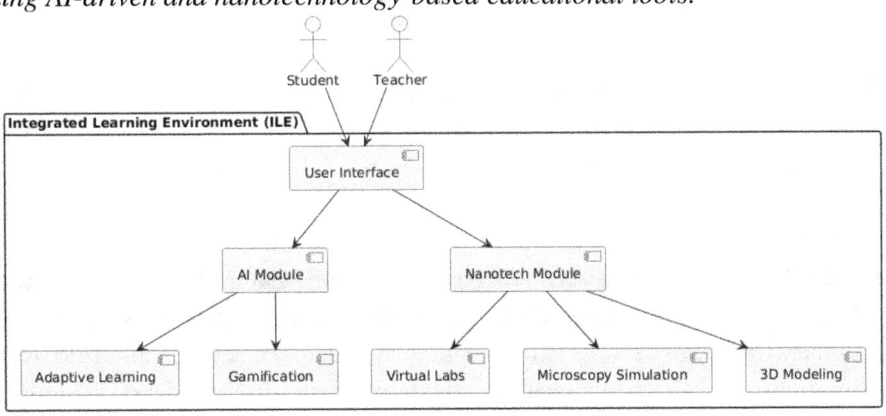

7.1 Interdisciplinary Curriculum Design

AIs and Nanotechnology are intrinsically interdisciplinary sciences and cross over traditional disciplines such as biology, chemistry, physics, computer science, and engineering. Traditional silo-type education systems have proved not to match the required dynamism of these subjects, and such systems will most likely result in disjunctive learning and less real-life utilization (Klein, 2017).

The intersection of interdisciplinary courses integrates knowledge and techniques from several disciplines, aiding students in developing a more comprehensive understanding and enhanced problem-solving skills. By infusing concepts from several fields, learners will be in a position to face problems from different perspectives. To offer a specific example, a course in AI applied to biomedical nanotechnology could have several components related to AI (such as neural networks, nanomaterial-based drug delivery, data analysis, and molecular modeling) as well as several components related to nanotechnology (such as biomaterials, nanofabrication processes nanoparticle fundamental properties), providing students with a theoretical and practical foundation in both fields as applied to aspects of biomedicine. It not only deepens concepts but also prepares students to complete collaborative research and cross-disciplinary innovation (Repko et al., 2020).

Effective interdisciplinary design requires:

- Modular learning structures allowing for flexible advancement
- Co-teaching frameworks that incorporate subject matter experts in multiple fields
- Project-based learning (PBL) focused on real-world impact
- Modules of ethics and policy to explore the societal consequences

These curricula foster creativity, innovation, and systems thinking needed to work in complex, technologically driven fields.

Universities like ETH Zurich and Singapore's NUS have constructed interdisciplinary “fusion tracks” where students can major in AI with a focus on nanotech applications, or nanotech with an emphasis on AI. These programs incorporate team-based projects where AI models forecast nanomaterial behavior, or where nanotech devices gather data to use in machine learning analysis. This kind of cross-disciplinary framework serves to make students realize the actual real-world overlap of disciplines, developing innovation, adaptability, and lifelong learning ability.

7.2 Hybrid Platforms for Teaching AI & Nanotech

Hybrid learning platforms integrate physical classroom interactions with online content delivery, providing a flexible and responsive learning model. These platforms are particularly well-suited for the instruction of AI and Nanotechnology, as they accommodate simulation-based learning, coding practice, and virtual experimentation in conjunction with in-person mentoring. (Lyshevski et al., 2006).

Examples of effective hybrid platforms include:

- **Moodle and Canvas**: Learning management systems that integrate AI coding labs and simulation modules
- **Google Classroom** with Jupiter Notebooks and Collab integration
- **Nano Hub and MIT Open Courseware,** which provide asynchronous content supported by real-time faculty interaction

Hybrid platforms provide for asynchronous exploration of challenging concepts, individualized feedback, and collaborative projects beyond traditional geographical locations (Garrison & Vaughan, 2008). They also ensure learning continuity in situations where direct access to lab resources is constrained (Hamimi et al., 2025). Research has confirmed that hybrid models substantially enhance learning outcomes in technical fields through repeated exposure to complex simulations and interactive exercises (Bernard et al., 2009).

7.3 Use of Augmented and Virtual Reality

Virtual and Augmented Reality technologies are transforming science and engineering education by allowing learners to be fully immersed in the learning process. For learning nanotechnology, VR enables students to “enter” molecular environments, see atomic structures, and manipulate nanoscale phenomena that are otherwise not visible. For AI, AR/VR can depict neural network architecture, data flows, and robotics simulations.(Kamali-Sarvestani et al., 2020).

Applications include:

- **Nano view VR:** Facilitates viewing of atomic lattices and crystal structures
- **Labster:** Provides VR-based nanotech experiments like nanoparticle synthesis or quantum tunneling
- **CoSpaces Edu:** Utilized for simulating AI-driven robots and algorithms in 3D spaces
- **Merge Cube AR:** Facilitates manipulation of molecular models through mobile devices

These tools help boost student interest, improve spatial understanding, and strengthen memory by involving multiple senses during the learning process (Makransky & Mayer, 2021). They also support inquiry-based learning, where students interact with detailed models, ask questions, and conclude hands-on exploration(Hemme et al., 2023).

8. WORKFORCE ALIGNMENT AND COMPETENCY MAPPING

Artificial Intelligence (AI) and Nanotechnology must meet the needs of modern workers by converging in the area of education. However, this convergence should be based not only on conceptual understanding but also correspond to the competencies needed in the contemporary workplace. With the increasing use of AI and nanoscale technologies in industries such as healthcare, manufacturing, autonomous systems, and advanced materials, educational outcomes should align with global and national skill maps to ensure employability and professional preparedness. (Friedersdorf, 2020a).

8.1 Industry Skill Needs

- Skills connected to AI include data analysis, developing machine learning, processing natural language, and ethically using AI.
- Skills connected to nanotechnology include nanoscale fabrication, material characterization, safe handling of nanomaterials, and solving problems across disciplines (Yawson & Greiman, 2017a).
- Skills that cross disciplines include systems thinking, being able to use simulations, working together on projects, and being able to use digital tools (Yawson, 2010).

8.2 Competency Frameworks

- Several established frameworks can guide alignment between curriculum and workforce needs:
- National Science Foundation (NSF) Advanced Technological Education (ATE) model - technical workforce training (Yawson & Greiman, 2017b).
- European Skills, Competences, Qualifications and Occupations (ESCO) - mapping skills on occupational demand.
- NIST Nano-EHS Competency Guidelines - safe and ethical management of nanomaterials.

- Industry 4.0 skill areas of the World Economic Forum - Robotics, IoT, edge AI, and nanofabrication (Yawson, 2012).

8.3 Educational Implementation

- Curriculum design: every learning outcome must be connected with one or several workforce competencies (e.g., a VR-based nano-lab activity → "materials characterization competency).
- Assessment mapping: Rubrics need to measure both theoretical knowledge and practice of professional skills (e.g., breaking safety rules, using AI tools to diagnose).
- Industry networks: Student competencies need to be tested in the field through internships, apprenticeships, and joint projects.

8.4 Benefits of Alignment

- Improves the employability of graduates through linking learning with real job positions.
- Gives specific standards to teachers and policymakers.
- Rationalizes investment in AI and nanotechnology education as an economic and labor source.

9. EVALUATION AND ASSESSMENT TOOLS

As Nanotechnology and Artificial Intelligence become core topics for tertiary education, conventional methods of testing often fail to capture the full extent of what students know. More than mere recitation of facts, these disciplines include hands-on skills, sophisticated ideas, and bridging connections among different bodies of ideas (Ghattas & Carver, 2012c). Thus, tests need to change to better measure not only what the students know but also how competently they can utilize knowledge and make connections across different bodies of knowledge. This section discusses how AI-driven assessments, adaptive feedback systems, and cross-disciplinary models are transforming educational evaluation in these multidisciplinary domains. (Mandrikas et al., 2020b).

9.1 Learning Analytics and AI-Based Assessments

Learning analytics uses data from online learning systems to track students' behavior, activity, and patterns of performance. In the teaching of AI and nanotechnology, these analytics can monitor students' learning pathways through simulations, coding interfaces, and virtual laboratories to identify conceptual roadblocks or disaffection(Ajayi & Pub, 2024).

AI-based tests utilize algorithms to assess student performance in real time. Grade scope, Coursera's auto-grading tools, and Edmentum utilize machine learning to mark open-ended responses, code, and design work effectively and uniformly (Baker & Inventado, 2014). These tools not only lighten the workload of instructors but also provide timely feedback—a critical success factor for technical courses(Sajja et al., 2025).

Besides, natural language processing (NLP) is applied to assess written reflections, research reports, and lab journals. AI can analyze structure, argument, and content relevance in student work and provide summative scores and formative feedback (Roll & Wylie, 2016).

These analytics-based evaluations make remediation individualized, pointing out misconceptions early and enabling instructors to make course strategy changes accordingly. (Hassan et al., 2023b).

9.2 Adaptive Learning and Feedback Systems

Adaptive learning systems modify course materials and quizzes based on each student's learning style, pace, and level of knowledge. This customized method is particularly useful for subjects such as AI and nanotechnology, where students tend to have different levels of knowledge within different topics. (Díaz-Marcos & Mendoza Gonzálvez, 2025b). These tools, such as Knewton, Smart Sparrow, and Squirrel AI, use student response data to personalize the order, difficulty, and manner of content delivery, enabling learners to focus on their areas of weakness and develop a stronger overall understanding (Pane et al., 2017).

For instance, a student who is having trouble with nanomaterial simulations can be required to study molecular bonding concepts first before moving to lab assignments. (Hue-Shin & Dine, 2025). Feedback systems within adaptive systems are intelligent and operate in real-time. They not only tell learners what is wrong but also why, leading them to correct thinking. With project-based evaluations, AI-based feedback can review design phases, code reasoning, and even mimic experimental outcomes for comparison (Woolf, 2010). These systems are conducive to competency-based learning, where learners advance by proving mastery—more than seat time—making them well-suited for modular, interdisciplinary learning settings (Haq et al., 2025).

9.3 Cross-Disciplinary Innovations and Future Tools

Future evaluation methods in AI and nanotechnology teaching will likely integrate augmented reality (AR), gamification, and blockchain to enhance credibility and engagement. For instance:

Gamified tests in AI-driven simulations can check problem-solving skills in actual nanotech or AI-driven settings.

AR-enabled quizzes enable learners to explore interactive models of molecules or neural networks and respond to spatial reasoning queries.(Jeffcoat et al., 2024).

Blockchain-based certificates can securely preserve competency records and micro-credentials, which are beneficial for use in modular, skill-focused learning environments (Sharples & Domingue, 2016).

Cross-disciplinary assessment frameworks are being developed as well. For example, STEAM rubrics measure creativity, systems thinking, and technical knowledge together, providing a more comprehensive evaluation of AI-nanotech integration projects (Henriksen, 2014).

In the future, tools will prioritize authentic assessment, gauging students' capacity to translate knowledge across subjects and address genuine real-world challenges—going beyond memorization for the sake of memorization and instead concentrating on application, design, and innovation.(Van de Voorde & Fitzpatrick, 2019).

10. FUTURE RESEARCH DIRECTIONS

As Artificial Intelligence (AI) and Nanotechnology increasingly transform world industries, their applications in education will evolve as well. Future research will need to address the pedagogical, technological, and equity issues surrounding the instruction of these advanced fields. The following section identifies emerging areas of research that can have a substantial impact on instructional methods, learning environments, and student outcomes in AI and nanotechnology education.(Yıldız et al., 2025).

Future frameworks will also need to consider the addition of neurotechnology as an auxiliary sector in addition to AI and Nanotechnology. Brain-computer interfaces powered by AI are most likely to become a vital educational tool, possibly enabling direct neural feedback regarding cognitive load, emotional interest, and knowledge retention. Research in this area has the potential to create ultra-personalized educational tools.(Tawiah et al., 2024). It becomes increasingly important to research how AI-generated learning materials impact critical thinking and media literacy in learners—a measure that is generally disregarded in tech-focused models of education.(Díaz-Marcos & Mendoza Gonzálvez, 2025c)

10.1 AI-Powered Pedagogical Models

One promising area of research is developing pedagogical theories grounded in AI that can model and predict student learning patterns across interdisciplinary settings. Emerging technology may include affective computing to assess emotional engagement and cognitive modeling to tailor interventions (D'Mello & Graesser, 2015) dynamically. The frameworks can establish innovative, compassionate learning environments that adapt not only to knowledge deficits but to motivation and emotions as well.

10.2 Integration of Quantum Computing in Curricula

With the evolution of quantum computing, interest in its overlap with AI and nanotechnology is growing. Curricula of the future might necessitate students to work with quantum algorithms, qubit modeling, and quantum machine learning (QML). Studies must be conducted to create age-suitable learning paths that expose quantum concepts early using gamified or visual models (Cao et al., 2019)

10.3 Data Ethics and Responsible Innovation

Both AI and nanotechnology pose fundamental ethical issues—privacy, safety, and bias—and future learning resources need to include ethics-by-design. Researchers will need to research how to build ethical reasoning structures, inclusive design principles, and regulatory literacy into both learning materials and assessment materials (Binns, 2018). There will be a need for multidisciplinary teamwork between ethicists, educators, and technologists.

10.4 Immersive and Haptic Learning Interfaces

Aside from AR/VR, future studies will probably revolve around haptic interfaces and multisensory environments that mimic the tactile qualities of nanomaterials or the dynamic behavior of AI. These instruments may further advance kinesthetic learning and accessibility of children with disability (Minogue & Jones, 2006).

10.5 Learning Analytics and Longitudinal Studies

There must be longitudinal studies to assess the effects of AI and nanotechnology teaching on career trajectories, innovation capacity, and public value. Researchers should employ longitudinal learning analytics to monitor knowledge retention, skill acquisition, and ethical consciousness over years, rather than semesters.

10.6 Global Collaboration and Open Innovation

Global research should prioritize developing open-source learning platforms that help close the gap between more developed (Global North) and less developed (Global South) regions. To make education in AI and nanotechnology more inclusive, these platforms need to be adaptable to local needs. This includes providing material in different languages, creating mobile-compatible applications in regions with low infrastructure, and offering culturally relatable examples and topics that appeal to local people (UNESCO, 2021).

11. GLOBAL CASE STUDIES AND IMPLEMENTATIONS

Since most schools globally have recognized the importance of Artificial Intelligence (AI) and Nanotechnology in education to equip students with future-readiness, real-life applications have started emerging on a large scale. Such case studies are an expression of best practices, pitfalls, and situational solutions that have made AI and nanotechnology accessible to various learners residing in divergent geographies.

Table 4. Global Case Implementations of AI and Nanotechnology Education

Country	AI Implementation	Nanotech Implementation	Unique Feature
India	CBSE curriculum, AI for Youth	Virtual lab kits in IITs	Community-based innovation
USA	AI4K12, Chatbots, MOOCs	NNCI virtual access	Integration with maker spaces
Finland	'Elements of AI' in schools	Early curriculum exposure	No-code AI learning
Germany	Ethical AI initiatives	NanoTruck project	Mobile outreach lab
South Korea	AI debate platforms	Nano VR in rural schools	Moral engagement through tech

11.1 India: Integrating AI and Nanotech in CBSE and Higher Education

In India, it is optional and taught in Classes 9th to 12th since 2019 by the Central Board of Secondary Education (CBSE). This program is handled with the help of AICTE, Intel, and IBM. The curriculum aims to provide gamified experiential learning through modules on understanding open data sets, AI, and Python programming for novices. In order to generate interest and relevance, the AI curriculum integrates practical themes such as traffic security, environmental sustainability, and language translation (CBSE, 2020).

Nanotechnology has been introduced into undergraduate engineering curricula at some institutes in India, including the IIT Bombay and Amity University. Students at the Amity Institute of Nanotechnology practice nanoscale phenomena virtually using software tools like Nano Hub before having the opportunity to operate costly physical instrumentation. Likewise, IIT Kanpur's "Nanoscience Outreach" program offers high school teachers free simulation packages and training modules for classroom application. One of the significant achievements is Intel India's "AI for Youth" program, which enabled 1,000 teachers and 10,000 students in 12 states. The scheme promotes student-designed AI solutions based on the community, such as innovative irrigation systems and traffic controllers.

11.2 United States: AI Integration through Public School Curriculum and Maker Spaces

The United States has seen excellent integration of AI and nanotechnology, particularly in K-12 and undergraduate schooling. Groups such as AI4K12, funded by the National Science Foundation (NSF), created five grand ideas of AI to help primary and secondary schooling. Perception, Representation, Learning, Natural Interaction, and Societal Impact are among these (Touretzky et al., 2019). Several public high schools across states such as California, Texas, and New York have AI integrated in STEM education with platforms such as Google Teachable Machine, MIT App Inventor, and Scratch AI extensions. The students construct voice assistants, chatbots, and basic machine learning models woven within real-life contexts.

Nanotechnology training in the United States is backed by NNCI (National Nanotechnology Coordinated Infrastructure), including institutions such as Stanford, Cornell, and Northwestern. The universities provide remote access to nanofabrication facilities, virtual SEM simulators, and gratis curriculum modules for high school science instructors. An excellent example of implementation is the "Nano@Stanford" virtual lab outreach, where students from across the country can remotely access real instrumentation to observe nanomaterials and experiment with fabrication methods. Additionally, maker spaces and community colleges provide AI/Nano bootcamps on Arduino-based sensors, Raspberry Pi, and open-source platforms such as TensorFlow Lite. These bootcamps are particularly in demand from underrepresented and rural communities to lower STEM barriers.

11.3 Europe: Cross-National Collaboration and Early Introduction

European nations have prioritized cross-border education cooperation to harmonize the learning of AI and nanotechnology under the Erasmus+ and Horizon Europe programs. In Finland, AI learning begins as early as age 10. The "Elements of AI" course of the University of Helsinki has been taken by more than 750,000 learners worldwide and is included in Finland's national curriculum. It covers the fundamental concepts of machine learning, bias, ethics, and human-AI interaction—without the need to write code. In Germany, the Nanoparticles in Schulunterricht (Nano in School) program equips high school students with lab kits, AR tools, and 3D modeling software to investigate nanoscale materials. The program, funded by the Federal Ministry of Education and Research (BMBF), focuses on the safe and social handling of nanomaterials. The Alan Turing Institute is spearheading the UK's AI Roadmap for Education, which aims to develop inclusive and ethical AI solutions for classroom use. Through initiatives like Teen Tech AI and STEM Learning AI competitions, students participate in interdisciplinary activities to develop solutions that benefit society.

11.4 Common Success Factors Across Countries

Throughout these geographies, some shared success drivers arise:

Localized Content: Examples and data align with the learners' actual contexts (Indian agriculture, European health, U.S. smart home).

Partnerships: Governments, universities, NGOs, and technology companies join forces to share open resources, teacher training, and platform access.

Hands-On Learning: All successful deployments feature lab simulations, coding exercises, and project-based evaluation.

Inclusivity and Ethics: Initiatives are incorporating digital equity, data privacy, and AI bias awareness into learning systems.

11.5 Challenges and Considerations

There are various Challenges and Considerations to follow up:

- **Scalability**: Due to financial limitations or linguistic barriers, the majority of potential initiatives are still only available as pilot programs, which prevents widespread adoption.

- **Teacher Training**: The vast majority of educators lack the confidence or feel unprepared to teach abstract, multidisciplinary subjects like artificial intelligence and nanotechnology.
- **Assessment Gaps:** Current evaluation systems often emphasize rote memorization rather than assessing creative thinking, practical application, or interdisciplinary understanding.

To overcome these hurdles, efforts are being made to develop adaptive learning tools, provide policy-level backing, and create open-access repositories for curriculum materials. These initiatives aim to make advanced STEM education more inclusive, flexible, and future-ready.

Table 5. Trade-off Box 1: Virtual Reality (VR/AR) in STEM Education

Pros	Cons	Cost/Support Drivers
Enhances spatial understanding and immersive learning	High equipment cost (VR headsets, PCs)	Hardware procurement and maintenance
Increases student motivation and engagement	Motion sickness, accessibility barriers	Teacher training requirements
Enables remote/virtual lab experiences	Requires strong bandwidth & technical support	Reliable internet and IT staff

Table 6. Trade-off Box 2: Wet-labs vs Virtual Labs

Wet-labs (Hands-on)	Virtual Labs (Digital)
Real-world skill practice (chemical handling, instrumentation)	Safe, low-cost, scalable
Develops tactile and procedural skills	Accessible anytime, anywhere
Requires regulatory approval, lab infrastructure, and PPE	Limited exposure to physical handling
High recurring cost (consumables, waste disposal)	One-time development/maintenance cost

Table 7. Trade-off Box 3: AI-based Assessment Systems

Pros	Cons	Cost/Support Drivers
Automated grading saves time.	Risk of algorithmic bias	Need for bias audits & human oversight
Adaptive tests personalize learning.	Lack of transparency in scoring	Development of explainable models
Scales easily to large classrooms	Over-reliance may reduce teacher-student interaction	Ongoing software licensing & updates

12. POLICY AND INDUSTRY COLLABORATION FOR SCALING EDUCATIONAL TOOLS

The simultaneous and widespread availability of AI and Nanotechnology teaching resources requires pedagogic creativity and firm structural support from the public and business sectors.(Friedersdorf, 2020b). Technology can create new instruments and platforms, but their utilization and effects will be limited unless policy, finances, and inter-sector coordination deliberately support them. This chapter focuses on the role of government policies, active industry-led public-private partnerships, and strategic public-private collaborations to create sustainable, scalable education spaces for Nanotechnology and AI(Marcos, n.d.).

12.1 Government Policies and National Initiatives

Educational policy plays the key role in determining how future technologies are taught and utilized across learning facilities. In the last couple of years, several countries have instituted national programs to support education in AI and Nanotechnology. For example, India's National Education Policy (NEP) 2020 prioritizes digital education, vocational training in emerging technology, and an interdisciplinary curriculum, aligning with the integration of AI and Nanotechnology into secondary and higher education (Ministry of Education, 2020).

At the same time, the Digital Education Action Plan prepared by the European Commission (2021-2027) encourages spending on digital infrastructure, educating teachers, and collaborating in research to promote the use of technology in classrooms (European Commission, 2020). In the UK, USA, Canada, and Singapore, in circumstances stressed by national approaches in the field of AI, there is an emphasis on human-friendly AI, workforce upskilling, and fair science, technology, engineering, and mathematics (STEM) education (OECD, 2021), with an indirect promotion of AI tool development in an educational environment.

Well-defined implementation roadmaps should be developed to achieve the effectiveness of these policies, which should lay down:

- Tool development funding allocations
- Schedules for modernization of the curriculum
- Incentives for the adoption of modules of emerging technologies by schools and universities
- Teacher upskilling requirements
- Designing evaluation frameworks for tools and results

The integration of AI and Nanotechnology tools should also consider international development agendas, such as UN Sustainable Development Goal 4: Quality Education, and address issues of inclusivity and equity concerning access to high-quality STEM education.

12.2 Public-Private Partnerships in Education Technology

Public-private partnerships (PPPs) offer a significant means of combining state scrutiny and personal investment in the hands of a business association. With PPPs, schools and universities can access new AI and nanotech toolkits, software packages, training modules, and mentoring initiatives.

In India, the AI for Youth Program by Intel, in collaboration with the Ministry of Education and CBSE, has enabled thousands of students to build AI-driven community solutions. Similarly, IBM AI Curriculum for Schools, in association with the United States Department of Education, offers online resources, virtual lab access, and project-based learning frameworks to schools in underprivileged communities (Intel, 2021; IBM, 2022).

Examples of PPP-based models are:

- **EdTech Sandboxes**: Where AI firms test educational software in concert with schools, sharing real-time feedback and optimizing algorithms.
- **Joint Certification Programs:** University-industry partnerships for co-crafting short-term nanotech or AI courses, tested by both entities for use in industry.
- **Innovation Labs and Hackathons:** Students engage in hands-on problem-solving with state-of-the-art technologies such as nanotechnology and artificial intelligence during innovation labs and hackathons, typically guided by industry experts. These types of platforms foster innovation, collaboration, and application of knowledge in real-life situations.

PPPs can be cost-saving, promote tool co-design, and provide field-tested content. However, such collaboration must be operated ethically to ensure student data privacy, avoid commercialization of the classroom, and uphold academic impartiality.

12.3 Industry-Academia Collaboration Models

Deep collaboration with industry and academia ensures education is conducted using tools reflective of contemporary industry practice. Technology continues to evolve, but universities often struggle to keep up with curriculum updates or expose students to practical systems. Collaboration with industry bridges this gap by exposing students to case studies, internal software, internships, and guest lectures(Vetrivel et al., 2025). The Capstone Model is an effective model where engineering or applied sciences third-year students apply AI and nanotechnology tools provided by collaborating companies to resolve actual business problems. These projects enhance student skills at the same time and offer companies innovative concepts as well as talent identification (Esangbedo et al., 2023).

Some of the other collaboration formats are:

- **Faculty Immersion Programs:** Where teachers spend time with technology companies in order to learn about tool development and bring best practices back home.
- **Sponsored Research Chairs:** Where businesses sponsor a professor or research group dedicated to education tool innovation.
- **Virtual Labs from Industry:** E.g., Bosch, Siemens, and Samsung, providing simulation tools to engineering colleges.

Such interactions also allow firms to shape curriculum trends, such that graduates possess employment-ready skills in AI modeling or nanofabrication methods.

12.4 Policy-Level Barriers and Recommendations

Despite best efforts, substantial obstacles to growing industry-university collaborations persist:

- Insufficient regulatory guidance regarding EdTech collaborations in public institutions.
- Insufficient funding for inter-sector research, especially in non-metropolitan or resource-constrained institutions.
- Privacy issues of student data, particularly if they engage with AI-based assessment software (Akinsipo (Oyelaja) & Anselm, 2025).
- Resistance from mainstream educators who tend to perceive technology as disruptive instead of complementary.

For the accomplishment of these challenges, the following are proposed:

1. **Policy Alignment:** Provincial and national education policy needs to be integrated with industrial policy, with room for adaptive relations and experimental methods.

2. **Tool Accreditation Framework:** Governments and accrediting agencies need to define standards for learning tools so that their reliability, availability, inclusiveness, and pedagogic content can be assessed.
3. **Incentivizing Innovation:** Relief from taxes, research grants, and ranking rewards can encourage universities to invest in AI/nano education infrastructure.
4. **Local Innovation Ecosystems:** Foster the creation of AI/Nano hubs or incubators in universities wherein industry can partner with academia to create education tools appropriate to local needs.
5. **Open-Source Resource Banks:** Encourage institutions to open-source co-developed tools for broader use(Berardi et al., 2024).

12.5 Success Stories of Collaboration

Several international case studies illustrate the value of industry-academic collaboration:

- **Stanford's Bio-X Program:** Interdepartmental research center combining AI, nano biotech, and education to create new learning tech and spin-offs.
- **MIT's J-WEL (Abdul Latif Jameel World Education Lab):** Works with governments and industry to create AI/nanotech-enabled curricula for low-income nations.
- **Germany's Nano Truck Initiative**: An exhibition, lab-on-wheels-based truck set up by German federal ministries of industry and education to bring nanotechnology education to distant schools (BMBF, 2019).

Such success stories demonstrate that increasing access to quality and contextualized learning content to support the development of future-proof competencies can be achieved through government-industry-academia partnerships.

CONCLUSION

Artificial Intelligence (AI) and Nanotechnology are two important fields whose convergence marks a revolutionary period in scientific and technological progress. Learning resources are quite important in promoting learning and creativity in these disciplines. Learning algorithms, data structures, and predictive models have been revolutionized over machine-learning platforms, coding environments, and data-visualization software (Şenocak et al., 2021). Simulations, 3D modeling, and virtual laboratories have improved nanotechnology education and encouraged interdisciplinary learning and cross-field collaboration. (Bauer, 2021). Nevertheless, developing and introducing educational tools remains a complicated process due to issues of accessibility, cost, and the necessity for constant curriculum updates. Democratization of the advantages of AI and nanotechnology relies on inclusivity and increased access to high-quality educational material. Nanotechnology and AI tools for instruction are promising for the future, as the introduction of Augmented Reality, Virtual Reality, and Mixed Reality is pioneering the learning world on a whole new level. The process of learning can be brought to a new level of depth and retention by using AI-based personalized learning tools, and interactive storytelling can help make the abstract concepts of nanotechnology more interesting (Mandrikas et al., 2020c). The possibilities require interdisciplinary cooperation between educators and technologists, policymakers, and industry stakeholders to be realized. Research, development, teacher training, and infrastructural investment will play a critical role in multiplying the reach and effectiveness of education tools (Pelleg et al., 2011).

REFERENCES

Adams, J. D., Rogers, B. S., & Leifer, L. J. (2004). Microtechnology, nanotechnology, and the scanning-probe microscope: An innovative course. *IEEE Transactions on Education*, *47*(1), 51–56. DOI: 10.1109/TE.2003.817623

Ahn, J. J., Kim, Y., Corley, E. A., & Scheufele, D. A. (2016). Laboratory Safety and Nanotechnology Workers: An Analysis of Current Guidelines in the USA. *NanoEthics*, *10*(1), 5–23. DOI: 10.1007/s11569-016-0250-9

Ajayi, A., & Pub, A. (2024). Creating a Conceptual Framework for AI-Powered STEM Education Analytics to Enhance Student Learning Outcomes. *Journal of Frontiers in Multidisciplinary Research*, *05*, 157–167.

Akinsipo (Oyelaja), O. B., & Anselm, O. H. (2025). Challenges and Opportunities for Implementing Green Chemistry in Nigerian Universities: Educational and Policy Perspectives. *Sustainability & Circularity NOW, 02*. https://doi.org/DOI: 10.1055/a-2534-1903

Alam, A., & Mohanty, A. (2023). Educational technology: Exploring the convergence of technology and pedagogy through mobility, interactivity, AI, and learning tools. *Cogent Engineering*, *10*(2), 2283282. DOI: 10.1080/23311916.2023.2283282

Baker, R. S., & Inventado, P. S. (2014). Educational data mining. In Larusson, J. A., & White, B. (Eds.), *Learning analytics: From research to practice* (pp. 61–75). Springer.

Baker, T., & Smith, L. (2019). *Educ-AI-tion Rebooted?* Nesta. https://www.nesta.org.uk/report/education-rebooted/

Bauer, J. (2021). Teaching Nanotechnology through Research Proposals. *Journal of Chemical Education*, *98*(7), 2347–2355. DOI: 10.1021/acs.jchemed.0c01251

Berardi, C., Antonini, M., Jordan, Z., Wechtler, H., Paolucci, F., & Hinwood, M. (2024). Barriers and facilitators to the implementation of digital technologies in mental health systems: A qualitative systematic review to inform a policy framework. *BMC Health Services Research*, *24*(1), 243. DOI: 10.1186/s12913-023-10536-1 PMID: 38408938

Bernard, R. M., Borokhovski, E., Schmid, R. F., Tamim, R. M., & Abrami, P. C. (2009). A meta-analysis of blended learning and technology use in higher education: From the general to the applied. *Journal of Computing in Higher Education*, *21*(3), 95–123.

Binns, R. (2018). Fairness in machine learning: Lessons from political philosophy. In *Proceedings of the 2018 Conference on Fairness, Accountability and Transparency* (pp. 149–159).

Cao, Y., Guo, Y., & Liu, Y. (2019). Quantum machine learning: A comprehensive review. *Nature Communications, 10*(1), 1–13. PMID: 30602773

CBSE. (2020). *Curriculum for Artificial Intelligence: Class IX–XII*. Central Board of Secondary Education.

Chatterjee, S., Nath, A., & Ghosh, S. K. (2021). Leveraging AI-driven educational tools for digital transformation in higher education. *Education and Information Technologies, 26*(3), 3297–3315. DOI: 10.1007/s10639-020-10392-2

Chichi. (n.d.). *Nanotechnology and Education: Preparing the Next Generation of Future Leaders | SpringerLink*. Retrieved September 2, 2025, from https://link.springer.com/chapter/10.1007/978-981-97-6184-5_7

Curreli, M., & Rakich, S. S. (2020). An integrated approach to teaching and learning nanotechnology: The Omni Nano model. *Journal of Research in Innovative Teaching & Learning, 13*(1), 141–145. DOI: 10.1108/JRIT-02-2020-0012

D'Mello, S. K., & Graesser, A. C. (2015). Feeling, thinking, and computing with affect-aware learning technologies. In *The Oxford Handbook of Affective Computing* (pp. 419–434). Oxford University Press.

Díaz-López, M. M., Bustos, R.-H., Guevara-Ramírez, J. F., & García, J.-C. (2025). Understanding Barriers to Nanotechnology Education in Medical Specialty Programs: A Qualitative Nominal Group Study Among Postgraduate Students. *Advances in Medical Education and Practice, 16*, 1481–1499. DOI: 10.2147/AMEP.S523258 PMID: 40861025

Díaz-Marcos, J., & Mendoza Gonzálvez, J. (2025a). Shaping the future of nanotechnology: A perspective on education, dissemination and ethics. *Frontiers in Nanotechnology, 7*. Advance online publication. DOI: 10.3389/fnano.2025.1524578

Dorouka, P., & Kalogiannakis, M. (2024a). Teaching nanotechnology concepts in early-primary education: An experimental study using digital games. *International Journal of Science Education, 46*(13), 1311–1338. DOI: 10.1080/09500693.2023.2286299

Esangbedo, C. O., Zhang, J., Esangbedo, M. O., Kone, S. D., & Xu, L. (2023). The role of industry-academia collaboration in enhancing educational opportunities and outcomes under the digital driven Industry 4.0. *Journal of Infrastructure. Policy and Development, 8*(1). Advance online publication. DOI: 10.24294/jipd.v8i1.2569

European Commission. (2020). *Digital education action plan 2021–2027*. https://education.ec.europa.eu

Friedersdorf, L. E. (2020a). Developing the Workforce of the Future: How the National Nanotechnology Initiative Has Supported Nanoscale Science and Engineering Education in the United States. *IEEE Nanotechnology Magazine*, *14*(4), 13–20. DOI: 10.1109/MNANO.2020.2994799

Ganesan, T. (n.d.). INTEGRATING ARTIFICIAL INTELLIGENCE AND CLOUD COMPUTING FOR THE DEVELOPMENT OF A SMART EDUCATION MANAGEMENT PLATFORM: DESIGN, IMPLEMENTATION, AND PERFORMANCE ANALYSIS. *International Journal of Engineering, 11*(2).

Garrison, D. R., & Vaughan, N. D. (2008). *Blended learning in higher education: Framework, principles, and guidelines*. Jossey-Bass.

Ghattas, N. I., & Carver, J. S. (2012a). Integrating nanotechnology into school education: A review of the literature. *Research in Science & Technological Education*, *30*(3), 271–284. DOI: 10.1080/02635143.2012.732058

Ghosh, M., & Kumar, R. (2024a). Regulatory Issues in Nanotechnology. In Prasad, M., Kumar, R., Ghosh, M., Syed, S. M., & Chakravarti, S. (Eds.), *Nanotechnology Theranostics in Livestock Diseases and Management* (pp. 765–788). Springer Nature., DOI: 10.1007/978-981-16-1610-5_31

Ghosh, S., Das, N., & Nath, A. (2020). Educational platforms for applied AI: A comparative study. [IJEAT]. *International Journal of Engineering and Advanced Technology*, *9*(5), 18–23.

Goel, A., & Polepeddi, L. (2016). *Jill Watson: A virtual teaching assistant for online education*. Georgia Institute of Technology.

Hamimi, E., Ibrohim, I., Kusumaningrum, I. K., Putra, W. E., Sudrajat, A. K., Hayuana, W., Kundariati, M., Hajar, M. U., & Malek, N. A. N. N. (2025). Developing Hi Nano mobile apps as integrated education instrument on nanotechnology and cancer treatment learning. *International Journal of Mobile Learning and Organisation*, *19*(3), 279–298. DOI: 10.1504/IJMLO.2025.147176

Haq, I. U., Mohiz, A., Spadafora, G., Palermo, A. M., Bilotta, E., Felicetti, F., & Lamonaca, F. (2025). AI in the Network of Extended Reality-Enabled Laboratories for STEM Education: Current Applications and Future Potential for Adaptive Learning. In G. Carbone & G. Quaglia (Eds.), *Proceedings of I4SDG Workshop 2025—IFToMM for Sustainable Development Goals* (pp. 373–382). Springer Nature Switzerland. https://doi.org/DOI: 10.1007/978-3-031-91179-8_39

Hassan, S. A.-D. H., Almaliki, M. N. S., Hussein, Z. A., Albehadili, H. M., Rabeea Banoon, S., Abboodi, A., & Al-Saady, M. (2023a). Development of Nanotechnology by Artificial Intelligence: A Comprehensive Review. *Journal of Nanostructures*, *13*(4), 915–932. DOI: 10.22052/JNS.2023.04.002

He, J.-H. (2025). *AI-based Nanotechnology: Breakthroughs*. Applications, Challenges, and the Road Ahead., DOI: 10.2174/0118722105384013250623094407

Hemme, C. L., Carley, R., Norton, A., Ghumman, M., Nguyen, H., Ivone, R., Menon, J. U., Shen, J., Bertin, M., King, R., Leibovitz, E., Bergstrom, R., & Cho, B. (2023). Developing Virtual and Augmented Reality Applications for Science, Technology, Engineering and Math Education. *BioTechniques*, *75*(1), 11–20. DOI: 10.2144/btn-2023-0029 PMID: 37291856

Henriksen, D. (2014). Full STEAM ahead: Creativity in excellent STEM teaching practices. *The STEAM Journal*, 1(2), Article 15.

Hidayatulloh, I., Pambudi, S., Surjono, H., & Sukardiyono, T. (2021). Gamification on Chatbot-Based Learning Media: A Review and Challenges. *Elinvo (Electronics, Informatics, and Vocational Education)*, *6*(1), 71–80. https://doi.org/DOI: 10.21831/elinvo.v6i1.43705

Hue-Shin, L., & Dine, P. (2025). Adaptive Learning Technologies in Structural Engineering Education: Immediate and Contextual Evaluations. *Innovative Journal of Educational Research and Insights*, 77–84.

Humphrey, W., Dalke, A., & Schulten, K. (1996). VMD: Visual molecular dynamics. *Journal of Molecular Graphics*, *14*(1), 33–38. PMID: 8744570

Husseiny, F. (n.d.). *Education's Role in Integrating Nanotechnology in the Curriculum for the Future: Education Book Chapter | IGI Global Scientific Publishing*. Retrieved September 2, 2025, from https://www.igi-global.com/chapter/educations-role-in-integrating-nanotechnology-in-the-curriculum-for-the-future/348778

IBM. (2022). *IBM AI education program*. https://www.ibm.com

Imamguluyev, R., Hasanova, P., Imanova, T., Mammadova, A., Hajizade, S., & Samadova, Z. (2024). AI-POWERED EDUCATIONAL TOOLS: TRANSFORMING LEARNING IN THE DIGITAL ERA. *International Research Journal of Modernization in Engineering Technology and Science*, *6*, 920–929. DOI: 10.56726/IRJMETS65040

Intel. (2021). *AI for youth: Global impact report*. Intel Corporation.

Jeffcoat, P., Di Lernia, C., Hardy, C., New, E. J., & Chrzanowski, W. (2024). (Re) imagining purpose: A framework for sustainable nanotechnology innovation. *NanoImpact, 35*, 100511. DOI: 10.1016/j.impact.2024.100511 PMID: 38750963

Jordan, K. (2015). Massive open online course completion rates revisited: Assessment, length and attrition. *International Review of Research in Open and Distance Learning, 16*(3), 341–358.

Kamali-Sarvestani, R., Weber, P., Clayton, M., Meyers, M., & Slade, S. (2020). Virtual Reality to Improve Nanotechnology Education: Development Methods and Example Applications. *IEEE Nanotechnology Magazine, 14*(4), 29–38. DOI: 10.1109/MNANO.2020.2994802

Khan, A. S. (2020a). Application of Simulation for Emulating and Visualizing Nanotechnology. In Vucinic, D., Rodrigues Leta, F., & Janardhanan, S. (Eds.), *Advances in Visualization and Optimization Techniques for Multidisciplinary Research: Trends in Modelling and Simulations for Engineering Applications* (pp. 307–316). Springer., DOI: 10.1007/978-981-13-9806-3_11

Khan, A. S. (2020b). Application of Simulation for Emulating and Visualizing Nanotechnology. In Vucinic, D., Rodrigues Leta, F., & Janardhanan, S. (Eds.), *Advances in Visualization and Optimization Techniques for Multidisciplinary Research: Trends in Modelling and Simulations for Engineering Applications* (pp. 307–316). Springer., DOI: 10.1007/978-981-13-9806-3_11

Kilani, M., Torabi, K., & Mao, G. (2018). Application of virtual laboratories and molecular simulations in teaching nanoengineering to undergraduate students. *Computer Applications in Engineering Education, 26*(5), 1527–1538. DOI: 10.1002/cae.21940

Klein, J. T. (2017). Beyond interdisciplinarity. In Frodeman, R. (Ed.), *The Oxford handbook of interdisciplinarity* (pp. 35–49). Oxford University Press.

Klimeck, G., McLennan, M., & Brophy, S. P.. (2008). nanoHUB.org: Advancing education and research in nanotechnology. *Computing in Science & Engineering, 10*(5), 17–23. DOI: 10.1109/MCSE.2008.120

Kulkarni, S. K. (2015). *Nanotechnology: Principles and Practices*. Springer International Publishing., DOI: 10.1007/978-3-319-09171-6

Lee, J. (2019). Equity and access in STEM education: Insights and recommendations. *Journal of Educational Technology Development and Exchange, 12*(1), 1–15.

Luckin, R., & Cukurova, M. (2019). Designing educational technologies in the age of AI: A learning sciences-driven approach. *British Journal of Educational Technology*, *50*(6), 2824–2838. DOI: 10.1111/bjet.12861

Lundstrom, M., & Klimeck, G. (2009). Nanoscience: A pathway for STEM education. *Science*, *324*(5933), 325–326.

Lyshevski, S. E., Andersen, J. D., Boedo, S., Fuller, L., Raffaelle, R., Savakis, A., & Skuse, G. R. (2006). Multidisciplinary Undergraduate Nano-Science, Engineering and Technology Course. *2006 Sixth IEEE Conference on Nanotechnology, 1*, 399–402. https://doi.org/DOI: 10.1109/NANO.2006.247661

Makransky, G., & Mayer, R. E. (2021). Cognitive and motivational effects of students' emotional design in virtual reality learning environments. *Educational Psychology Review*, *33*(4), 1671–1690.

Mandrikas, A., Michailidi, E., & Stavrou, D. (2020a). Teaching nanotechnology in primary education. *Research in Science & Technological Education*, *38*(4), 377–395. DOI: 10.1080/02635143.2019.1631783

Marcos. (n.d.). *Frontiers | Shaping the future of nanotechnology: A perspective on education, dissemination and ethics*. Retrieved September 2, 2025, from https://www.frontiersin.org/journals/nanotechnology/articles/10.3389/fnano.2025.1524578/full

Mendoza, K. Z. V., Zambrano, M. M. B., & Baquerizo, A. S. M. (2024). Emerging Technologies and its Role in Modern Education. *Estudios y Perspectivas Revista Científica y Académica*, *4*(3), 3052–3073. DOI: 10.61384/r.c.a.v4i3.597

Ministry of Education. (2020). National [Government of India.]. *Educational Policy*, •••, 2020.

Minogue, J., & Jones, M. G. (2006). Haptics in education: Exploring an untapped sensory modality. *Review of Educational Research*, *76*(3), 317–348.

Moeck, P., Stone-Sundberg, J., Snyder, T. J., & Kaminsky, W. (n.d.). Enlivening 300 level general education classes on nanoscience and nanotechnology with 3D printed crystallographic models. *The Journal of Materials Education.*

Neto, O. P. V. (2014). Intelligent Computational Nanotechnology: The Role of Computational Intelligence in the Development of Nanoscience and Nanotechnology. *Journal of Computational and Theoretical Nanoscience*, *11*(4), 928–944. DOI: 10.1166/jctn.2014.3446

Neumann, A. T., Conrardy, A. D., Decker, S., & Jarke, M. (2023). Motivating Learners with Gamified Chatbot-Assisted Learning Activities. In H. Xie, C.-L. Lai, W. Chen, G. Xu, & E. Popescu (Eds.), *Advances in Web-Based Learning – ICWL 2023* (pp. 189–203). Springer Nature. https://doi.org/DOI: 10.1007/978-981-99-8385-8_15

Nur Fitria, T. (2021). *Artificial Intelligence (AI) In Education: Using AI Tools for Teaching and Learning Process.*

OECD. (2021). *National AI strategies.* https://www.oecd.org

Ofozoba, C. A., Okafor, P. C., Ikegbusi, N. G., Manafa, F. U., Egwu, J. U., Nwobu, C. M., Okafor, S. O., Onafowope, M. A., Olofinkua, V. K., Adeagbo, J. O., Gbaeprekumo, O. V., Ayoko, V. O., Udogu, A. N., Uwannah, N. C., Okudo, O. C., Onyiorah, B. O., Ayodele, M. F., Nnoyelu, O. G., Ekwugha, J. C., & Nnalue, O. H. (2025). AI-driven service-learning to enhance students' understanding of green nanomaterials in sustainability education. *Multidisciplinary Reviews*, *8*(12), 2025429–2025429. DOI: 10.31893/multirev.2025429

Oliveira, A., Feyzi Behnagh, R., Ni, L., Mohsinah, A. A., Burgess, K. J., & Guo, L. (2019). Emerging technologies as pedagogical tools for teaching and learning science: A literature review. *Human Behavior and Emerging Technologies*, *1*(2), 149–160. DOI: 10.1002/hbe2.141

Pane, J. F., Steiner, E. D., & Baird, M. D.. (2017). *Informing progress: Insights on personalized learning implementation and effects*. RAND Corporation.

Pelleg, B., Figueroa, M., VanKouwenberg, M., Fontecchio, A., & Fromm, E. (2011). Implementing nanotechnology education in the high school classroom. *2011 Frontiers in Education Conference (FIE)*, F4D-1-F4D-6. https://doi.org/DOI: 10.1109/FIE.2011.6142886

Repko, A. F., Szostak, R., & Buchberger, M. P. (2020). *Introduction to interdisciplinary studies* (3rd ed.). SAGE Publications.

Roco, M. C. (2023a). National Nanotechnology Initiative at 20 years: Enabling new horizons. *Journal of Nanoparticle Research*, *25*(10), 197. DOI: 10.1007/s11051-023-05829-9

Roco, M. C. (2023b). National Nanotechnology Initiative at 20 years: Enabling new horizons. *Journal of Nanoparticle Research*, *25*(10), 197. DOI: 10.1007/s11051-023-05829-9

Roco, M. C., & Bainbridge, W. S. (2005). Societal implications of nanoscience and nanotechnology. In Roco, M. C., & Bainbridge, W. S. (Eds.), *Nanotechnology: Societal implications* (pp. 3–28). Springer.

Roll, I., & Wylie, R. (2016). Evolution and revolution in artificial intelligence in education. *International Journal of Artificial Intelligence in Education, 26*(2), 582–599.

Russell, S., & Norvig, P. (2020). *Artificial intelligence: A modern approach* (4th ed.). Pearson.

Sajja, R., Sermet, Y., Cwiertny, D., & Demir, I. (2025). *Integrating AI and Learning Analytics for Data-Driven Pedagogical Decisions and Personalized Interventions in Education*. Technology, Knowledge and Learning., DOI: 10.1007/s10758-025-09897-9

Sanders, W. (2018). *Basic Principles of Nanotechnology*. CRC Press., DOI: 10.1201/9781351054423

Şenocak, E., Ozdemir, T., & Yilmaz, F., Tayhan, & Mcnally. (2021). A NANOSCIENCE AND NANOTECHNOLOGY EDUCATIONAL PROGRAM FOR HIGH SCHOOL STUDENTS: EXAMINATION OF PRELIMINARY TRIALS. *The Journal of Materials Education, 43*, 93–108.

Sharma, P., Singh, V., Thakur, A., & Malik, A. K. (2024). Nanotechnology, Artificial Intelligence, and the Future of Manufacturing. In George, S. C., & Tawiah, B. (Eds.), *Nanotechnology in Societal Development* (pp. 307–332). Springer Nature., DOI: 10.1007/978-981-97-6184-5_9

Sharples, M., & Domingue, J. (2016). The blockchain and kudos: A distributed system for educational record, reputation and reward. In *International Conference on Web-Based Learning* (pp. 490–496). Springer.

Spector, J. M. (2013). Emerging Educational Technologies and Research Directions. *Journal of Educational Technology & Society, 16*(2), 21–30.

Srikantaiah, K., & Muralidhar, S. (2020). Pedagogical challenges in nanotechnology education in India. *Indian Journal of Engineering Education, 36*(2), 489–498.

Sweeney, A. E., & Seal, S. (2018). Nanotechnology education: Addressing the issues of workforce preparation and public understanding. *Journal of Nano Education, 10*(1), 23–32.

Tawiah, B., Seidu, R. K., Fobiri, G. K., & Chichi, C. A. (2024). Nanotechnology and Education: Preparing the Next Generation of Future Leaders. In George, S. C., & Tawiah, B. (Eds.), *Nanotechnology in Societal Development* (pp. 247–270). Springer Nature., DOI: 10.1007/978-981-97-6184-5_7

Touretzky, D. S., Gardner-McCune, C., Martin, F., & Seehorn, D. (2019). Envisioning AI for K-12: What should every child know about AI? *Proceedings of the AAAI Conference on Artificial Intelligence*, *33*(1), 9795–9799.

UNESCO. (2022). *Technology in education: A tool on the road to equality*. https://www.unesco.org/en/education/technology

Ünlütabak, B., Ozdil, Z. C. C., & Cirit, E. S. (2023). *Exploring the Use of ChatGPT as a Learning and Teaching Tool in Material Science and Nanotechnology Engineering Education*. *ChemRxiv*. https://doi.org/DOI: 10.26434/chemrxiv-2023-4d99t

Van de Voorde, M., & Fitzpatrick, M. E. (2019). Discovery and Innovation in Science and Technology: Inter- and trans-disciplinary in cognitive learning. *2019 IEEE 18th International Conference on Cognitive Informatics & Cognitive Computing (ICCI*CC)*, 261–264. https://doi.org/DOI: 10.1109/ICCICC46617.2019.9146083

Vega-Baudrit, J.-R., Camacho, M., Araya, A., & Camacho, M. (2023). Regulating Nanotechnology: Ensuring Responsible and Safe Innovation in the Advancement of Science and Technology. *Ciencia. Tecnología y Salud*, *10*(2), 177–191. DOI: 10.36829/63CTS.v10i2.1581

Women, U. N. (2021). *Gender equality in STEM education*. https://www.unwomen.org

Yaseen, H., Mohammad, A. S., Ashal, N., Abusaimeh, H., Ali, A., & Sharabati, A.-A. A. (2025). The Impact of Adaptive Learning Technologies, Personalized Feedback, and Interactive AI Tools on Student Engagement: The Moderating Role of Digital Literacy. *Sustainability*, *17*(3), 1133. DOI: 10.3390/su17031133

Yawson, R. M. (2010). Skill needs and human resources development in the emerging field of nanotechnology. *Journal of Vocational Education and Training*, *62*(3), 285–296. DOI: 10.1080/13636820.2010.499474

Yawson, R. M. (2012). An epistemological framework for nanoscience and nanotechnology literacy. *International Journal of Technology and Design Education*, *22*(3), 297–310. DOI: 10.1007/s10798-010-9145-1

Yawson, R. M., & Greiman, B. C. (2017a). Strategic flexibility analysis of agrifood nanotechnology skill needs identification. *Technological Forecasting and Social Change*, *118*, 184–194. DOI: 10.1016/j.techfore.2017.02.019

Yıldız, M., Kaltakçı Gürel, D., Salmankurt, B., & Gürel, H. H. (2025). Exploring the Evolution of Nanotechnology Education: Insights from Bibliometric Analysis. *Journal of Chemical Education*, *102*(1), 253–269. DOI: 10.1021/acs.jchemed.4c01120

Zafar, S., & Rana, N. (2025). *The Convergence of Nanotechnology and Artificial Intelligence: Unlocking Future Innovations*. https://doi.org/DOI: 10.2174/012405 520435921525 0127071256

APPENDIX A – IMPLEMENTATION TOOLKIT

Table 8. Teaching Module (6-week plan)

Week	Learning Objective	Activities	Assessment Method	Resources/Cost
Week 1	Introduction to AI & Nanotechnology	Lecture + VR demo	Quiz	Free online tools
Week 2	AI in everyday life	Case study + group discussion	Reflection note	Open datasets
Week 3	Basics of Nanotechnology	Virtual lab activity	Short test	Simulation software
Week 4	AI + Nano convergence	Project brainstorming	Peer review	Whiteboard tools
Week 5	Hands-on implementation	Arduino/Raspberry Pi + sensors	Lab report	~$50 kit
Week 6	Project showcase & evaluation	Presentations	Rubric-based grading	Classroom setup

Table 9. Project-Based Assessment Rubric

Criteria	Excellent (4)	Good (3)	Fair (2)	Poor (1)
Technical Understanding	Profound conceptual clarity, correct application	Mostly correct with minor gaps	Partial understanding	Major misconceptions
Innovation & Creativity	Highly original, practical idea	Some originality	Limited novelty	No originality
Implementation	Fully functional, well-tested	Functional with minor bugs	Partially functional	Not functional
Presentation	Clear, engaging, professional	Adequate clarity	Needs improvement	Confusing, unclear

Table 10. Procurement Guide (Low/Medium/High-Cost Options)

Item	Low-Cost Option	Medium-Cost Option	High-Cost Option
VR/AR Tools	Google Cardboard (~$10)	Oculus Quest (~$300)	HTC Vive Pro (~$800)
Microscopes	Basic USB microscope (~$30)	Student-grade (~$150)	Research-grade (~$1000)
Computing Devices	Refurbished laptops (~$100)	Raspberry Pi 4 (~$70)	High-end workstation (~$1500)
Sensors/Kits	Arduino kit (~$25)	Arduino + sensors bundle (~$80)	Full lab kit (~$500+)

Chapter 8
Forecasting E-Waste Generation Using NARNET:
A Case Study of Germany

Mikail Pehlivan
https://orcid.org/0000-0002-7273-333X
Trakya University, Turkey

Umut Akduğan
https://orcid.org/0000-0001-8659-541X
Trakya University, Turkey

ABSTRACT

Electronic waste (e-waste) is the fastest-growing waste stream worldwide, posing environmental, economic, and health challenges. Despite containing valuable precious and rare metals, less than 22% of e-waste is recycled, worsening resource loss and pollution. This study estimates and forecasts Germany's e-waste generation from 1960 to 2040. Using Robinson's (2008) method and complying with the EU WEEE Directive, total e-waste was estimated until 2025. After testing nonlinearity, a Nonlinear Autoregressive Neural Network (NARNET) forecasted trends through 2040, showing steady growth. Results highlight trade-offs: pre-ElektroG models recovered value but were costly, while post-ElektroG bulk processing cut costs but limited reuse. We propose a hybrid approach—value-focused collection for high-yield items and bulk processing for low-value waste—to optimize economic and environmental outcomes. Policy recommendations stress collaboration, awareness, and adaptive frameworks to foster a circular economy.

DOI: 10.4018/979-8-3373-2883-6.ch008

INTRODUCTION

Electronic waste (e-waste) is the fastest-growing category of solid waste worldwide (Guo and Yan, 2017; Yang et al., 2017; Zeng et al., 2017; Qu et al., 2019). The rise in related publications over the past two decades highlights e-waste management as a global issue affecting both developing and industrialized nations (Sthiannopkao and Wong 2013; Li et al. 2015; Awasthi and Li 2017). The growing volume of e-waste plays a critical role in shaping policies and advancing treatment processes. Experts view e-waste generation as a natural result of technological progress, particularly in developed countries (Erumban and Das 2016; Jorgenson et al. 2016; Song et al. 2016). This trend also reflects increasing economic prosperity, as waste generation often correlates with material wealth (Kusch and Hills, 2017; Awasthi et al. 2018).

Building on this foundation, recycling not only mitigates environmental degradation but also serves as a strategic approach to sustainable development. As a core element of the circular economy, recycling emphasizes minimizing ecological harm by recovering valuable materials from waste streams (Saavedra et al. 2018; Forti et al. 2020; Oughton et al. 2022). The recycling rate commonly indicates the share of waste redirected toward material reclamation, serving as a benchmark for environmental progress and efficient resource utilization. Nations exhibiting higher recycling performance typically generate less overall waste (Antoni and Marzetti 2019; Islam et al. 2019), contribute to lower greenhouse gas (GHG) outputs (Magazzino et al. 2020; Turner et al. 2015), and often witness enhanced economic performance and employment opportunities (Di Vita 2001; George et al. 2015; Liu et al. 2020).

Recycling offers three primary advantages: economic, environmental, and public health-related. First, economically, e-waste represents a valuable resource due to its high content of precious and rare metals. Yet, less than 22% of global e-waste is currently recycled, and only about 1% of the global demand for rare earth elements is met through this process (Baldé et al. 2024). Inefficient e-waste management not only leads to the loss of recoverable materials but also increases reliance on primary extraction, thereby intensifying mining activities. The concentration of metals in e-waste significantly exceeds that found in traditional ores. With global ore grades steadily declining, mining operations are increasingly forced to extract more complex and lower-grade deposits to meet rising metal demand (Lèbre and Corder 2015). Research comparing the metal content of electronic devices with average ore grades has shown that electronics contain far higher concentrations of copper, gold, silver, and palladium (Hagelüken 2006, 2014; Cui and Zhang 2008; Namias 2013; Desjardins 2014; McLeod 2014; Pistilli 2016).

Beyond precious metals, electronic devices also contain a range of rare elements such as terbium (Tb), yttrium (Y), praseodymium (Pr), europium (Eu), lanthanum (La), neodymium (Nd), gadolinium (Gd), gallium (Ga), and indium (In) (Schüler et al. 2011; Cucchiella et al. 2015; Li et al. 2015a). These metals are difficult and hazardous to extract, limiting their production to a few geographic areas—predominantly China. China's implementation of export quotas on these critical materials raises concerns over future supply security (Wilburn 2012). Consequently, such metals are classified as strategic resources essential for national economic development and technological innovation (Wang et al. 2017).

Second, beyond the increasing volume of e-waste, its environmental and health impacts are significant due to the presence of toxic heavy metals and organic contaminants (Tsydenova and Bengtsson 2011; Kiddee et al. 2013). In addition to primary emissions from hazardous substances within e-waste, improper disposal and informal recycling can also lead to secondary and tertiary emissions, further exacerbating pollution (Schluep et al. 2009). Recycling e-waste significantly reduces the global demand for primary metal extraction, thereby contributing to lower GHG emissions. Manufacturing a single computer with a monitor, for instance, requires approximately 240 kg of fossil fuels, 22 kg of chemicals, and 1.5 tons of water (Kuehr et al. 2003; Electronics TakeBack Coalition 2016). Recovering metals from e-waste is far more energy-efficient than using raw materials—recycling saves up to 95% of energy for aluminum, 85% for copper, 74% for iron and steel, 65% for lead, 60% for zinc, 64% for paper, and over 80% for plastics (Cui and Forssberg 2003). These substantial energy savings directly translate into reduced emissions associated with metal production (Kumar et al. 2017).

Third, e-waste contains hazardous metals and chemicals that pose risks not only to the environment but also to human health. Exposure occurs primarily in two ways: through emissions that contaminate soil, water, and air—introducing toxins into the food chain—and through direct contact during disposal and informal recycling processes, which can lead to both short- and long-term health effects (Kuper and Hojsik 2008; Kiddee et al. 2013; Garlapati 2016; Scruggs et al. 2016). Informal recovery methods, such as open burning of polyvinyl chloride-coated cables or chemical leaching of circuit boards, are particularly harmful (Velis and Mavropoulos 2016). For instance, cadmium from mobile phone batteries can pollute up to 600 m^3 of water (Garlapati 2016). Toxic substances in e-waste have been linked to impaired cognitive development, liver and kidney damage, and the release of carcinogens that contribute to respiratory illnesses (Baldé et al. 2015).

To the best of our knowledge, this is the first study to estimate and forecast e-waste generation in Germany. The initial dataset used in the analysis was derived from the authors' calculations based on Robinson (2009). This is represented by the equation $TEW = (M_t . N)/L$, where TEW (kg/year) depends on the mass of the item in kilograms (M), the number of units in service (N), and L (years) is the average lifespan. According to this, the information related to the electronic products used in the creation of the TEW (total e-waste) series, as determined by the six classification categories, is compliant with the EU Directive 2012/19/EU on waste electrical and electronic equipment (WEEE Directive), and is provided in Appendix Table 1A. This approach allowed for the use of a longer sample, spanning from 1960 to 2025, which provided more consistent results for e-waste forecasting in Germany. The objective of this paper is to estimate e-waste generation and forecast trends for Germany from 2026 to 2040. Based on the findings, the paper aims to emphasize the significance of e-waste and propose a policy framework for improved e-waste management.

Given the scale and complexity of e-waste generation, accurate forecasting is essential for informing policy, optimizing collection systems, and aligning recycling infrastructure with future demand. While international studies have applied a range of statistical and AI-based approaches, the literature on Germany reveals a notable gap in comprehensive, long-term forecasting that accounts for both nonlinear trends and policy frameworks. Addressing this gap requires evaluating existing methodologies and selecting models capable of handling the variability inherent in e-waste data. This need for robust, context-specific forecasting forms the basis for the following section, which reviews Germany's e-waste management framework and relevant forecasting literature.

In light of the discussion above, the structure of the study is organized as follows: Section 2 outlines e-waste management in Germany, Section 3 describes the methodology, Section 4 presents the empirical results, Section 5 offers a discussion, and Section 6 concludes the paper.

CONTEXT OF GERMANY AND REVIEW OF THE LITERATURE

Germany is the largest generator of e-waste among developed countries in Europe. In total, 1,767 kilotonnes (Kt) of e-waste were produced, corresponding to approximately 21.2 kilograms per capita. Of this amount, 956.6 Kt—or 54%—was officially collected and recycled (Baldé et al. 2024). As shown in Figure 1, the amount of e-waste in Germany has increased rapidly over time, closely correlating with rising income levels.

Figure 1. The trend of e-waste generation in Germany (1960-2023)

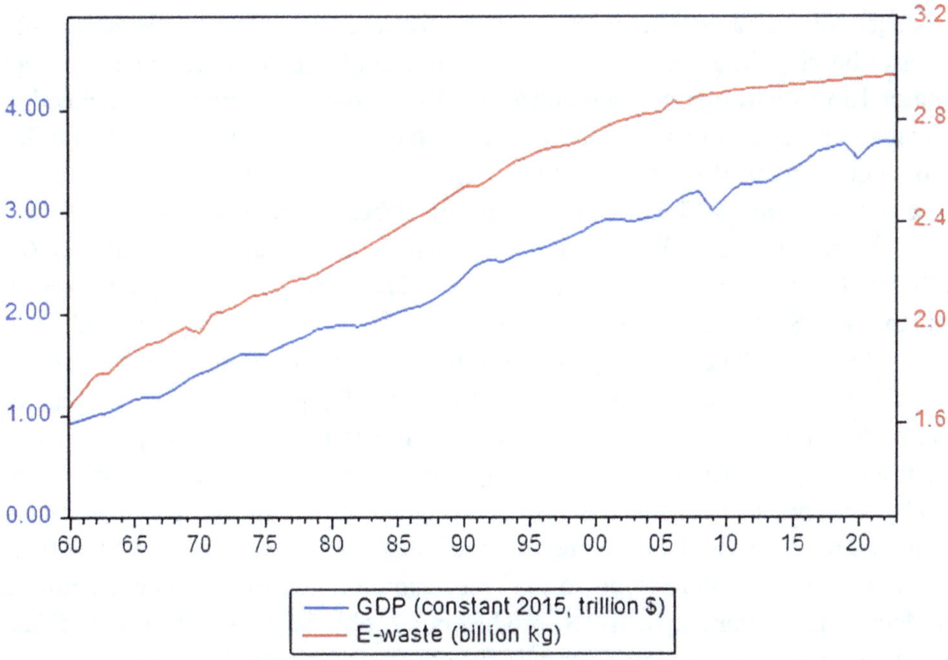

Despite the existence of a formal e-waste collection system in Germany, some consumers continue to dispose of their electronic waste alongside regular household garbage (Dimitrakakis et al. 2009). The European Union (EU) introduced an Extended Producer Responsibility (EPR) framework through the WEEE Directive to regulate e-waste. This directive imposes economic, physical, and informational responsibilities on producers, aiming first to prevent e-waste generation, and then to prioritize reuse, recycling, and other recovery methods (Walther et al. 2009). However, national implementation of the directive has varied across EU member states, resulting in different operational and economic structures based on the type of EPR model adopted (Magalini and Huisman 2007).

In Germany, the WEEE and Restriction of Hazardous Substances (RoHS) Directives were transposed into national law through the Electrical and Electronic Equipment Act (ElektroG), enacted in 2005. This legislation outlines responsibilities across several areas: product design, collection, take-back obligations, treatment and recycling, financing, monitoring and reporting, as well as labeling and consumer information. The entire process—from collection to meeting recovery targets—is coordinated with public authorities. Once collected, e-waste is transferred to private or state-subsidized dismantling facilities, where devices are tested for reuse or processed for material recovery, including the extraction of valuable resources and safe removal of hazardous substances. In addition to municipal collection schemes, producers may establish individual or collective take-back systems for household e-waste, bearing the associated collection costs. Retailers can also voluntarily accept returned e-waste from consumers and forward it to producers or municipal collection centers, as producers may not have access to all categories of discarded electronics (Rotter et al. 2009; Walther et al. 2009; Ongondo et al. 2011).

Under Germany's ElektroG, producers are offered three options for organizing the return of e-waste: (i) individual brand-selective take-back schemes (IBTS), (ii) individual non-selective take-back schemes (INTS), and (iii) collective take-back schemes (CTS). In IBTS, producers are responsible for collecting only their own branded e-waste. To do so, they must establish agreements with public authorities and distributors to identify and separate their branded products from the general waste stream. Once separated, the e-waste is directed to the service providers of the respective producers. The quantity of e-waste collected must correspond to the producer's market share, which is calculated and monitored by the Stiftung Elektro-Altgeräte Register (EAR). EAR regularly evaluates whether producers meet their collection quotas; failure to do so may result in a ban on placing new electronic products on the market. In INTS, producers are obligated to collect a specified amount of e-waste regardless of brand. The containers supplied by the government contain mixed-brand electronic waste. Producers may contract different companies for logistics and collection. In CTS, producers jointly organize and finance the collection of electronic products (Chaudhary and Vrat 2018).

E-waste management involves both costs and potential revenues associated with collection, transport, processing, and the resale of recovered materials. Revenues and expenses vary depending on the transportation model used and the degree of product sorting prior to collection. For instance, pre-sorted e-waste may incur additional costs due to the need for *value-conserving transport*, which requires more containers and space, as well as increased labor to assess product reusability. After this, products are tested for reuse potential. In contrast, *bulk transportation* is less costly due to reduced sorting, though it may damage products during transit, increasing the risk of hazardous substance leakage and contamination of entire loads. This also raises environmental concerns. As such, the choice of transport model plays a crucial role in e-waste management. The three main recovery options—product reuse, component extraction, and material recycling—are mutually exclusive to varying degrees. A reused product cannot yield components or materials. If components are extracted, the product can no longer be reused, and when materials are recycled, neither reuse nor component recovery is possible. Due to varying material values across products and components, the chosen recovery pathway significantly affects the economic outcome of the process (Walther et al. 2009).

Given the increasing importance of accurate data for effective e-waste management, a growing number of estimation studies began to emerge just before the 2000s (Matthews et al. 1997; Lohse et al. 1998), utilizing a range of methodological approaches. In this study, the literature is reviewed in two parts: first, by examining e-waste quantification studies based on varying criteria such as method and regional scope; and second, by focusing specifically on research related to e-waste estimation in Germany. While there is extensive literature addressing global and regional e-waste trends, a notable gap exists in detailed forecasting studies specifically focused on Germany (Table 1). This highlights the need for country-specific models that can inform national policy and improve the accuracy of future e-waste projections.

Table 1. E-waste forecasting-related studies in the literature

Study	Method	Region	Unit	Estimation Period
Tasaki et al. (2004)	DD, LM	Japan	kg	1995–2020
Streicher-Porte et al. (2005)	SD	India/Delhi	units	2003–2010
Peralta and Fontanos (2006)	CM	Philippines	units	1995, 2000, 2005, 2010
Jain and Sareen (2006)	SD	India/Delhi	units	-2010
Kumar and Shrihari (2007)	SD, DD, C, and U	India/Mangalore	kg, tonnes	2007–2015, 2007
Oguchi et al. (2008)	DD	Japan	units	2003
Yang and Williams (2009)	LM	United States	units	1980–2050
Walk (2009)	DD	Germany/Baden-Württemberg	tonnes	2000–2015
Robinson (2009)	C and U	General	kg	2009
Dwivedy and Mittal (2010a)	CM	India	units	2007–2011
Steubing et al. (2010)	CM	Chile	kg	1996–2020
Dwivedy and Mittal (2010)	LM, CM	India	units	1994–2026
Chung et al. (2011)	C and U	China/Hong Kong	kg	2009
Polak and Drapalova (2012)	DD	Czech Republic	units	1995–2020
Araújo et al. (2012)	C and U, TS	Brazil	tonnes	2000–2008 and 2008
Wang et al. (2013)	SD, DD, CM, C and U, TS	Finland	kg	1990–2011
Kim et al. (2013)	LM	South Korea	units	2000–2020
Pant (2013)	Math. expression	India/Dehradun	units	Equation depends on the year
Öztürk (2014)	C and U	Turkey	tonnes	2002–2012, 1999–2012 (TV), 2005–2012 (PC)

continued on following page

Table 1. Continued

Study	Method	Region	Unit	Estimation Period
Yazici et al. (2014)	DD	Turkey	tonnes, units	2013
Li et al. (2015b)	DD, C and U, TS	China	units	1998–2012
Kalmykova et al. (2015)	DD	Sweden	tonnes, units	2014–2040
Alavi et al. (2015)	–	Iran/Ahvaz	tonnes	2011
Ikhlayel (2016)	SD, CM, C and U, SM, TS	Jordan/Middle East-North Africa	kg	2015
Petridis et al. (2016)	LM and Time Series Models	East-West Euro, Asia/Pacific, Japan/ Aust./NZ, Americas	units	2013–2030
Zhao et al. (2016)	GM	China	units	2014–2040
Salihoğlu and Kahraman (2016)	C and U	Turkey/Bursa	kg	2015
Golev et al. (2016)	SSL	Australia	tonnes	2010–2024
Guo and Yan (2017)	LM	China	kg, units	1997–2025
Gusukuma and Kahhat (2018)	MFA	Peru	units	2005–2017, 2025
Tran et al. (2018)	DD, MFA	Vietnam	units	1966–2030
Wang et al. (2018)	DD, LM	China	units	1992–2040
Kececi et al. (2018)	CM	Turkey/Ankara, Cankaya District	tonnes	2011–2016
Thiébaud-Müller et al. (2018)	MFA	Swiss	–	2014
Duman et al. (2019)	GM	USA/Washington	units	2018–2030
Sajid et al. (2019)	MFA	Pakistan	units	2005–2020
Ravindra and Mor (2019)	Average Household	India/Chandigarh	kg, units	2016
Abbondanza and Souza (2019)	DD	Brazil/Sao Jose dos Campos	kg, units	2009–2017, 2014–2025
Golev et al. (2019)	MFA	Australia	tonnes	2010–2024
Gonda et al. (2019)	C and U, MFA	Belgium/Brussels, Walloon Region	tonnes	2013
Islam and Huda (2019)	Holt's	Australia	tonnes	2018–2047
Duman et al. (2020)	GM	USA	tonnes	2016–2023
Kosai et al. (2020)	Bass	Vietnam	units	2015–2050
Kazancoglu et al. (2020)	GM	Turkey	tonnes	2019–2021
Koshta et al. (2021)	DD, S-shaped Curves	India/Delhi	units	1995–2030

continued on following page

Table 1. Continued

Study	Method	Region	Unit	Estimation Period
Ryan-Fogarty et al. (2021)	Analysing sites data	Ireland	units, kg	2019
Wang et al. (2021)	GM	USA	tonnes	2016–2025
Zeng et al. (2021)	Bass and Kuznets curve	China	units, tonnes	2010–2050
Zhang et al. (2021)	DD	China	units, tonnes	1998–2035
Kastanaki and Giannis (2022)	LM, DD	Greece	units, tonnes	1995–2035
Sakthivel et al. (2022)	MFA, DD	India	units	2010–2030
Li et al. (2022)	SM	China	units, tonnes	1998–2020
Owusu-Sekyere et al. (2022)	Analysing Input-output materials	Ghana	tonnes	2021
Bogar and Gungor (2023)	DD and time series forecasting techniques	Turkey	units, tonnes	2001–2035
This Study	**NARNET**	**Germany**	**kg**	**2026–2040**

Notes: Sales Method (SM), Simple Delay Method (SD), Distribution Delay Method (DD), Carnegie Mellon Method (CM), Time Step Method (TS), Consumption and Use Method (C and U), Material Flow Analysis (MFA), Grey Model (GM), Logistic Model (LM), and Sales-Stock-Lifespan (SSL).

While the descriptive context remains central, it is equally important to critically examine the forecasting methods used in prior studies. Existing approaches such as Grey Models (GM), Logistic Models (LM), Material Flow Analysis (MFA), ARIMA and hybrid statistical models each have inherent strengths and weaknesses. GM and LM, for example, perform well for short- to medium-term forecasts with stable trends but may underperform in scenarios with structural breaks or abrupt technological shifts. MFA provides detailed material composition insights but requires extensive input data, which may limit its forecasting horizon. ARIMA and other linear time-series models are less capable of capturing nonlinearities typical of e-waste generation, especially in contexts where policy, technology adoption, and consumer behavior interact in complex ways. In contrast, NARNET, with its recurrent structure and feedback connections, has demonstrated superior ability to model nonlinear, stochastic patterns over extended periods, making it well-suited for e-waste forecasting where collection rates, product lifespans, and technological obsolescence introduce nonlinearity into the data.

DATA AND METHODOLOGY

While Nonlinear Autoregressive Neural Network (NARNET) is the central forecasting tool in this study, alternative AI-based approaches were evaluated conceptually. Long Short-Term Memory (LSTM) networks, Convolutional Neural Network–LSTM hybrids (CNN-LSTM), NAR-GMDH, and Online Sequential Extreme Learning Machines (OS-ELM) all offer capabilities for nonlinear modeling. LSTM networks excel in long-range dependency capture but typically require larger datasets and more computational resources. CNN-LSTM models can detect spatial–temporal patterns but are more complex to train for purely time-series data. NAR-GMDH offers automated model structure discovery but has shown variable forecasting accuracy in small datasets. OS-ELM is fast to train but can underperform when data exhibit strong autocorrelation. NARNET, by contrast, offers a balance of training efficiency and accuracy, particularly when implemented with the Levenberg–Marquardt algorithm, which is well-suited for moderate-sized datasets. Comparative evidence from environmental and resource forecasting studies indicates that NARNET often achieves lower Mean Absolute Percentage Error (MAPE) and Root Mean Square Error (RMSE) than these alternatives when the dataset length is limited but exhibits pronounced nonlinearity (Zeinali et al. 2021).

Harvey Et Al. (2008) Linearity Test

In empirical studies that use time series, it is crucial to examine the characteristics of the series, such as whether the series is stationary, linear, or exhibits seasonality. Determining the linearity/nonlinearity of a series, which is one of its characteristic features, is necessary for deciding which type of unit root test (linear or nonlinear unit root tests) to apply. In this context, although many linearity tests have been developed to determine whether the series is linear, the linearity test developed by Harvey et al. (2008) is one of the most advanced. The advantage of this test over previous ones is that it does not require determining the order of integration of the series and can be applied to both stationary (I(0)) and first-differenced stationary (I(1)) variables (Güriş et al. 2017). The test proposed by Harvey et al. (2008) suggests a Wald test statistic, which is a weighted average of the Wald tests for the stationary and unit root cases of the series when the order of integration is unknown (Cuestas et al. 2012). The corresponding Wald test statistic is as follows:

$$W_\lambda = (1 - \lambda)W_0 + \lambda W_1 \tag{1}$$

where W_0 and W_1 denote the Wald tests when the variable is I(0) and I(1), respectively. W_λ is asymptotically distributed as χ_2^2 and W_0 and W_1 follow the standart

χ_2^2. λ is a function that converges in probability to 1 when the variable is I(1) and to 0 when the variable is I(0). Harvey et al. (2008) state that an appropriate function should be selected for λ. Therefore, the following functional form is suggested (Khraief et al. 2020):

$$\lambda(U,S) = \exp(-g\left(\frac{U}{S}\right)^2 \tag{2}$$

In this context, g represents a finite positive constant that does not influence the asymptotic properties of W_λ. U and S refer to the selected unit root and stationarity statistics, respectively. Harvey et al. (2008) suggest using the standart Dickey-Fuller unit root statistic for U and the nonparametric stationarity statistic introduced by Harris et al. (2003) for S. When the time series is stationary, $(U/S)^2$ diverges and λ converges to 0, and when the time series is unit root, $(U/S)^2$ converges to 0 resulting in λ converging to 1 (Khraief et al., 2020).

Nonlinear Autoregressive Neural Network (NARNET)

In fact, most time series exhibit significant fluctuations and rapid transient behavior, making nonlinear modeling approaches more suitable for capturing their dynamics. A time series can be effectively represented using a Nonlinear Autoregressive (NAR) model (Ruiz et al. 2016). On the other hand, due to their high potential and flexibility in capturing structural breaks and variations in nonlinear time series, artificial neural networks (ANNs) are widely regarded as effective function approximators in predictive modeling. This is particularly relevant because traditional Recurrent Neural Networks (RNNs) often struggle to capture long-term dependencies—a common challenge in time series data. This limitation primarily arises from gradient-related issues, a problem shared by most recurrent architectures except for the Nonlinear Autoregressive Neural Network (NARNET) (Ibrahim et al. 2016). As an autoregressive network model, NARNET explicitly accounts for stochastic nonlinear relationships. Consequently, it has demonstrated robust performance in forecasting nonlinear time series (He and Jin 2018). NARNET employs a multi-layered recurrent architecture with feedback connections, allowing it to predict future values based on past observations. This structure makes NARNET particularly well-suited for modeling and forecasting time series with complex nonlinear patterns.

As a type of RNNs, the NARNET represents a discrete, nonlinear autoregressive system driven by endogenous inputs. In the NARNET method, the following equation can express the output of the NARNET model (Yt) (Kheimi et al. 2024):

$$\hat{Y}_t = f(Y_{t-1} + Y_{t-2} + \ldots + Y_{t-d}) + \varepsilon_t \quad (3)$$

Where $\hat{Y}_t$ is the modeled value, f presents a nonlinear function, d denotes the number of earlier values (input delay) of the considered parameter, and ε_t is the error term. The equation illustrates how NARNET can be used to predict the value of the series y at time t, denoted as y(t), based on the previous (p) values of the series. The function f is not known beforehand, and the goal of training the NARNET is to approximate this function by optimizing its weights and biases. The error term ε_t stands for the error of the approximation of the series y at time t. NARNET is a multilayer, recurrent, dynamic network, with feedback connections. In a NARNET the terms y(t−1), y(t−2), . . ., y(t−p), are known as feedback delays (Işıkdağ et al. 2023). Like a multilayer perceptron neural network, the NARNET requires the optimization of synaptic weights and biases during the training phase, following exposure to the input dataset. The general schematic structure of the NARNET model is shown in Figure 2.

Figure 2. The NARNET structure

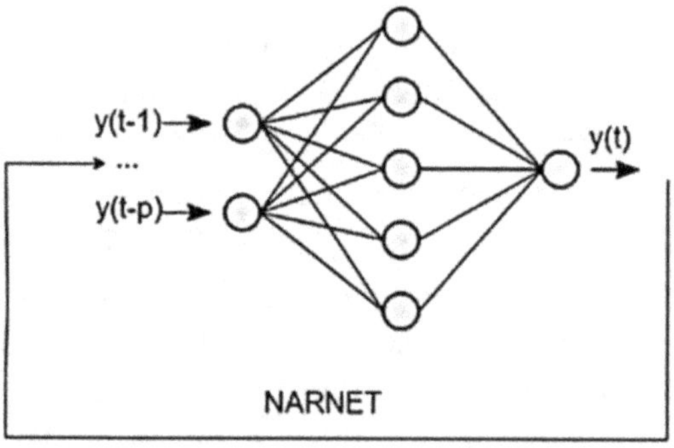

Typically, the NARNET is trained using the backpropagation method. In this study, the Levenberg–Marquardt (LM) backpropagation algorithm is employed for training, with the "logsig" activation function applied in the hidden layers. The LM algorithm is the most commonly used learning rule for NARNET. It integrates two numerical optimization techniques: the gradient descent method and the Gauss-Newton method. The gradient descent approach reduces the sum of squared errors by updating parameters in the direction of the steepest descent. In contrast, the Gauss-Newton method assumes that the error function behaves quadratically near the minimum and finds its optimum accordingly. The LM algorithm transitions between these two methods—behaving like gradient descent when parameters are far from optimal, and like the Gauss-Newton method when they are near their optimal values (Gavin

2020). The LM algorithm is designed to achieve the speed of second-order training methods without explicitly computing the Hessian matrix. It is considered one of the fastest training algorithms for moderately sized feedforward neural networks, typically those with up to several hundred weights (MathWorks 2022).

In this study, we propose and implement a grid search algorithm to identify the most accurate NARNET—defined as the network with the lowest prediction error—by optimizing its hyperparameters. The LM algorithm is used as the training function, given its widespread use in training NARNETs. The hyperparameters considered in our objective function are: (i) the number of neurons in the hidden layer, and (ii) the number of feedback delays. To improve accuracy, the training process can be repeated N times. The number of epochs for each training session is automatically determined by the LM algorithm. Model accuracy is evaluated using the test dataset (Işıkdağ et al. 2023).

After optimizing the NARNET's hyperparameters through grid search, the next critical step is evaluating its forecasting performance using robust accuracy metrics. Performance metrics, or accuracy measures, are essential statistical tools for selecting the most suitable model for forecasting a given time series. The proposed NARNET model will be evaluated using these metrics to ensure both qualitative and quantitative assessment of its forecasting capabilities. To accurately gauge the model's prediction precision, Hyndman and Athanasopoulos (2018) recommend using validation data that is distinct from the training data. Various model validation techniques exist, and among them, examining the values of key performance indices is a common approach to assess forecasting accuracy (Brosilow 2002).

FINDINGS

Determining the linearity of the time series constitutes an essential prerequisite for NARNET analysis. The linearity test developed by Harvey et al. (2008) was employed for this purpose, with the test results presented in Table 2. Analysis of the linearity test results indicates that the series exhibits nonlinear characteristics. This confirmed nonlinearity necessitates the application of NARNET's specialized architecture to effectively model the series' behavior.

Table 2. The results of linearity test

Variable	Test Statistics W	Result
TEW	12.4569*	Nonlinear

*** denotes the rejection of the null hypothesis which define linearity of the series at the 1%.** χ_2^2 table critical values of Harvey et al. (2008) test are 9.21, 5.99, 4.60 at the 1, 5 and 10%, respectively.

Training Results

Once the NARNET architecture is configured, training is initiated within the workflow or training application by clicking the “Train” button and waiting for the process to complete. The output from the training includes several key parameters that are essential for optimizing the neural network's performance.

In Figure 3, the Progress box labeled “Trained Output” displays the network's error performance, which begins at 295 MSE (Mean Squared Error) and decreases to 103 MSE by the end of training. Figure 4, showing the training performance window, indicates that the network successfully avoids both overfitting and under-fitting, as evidenced by the parallel alignment of the training, validation, and testing performance curves.

Figure 3. The progress box

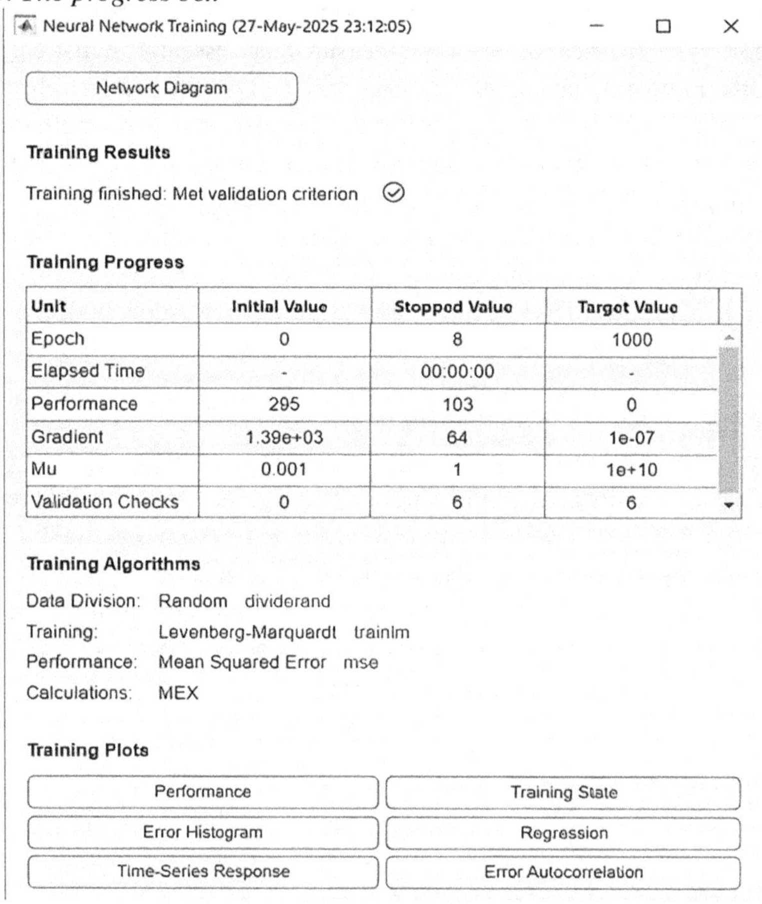

Unit	Initial Value	Stopped Value	Target Value
Epoch	0	8	1000
Elapsed Time	-	00:00:00	-
Performance	295	103	0
Gradient	1.39e+03	64	1e-07
Mu	0.001	1	1e+10
Validation Checks	0	6	6

Figure 4. Training performance

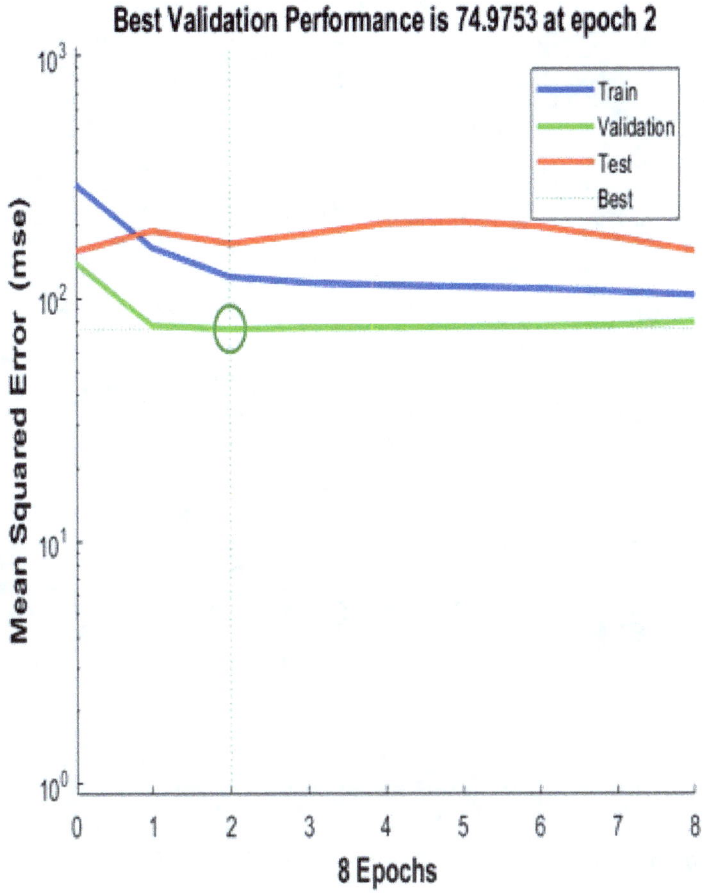

The R value, which represents the strength of the linear relationship between the model's outputs and target values (i.e., the goodness of fit), exceeds 72% for the training dataset. As shown in Figure 5, the model performs significantly better on the validation and testing datasets, with R values above 93%. These results indicate that the model generalizes well and demonstrates strong predictive performance on unseen data. The fitted model for the residual is presented in Figure 6.

Figure 5. Model Summary

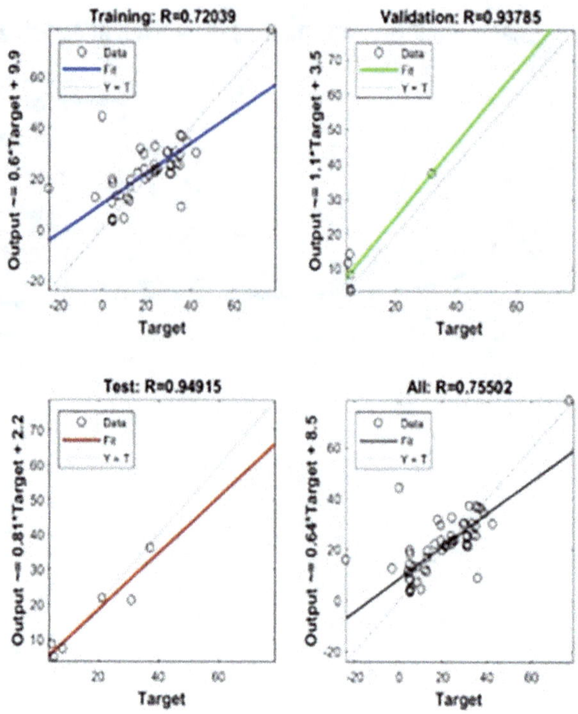

Figure 6. Response of output element for residual time-series

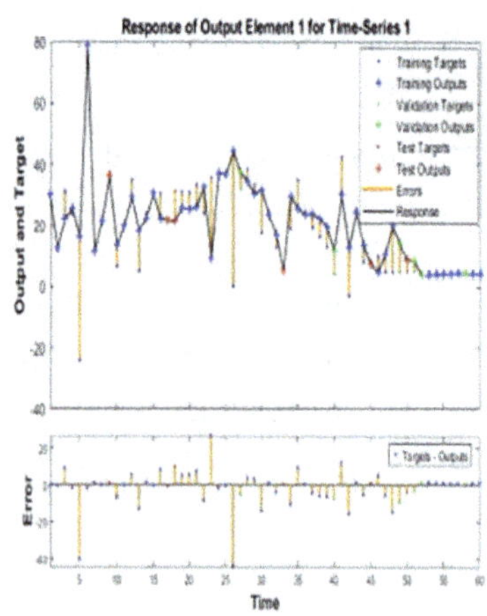

While the fitted residual model is illustrated in Figure 6, Figure 7 displays the errors associated with the neural network model. The adequacy of the model is evaluated through the autocorrelation analysis of these errors. As shown in Figure 8, there is no significant autocorrelation in the error term (Error1), suggesting that the residuals are randomly distributed.

Figure 7. Errors of the gitted residual neaural network

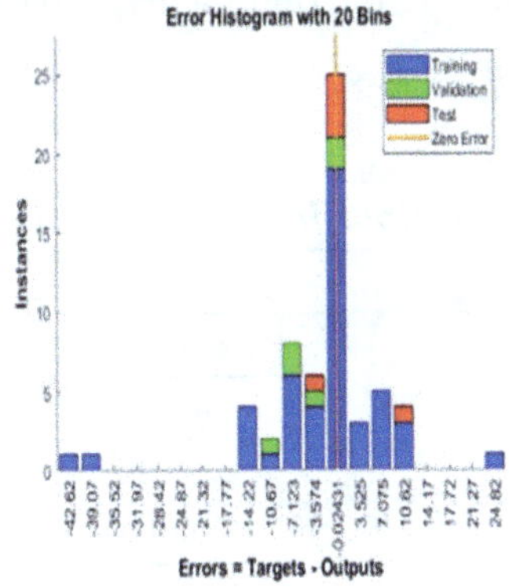

Figure 8. Autocorrelation of Error 1

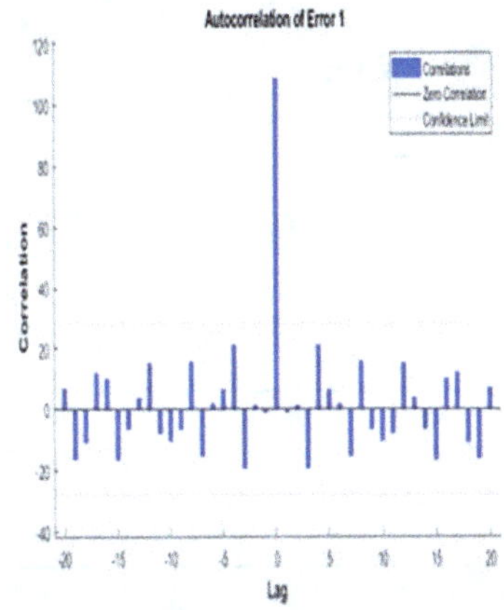

The Outputs of Optimal NARNET Models

All indicators were modeled and estimated using the optimal NARNET architectures, identified through the MATLAB App developed as part of this study.

According to the ADF unit root test, the variable was not stationary at level (Table 3); therefore, the first difference of the variable was used to construct the training and test dataset. The training dataset spans from 1960-2012 (53 observations), while the test dataset covers the period from 2013-2025 (13 observations). A static in-sample estimation method was applied during model training. The model's hyperparameters—specifically, feedback delays and the number of hidden layer neurons—were evaluated and optimized using a grid search algorithm. A summary of the estimation results, including the optimal hyperparameters for the best NARNET model, is provided in Table 4.

Table 3. The Result of ADF Unit Root Test

			TEW
ADF Unit Root Test	**Level**	Test Statistic	0.4450
		p-value	0.9989
		Optimal lag	7
	First Difference	Test Statistic	-1.9866**
		p-value	0.0459
		Optimal lag	5

** denotes the rejection of the null hypothesis at 5%.

Table 4. Estimation results of NARNET

Parameter	**Value**
R^2	0.92
MAPE (%)	3.37
SMAPE (%)	3.44
MSE	8969.82
RMSE	94.70
MAE	81.93
Delay	[1 2 3 4 5]
Hidden Size	10

The selected NARNET model, configured with feedback delays of [1 2 3 4 5] and a hidden layer size of 10, demonstrates strong predictive performance. The model achieved a Mean Absolute Percentage Error (MAPE) of 3.37% and a Symmetric Mean Absolute Percentage Error (SMAPE) of 3.44%, indicating high accuracy and consistency in its forecasts. Additionally, the coefficient of determination (R^2) is 0.92, suggesting that approximately 92% of the variance in the actual data is explained by the model. The error metrics—Root Mean Square Error (RMSE) of 94.70, Mean Absolute Error (MAE) of 81.93, and Mean Squared Error (MSE) of 8969.82—are all within acceptable ranges, reflecting low average deviations between predicted and actual values (Legates 1999; Pinheiro et al. 2020). Overall, these results indicate that the model provides a reliable and well-fitted representation of the underlying time series and is well-suited for forecasting purposes. The forecasting process continues using this method through to the year 2040. Figure 9 presents the estimated and predicted e-waste quantities from 1960 to 2040. The results indicate an upward trend in e-waste generation in the coming years. Notably, forecasts suggest that e-waste will surpass 3.1 million kg by 2040.

Figure 9. Forecasted e-waste for the period 2026-2040

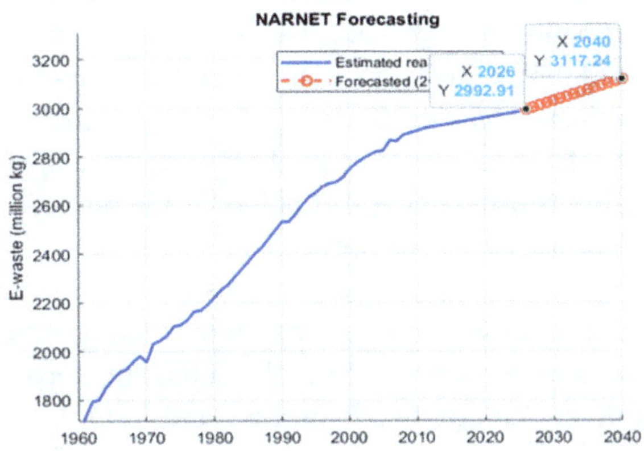

DISCUSSION

The primary legislative instrument governing e-waste in Germany is the ElektroG, which transposes the EU WEEE and RoHS Directives into national law. The most recent amendment (2023) introduced obligations for online marketplaces and fulfillment service providers to verify producer registration before offering products for sale, closing regulatory gaps that previously allowed unregistered products to enter the market (ElektroG 2024). Producer responsibilities for financing collection, treatment, and recovery have also been intensified, aligning with the EU's broader emphasis on full lifecycle accountability. In addition, Germany's waste management framework incorporates updated national recycling targets. According to the Federal Ministry for the Environment (BMUV 2023), municipal recycling goals have been revised in line with the amended EU Waste Framework Directive, with specific targets for e-waste collection and recovery set to increase over the next decade. These measures aim to reinforce circular economy principles by ensuring recovered materials, especially critical raw materials, are reintegrated into production cycles.

Before the implementation of the ElektroG, e-waste was collected by local public authorities and typically processed by contracted dismantling or recycling firms. Due to the lack of legal obligation to separate e-waste from household waste, many small electronic devices were not collected. To reduce transport costs, many authorities worked with regional contractors. Under the *value-conserving transport* model, only about 10% of collected items and components were suitable for reuse. Revenues came from both reuse and the recovery of valuable materials like metals. However, when processing costs exceeded these revenues, public authorities had to cover the deficit, financed through consumer waste management fees. With the introduction of ElektroG, a more centralized system was established, replacing regional contracts with national coordination through the EAR. Although this model improved cost-efficiency, it lowered the quality of collected e-waste. *Value-conserving transport* was no longer feasible, and *bulk transport* significantly limited reuse opportunities, reducing the reuse rate from 10% to around 3%. Consumers can now dispose of e-waste free of charge at official drop-off points, while producers bear the cost of collection and transportation (Walther et al. 2009).

ElektroG mandates that multiple authorized dismantling companies handle each e-waste category to prevent monopolies and promote competition, particularly in high-value waste streams (Walther and Spengler 2005). Supporting legislation, such as the Federal Immission Control Act (BImSchG) and the Ordinance on Specialized Waste Management Companies (EfbV), further enhances market competition while addressing issues like free-riding and cherry-picking. The EPR system requires all producers to contribute to e-waste processing and financing based on market share, discouraging free-rider behavior. Additionally, by obligating producers to cover the entire national territory for collection, the system prevents cherry-picking practices—such as targeting only urban areas and avoiding less profitable rural regions (Deubzer 2011). This system aligns with Germany's growing circular economy, as reflected in both Figure 10 and empirical forecasts, which project an approximate 4.2% increase in e-waste generation over the next 15 years. The success of the system is closely tied to the increase in the e-waste recycling rate.

Figure 10. Turnover of the waste collection industry in Germany (million Euro).

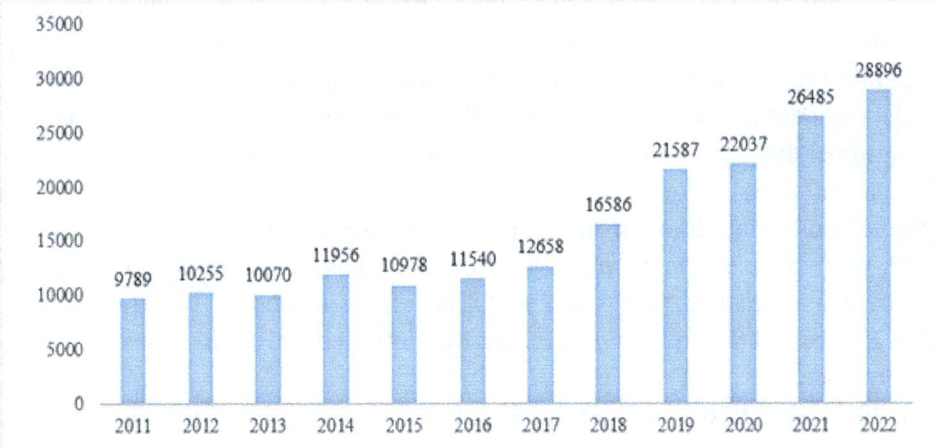

Source: Statista database

A review of the literature reveals that Walk (2009) remains the only published study specifically focused on e-waste forecasting in Germany, using a regional case study to project cathode ray tube (CRT) appliance disposal in Baden-Württemberg. This significant gap highlights the need for updated, nationwide forecasts that reflect modern e-waste streams and current policy frameworks such as the ElektroG system. Therefore, this study extends Walk's (2009) regional CRT waste forecasting by analyzing Germany's nationwide e-waste trends from 2026 to 2040 within the ElektroG framework, utilizing NARNET for AI-driven predictions. While Walk (2009) emphasized localized data and peak disposal periods, this research introduces a hybrid collection model that preserves value for high-value e-waste (e.g., smartphones) and uses bulk collection for low-value items, balancing cost efficiency with reuse potential. It also critically assesses the impact of ElektroG's centralized system, which has reduced reuse rates from 10% to 3%, and proposes policy adjustments to reintegrate circular economy principles. Unlike Walk's (2009) short-term, CRT-focused analysis, this study offers a long-term perspective and highlights public awareness as a key driver of sustainable e-waste management.

The interplay between regulatory design and forecasting accuracy is non-trivial. Changes in collection schemes, recovery targets, or compliance monitoring can shift both the level and volatility of collected e-waste, thereby affecting the parameters and residual patterns of forecasting models. NARNET's flexibility allows it to adapt to such shifts, but accuracy depends on the timely integration of updated policy-driven changes in collection behavior. This link between regulation and model performance underscores the importance of maintaining high-quality, granular data within the policy framework to support reliable long-term forecasts.

This study is subject to several limitations that should be acknowledged. First, the forecasting relies on historical e-waste generation data. It assumes that past patterns—adjusted for identified trends—will continue, which may not fully capture abrupt market shifts, policy changes, or technological disruptions. Second, the model does not explicitly integrate real-time behavioral data, such as fluctuations in consumer disposal habits or informal collection volumes, which can significantly affect actual e-waste flows. Third, the composition of e-waste is assumed to remain broadly consistent, although emerging device categories (e.g., Internet of Things (IoT), wearable electronics) may alter material recovery potential and processing requirements. Finally, the environmental benefit estimates are based on existing LCA studies and thus reflect general recovery efficiencies rather than site-specific performance.

Future research could address these limitations by integrating dynamic behavioral datasets, linking forecasting models with economic and policy simulation tools, and incorporating device-level technological trends into projection algorithms. Additionally, expanding the analysis to include scenario-based forecasts under varying regulatory, technological, and market conditions would provide policymakers with a more flexible decision-support tool. Applying hybrid AI models that combine NARNET with other machine learning architectures, such as LSTM or ensemble methods, could also enhance predictive accuracy and adaptability in rapidly evolving e-waste landscapes.

CONCLUSION

Different implementation models before and after the introduction of the ElektroG resulted in varying costs and revenues. From an ecological perspective, waste prevention is essential, whereas from an economic standpoint, overall cost efficiency is prioritized. To realize the economic potential of reuse, the distribution of costs among stakeholders plays a critical role. Prior to ElektroG, value-conserving transport of e-waste incurred high costs but enabled product reuse from an environmental standpoint. Under ElektroG, bulk collection and processing reduced overall costs but eliminated the possibility of reuse and value preservation. Consequently, a more balanced model has been proposed—one that is both economically and ecologically viable—based on cost-sharing among stakeholders and collaboration between public authorities and private companies.

The present study examines the e-waste forecast for the 2026-2040 period by using intelligence of NARNET. As it seems for Germany, e-waste will continue into the forecasted future. Moreover, by combining this idea with the current management system, for the sake of the circular economy and to increase the potential revenue gathered from e-waste, the best suggestion is to collect some e-waste that contains valuable materials (e.g., mobile phones and computers) with a value-preserving type of collection, and the other one, less valuable, with the bulk collection.

Technological trends will also influence the forecasted trajectory. Shorter product lifecycles, driven by rapid innovation, frequent model releases, and planned obsolescence, accelerate turnover and increase waste generation. The spread of multifunctional devices may reduce the total number of gadgets but can create larger, more complex products that are harder to repair or recycle. Conversely, advances in modular design, durability, and standardized components could slow replacement cycles and improve material recovery rates. The growing adoption of IoT and wearable electronics, although individually small, may collectively add new waste streams requiring adapted collection and processing systems. Incorporating such technological dynamics into forecasting models will refine projections by aligning them with shifts in product composition, lifespan, and recyclability.

Lastly, increasing public awareness is a key catalyst for sustainable e-waste management. Beyond regulatory and technical measures, consumer behavior strongly shapes e-waste flows. Factors such as awareness, convenience of collection points, and trust in recycling systems determine whether electronics enter formal channels or are diverted to informal disposal, resale, or storage. Informal practices reduce official collection volumes, lower material quality, and create unrecorded environmental impacts. Incorporating behavioral dynamics into forecasting models can improve accuracy by accounting for changes in participation rates and seasonal patterns. Targeted interventions—such as awareness campaigns, take-back incentives, and integrating informal collectors—can significantly influence the e-waste trajectory predicted by the model.

REFERENCES

Abbondanza, M. N. M., & Souza, R. G. (2019). Estimating the generation of household e-waste in municipalities using primary data from surveys: A case study of Sao Jose dos Campos, Brazil. *Waste Management (New York, N.Y.)*, *85*, 374–384.

Araújo, M. G., Magrini, A., Mahler, C. F., & Bilitewski, B. (2012). A model for estimation of potential generation of waste electrical and electronic equipment in Brazil. *Waste Management (New York, N.Y.)*, *32*, 335–342.

Alavi, N., Shirmardi, M., Babaei, A., Takdastan, A., & Bagheri, N. (2015). Waste electrical and electronic equipment (WEEE) estimation: A case study of Ahvaz City, Iran. *Journal of the Air & Waste Management Association*, *65*, 298–305.

Antoni, G., & Marzetti, G. (2019). Recycling and waste generation: An estimate of the source reduction effect of recycling programs. *Ecological Economics*, *161*, 321–329.

Awasthi, A., & Li, J. (2017). Management of electrical and electronic waste: A comparative evaluation of China and India. *Renewable & Sustainable Energy Reviews*, *76*, 434–447.

Awasthi, A., Cucchiella, F., D'Adamo, I., Li, J., Rosa, P., Terzi, S., & Zeng, X. (2018). Modelling the correlations of e-waste quantity with economic increase. *The Science of the Total Environment*, *613–614*, 46–53.

Baldé, C. P., Kuehr, R., Yamamoto, T., McDonald, R., D'Angelo, E., Althaf, S., & Wagner, M. (2024). *Global E-waste Monitor 2024. International Telecommunication Union (ITU) and United Nations Institute for Training and Research*. UNITAR.

Baldé, K., Wang, F., Huisman, J., & Kuehr, R. (2015). The Global E-Waste Monitor. United Nations University, IAS – SCYCLE.

BMUV. (Federal Ministry for the Environment, Nature Conservation, Nuclear Safety and Consumer Protection), (2023). Waste Management in Germany 2023: Facts, data, figures. Berlin: BMUV Printing Office

Brosilow, C., & Joseph, B. (2002). *Techniques of model-based control*. Prentice Hall.

Bogar, Z. O., & Gungor, A. (2023). Forecasting waste mobile phone (WMP) quantity and evaluating the potential contribution to the circular economy: A case study of Turkey. *Sustainability*, *15*(4), 3104.

Chaudhary, K., & Vrat, P. (2018). Case study analysis of e-waste management systems in Germany, Switzerland, Japan and India: A RADAR chart approach. *Benchmarking*, *25*(9), 3519–3540.

Cucchiella, F., D'Adamo, I., Koh, S., & Rosa, P. (2015). Recycling of WEEEs: An economic assessment of present and future e-waste streams. *Renewable & Sustainable Energy Reviews*, *51*, 263–272.

Cuestas, J., Gil-Alana, L., & Taylor, K. (2012). Inflation convergence in Central and Eastern Europe with a view to adopting the Euro. Sheffield: Sheffield Economics Research Papers no 2012005, Department of Economics, University of Sheffield.

Cui, J., & Forssberg, E. (2003). Mechanical recycling of waste electric and electronic equipment: A review. *Journal of Hazardous Materials*, *99*(3), 243–263.

Cui, J., & Zhang, L. (2008). Metallurgical recovery of metals from electronic waste: A review. *Journal of Hazardous Materials*, *158*(2–3), 228–256.

Chung, S. S., Lau, K. Y., & Zhang, C. (2011). Generation of and control measures for e-waste in Hong Kong. *Waste Management (New York, N.Y.)*, *31*, 544–554.

Desjardins, J. (2014). Global gold mine and deposit rankings 2013. *Visual Capitalist*. https://www.visualcapitalist.com/global-gold-mine-and-deposit-rankings-2013/

Deubzer, O. (2011). *E-waste management in Germany*. United Nations University Institute for Sustainability and Peace (UNU-ISP).

Di Vita, G. (2001). Technological change, growth and waste recycling. *Energy Economics*, *23*(5), 549–567.

Dimitrakakis, E., Janz, A., Bilitewski, B., & Gidarakos, E. (2009). Small WEEE: Determining recyclables and hazardous substances in plastics. *Journal of Hazardous Materials*, *161*(2–3), 913–919.

Duman, G. M., Kongar, E., & Gupta, S. M. (2019). Estimation of electronic waste using optimized multivariate grey models. *Waste Management & Research*, *37*(3), 241–249.

Duman, G. M., Kongar, E., & Gupta, S. M. (2020). Predictive analysis of electronic waste for reverse logistics operations: A comparison of improved univariate grey models. *Soft Computing*, *24*, 15747–15762.

Dwivedy, M., & Mittal, R. K. (2010a). Estimation of future outflows of e-waste in India. *Waste Management (New York, N.Y.)*, *30*, 483–491.

Dwivedy, M., & Mittal, R. K. (2010). Future trends in computer waste generation in India. *Waste Management (New York, N.Y.)*, *30*, 2265–2277.

Electronics TakeBack Coalition. (2016). Facts and figures on e-waste and recycling. https://blog.midwestind.com/wp-content/uploads/2017/11/Facts_and_Figures_on_EWaste_and_Recycling1.pdf

ElektroG. (2024). *Act on the placing on the market, return and environmentally sound disposal of electrical and electronic equipment (Electrical and Electronic Equipment Act)*. https://www.umweltbundesamt.de/en/topics/waste-resources/product-stewardship-waste-management/electrical-electronic-waste/electrical-electronic-equipment-act#purpose-of-the-elektrog

Erumban, A., & Das, D. (2016). Information and communication technology and economic growth in India. *Telecommunications Policy*, *40*(5), 412–431.

Forti, V., Baldé, C., Kuehr, R., & Bel, G. (2020). *The Global E-waste Monitor 2020: Quantities, flows and the circular economy potential. United Nations University (UNU)/International Telecommunication Union (ITU)/International Solid Waste Association*. ISWA.

Garlapati, V. (2016). E-waste in India and developed countries: Management, recycling, business and biotechnological initiatives. *Renewable & Sustainable Energy Reviews*, *54*, 874–881.

Gavin, H. P. (2020). The Levenberg-Marquardt algorithm for nonlinear least squares curve-fitting problems. https://people.duke.edu/~hpgavin/ce281/lm.pdf

George, D., Chiang Lin, B., & Chen, Y. (2015). A circular economy model of economic growth. *Environmental Modelling & Software*, *73*, 60–63.

Golev, A., Schmeda-Lopez, D. R., Smart, S. K., Corder, G. D., & McFarland, E. W. (2016). Where next on e-waste in Australia? *Waste Management (New York, N.Y.)*, *58*, 348–358.

Golev, A., Corder, G. D., & Rhamdhani, M. A. (2019). Estimating flows and metal recovery values of waste printed circuit boards in Australian e-waste. *Minerals Engineering*, *137*, 171–176.

Gonda, L., D'Ans, P., & Degrez, M. (2019). A comparative assessment of WEEE collection in an urban and rural context: Case study on desktop computers in Belgium. *Resources, Conservation and Recycling*, *142*, 131–142.

Guo, X., & Yan, K. (2017). Estimation of obsolete cellular phones generation: A case study of China. *The Science of the Total Environment*, *575*, 321–329.

Gusukuma, M., & Kahhat, R. (2018). Electronic waste after a digital TV transition: Material flows and stocks. *Resources, Conservation and Recycling*, *138*, 142–150.

Güriş, B., Tiftikçigil, B., & Tıraşoğlu, M. (2017). Testing for unemployment hysteresis in Turkey: Evidence from nonlinear unit root tests. *Quality & Quantity*, *51*, 35–46.

Hagelüken, C. (2006). Improving metal returns and eco-efficiency in electronics recycling—A holistic approach for interface optimisation between. *2006 IEEE International Symposium on Electronics and the Environment* (pp. 218–223). IEEE.

Hagelüken, C. (2014). Recycling of (critical) metals. In Gunn, G. (Ed.), *Critical Metals Handbook* (pp. 41–69). Wiley.

Harris, D., McCabe, B., & Leybourne, S. (2003). Some limit theory for autocovariances whose order depends on sample size. *Econometric Theory*, *19*, 829–864.

Harvey, D., Leybourne, S., & Xiao, B. (2008). A powerful test for linearity when the order of integration is unknown. (Article 2). *Studies in Nonlinear Dynamics and Econometrics*, *12*(3), 1–24.

He, A., & Jin, X. (2018). NARNET-based prognostics modeling for deteriorating systems under dynamic operating conditions. In *Proceedings of the 14th IEEE International Conference on Automation Science and Engineering (CASE)* (pp. 1–6). IEEE.

Hyndman, R. J., & Athanasopoulos, G. (2018). *Forecasting: Principles and practice* (2nd ed.). OTexts.

Ibrahim, M., Jemei, S., Wimmer, G., & Hissel, D. (2016). Nonlinear autoregressive neural network in an energy management strategy for battery/ultra-capacitor hybrid electrical vehicles. *Electric Power Systems Research*, *136*, 262–269.

Ikhlayel, M. (2016). Differences of methods to estimate generation of waste electrical and electronic equipment for developing countries: Jordan as a case study. *Resources, Conservation and Recycling*, *108*, 134–139.

Islam, R., Nazifa, T., Yuniarto, A., Uddin, A., Salmiati, S., & Shahid, S. (2019). An empirical study of construction and demolition waste generation and implication of recycling. *Waste Management (New York, N.Y.)*, *95*, 10–21.

Islam, M. T., & Huda, N. (2019). E-waste in Australia: Generation estimation and untapped material recovery and revenue potential. *Journal of Cleaner Production*, *237*, 117787.

Işıkdağ, Ü., Hepsağ, A., Bıyıklı, S. İ., Öz, D., Bekdaş, G., & Geem, Z. W. (2023). Estimating construction material indices with ARIMA and optimized NARNETs. *Computers, Materials & Continua*, *74*(1), 113–129.

Jain, A., & Sareen, R. (2006). E-waste assessment methodology and validation in India. *Journal of Material Cycles and Waste Management*, *8*, 40–45.

Jorgenson, D., Ho, M., & Samuels, J. (2016). The impact of information technology on postwar US economic growth. *Telecommunications Policy*, *40*(5), 398–411.

Kalmykova, Y., Patricio, J., Rosado, L., & Berg, P. E. (2015). Out with the old, out with the new—The effect of transitions in TVs and monitors technology on consumption and WEEE generation in Sweden 1996–2014. *Waste Management (New York, N.Y.)*, *46*, 511–522.

Kastanaki, E., & Giannis, A. (2022). Forecasting quantities of critical raw materials in obsolete feature and smart phones in Greece: A path to circular economy. *Journal of Environmental Management*, *306*, 114566.

Kazancoglu, Y., Ozbiltekin, M., Ozen, Y. D. O., & Sagnak, M. (2020). A proposed sustainable and digital collection and classification center model to manage e-waste in emerging economies. *Journal of Enterprise Information Management*, *34*(1), 267–291.

Kececi, B., Dengiz, O., Dengiz, B., Sumer, E., Kilic, A., Ceki, E., Inan, B., & Cicek, S. (2018). WEEE estimation and determination of collection points: A case for the Municipality of Cankaya. *Pamukkale University Journal of Engineering Sciences*, *24*(5), 692–704.

Kheimi, M., Almadani, M., & Zounemat-Kermani, M. (2024). Stochastic (S[ARIMA]), shallow (NARnet, NAR-GMDH, OS-ELM), and deep learning (LSTM, Stacked-LSTM, CNN-GRU) models, application to river flow forecasting. *Acta Geophysica*, *72*(4), 2679–2693.

Khraief, N., Shahbaz, M., Heshmati, A., & Azam, M. (2020). Are unemployment rates in OECD countries stationary? Evidence from univariate and panel unit root tests. *The North American Journal of Economics and Finance*, *51*(100838), 1–15.

Kiddee, P., Naidu, R., & Wong, M. (2013). Electronic waste management approaches: An overview. *Waste Management (New York, N.Y.)*, *33*(5), 1237–1250.

Kim, S., Oguchi, M., Yoshida, A., & Terazono, A. (2013). Estimating the amount of WEEE generated in South Korea by using the population balance model. *Waste Management (New York, N.Y.)*, *33*, 474–483.

Kosai, S., Kishita, Y., & Yamasue, E. (2020). Estimation of the metal flow of WEEE in Vietnam considering lifespan transition. *Resources, Conservation and Recycling*, *154*, 104621.

Koshta, N., Patra, S., & Singh, S. P. (2021). Estimation of E-waste at micro level for reverse logistics: A case of Delhi. *Journal of Cleaner Production*, *314*, 128063.

Kumar, A., Holuszko, M., & Espinosa, D. (2017). E-waste: An overview on generation, collection, legislation and recycling practices. *Resources, Conservation and Recycling*, *122*, 32–42.

Kumar, P., & Shrihari, S. (2007). Estimation and material flow analysis of waste electrical and electronic equipment (WEEE)—A case study of Mangalore City, Karnataka, India. *Proceedings of the International Conference on Sustainable Solid Waste Management* (pp. 5–7). Chennai, India.

Kuehr, R., Velasquez, G., & Williams, E. (2003). Computers and the environment: Understanding and managing their impact. In Kuehr, R., & Williams, E. (Eds.), *Computers and the Environment* (pp. 1–17). Kluwer Academic Publishers.

Kuper, J., & Hojsik, M. (2008). *Poisoning the poor: Electronic waste in Ghana*. Greenpeace.

Kusch, S., & Hills, C. (2017). The link between e-waste and GDP—New insights from data from the pan-European region. *Resources*, *6*(1), 15.

Legates, D. R., & Davis, R. E. (1999). Evaluating the use of "goodness-of-fit" measures in hydrologic and hydroclimatic model validation. *Water Resources Research*, *35*(1), 233–241.

Lèbre, É., & Corder, G. (2015). Integrating industrial ecology thinking into the management of mining waste. *Resources*, *4*(4), 765–786.

Li, J., Yang, J., & Liu, L. (2015a). Development potential of e-waste recycling industry in China. *Waste Management & Research*, *33*(6), 533–542.

Li, J., Zeng, X., Chen, M., Ogunseitan, O., & Stevels, A. (2015). Control-Alt-Delete: Rebooting solutions for the e-waste problem. *Environmental Science & Technology*, *49*(12), 7095–7108.

Li, B., Yang, J., Lu, B., & Song, X. (2015b). Estimation of retired mobile phones generation in China: A comparative study on methodology. *Waste Management (New York, N.Y.)*, *35*, 247–254.

Li, A., Li, B., Lu, B., Yang, D., Hou, S., & Song, X. (2022). Generation estimation and material flow analysis of retired mobile phones in China. *Environmental Science and Pollution Research International*, *29*, 75626–75635.

Liu, Y., Park, S., Yi, H., & Feiock, R. (2020). Evaluating the employment impact of recycling performance in Florida. *Waste Management (New York, N.Y.)*, *101*, 283–290.

Lohse, J., Winteler, S., & Wulf-Schnabel, J. (1998). Collection targets for waste from electrical and electronic equipment (WEEE) the directorate general (DG XI) environment. In Nuclear safety and civil protection of the Commission of the European Communities. Ökopol Institut für Ökologie und Politik.

Magalini, F., & Huisman, J. (2007). Management of WEEE and cost models across the EU: Could the EPR principle lead us to a better environmental policy? In *2007 IEEE International Symposium on Electronics and the Environment* (pp. 143–148). IEEE.

Magazzino, C., Mele, M., & Schneider, N. (2020). The relationship between municipal solid waste and greenhouse gas emissions: Evidence from Switzerland. *Waste Management (New York, N.Y.)*, *113*, 508–520.

Matthews, H. S., McMichael, F. C., Hendrickson, C. T., & Hart, D. J. (1997). *Disposition and end-of-life options for personal computers*. Carnegie Mellon University.

McLeod, C. (2014). Top primary silver miners facing declining ore grades. *Investing News Network*. https://investingnews.com/daily/resource-investing/precious-metals-investing/silver-investing/top-primary-silver-miners-facing-declining-ore-grades/

MathWorks. (2022). MATLAB implementation of the Levenberg-Marquardt algorithm. MATLAB & Simulink Documentation. https://www.mathworks.com/help/deeplearning/ref/trainlm.html

Namias, J. (2013). The future of electronic waste recycling in the United States: Obstacles and domestic solutions [Master's thesis, Columbia University].

Oguchi, M., Kameya, T., Yagi, S., & Urano, K. (2008). Product flow analysis of various consumer durables in Japan. *Resources, Conservation and Recycling*, *52*, 463–480.

Ongondo, F., Williams, I., & Cherrett, T. (2011). How are WEEE doing? A global review of the management of electrical and electronic wastes. *Waste Management (New York, N.Y.)*, *31*(4), 714–730.

Oughton, C., Kurup, B., Anda, M., & Ho, G. (2022). Industrial symbiosis to circular economy: What does the literature reveal for a successful complex industrial area? *Circular Economy and Sustainability*, *2*, 1317–1344.

Öztürk, T. (2014). Generation and management of electrical–electronic waste (e-waste) in Turkey. *Journal of Material Cycles and Waste Management*, *17*, 411–421.

Owusu-Sekyere, K., Batteiger, A., Afoblikame, R., Hafner, G., & Kranert, M. (2022). Assessing data in the informal e-waste sector: The Agbogbloshie Scrapyard. *Waste Management (New York, N.Y.)*, *139*, 158–167.

Pant, D. (2013). E-waste projection using life-span and population statistics. *The International Journal of Life Cycle Assessment*, *18*, 1465–1469.

Peralta, G. L., & Fontanos, P. M. (2006). E-waste issues and measures in the Philippines. *Journal of Material Cycles and Waste Management*, *8*, 34–39.

Petridis, N. E., Stiakakis, E., Petridis, K., & Dey, P. (2016). Estimation of computer waste quantities using forecasting techniques. *Journal of Cleaner Production*, *112*, 3072–3085.

Pinheiro, E. A., Leal Júnior, J. B. V., de Lima, L. C., & de Melo, F. G. (2020). Nonlinear autoregressive neural networks for forecasting wind speed time series. *International Journal of Developmental Research*, *10*(9), 40336–40343.

Pistilli, M. (2016). World-class copper deposits. *Investing News Network*. https://investingnews.com/daily/resource-investing/base-metals-investing/copper-investing/what-makes-a-world-class-copper-deposit/

Polak, M., & Drapalova, L. (2012). Estimation of end of life mobile phones generation: The case study of the Czech Republic. *Waste Management (New York, N.Y.)*, *32*, 1583–1591.

Qu, Y., Wang, W., Liu, Y., & Zhu, Q. (2019). Understanding residents' preferences for e-waste collection in China—A case study of waste mobile phones. *Journal of Cleaner Production*, *228*, 52–62.

Ravindra, K., & Mor, S. (2019). E-waste generation and management practices in Chandigarh, India and economic evaluation for sustainable recycling. *Journal of Cleaner Production*, *221*, 286–294.

Robinson, B. (2009). E-waste: An assessment of global production and environmental impacts. *The Science of the Total Environment*, *408*(2), 183–191.

Rotter, V., Chancerel, P., & Schill, W. (2009). Implementing individual producer responsibility (IPR) under the European WEEE Directive – Experiences in Germany. In *2009 IEEE International Symposium on Sustainable Systems and Technology* (pp. 1–6). IEEE.

Ruiz, L. G. B., Cuéllar, M. P., Calvo-Flores, M. D., & Jiménez, M. D. C. P. (2016). An application of non-linear autoregressive neural networks to predict energy consumption in public buildings. *Energies*, *9*(9), 684.

Ryan-Fogarty, Y., Coughlan, D., & Fitzpatrick, C. (2021). Quantifying WEEE arising in scrap metal collections: Method development and application in Ireland. *Journal of Industrial Ecology*, *25*(4), 1021–1033.

Saavedra, Y., Iritani, D., Pavan, A., & Ometto, A. (2018). Theoretical contribution of industrial ecology to circular economy. *Journal of Cleaner Production*, *170*, 1514–1522.

Sajid, M., Syed, J. H., Iqbal, M., Abbas, Z., Hussain, I., & Baig, M. A. (2019). Assessing the generation, recycling and disposal practices of electronic/electrical-waste (E-Waste) from major cities in Pakistan. *Waste Management (New York, N.Y.)*, *84*, 394–401.

Sakthivel, U., Swaminathan, G., & Anis, J. (2022). Strategies for Quantifying Metal Recovery from Waste Electrical and Electronic Equipment (WEEE/E-waste) Using Mathematical Approach. *Process Integration and Optimization for Sustainability*, *6*(4), 781–790.

Salihoğlu, G., & Kahraman, A. E. (2016). Electrical and Electronical Waste Generation in Turkey: Bursa Case Study. *Uludağ University Journal of the Faculty of Engineering*, *21*(1), 95–106.

Schluep, M., Hagelüken, C., Kuehr, R., Magalini, F., Maurer, C., Meskers, C., & Wang, F. (2009). *Recycling: From e-waste to resources*. United Nations Environment Programme & United Nations University.

Schüler, D., Buchert, M., Liu, R., Dittrich, S., & Merz, C. (2011). Study on the review of the list of critical raw materials: Final report. Öko-Institut e.V.

Scruggs, C., Nimpuno, N., & Moore, R. (2016). Improving information flow on chemicals in electronic products and e-waste to minimize negative consequences for health and the environment. *Resources, Conservation and Recycling*, *113*, 149–164.

Song, Q., Li, J., Liu, L., Dong, Q., Yang, J., Liang, Y., & Zhang, C. (2016). Measuring the generation and management status of waste office equipment in China: A case study of waste printers. *Journal of Cleaner Production*, *112*(Part 5), 4461–4468.

Sthiannopkao, S., & Wong, M. (2013). Handling e-waste in developed and developing countries: Initiatives, practices, and consequences. *The Science of the Total Environment, 463–464*, 1147–1153.

Steubing, B., Boni, H., Schluep, M., Silva, U., & Ludwig, C. (2010). Assessing computer waste generation in Chile using material flow analysis. *Waste Management (New York, N.Y.), 30*, 473–482.

Streicher-Porte, M., Widmer, R., Jain, A., Bader, H.-P., Scheidegger, R., & Kytzia, S. (2005). Key drivers of the e-waste recycling system: Assessing and modelling e-waste processing in the informal sector in Delhi. *Environmental Impact Assessment Review, 25*, 472–491.

Tasaki, T., Takasuga, T., Osako, M., & Sakai, S. (2004). Substance flow analysis of brominated flame retardants and related compounds in waste TV sets in Japan. *Waste Management (New York, N.Y.), 24*, 571–580.

Thiébaud-Müller, E., Hilty, L. M., Schluep, M., Widmer, R., & Faulstich, M. (2018). Service Lifetime, Storage Time, and Disposal Pathways of Electronic Equipment: A Swiss Case Study. *Journal of Industrial Ecology, 22*(2), 196–208.

Tran, H. P., Schaubroeck, T., Nguyen, D. Q., Ha, V. H., Huynh, T. H., & Dewulf, J. (2018). Material flow analysis for management of waste TVs from households in urban areas of Vietnam. *Resources, Conservation and Recycling, 139*, 78–89.

Tsydenova, O., & Bengtsson, M. (2011). Chemical hazards associated with treatment of waste electrical and electronic equipment. *Waste Management (New York, N.Y.), 31*(1), 45–58.

Turner, D., Williams, I., & Kemp, S. (2015). Greenhouse gas emission factors for recycling of source-segregated waste materials. *Resources, Conservation and Recycling, 105*, 186–197.

Velis, C., & Mavropoulos, A. (2016). Unsound waste management and public health: The neglected link? *Waste Management & Research, 34*(4), 277–279.

Walk, W. (2009). Forecasting quantities of disused household CRT appliances—A regional case study approach and its application to Baden-Wurttemberg. *Waste Management (New York, N.Y.), 29*, 945–951.

Walther, G., Steinborn, J., Spengler, T., Luger, T., & Herrmann, C. (2009). Implementation of the WEEE–Directive – Economic effects and improvement potentials for reuse and recycling in Germany. *International Journal of Advanced Manufacturing Technology, 47*, 461–474.

Wang, F., Huisman, J., Stevels, A., & Balde, C. P. (2013). Enhancing e-waste estimates: Improving data quality by multivariate Input-Output Analysis. *Waste Management (New York, N.Y.)*, *33*, 2397–2407.

Wang, M., Tan, Q., Chiang, J., & Li, J. (2017). Recovery of rare and precious metals from urban mines—A review. *Frontiers of Environmental Science & Engineering*, *11*(5), 1–17.

Wang, M., You, X., Li, X., & Liu, G. (2018). Watch more, waste more? A stock-driven dynamic material flow analysis of metals and plastics in TV sets in China. *Journal of Cleaner Production*, *187*, 730–739.

Wang, F., Yu, L., & Wu, A. (2021). Forecasting the electronic waste quantity with a decomposition-ensemble approach. *Waste Management (New York, N.Y.)*, *120*, 828–838.

Wilburn, D. (2012). Byproduct metals and rare-earth elements used in the production of light-emitting diodes—Overview of principal sources of supply and material requirements for selected markets. *U.S. Geological Survey*.

Yang, Y., & Williams, E. (2009). Logistic model-based forecast of sales and generation of obsolete computers in the U.S. *Technological Forecasting and Social Change*, *76*, 1105–1114.

Yang, C., Tan, Q., Liu, L., Dong, Q., & Li, J. (2017). Recycling tin from electronic waste: A problem that needs more attention. *ACS Sustainable Chemistry & Engineering*, *5*(11), 9586–9598.

Yazici, B., Can, Z. S., & Calli, B. (2014). Prediction of future disposal of end-of-life refrigerators containing CFC-11. *Waste Management (New York, N.Y.)*, *34*, 162–166.

Zeinali, S., Dehghani, M., & Talebbeydokhti, N. (2021). Artificial neural network for the prediction of shoreline changes in Narrabeen, Australia. *Applied Ocean Research*, *107*, 102362.

Zeng, X., Yang, C., Chiang, J., & Li, J. (2017). Innovating e-waste management: From macroscopic to microscopic scales. *The Science of the Total Environment*, *575*, 1–5.

Zeng, X., Ali, S. H., & Li, J. (2021). Estimation of waste outflows for multiple product types in China from 2010–2050. *Scientific Data*, *8*(1), 15.

Zhang, S., Gu, Y., Tang, A., Li, B., Li, B., Pan, D., & Wu, Y. (2021). Forecast of future yield for printed circuit board resin waste generated from major household electrical and electronic equipment in China. *Journal of Cleaner Production*, *283*, 124575.

Zhao, M., Zhao, C., Yu, L., Li, G., Huang, J., Zhu, H., & He, W. (2016). Prediction and Analysis of WEEE in China Based on the Gray Model. *Procedia Environmental Sciences, 31*, 925–934.

KEY TERMS AND DEFINITIONS

Circular Economy: The concept of prolonging the usable life of goods, materials, or resources is known as the "circular economy". The use of resources inside closed-loop systems, lowering pollution or preventing resource leakage while maintaining economic growth, is a key component of the circular economy idea. Furthermore, this strategy is the most recent attempt to conceive how to combine economic activity and environmental well-being.

E-Waste: E-waste covers a wide and growing range of consumer electronics, from large household appliances to small appliances, such as refrigerators, air conditioners, cell phones, stereos, and computers, that are discarded by their users.

NARNET: NARNET (Nonlinear Autoregressive Neural Network) is a type of neural network model used for time series forecasting. It predicts future values by using past observations of the same series, capturing nonlinear relationships in the data. Unlike linear autoregressive models, NARNET can model complex patterns and dynamics through its neural network architecture, making it effective for applications such as weather prediction, financial forecasting, and energy demand estimation.

Nonlinearity: It refers to situations where the relationship between past and current values is not proportional or constant. This means the effect of past data on future values can change depending on factors like regime shifts, thresholds, or time-varying conditions, making the model more complex than linear ones.

APPENDIX

Table 5. Information on the e-products used in the creation of the TEW series

	Classification	E-products	Wt of item (kg)	Typical life (year)
1	**Temperature exchange equipments**			
		Air conditioning	30	15
		Dehumidifying	10	10
		Freezers	50	20
		Heat pumps	50	20
		Refrigerators	60	20
2	**Screens**			
		Laptops	1	7
		LCD photo frames	0.5	10
		Monitors	2	15
		Notebooks	1	7
		Screens	1	15
		Televisions	10	15
3	**Lamps**			
		Compact fluorescent lamps	0.2	15
		Fluorescent lamps	0.5	10
		High intensity discharge lamps	1	15
		Low pressure sodium lamps	0.5	10
		Straight fluorescent lamps	0.5	10
4	**Large equipments**			
		Appliances for knitting and weaving	5	20
		Clothes dryers	40	15
		Cookers	5	20
		Copying equipment	10	20
		Dish washing machines	50	15
		Electric hot plates	1	10
		Electric stoves	20	25
		Large appliances which automatically deliver products and money	100	20
		Large coin slot machines	100	30
		Large computer-mainframes	500	20

continued on following page

Table 5. Continued

	Classification	E-products	Wt of item (kg)	Typical life (year)
		Large monitoring and control instruments	10	25
		Large printing machines	200	20
		LED	0.05	25
		Luminaires	1	15
		Photovoltaic panels	10	30
		Washing machines	50	15
5	**Small equipments**			
		Appliances for sewing	5	20
		Calculators	0.1	10
		Carpet sweepers	2	10
		Clocks and Watches	0.1	10
		Computers for biking, diving, running, rowing, etc.	0.5	5
		Electric kettles	1	10
		Electric knives	0.5	10
		Electric shavers	0.5	10
		Hi-fi equipment	5	20
		Irons	1	20
		Microwaves	15	10
		Radio sets	2	15
		Small appliances which automatically deliver products	0.5	15
		Small equipment with integrated photovoltaic panels	0.1	25
		Small monitoring and control instruments	0.5	15
		Smoke detectors	0.1	10
		Sound recorder	1	10
		Thermostats	0.2	20
		Toasters	1	15
		Vacuum cleaners	4	15
		Video cameras	1	10
6	**Small IT and telecommunication equipments**			
		GPS	0.1	10
		Mobile phones	0.1	10

continued on following page

Table 5. Continued

Classification		E-products	Wt of item (kg)	Typical life (year)
		Personal computers	5	10
		Pocket calculators	0.1	10
		Printers	3	10
		Routers	0.5	10
		Telephones	1	10

Source: Authors' internet research based on WEEE Directive (2012/19/EU)

Chapter 9
AI-Optimized Nanoelectronics From Design to Fabrication:
Nanotechnology in Device Fabrication

Manjunatha Badiger
https://orcid.org/0000-0001-8073-0270
NMAM Institute of Technology, Nitte University, India

P. S. Sushma
NMAM Institute of Technology, Nitte University, India

Savidhan Shetty C. S.
https://orcid.org/0009-0007-3032-1736
NMAM Institute of Technology, Nitte University, India

Tanya Mendez
NMAM Institute of Technology, Nitte University, India

Jose Alex Mathew
AJ Institute of Engineering and Technology, India

Varuna Kumara
Moodlakatte Institute of Technology, India

ABSTRACT

Nanoelectronics development experiences a transformation through the union of artificial intelligence (AI) with nanotechnology, which leads to highly precise and high-performing devices. The chapter demonstrates how AI approaches enhance nanoelectronic device creation through the systematic optimization of each development stage, up to fabrication and testing phases. The article introduces nanoelectronics transistors, photonic devices, and biomedical implants as major applications while describing how AI handles design optimization, defect detection, performance predictions, and multi-scale simulations. The power of AI emerges in

DOI: 10.4018/979-8-3373-2883-6.ch009

specific cases, which show its ability to revolutionize transistor fabrication, photonic crystal structure development, and electronic defect analysis. At the end of the chapter, problems like limited data availability, scalability issues, and unreliability are talked about. Next, the authors look at the future possibilities for self-operating fabrication equipment and AI-quantum computing approaches that work together.

1. INTRODUCTION

The merger of artificial intelligence along with nanotechnology drives nanoelectronics research toward building efficient, miniature yet powerful electronic devices. Nanotechnology has transformed multiple industries through its work in semiconductor production, along with photonics and bio applications (Hossain et al., 2023). Traditional fabrication methods for nanoelectronics face several obstacles, including difficulties in achieving exactness, large-scale production, and managing manufacturing errors. AI-driven approaches resolve these limitations through powerful data analytics algorithms and optimized machine learning models, which upgrade device reliability and performance levels (Biswas et al., 2024). AI powers quick research and development (R&D) advancement in nanoelectronics through automated complex computations, predictive models, and optimized production methods. The integration of AI algorithms improves material choice processes, develops device architectures, and implements real-time error correction, which together enhance the efficiency and cost-effectiveness of fabrication processes (Soori et al., 2025). AI's ability to detect manufacturing defects and forecast component failures leads to higher output rates in the semiconductor industry by minimizing production losses and expanding capacity. AI-based nanoelectronics achieves several advantages, including the ability to perform multi-scale simulations that link atomic-level material interactions with large-scale device performance. Research predictability through simulation methods allows scientists to optimize the behavior of devices under different operating environments, thus enabling the development of future-generation transistors, photonic devices, and biomedical implants (Fu et al., 2024). Artificial intelligence supports innovative component design through its capability to discover the ultimate configurations for complex nanoelectronic elements. This section examines the various ways AI optimizes nanoelectronic devices, starting from design to fabrication and ending with performance forecasting. The main research areas in this field encompass AI-driven lithography solutions, nanoscale deposition methodologies, and defect spotting capabilities and simulation across multiple scales (Sharma et al., 2024). Our analysis includes specific examples that demonstrate how AI transforms the design of transistors, photonic crystal structures, and biomedical nanoelectronics. The present research also examines prominent obstacles such as limited data availability, unpredictable results, and growing requirements, as these substantially inhibit AI implementation in nanoelectronics on a broad scale.

The flowchart depicted in Figure 1 illustrates the role of AI and ML in optimizing nanoelectronics fabrication. Every step of the fabrication operation benefits from AI technology, which includes material seeking and defect finding alongside production parameter improvement.

1. **AI-Assisted Material Selection** –A large pool of electrical, mechanical, and thermal data feeds into AI systems, which selects optimum materials for use.
2. **AI-Optimized Nanolithography** –Deep learning technology enables improvements in nanoscale printing capabilities, thus improving transistor and circuit design performance.
3. **ML-Based Process Optimization** –AI implements dynamic parameter control through ML models, which enables uniformity in nanoscale deposition procedures.

Figure 1. Role of AI and ML in optimizing nanoelectronics fabrication

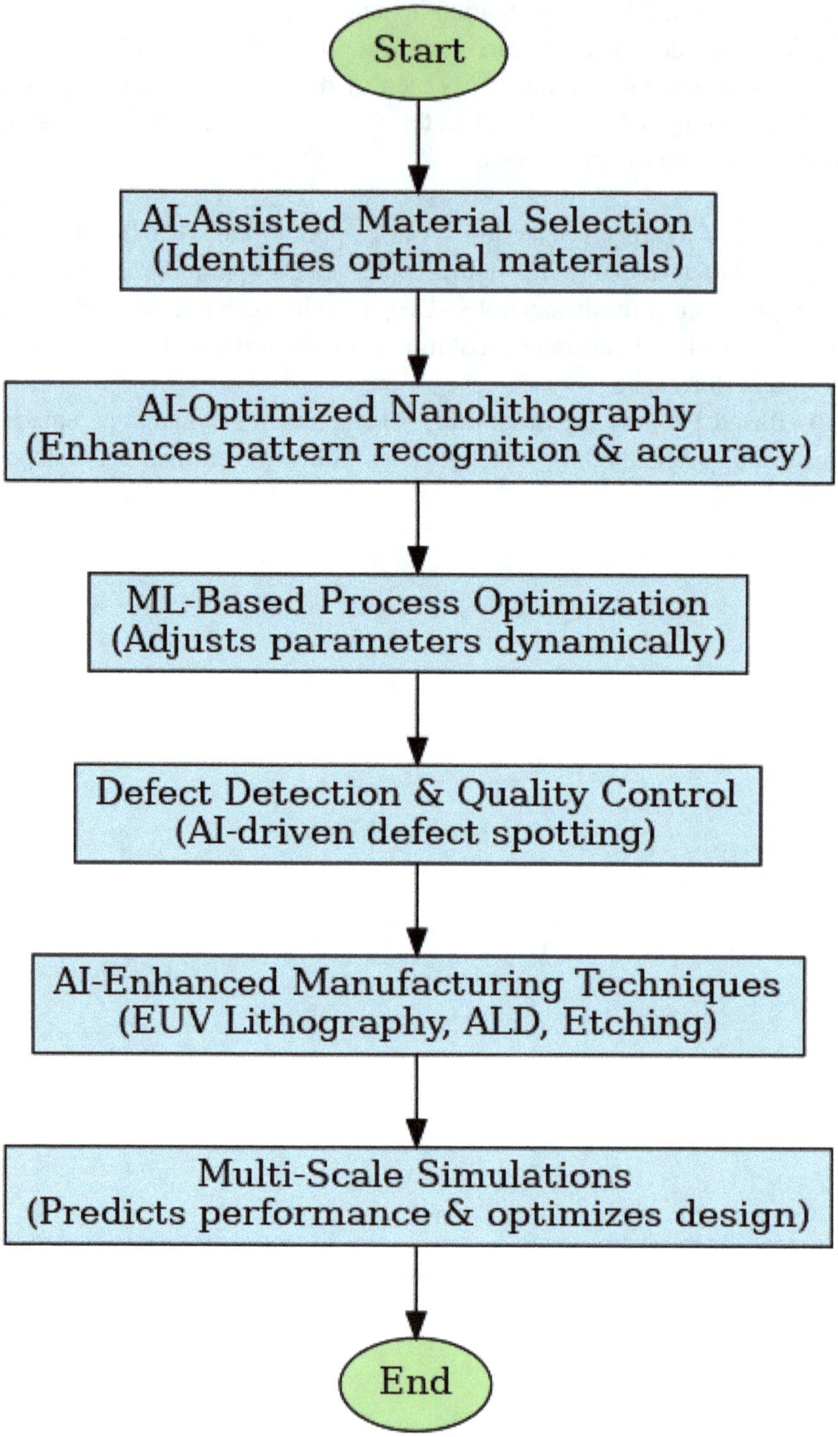

4. **Defect Detection & Quality Control** –The combination of AI and computer vision results in systems that identify atomic-scale defects, which enhances manufacturing yield for quality control purposes.
5. **AI-Enhanced Manufacturing Techniques** –The combination of AI and manufacturing enables optimized results for Extreme Ultraviolet (EUV) Lithography, Atomic Layer Deposition (ALD), and Nanoscale Etching fabrication methods.
6. **Multi-Scale Simulations** – Through predictive models controlled by AI researchers, optimize nanoelectronic devices by executing multi-scale simulations from atomic material interactions to large-scale device assessment.

2. AI IN NANOELECTRONIC DEVICE DESIGN

The advancement of nanoelectronic device design heavily relies on artificial intelligence, which plays essential roles in structure optimization, improved device performance, and new materials discovery (Olawade et al., 2024). In nanoelectronics, measurements are made in atoms and molecules, so it requires complex mathematical and computer methods to design and create devices. AI methodologies deliver strong solutions through automated optimization algorithms that reduce computational expenses while shortening innovation development periods. An AI application requires the initial development of fundamental competencies across multiple specific domains in the first place. In this field, nanotechnology, along with nanoelectronics, serves as an essential component, requiring a fundamental understanding of nanomaterials, transistor designs, and photonic equipment. Scientists working on electronic devices need to understand the behavior of quantum and atomic structures in electronic components, as traditional electrical principles become inadequate at this scale. AI technology explores and optimizes the unique electronic, mechanical, and optical properties of emerging materials like graphene, carbon nanotubes, and transition metal dichalcogenides (TMDs) (Bai & Zhang, 2025). Machine learning (ML) and artificial intelligence (AI) algorithms make decisions based on extensive data, which is very helpful for nanoelectronic design. One must understand deep learning alongside neural networks, reinforcement learning, and evolutionary algorithms to create AI-based optimization procedures. By implementing these methods, researchers explore intricate design areas to find perfect transistor designs while predicting how devices will operate across different situations. The analytical power of AI models permits large-scale analysis of experimental data and simulation results, allowing them to discover pattern connections that tend to hide from conventional techniques (Collins et al., 2021). The study of semiconductor physics is essential because all nanoelectronic devices require semiconductor materials for their construction. Knowledge about electrical conductivity, carrier mobility, thermal management, and fabrication techniques serves as essential criteria when designing high-performance devices. Semiconductor behavior at ultra-small sizes shows dissimilarities with bulk materials because of quantum mechanics effects, which need accurate simulation and development techniques. The analysis of material properties is conducted with AI, which predicts how materials behave and optimizes device operational efficiency.

The complete picture of material behavior, circuit optimization, and predictive device outcomes requires computational modeling. Scientific researchers who master simulation tools, such as density functional theory (DFT), molecular dynamics (MD), finite element analysis (FEA), and electronic circuit simulators, effectively gain AI-driven analytical capabilities. AI boosts modeling techniques through computational optimization that lowers operation costs and provides enhanced result predictions.

AI demonstrates its effectiveness in nanoelectronic device design by optimizing their design and selecting appropriate materials (Batool et al., 2024). The optimization of transistor structures and improvements in circuit performance due to AI are part of design optimization procedures, which also lead to decreased power consumption. Through artificial intelligence, researchers can make use of generative design, which evaluates diverse system configurations to identify their most optimized forms. The discovery of high-performance nanomaterials for next-generation semiconductor technologies is accelerated by AI-driven informatics and predictive modeling in material selection. Integrating artificial intelligence technologies into nanoelectronic systems allows researchers to extend innovation capabilities, reduce development expenses, and create devices that outperform traditional engineering standards (Krishnan et al., 2024).

2.1 AI-Driven Design Optimization

The field of nanoelectronic devices is undergoing a significant transformation due to AI-driven design optimization, which analyzes a wide range of design parameters to identify the most effective configurations. Traditional methods often struggle with the complex optimization challenges associated with component architectures, circuit layouts, and material properties, leading to difficulties in enhancing performance and reducing costs. Machine learning algorithms within AI offer a robust solution for design optimization and structural enhancement through diverse design exploration and efficiency enhancement (Pai et al., 2025). The fundamental structure of nanoelectronic systems benefits significantly from AI optimization techniques, which make up one of the key applications in this field.

Optimizing Device Architectures:

Artificial intelligence technology, through deep learning and reinforcement learning, enables the precise refinement of important nanoelectronic features, ranging from transistors to photonic crystal structures and nanoscale circuits. Through AI-driven techniques, engineers can examine massive design areas to discover performance-enhancing, reliable configurations. Multiple design parameters are analyzed through AI systems to enable engineers to create efficient architectures for specific applications (Yüksel et al., 2023). The main goal of AI-guided design optimization is to reduce power usage, as this directly supports the development of energy-efficient electronic systems. The reduction in transistor dimensions leads to two critical challenges: leakage currents and power consumption. Through AI algorithms, designers obtain power efficiency forecasts that help them identify the most energy-conserving layouts matching high-speed operational requirements. AI technology offers effective power performance solutions in ultra-low-power processors, benefiting wearable devices and IoT sensors due to their critical energy efficiency needs.

AI-based optimization enhances devices by refining transistor and circuit layouts (Murali, 2024). Large collections of design data feed machine learning algorithms, which generate predictions about the performance effects of altering transistor locations, interconnect patterns, and gate planning configurations. Through optimized parameter adjustment, AI enables engineers to develop circuits with faster switching capabilities, enhanced signal stability, and lower levels of electromagnetic interference. The technology brings significant value to high-performance computing systems, artificial intelligence accelerators, and upcoming semiconductors that require rapid processing capabilities (Gao & Adnan, 2025).

The structural integrity of nanoelectronic devices depends heavily on AI systems to protect their long-term operational reliability. The persistent shrinking of semiconductors creates two main types of stress, which can lead to equipment breakdown throughout device operation. Engineers design robust nanoscale structures that resist mechanical stress and thermal drift, using predictions from A.I. simulations of material behavior. Nanoelectronic components demonstrate longevity through these design methods, which allows their employment in aerospace systems together with medical instruments and equipment used in demanding environments (Khan et al., 2019). Engineers achieve better nanoelectronic device development through AI integration into design optimization procedures, which enables smaller components that operate at faster speeds and higher energy efficiency. AI serves as an important engineering tool for nanoelectronic device development through its ability to examine intricate design landscapes, forecast system output, and refine technical architectures with precision.

Generative Design and Evolutionary Algorithms:

AI-powered generative design techniques automate several aspects of nanoelectronics through design creation, performance evaluation, and optimization of multiple solutions. Before the advent of automation, engineers needed to perform multiple manual design simulations to optimize nanoelectronic devices. Generative AI algorithms swiftly generate thousands of design configurations while executing predefined performance evaluations for factors such as energy efficiency, signal integrity, and thermal stability (Shobanke et al., 2025). The development process of high-performance nanoelectronic components consumes fewer resources and takes less time due to this method, which also results in reduced material waste.

Evolutionary algorithms form a vital aspect of generative design because they apply natural selection principles to perfect both circuit configurations and tiny interconnects at the nanoscale level (Gupta et al., 2024). The design optimization method uses multiple evolving cycles to evaluate design populations, which start from random generation and advance through subsequent iterations. The selection process selects the most effective design elements based on power efficiency and transistor density metrics, using them to generate better versions in subsequent generations. Through successive rounds of refinement, electronic components become highly efficient, even though conventional design methods cannot achieve such results. Evolutionary algorithms demonstrate significant value in nanoscale transistor placement optimization, improving power leakage reduction and enhancing circuit integration capabilities.

The application of neural networks serves as an essential AI technique during nanoelectronic design to predict performance metrics and select optimal device structures (Abiodun et al., 2018). Neural networks take advantage of vast databases of previous design solutions to determine performance correlations through pattern recognition. The trained models deliver rapid assessments for newly proposed designs by predicting their performance behavior in actual operational settings, helping engineers understand them better. Research teams utilize neural networks to avoid lengthy trial-and-error testing, which speeds up the process (Bishnu et al., 2023).

Advantages of AI-Enhanced Optimization:

The incorporation of artificial intelligence in the optimization of nanoelectronic design presents a variety of advantages, significantly improving both device performance and efficiency. A key benefit is the enhancement of transistor density, enabling a greater number of transistors to be fitted into smaller spaces. This results in more compact and powerful electronic circuits, which are vital for applications in high-performance computing, artificial intelligence (AI) processors, and mobile technology. AI-driven optimization uncovers novel methods for arranging transistors at the nanoscale, thereby maximizing space utilization while ensuring functionality and reliability.

Additionally, another significant advantage is the reduction of signal interference, which is crucial for enhancing the reliability of nanoelectronic devices. As circuit density increases, the risk of unwanted electromagnetic interference and signal crosstalk rises, potentially impairing performance (Barzdenas & Vasjanov, 2023). AI-assisted layout optimization strategically positions signal paths to minimize interference, thereby enhancing signal integrity and creating more stable electronic systems. This is especially important in communication devices, high-speed processors, and sensor networks, where accurate signal transmission is essential.

Benefits of AI-Driven Optimization:

The incorporation of artificial intelligence into nanoelectronic design optimization offers numerous advantages that enhance both device efficiency and performance metrics. The main benefit of this approach is higher transistor density, which allows designers to place more transistors within confined spaces (Wei et al., 2022). These more condensed and efficient electronic systems enable vital operations in high-performance computing, AI processing, and mobile applications. AI optimization systems find original transistor placement solutions at the nanoscale to fully utilize space without affecting performance stability.

Nanoelectronic devices become more reliable because signal interference is significantly minimized through this approach. Performance degradation occurs when circuits become closer together due to unwanted electromagnetic interference and signal crosstalk. AI layout optimization algorithms place signals in precise positions to reduce interference, creating more reliable electronic systems with superior signal integrity. The accurate transmission of signals remains crucial for devices operating in communication systems, high-speed processing, and sensor networks (Swain et al., 2022).

2.2 Machine Learning for Material Selection

A key step in nanoelectronic device development is selecting suitable materials, as their properties directly determine the performance, efficiency, and reliability of electronic components. The traditional approach to materials discovery through experimental trials took an exceptionally long time before discovering suitable materials. Researchers utilize AI technology to review enormous datasets and create fast, precise predictions about material performance while simultaneously finding optimal materials. Modern nanoelectronics discovers materials with the help of AI-powered material informatics and molecular modeling, which speeds up innovation while improving fabrication processes (Bai & Zhang, 2025).

AI-Driven Material Informatics

The utilization of AI algorithms in material informatics enables researchers to study extensive material properties databases, leading to more accurate suitability predictions for nanoelectronic applications. Artificial intelligence analyzes substantial datasets of experimental and simulated data to help researchers uncover materials that match predetermined performance criteria (Mobarak et al., 2023). Nanoelectronic applications require materials from a selected group of properties, which include:

- **Electrical conductivity:** The electrical conductivity of materials must have high carrier mobility and low resistance for application in fast electronic components like transistors, memory devices, and logic circuits.
- **Thermal stability:** The thermal stability of materials functions to protect device performance from temperature fluctuations in semiconductor devices with tight components because hot conditions can reduce both device lifetime and efficiency.
- **Mechanical robustness:** The durability and reliability of nanomaterials need improvement through better mechanical resistance to maintain stability throughout manufacturing and operational phases.
- **Integration with semiconductor processes:** The technology needs to integrate seamlessly with semiconductor production methods to allow its use within current semiconductor manufacturing environments.

The prediction of material properties achieves high accuracy through processing with machine learning technologies, particularly support vector machines (SVMs) and convolutional neural networks (CNNs), which handle high-dimensional material data. The SVM classification method relies on electronic and mechanical properties of materials, whereas CNNs utilize complex material structure image data to find desirable property correlations(Gola et al., 2019). AI-driven methods enable quick testing of multiple thousands of possible materials for nanoelectronics, thus speeding up the search for superior performance materials.

Molecular Modeling and Predictive Analysis:

Artificial intelligence solutions for molecular modeling deliver atomic-level analysis of nanomaterials, which enables scientists to forecast their behavior before production takes place. Artificial intelligence-powered simulations now allow researchers to conduct effective material discovery through a computational method that overcomes the time-consuming and expensive nature of traditional material examination (Badini et al., 2023). The predictive capabilities generate multiple important benefits.

- Exploration of 2D materials: Research simulations run by artificial intelligence help scientists find valuable two-dimensional (2D) materials, including graphene and transition metal dichalcogenides (TMDs), because they possess outstanding electrical, mechanical, and optical properties. Scientists are developing these materials because they have potential applications in next-generation transistors, flexible electronic technology, and photonic devices.
- Optimization of dopants and defect engineering: The optimization of dopants alongside defect engineering can be achieved through AI, which predicts the impacts of dopants (intentional impurities) and defects on the electrical and optical behavior of nanomaterials. The optimized results from artificial intelligence improve device performance by enhancing bandgap properties, charge carrier mobility, and overall device characteristics.
- Virtual prototyping and validation: AI simulations enable researchers to perform digital material evaluations, which help them validate concepts before creating physical fabrication prototypes. Virtual prototyping through AI simulations leads to faster development of nanoelectronic components by reducing experimental prototyping expenses and duration.

The incorporation of artificial intelligence for material selection enables investigators to find and improve advanced materials that extend the possibilities of nanoelectronics (Ori et al., 2023). The combination of AI technology improves material discovery practices and enables advanced material design with precision, leading to future-generation devices and efficient electronic elements. The evolution of AI technology will enhance its importance in material selection, fueling revolutionary progress in semiconductor and nanotechnology research.

2.3 AI-Assisted Transistor Design Case Study

Modern semiconductor technologies push designers toward creating more dependable transistor innovations that deliver both high performance and improved reliability. Modern transistor fabrication faces a key challenge in optimizing wafer-level packaging because this essential technique directly affects circuit integration and thermal and mechanical reliability (X. Liu et al., 2025). The standard design practice based on physical simulations requires repeated experimental trials, which produce expensive, time-consuming outcomes and yield unreliable predictions regarding durability failures in the long run. The challenges are addressed by AI-assisted design tools, which offer fast development times, reduced failure risks, and superior overall results.

AI technology is used in a real-world application to improve advanced transistor development by analyzing WLP technology stresses and temperature effects while studying material characteristics. Machine learning trained its models using various datasets, which contained experimental historical data, simulation outputs, and measurement data from operating environments. Artificial intelligence models analyzed device patterns to forecast three categories of transistor defects that traditional testing methods could not identify. The design process received a significant advantage from artificial intelligence since it could analyze multiple parameter relationships that impacted transistor functioning at high speed (Amuru et al., 2023). With the implementation of neural networks along with optimized algorithms, engineers can adjust design parameters, including interconnect materials, packaging structures, and stress reduction techniques, which reduces the requirement of physical prototypes. AI-driven models gave predictions to anticipate future reliability problems while ensuring that transistor designs surpassed their performance and longevity standards. Engineers received adaptive models through AI integration to make real-time decisions using data streams that continually learned from new input. The adaptable methodology protected the transistor development process from emerging material advances, shifting production boundaries, and developing technological demands. Semiconductor development times decreased along with manufacturing price reductions, which led to superior product reliability (Mohammad et al., 2022).

Semiconductor engineering has been transformed by AI technologies, which have produced transistor designs that optimize manufacturing speed and cost while enhancing performance capabilities. AI approaches in this field have proven successful and indicate upcoming opportunities for innovation through efficient semiconductor technology developments across other electronic production areas.

3. AI-ENHANCED FABRICATION TECHNIQUES FOR NANOELECTRONICS

The technological progress in nanoelectronic fabrication has reached an essential stage where manufacturing success depends on precision, defect elimination, and process optimization. The shrinking size of device architectures creates difficulties for traditional fabrication approaches to sustain consistent nanoscale patterning because it causes yield variations and higher production expenses (Bi et al., 2025). Through the implementation of AI-driven procedures, manufacturers can make real-time manufacturing choices, utilize predictive process control, and implement intelligent defect prevention. Manufacturers achieve productivity improvements and minimize waste while ensuring strict quality and dependable nanoelectronic devices through AI-powered fabrication technologies.

3.1 AI in Nanoscale Lithography and Deposition

The fundamental building blocks of semiconductor fabrication consist of nanoscale lithography and material deposition processes, which determine circuit design elements and module composition. The three fabrication techniques, EUV lithography, along with CVD and ALD, need strict control systems to guarantee proper uniformity and defect-free operation. The fair execution of these processes faces substantial difficulties because transistor components need to measure at sub-5nm nodes (Gul et al., 2022).

AI-driven models act as vital process optimization tools through data analysis of large process datasets which include optical proximity measurements alongside exposure dose fluctuations and etching reaction dynamics. The combination of experimental process data with high-resolution SEM images shapes machine learning algorithms which deliver the following system improvements:

- **Predictive Process Modeling:** Certain AI-generated process simulations assist with determining optimal photoresist exposure durations and etching profile depths while defining thin-film deposition sequences, which lead to regular nanoscale pattern formation. The predictive modeling system reduces the need for experimental methods, thereby accelerating process advancement.
- **Reduction of Line-Edge Roughness (LER) and Line-Width Variations:** Through an AI algorithm, monitoring the dimensions of features, the system automatically adjusts exposure settings and etching parameters, which produce precise and uniform patterns that deliver superior device functionality.
- **Enhanced Atomic Precision in Deposition Techniques:** AI controllers optimize atomic-level parameters of deposition processes by adjusting precursor flow rates together with substrate temperatures and reaction durations through ALD and CVD technologies. The optimized processes prevent material defects, which include film disruptions together with impurities and thickness abnormalities, for creating outstanding material quality.
- **Adaptive Lithography Calibration:** A feedback system enhanced through artificial intelligence positions itself to continuously optimize mask alignments along with exposure parameters and processing steps for resists, depending on measurements taken in real time. The method preserves full pattern accuracy throughout the entire wafer production despite shrinking feature dimensions.

Semiconductor manufacturers using AI in lithography and deposition workflows gain a market advantage by increasing fabrication yields, reducing material usage, and achieving better device performance results (A. Liu et al., 2024). Through integration with AI systems manufacturers utilize optimizations that enable both expandable and economical nanoelectronic manufacturing.

3.2 AI for Defect Detection and Correction

Semiconductor manufacturing defects directly affect yield rates, which in turn produce functional failures as well as elevated production costs and diminished reliability in nanoelectronic devices. Detection of nanoscale defects within traditional inspections requires intensive manual effort and slow procedures and fails to recognize complicated defects because of rule-based algorithms. Deep learning and computer vision, combined with real-time adaptive control systems, make up an AI-powered defect detection system that enhances semiconductor manufacturing quality assurance.

AI-Driven Defect Detection:

Artificial intelligence applies high-resolution SEM and atomic force microscopy (AFM) and optical inspection images through algorithms to discover minute nanometer-scale defects as well as pattern distortions along with particle contaminants and etching anomalies (Botifoll et al., 2022). The deep learning architecture with convolutional neural networks (CNNs) together with generative adversarial networks (GANs) enables:

- **Automated Real-Time Monitoring:** Through automated real-time monitoring AI models swiftly conduct fast wafer examinations to identify defects while eliminating human involvement and perform automated classification and identification of defects and their quantities. AI-based detection speeds up the process of identifying defects which lowers expenses for subsequent production inspections.
- **Sub-Pixel Accuracy in Defect Classification:** A system based on Sub-Pixel Accuracy helps manufacturing companies classify important defects which allows them to quickly solve the most vital problems.
- **Defect Localization and Root-Cause Analysis**: AI uses historical defect trends to determine recurring problem locations from mask alignment problems and contamination issues as well as process drifts.

4. AI-GUIDED DEFECT CORRECTION

The application of AI helps manufacturers improve defect correction by implementing adaptive control systems, which change parameters during fabrication automatically. Pre-emptive defect minimization occurs within these systems thanks to their predictive analytical capabilities that prevent defect propagation in the manufacturing process (Gunasegaram et al., 2024). Key corrective measures include:

- **Predictive Failure Prevention:** The assessment of wafer-level data by AI algorithms helps predict which regions will develop defects so that early adjustments in deposition, etching, or doping processes become possible.
- **Intelligent Process Tuning:** AI-based feedback systems create real-time adjustments of critical parameters, including plasma etch condition adjustments to accommodate material failings and exposure mask modifications to stop edge distortions.
- **Autonomous Self-Healing Processes:** Next-generation fabrication lines utilize AI-controlled systems to make automatic process adjustments while operating including dynamic rate adjustments to balance thickness variations or electrical path redirection around faulty areas.

Semiconductor manufacturers who implement AI into their defect detection and correction operations experience improved yield rates while cutting down expenses from reworks, and improving nanoelectronic device reliability. Through sustainable semiconductor manufacturing, AI-driven improvements allow miniaturization of transistors to continue, which enables future development of high-performance computing and 5G technology while supporting AI applications.

4.1 Machine Learning for Performance Prediction

Nanoelectronic devices need AI models to forecast their electrical along with thermal and mechanical responses when performing under different operational scenarios. Computational modeling through finite element analysis (FEA) together with density functional theory (DFT) remains limited in its performance when used to analyze complex nanoelectronic systems because of their extensive computing requirements (Laycock et al., 2024). Deep neural networks with machine learning algorithms serve as data-driven systems which produce fast and accurate performance predictions.

Machine learning models find their primary use in performance prediction through these applications:

- **Electrical Property Forecasting:** Deep learning models built from experimental and simulated data provide forecasts for electrical property metrics that consist of conductivity speed and mobility trends along with voltage limits and current leakage effects. Before fabrication engineers use this technology to optimize transistor geometries alongside material compositions.
- **Failure Rate Estimation:** The evaluation of historical reliability test results combined with stress conditions enables AI systems to predict Depression points that occur within nanoelectronic devices between interconnect electromigration and gate oxide dielectric breakdown. Predictive maintenance tools detect product deterioration patterns early to extend device operating life.
- **Process-Structure-Property Relationship Modeling:** AI creates a link that connects process conditions to device structure while modeling electrical properties. The technique allows researchers to make exact adjustments to fabrication parameters, which lead to specific device specifications.
- **Adaptive Optimization for Real-Time Design Adjustments:** The combination of AI-driven optimization frameworks uses real-time simulation outputs to modify device parameters that achieve specific performance targets automatically.

Engineers benefit from machine learning integration in nanoelectronic performance prediction workflows to inspect extensive design spaces quicker through efficient exploration while minimizing production cycles and strengthening semiconductor device stability (Gao & Adnan, 2025).

4.2 Multi-Scale Simulations Integrating AI

Similar to atomic-scale material interactions, the performance of macroscopic devices depends heavily on multi-scale simulations. The operation of traditional computational approaches occurs at separate scales since quantum mechanics controls atomic interactions, molecular dynamics reproduces material properties, and continuum models deliver device-level predictions. Making these separate scales work together in a unified structure proves to be both computationally complex and time demanding(Ingólfsson et al., 2023).

Machine learning algorithms within multiscaling simulations enable efficient data transmission between modeling scales, aiding in the understanding of intricate multi-scale relationships. Figure 2 represents the benefits of AI Integration in Quantum and Molecular Simulations.

Figure 2. Benefits of AI Integration in Quantum and Molecular Simulations

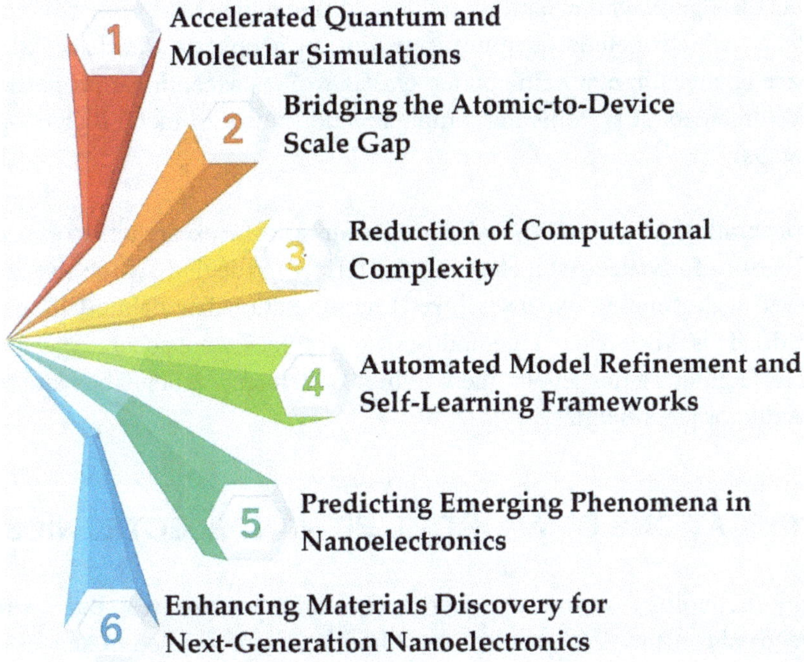

1. **Accelerated Quantum and Molecular Simulations:** Training AI models to handle quantum mechanical calculations based on density functional theory (DFT) produces accurate material property predictions using reduced computational expenses. The rapid material screening process becomes possible through this method.
2. **Bridging the Atomic-to-Device Scale Gap:** AI-based surrogate models provide fluid data exchange between quantum-level simulations and huge device modeling that accurately connects atomic-force to device-operation predictions.
3. **Reduction of Computational Complexity:** Traditional multi-scale simulations need very powerful computational resources because they require high-performance computing clusters to function optimally. The integration of reduced-order models created with AI reduces computational expenses and simulation durations through basic machine learning physics estimations.
4. **Automated Model Refinement and Self-Learning Frameworks:** AI-driven simulation frameworks employ automated model refinement mechanisms to learn from experimental data along with simulation data, which makes models accurate toward emerging fabrication techniques.

5. **Predicting Emerging Phenomena in Nanoelectronics:** AI-enhanced simulations through machine learning models combine to identify emerging physical effects, which include quantum tunneling in nanoscale transistors alongside phase changes in new semiconductors, as well as exceptional properties from two-dimensional systems like graphene and transition metal dichalcogenides (TMDs).

Incorporating AI into multi-scale simulations allows researchers to expedite the identification of advanced nanoelectronic materials and devices, while also considerably lowering the time and expenses linked to computational modeling(Chávez-Angel et al., n.d.). This AI-enhanced methodology guarantees more accurate performance forecasts, facilitating the development of highly efficient, dependable, and scalable semiconductor technologies.

5. APPLICATIONS OF AI-OPTIMIZED NANOELECTRONICS

Many technology sectors experience revolutionary progress due to the AI-nanoelectronics technology integration. AI optimization applied to nanoelectronic components achieves better device performance together with decreased power requirements and develops previously impossible capabilities. This section examines main applications of AI-optimized nanoelectronics as it transforms current technological processes.

5.1 Next-Generation Transistors and Semiconductor Devices

The semiconductor sector undergoes a transformation due to AI-based innovations in transistor development, which leads to higher-performing devices along with smaller size and greater efficiency. AI ensures the exploration of new architectures and materials to maintain transistor scaling at its present scope because traditional thresholds are closing in (Elahi et al., 2023).

AI provides essential contributions that enable the development of advanced semiconductor technologies through its following achievements:

- **Optimized Transistor Architectures:** Machine learning algorithms study extensive datasets to develop most efficient designs for FinFETs and GAA transistors along with 2D-material-based transistors that provide superior speed and energy efficiency.
- **Adaptive Process Control in Semiconductor Fabrication:** AI analytical tools in semiconductor production control processes optimize dopant dos-

age levels, annealing protocols, and etching methods to boost manufacturing yield and reduce product flaws.

- **AI-Assisted Electronic Design Automation (EDA):** The combination of AI with Electronic Design Automation (EDA) through assisted algorithms deducts development duration while decreasing computational expenses and automates tasks like layout optimization and power distribution and signal integrity evaluation.
- **Energy-Efficient Computing:** The optimization of neuromorphic computing systems using AI power enables the development of energy-efficient processors that serve AI workloads and support mobile devices as well as edge computing applications.

Semiconductor performance limits must be pushed beyond current boundaries through AI-based advancements because this approach keeps energy efficiency high as it enables better next-generation computing and mobile technology.

5.2 Photonic Devices and Quantum Computing Applications

The advancement of both photonic devices and quantum computing technologies progresses quickly because of AI-optimized nanoelectronics (Xu et al., 2025). Nanophotonic structures facilitate high-speed data transmission as artificial intelligence applications drive research toward next-generation quantum computing paradigms.

AI in Photonic Devices:

Artificial intelligence enhances the operation of photonic waveguides, optical switches, and plasmonic structures through optimized light-matter interactions, which create ultra-fast data transmission in fiber-optic networks and photonic integrated circuits (PICs). Deep learning algorithms discover and optimize nanophotonic metamaterials through an AI process, leading to remarkable optical properties for holographic imaging and sensing purposes. The integration of AI techniques allows for maximal efficiency in nano-scale LEDs and photodetectors and solar cells to elevate performance levels in optoelectronic products (Mishra et al., 2024).

AI in Quantum Computing:

The discovery process of quantum dot materials and superconducting qubit geometries receives acceleration through AI, which leads to superior error rate performance and longer coherence times. Machine learning employs techniques that enhance quantum error correction protocols through predicting and mitigating decoherence effects that occur in quantum circuits. Through AI assistance, quantum algorithms can be developed in three key areas: material science applications, cryptography, and complex problem-solving and scalability refinement. Photonic and quantum computing technologies advance rapidly through AI-driven optimization which leads to improved ultra-fast communication and next-generation computational abilities.

5.3 Biomedical Nanoelectronics: Implants, Sensors, and Drug Delivery

The fusion of AI with nanoelectronics technology advanced biomedical functions by allowing continuous healthcare supervision alongside detailed medical diagnostics and targeted medical treatment(Vora et al., 2023). The advancement of artificial intelligence technology into nanoelectronic devices has helped enhance medical care through compact implanted devices as well as biosensors and micro-droplet drug delivery systems.

AI-Powered Implantable Biosensors:

Thanks to permeable biosensors embedded in the human body the combination of AI technology enables nonstop tracking of biomarkers including glucose levels and blood pressure and neural signals which furnish immediate health assessments. Biosensor data analysis through AI algorithms allows for the discovery of early disease indications including cancer alongside neurodegenerative conditions and cardiovascular disorders which leads to proactively treating patients. AI-based bio-electronics use patient-specific biological data to personalize therapeutic responses which leads to better treatment outcomes(Parekh et al., 2023).

AI-Optimized Drug Delivery Systems:

Artificial Intelligence systems improve nano-based drug delivery platforms by creating targeted medication carriers that transport therapeutic agents directly into disease-specific areas and concurrently reduce adverse effects on the patient. Nanocarriers which use AI-controlled adaptive drug release mechanisms respond to biological environments by delivering medications at their best quantities and moments. Machine learning techniques speed up the discovery process of biocompatible nanomedicines, which possess advanced therapeutic properties for modern treatment methods.

The combination of artificial intelligence with biomedical nanoelectronics devices drives healthcare advancement through improved diagnosis and earlier treatment decisions along with personalized treatment plans for better patient results.

6. CHALLENGES IN AI-DRIVEN NANOELECTRONICS

AI has greatly improved nanoelectronics, yet multiple essential obstacles prevent its broad commercial utilization along with smooth industrial implementation. AI models require large datasets of high availability and superb quality for fundamental operation, as they need extensive, reliable information to make valuable predictions and extract helpful insights. Products developed in controlled laboratory conditions encounter significant problems when manufacturers attempt to implement them on an industrial scale, which reduces both performance reliability and manufacturing quality. Establishing reproducibility requires addressing various challenges, as fabrication methods vary and experimental setups and measurement practices differ across facilities, undermining the general applicability and stability of AI-based solutions. Industrial-scale AI implementation faces significant barriers due to ethical challenges and intricate regulations. These issues include intellectual property rights, data privacy risks, cybersecurity threats, environmental concerns, and the widespread effects of automated workforces (Rane et al., 2024). The complete utilization of AI revolutionary capabilities in nanoelectronic manufacturing and design needs purposeful strategies and collaborative interdisciplinary approaches to resolve and handle multiple complex challenges together with appropriate regulatory systems.

6.1 Data Scarcity and High-Quality Datasets

AI algorithms need substantial quality data collections to perform their training and optimization functions. The collection of required datasets represents substantial difficulties in nanoelectronics research because of these factors:

- **Limited Experimental Data:** Educational research about nanoelectronics produces complex proprietary experimental data which remains difficult to access by the wider scientific community. Both the rare availability and the lack of labels from large datasets prevent the development of stable AI modeling systems.
- **High Cost of Data Collection:** Generation of experimental data using advanced fabrication methods together with microscopy technology and device assessment methods requires expensive long-term research efforts.
- **Variability in Measurement Techniques:** Each fabrication laboratory operates using its unique set of characterization devices alongside measurement techniques that produce inconsistent results in their data collections.

These challenges can be addressed through synthetic data generation, collaborative data-sharing platforms, and hybrid AI models that combine physics-based techniques with machine learning.

- **Synthetic Data Generation:** Generative adversarial networks (GANs) operated by AI create artificial nanoelectronic datasets through synthetic data generation while mimicking actual experimental findings to lower the expense of physical experimentation.
- **Collaborative Data-Sharing Platforms:** Research institutions and manufacturers can join collaborative platforms for data sharing since these platforms provide protected access to anonymized datasets through open-access repositories and federated learning frameworks.
- **Hybrid AI Models:** Hybrid AI systems, which merge physics-based modeling techniques with machine learning technology, help overcome scarce experimental evidence to achieve better model precision.

These techniques work to reduce data insufficiencies, which will boost both the accuracy and universal application of AI-based nanoelectronics research and applications.

6.2 Scalability and Reproducibility

Scalability and reproducibility represent major obstacles for AI-driven nanoelectronics research because research lab findings need to scale up for commercial manufacturing while retaining production reliability (Sharmile et al., 2025). The effective reproduction of controlled research results into manufacturing processes poses many challenges for trained AI models because of the following reasons:

- **Variability in Fabrication Conditions:** Semiconductor fabrication requires multiple fabrication variables, including temperature, pressure, and material properties, which differ between production sites, hence affecting the AI-optimized process reproducibility.
- **Generalization of AI Models:** The wide application of AI models faces limitations when their training data comes from particular fabrication setups and specialized datasets, which proves challenging for operation across different manufacturing locations.
- **Computational Costs and Resource Constraints:** Large-scale nanoelectronic fabrication deployment using AI requires substantial computational processing power, and such demands are challenging for smaller labs and companies because of associated costs.

The solution combines transfer learning with domain adaptation techniques to enhance AI generalization capabilities. Meanwhile, standardized AI protocols improve consistency, and edge AI implementations enable real-time process control.

- **Transfer Learning and Domain Adaptation:** AI models gain enhanced ability to adapt and generalize through fine-tuning them with minor datasets from various fabrication settings.
- **Standardized AI Protocols:** Standard AI Protocols should be developed with benchmark datasets to create consistency throughout different fabrication environments.
- **Edge AI for Real-Time Adaptation:** The application of AI models as edge computations allows real-time updates of fabrication protocol adjustments, which enhances both adaptability and scalability.

To progress from laboratory research to commercial adoption in semiconductor manufacturing the expansion and predictable results of AI technologies in nanoelectronics need to be ensured.

6.3 Ethical and Regulatory Considerations

The adoption of AI within nanoelectronics demands thorough investigation of ethical and regulatory problems to build innovation with responsible practices and public confidence.

- **Intellectual Property (IP) Rights:** Patent laws become difficult to enforce because AI produces nanoelectronic designs. The current framework for intellectual property requires modifications in order to establish ownership rights for semiconductor designs developed through AI optimization. Organizations and research groups must overcome present and future legal roadblocks when creating AI innovations since they face conflicts regarding computer program authorship and ownership of system-generated data.
- **Data Privacy and Security:** The design and fabrication of semiconductors through AI systems depend on confidential elements such as intellectual property protocols and measurement standards of device performance. Protecting data privacy and maintaining cybersecurity measures is essential to prevent both industrial espionage and unauthorized system entries. Secure data protection methods, such as federated learning and encrypted artificial intelligence models, allow teams to collaborate safely, ensuring confidential information remains secure.

- **Environmental and Social Impact:** The energy requirements of AI-based semiconductor manufacturing facilities create long-term sustainability issues. Optimizing nanoelectronics using AI models requires specific attention to reduce manufacturing emissions and the effects of waste disposal. The deployment of AI-driven nanoelectronics needs to address workforce challenges resulting from automation, and its social effects on workers must be understood. Education programs that teach new skills and improve existing skills for semiconductor industry workers can reduce the total number of job losses in this sector.
- **Regulatory Frameworks:** Nanoelectronics demand clear regulatory and governmental policy for the proper ethical utilization of AI through development and implementation guidelines. AI-driven semiconductor design should have policies focusing on algorithmic bias standards, as well as measures to achieve transparency and accountability. Standardization of international semiconductor market AI ethics guidelines will create fair competition environments while promoting responsible technological innovation.

The nanoelectronics industry addresses ethical problems and regulatory requirements before AI-driven innovations begin, preventing misalignment between innovations and societal, environmental, and legal standards.

7. FUTURE PROSPECTS IN AI-DRIVEN NANOELECTRONICS

AI-driven nanoelectronics will bring revolutionary changes through combining leading technologies, which merge autonomous fabrication with advanced AI-quantum computing hybrid systems and self-repairing nanorobotic components (Hassan et al., 2023). New innovations will significantly boost manufacturing efficiency, improve semiconductor precision, and provide enhanced scalability. Moore's Law is demonstrated through advanced device technology, which brings faster operation, superior energy conservation, and enhanced dependability, contributing to revolutionary computational power. The implementation of self-maintenance along with automatic repair capabilities in nanorobotics technology produces substantial longevity of devices while minimizing maintenance periods. The innovative breakthroughs in nanoelectronics development will create substantial changes to the upcoming nanoelectronic technology era while spurring innovation throughout multiple industries.

7.1 Autonomous Fabrication Systems

AI-driven nanoelectronics will bring revolutionary changes through combining leading technologies which merge autonomous fabrication with advanced AI-quantum computing hybrid systems and self-repairing nanorobotic components. New innovations will significantly boost manufacturing efficiency and will improve semiconductor precision as well as provide enhanced scalability. Moore's Law demonstrates itself through advanced device technology which brings faster operation along with superior energy conservation and enhanced dependability measures toward revolutionary computational power. The implementation of self-maintenance along with automatic repair capabilities in nanorobotics technology produces substantial longevity of devices while minimizing maintenance periods (Kong et al., 2023). The innovative breakthroughs in nanoelectronics development will create substantial changes to the upcoming nanoelectronic technology era while spurring innovation throughout multiple industries.

The essential characteristics of autonomously operated fabrication lines through artificial intelligence include:

Self-Correcting Production Lines: AI models inspect production line data in real-time to perform automated adjustments, which control etching depth as well as deposition rates and temperature parameters for optimized system performance.

Automated Quality Control: AI-powered defect detection systems watch wafers in continuous operation through automated quality control to lower the chances of market-defects reaching end consumers.

Predictive Maintenance: AI algorithms enable predictive maintenance through equipment monitoring, predicting system breakdowns so that planned maintenance can occur before failure points, enhancing production efficiency.

Autonomous fabrication technology enables the semiconductor industry to achieve higher production yields, reduce costs, and expedite delivery times for nanoelectronic products.

7.2 Hybrid AI-Quantum Computing Models

The combination of AI technology with quantum computing is expected to be a catalyst for nanoelectronics to achieve revolutionary computational possibilities. Quantum computers, built using principles from quantum mechanics, demonstrate the ability to process enormous quantities of data concurrently. As a result, they accelerate complex optimization and simulation tasks in nanoelectronic systems (Gill & Buyya, 2024).

Potential Impact on Nanoelectronics:

- Faster Multi-Scale Simulations: Faster discoveries of new semiconductor materials and innovative device designs follow from accelerated nanoscale atomic interaction and material behavior simulations driven by hybrid AI-quantum computer models. The technical innovations allow items to reach consumers with less delay by reducing the timeframes between idea definition and market introduction.
- Enhanced Circuit Design Optimization: The ability of quantum algorithms allows them to perform quick and effective optimization of circuit designs, which creates highly efficient transistor layouts and best interconnect solutions that lead to significant device performance improvements. The capability leads to improved nanoelectronic devices that are both smaller and more energy-efficient, delivering stronger performance beyond traditional constraints.
- Breakthroughs in Quantum Nanoelectronics: Major quantum nanoelectronics breakthroughs driven by artificial intelligence optimization of quantum dots and superconducting qubits will efficiently push nanoelectronic technology limits. Along with ultra-fast data processing, quantum hardware improvements will provide circumstances for applying quantum computing solutions throughout industrial sectors for secure communication networks and exact sensing technologies.
- Advanced Materials Discovery: AI predictive modeling linked with quantum computing processing enables researchers to explore new nanoscale materials at advanced levels of precision and speed. Scientists working together help achieve breakthroughs in material science which results in lasting improvements of electricity transmission and durability standards along with increased energy efficiency.
- Improved Predictive Maintenance: The implementation of quantum computing with artificial intelligence in nanoelectronics manufacturing helps improve predictive maintenance systems, which enables superior capability in equipment deterioration and failure prediction. The capability improves device reliability while lowering operational costs for high-performance nanoelectronics industries by reducing downtime.

The development of next-generation nanoelectronic devices and technological innovation will advance as quantum computing matures, integrating with AI and quantum simulations.

7.3 Emerging Trends: AI-Enhanced Nanorobotics and Self-Repairing Nano-Devices

Nanotechnology, enabled by AI-driven systems and combined with self-repairing nanoelectronic devices, brings revolutionary progress to nanoelectronics, enabling extreme durability extension and major cost reductions for maintenance. Nanorobots supported by artificial intelligence use complex algorithms for atom-to-molecule manipulation to build precise nanoelectronic components while ensuring high standardization (Nandipati et al., 2024). Nanorobots gain real-time anomaly detection from embedded AI technologies, which enables them to prevent failures from occurring in advance. Nanoelectronic devices that self-repair utilize AI-guided sensors to detect micro-level failures, such as material degradation and performance issues, allowing repairs to begin as needed. Nanomaterials engineered with specific functionality enable advanced systems to activate self-healing chemical processes, automatically regaining operational performance in transistors and circuits. The self-healing ability of extended critical electronic operations substantially reduces replacement expenses and maintenance costs, benefiting numerous industries such as aerospace operations, healthcare applications, and consumer electronics fields. This integration of AI-powered nanorobotic systems and self-healing capabilities develops nanoelectronic platforms that become exceptionally strong and intelligent while reducing operational costs (Naik & Jagtap, 2024).

8. CONCLUSION

The combination of AI with nanotechnology creates a disruptive shift in nanoelectronics, pushing transistor engineering and device construction methods toward improved performance. Intelligent innovations utilizing AI systems deliver superior efficiency in manufacturing by providing enhanced precision fabrication, automatic defect recognition, and advanced multi-scale simulation software. These platforms improve manufacturing performance by raising yield rates and improving semiconductor capability. Notwithstanding recent developments, several significant challenges still exist in the field of nanoelectronics. These challenges arise from data constraints and the pragmatic application of artificial intelligence models to nanoscale systems, which together raise important ethical issues about AI control, privacy, and safety concerns in AI implementation. The complete realization of AI's transformative power in nanoelectronics requires structured approaches for sharing collective data inventories, coupled with adaptable AI systems that offer explanations and comprehensive governance policies. The path to overcoming difficulties in AI and nanotechnology collaboration depends on establishing partnerships between researchers, specialists, stakeholders, and policymakers. The semiconductor industry is poised for revolutionary changes through standalone automated manufacturing systems integrated with self-maintaining nanotech devices, which operate with high precision to minimize human intervention, alongside quantum computing systems that optimize complex computational processes. AI implementation of real-time monitoring with predictive maintenance systems will boost process control capabilities while cutting down operational downtime. Academic institutions that implement AI-driven educational measures along with training programs for the workforce will prepare future professionals to manage novel technological solutions in the industry. Nanoelectronics will be fostered by continuous collaboration, ethical governance, and ongoing innovation to develop smarter and more sustainable electronic devices. These devices will transform diverse sectors, including computing, healthcare, and energy, while promoting environmental sustainability.

REFERENCES

Abiodun, O. I., Jantan, A., Omolara, A. E., Dada, K. V., Mohamed, N. A., & Arshad, H. (2018). State-of-the-art in artificial neural network applications: A survey. *Heliyon*, *4*(11), e00938. DOI: 10.1016/j.heliyon.2018.e00938

Amuru, D., Zahra, A., Vudumula, H. V., Cherupally, P. K., Gurram, S. R., Ahmad, A., & Abbas, Z. (2023). AI/ML algorithms and applications in VLSI design and technology. *Integration (Amsterdam)*, *93*, 102048. DOI: 10.1016/j.vlsi.2023.06.002

Badini, S., Regondi, S., & Pugliese, R. (2023). Unleashing the Power of Artificial Intelligence in Materials Design. *Materials (Basel)*, *16*(17), 17. Advance online publication. DOI: 10.3390/ma16175927

Bai, X., & Zhang, X. (2025). Artificial Intelligence-Powered Materials Science. *Nano-Micro Letters*, *17*(1), 135. DOI: 10.1007/s40820-024-01634-8

Barzdenas, V., & Vasjanov, A. (2023). Crosstalk Reduction in High-Density Radio Frequency Printed Circuit Boards: Leveraging FR4 Coating Layers. *Coatings*, *13*(10), 10. Advance online publication. DOI: 10.3390/coatings13101801

Batool, M., Sanumi, O., & Jankovic, J. (2024). Application of artificial intelligence in the materials science, with a special focus on fuel cells and electrolyzers. *Energy and AI*, *18*, 100424. DOI: 10.1016/j.egyai.2024.100424

Bi, Y., Xie, S., Li, Z., Dong, S., & Teng, L. (2025). Precise nanoscale fabrication technologies are the "last mile" of medicinal development. *Acta Pharmaceutica Sinica. B*. Advance online publication. DOI: 10.1016/j.apsb.2025.03.040

Bishnu, S. K., Alnouri, S. Y., & Al-Mohannadi, D. M. (2023). Computational applications using data driven modeling in process Systems: A review. *Digital Chemical Engineering*, *8*, 100111. DOI: 10.1016/j.dche.2023.100111

Biswas, P., Rashid, A., Biswas, A., Nasim, M. A. A., Chakraborty, S., Gupta, K. D., & George, R. (2024). AI-driven approaches for optimizing power consumption: A comprehensive survey. *Discover Artificial Intelligence*, *4*(1), 116. DOI: 10.1007/s44163-024-00211-7

Botifoll, M., Pinto-Huguet, I., & Arbiol, J. (2022). Machine learning in electron microscopy for advanced nanocharacterization: Current developments, available tools, and future outlook. *Nanoscale Horizons*, *7*(12), 1427–1477. DOI: 10.1039/D2NH00377E

Chávez-Angel, E., Eriksen, M. B., Castro-Alvarez, A., Garcia, J. H., Botifoll, M., Avalos-Ovando, O., Arbiol, J., & Mugarza, A. (n.d.). Applied Artificial Intelligence in Materials Science and Material Design. *Advanced Intelligent Systems, n/a*(n/a), 2400986. https://doi.org/DOI: 10.1002/aisy.202400986

Collins, C., Dennehy, D., Conboy, K., & Mikalef, P. (2021). Artificial intelligence in information systems research: A systematic literature review and research agenda. *International Journal of Information Management*, *60*, 102383. DOI: 10.1016/j.ijinfomgt.2021.102383

Elahi, M., Afolaranmi, S. O., Martinez Lastra, J. L., & Perez Garcia, J. A. (2023). A comprehensive literature review of the applications of AI techniques through the lifecycle of industrial equipment. *Discover Artificial Intelligence*, *3*(1), 43. DOI: 10.1007/s44163-023-00089-x

Fu, Y., Liu, T., Wang, H., Wang, Z., Hou, L., Jiang, J., & Xu, T. (2024). Applications of nanomaterial technology in biosensing. *Journal of Science. Advanced Materials and Devices*, *9*(2), 100694. DOI: 10.1016/j.jsamd.2024.100694

Gao, P., & Adnan, M. (2025). Overview of emerging electronics technologies for artificial intelligence: A review. *Materials Today Electronics*, *11*, 100136. DOI: 10.1016/j.mtelec.2025.100136

Gill, S. S., & Buyya, R. (2024). Transforming Research with Quantum Computing. *Journal of Economy and Technology*. https://doi.org/DOI: 10.1016/j.ject.2024.07.001

Gola, J., Webel, J., Britz, D., Guitar, A., Staudt, T., Winter, M., & Mücklich, F. (2019). Objective microstructure classification by support vector machine (SVM) using a combination of morphological parameters and textural features for low carbon steels. *Computational Materials Science*, *160*, 186–196. DOI: 10.1016/j.commatsci.2019.01.006

Gul, W., Shams, M., & Al-Khalili, D. (2022). SRAM Cell Design Challenges in Modern Deep Sub-Micron Technologies: An Overview. *Micromachines*, *13*(8), 8. Advance online publication. DOI: 10.3390/mi13081332

Gunasegaram, D. R., Barnard, A. S., Matthews, M. J., Jared, B. H., Andreaco, A. M., Bartsch, K., & Murphy, A. B. (2024). Machine learning-assisted in-situ adaptive strategies for the control of defects and anomalies in metal additive manufacturing. *Additive Manufacturing*, *81*, 104013. DOI: 10.1016/j.addma.2024.104013

Gupta, P., Ding, B., Guan, C., & Ding, D. (2024). Generative AI: A systematic review using topic modeling techniques. *Data and Information Management*, *8*(2), 100066. DOI: 10.1016/j.dim.2024.100066

Hassan, S. A.-D. H., Almaliki, M. N. S., Hussein, Z. A., Albehadili, H. M., Rabeea Banoon, S., Abboodi, A., & Al-Saady, M. (2023). Development of Nanotechnology by Artificial Intelligence: A Comprehensive Review. *Journal of Nanostructures*, *13*(4), 915–932. DOI: 10.22052/JNS.2023.04.002

Hossain, N., Mobarak, M. H., Mimona, M. A., Islam, M. A., Hossain, A., Zohura, F. T., & Chowdhury, M. A. (2023). Advances and significances of nanoparticles in semiconductor applications – A review. *Results in Engineering*, *19*, 101347. DOI: 10.1016/j.rineng.2023.101347

Ingólfsson, H. I., Bhatia, H., Aydin, F., Oppelstrup, T., López, C. A., Stanton, L. G., Carpenter, T. S., Wong, S., Di Natale, F., Zhang, X., Moon, J. Y., Stanley, C. B., Chavez, J. R., Nguyen, K., Dharuman, G., Burns, V., Shrestha, R., Goswami, D., Gulten, G., & Streitz, F. H. (2023). Machine Learning-Driven Multiscale Modeling: Bridging the Scales with a Next-Generation Simulation Infrastructure. *Journal of Chemical Theory and Computation*, *19*(9), 2658–2675. DOI: 10.1021/acs.jctc.2c01018

Khan, I., Saeed, K., & Khan, I. (2019). Nanoparticles: Properties, applications and toxicities. *Arabian Journal of Chemistry*, *12*(7), 908–931. DOI: 10.1016/j.arabjc.2017.05.011

Kong, X., Gao, P., Wang, J., Fang, Y., & Hwang, K. C. (2023). Advances of medical nanorobots for future cancer treatments. *Journal of Hematology & Oncology*, *16*, 74. DOI: 10.1186/s13045-023-01463-z

Krishnan, A., Sundaram, T., Nagappan, B., & Devarajan, Y., & Bhumika. (2024). Integrating artificial intelligence in nanomembrane systems for advanced water desalination. *Results in Engineering*, *24*, 103321. DOI: 10.1016/j.rineng.2024.103321

Laycock, B. G., Chan, C. M., & Halley, P. J. (2024). A review of computational approaches used in the modeling, design, and manufacturing of biodegradable and biobased polymers. *Progress in Polymer Science*, *157*, 101874. DOI: 10.1016/j.progpolymsci.2024.101874

Liu, A., Zhang, X., Liu, Z., Li, Y., Peng, X., Li, X., Qin, Y., Hu, C., Qiu, Y., Jiang, H., Wang, Y., Li, Y., Tang, J., Liu, J., Guo, H., Deng, T., Peng, S., Tian, H., & Ren, T.-L. (2024). The Roadmap of 2D Materials and Devices Toward Chips. *Nano-Micro Letters*, *16*(1), 119. DOI: 10.1007/s40820-023-01273-5

Liu, X., Pei, W., Zhao, J., Xu, R., Zhong, Y., & Yu, D. (2025). A Review of Wafer-Level Packaging Technology for SAW and BAW Filters. *Micromachines*, *16*(3), 3. Advance online publication. DOI: 10.3390/mi16030320

Mishra, R. K., Verma, K., Chianella, I., Goel, S., & Nezhad, H. Y. (2024). Nanomaterial integration in micro LED technology: Enhancing efficiency and applications. *Next Nanotechnology*, *5*, 100056. DOI: 10.1016/j.nxnano.2024.100056

Mobarak, M. H., Mimona, M. A., & Islam, Md. A., Hossain, N., Zohura, F. T., Imtiaz, I., & Rimon, M. I. H. (2023). Scope of machine learning in materials research—A review. *Applied Surface Science Advances*, *18*, 100523. DOI: 10.1016/j.apsadv.2023.100523

Mohammad, W., Elomri, A., & Kerbache, L. (2022). The Global Semiconductor Chip Shortage: Causes, Implications, and Potential Remedies. *IFAC-PapersOnLine*, *55*(10), 476–483. DOI: 10.1016/j.ifacol.2022.09.439

Murali, S. L. (2024). *Artificial Intelligence in VLSI Physical Design of Circuits to Optimize Power, Performance and Area (PPA)*. *11*(10).

Naik, G. G., & Jagtap, V. A. (2024). Two heads are better than one: Unravelling the potential Impact of Artificial Intelligence in nanotechnology. *Nano TransMed*, *3*, 100041. DOI: 10.1016/j.ntm.2024.100041

Nandipati, M., Fatoki, O., & Desai, S. (2024). Bridging Nanomanufacturing and Artificial Intelligence—A Comprehensive Review. *Materials (Basel)*, *17*(7), 7. Advance online publication. DOI: 10.3390/ma17071621

Olawade, D. B., Ige, A. O., Olaremu, A. G., Ijiwade, J. O., & Adeola, A. O. (2024). The synergy of artificial intelligence and nanotechnology towards advancing innovation and sustainability—A mini-review. *Nano Trends*, *8*, 100052. DOI: 10.1016/j.nwnano.2024.100052

Ori, M. O., Ekpan, F. M., Samuel, H. S., & Egwuatu, O. P. (2023). Integration of Artificial Intelligence in Nanomedicine. *Eurasian Journal of Science and Technology, Online First*. https://doi.org/DOI: 10.48309/ejst.2024.422419.1105

Pai, S. M., Shah, K. A., Sunder, S., Albuquerque, R. Q., Brütting, C., & Ruckdäschel, H. (2025). Machine learning applied to the design and optimization of polymeric materials: A review. *Next Materials*, *7*, 100449. DOI: 10.1016/j.nxmate.2024.100449

Parekh, A.-D. E., & Shaikh, O. A., Simran, Manan, S., & Hasibuzzaman, Md. A. (2023). Artificial intelligence (AI) in personalized medicine: AI-generated personalized therapy regimens based on genetic and medical history: short communication. *Annals of Medicine and Surgery (London)*, *85*(11), 5831–5833. DOI: 10.1097/MS9.0000000000001320

Rane, N. L., Desai, P., & Choudhary, S. (2024). *Challenges of implementing artificial intelligence for smart and sustainable industry: Technological, economic, and regulatory barriers*. Deep Science Publishing., DOI: 10.70593/978-81-981271-1-2

Sharma, P., Singh, V., Thakur, A., & Malik, A. K. (2024). Nanotechnology, Artificial Intelligence, and the Future of Manufacturing. In George, S. C., & Tawiah, B. (Eds.), *Nanotechnology in Societal Development* (pp. 307–332). Springer Nature., DOI: 10.1007/978-981-97-6184-5_9

Sharmile, N., Chowdhury, R. R., & Desai, S. (2025). A Comprehensive Review of Quality Control and Reliability Research in Micro–Nano Technology. *Technologies*, *13*(3), 3. Advance online publication. DOI: 10.3390/technologies13030094

Shobanke, M., Bhatt, M., & Shittu, E. (2025). Advancements and future outlook of Artificial Intelligence in energy and climate change modeling. *Advances in Applied Energy*, *17*, 100211. DOI: 10.1016/j.adapen.2025.100211

Soori, M., Jough, F. K. G., Dastres, R., & Arezoo, B. (2025). Additive Manufacturing Modification by Artificial Intelligence, Machine Learning, and Deep Learning: A Review. *Additive Manufacturing Frontiers*, *200198*. Advance online publication. DOI: 10.1016/j.amf.2025.200198

Swain, A., Abdellatif, E., Mousa, A., & Pong, P. W. T. (2022). Sensor Technologies for Transmission and Distribution Systems: A Review of the Latest Developments. *Energies*, *15*(19), 19. Advance online publication. DOI: 10.3390/en15197339

Vora, L. K., Gholap, A. D., Jetha, K., Thakur, R. R. S., Solanki, H. K., & Chavda, V. P. (2023). Artificial Intelligence in Pharmaceutical Technology and Drug Delivery Design. *Pharmaceutics*, *15*(7), 1916. DOI: 10.3390/pharmaceutics15071916

Wei, T., Han, Z., Zhong, X., Xiao, Q., Liu, T., & Xiang, D. (2022). Two-dimensional semiconducting materials for ultimately scaled transistors. *iScience*, *25*(10), 105160. DOI: 10.1016/j.isci.2022.105160

Xu, D., Ma, Y., Jin, G., & Cao, L. (2025). Intelligent Photonics: A Disruptive Technology to Shape the Present and Redefine the Future. *Engineering*, *46*, 186–213. DOI: 10.1016/j.eng.2024.08.016

Yüksel, N., Börklü, H. R., Sezer, H. K., & Canyurt, O. E. (2023). Review of artificial intelligence applications from an engineering design perspective. *Engineering Applications of Artificial Intelligence*, *118*, 105697. DOI: 10.1016/j.engappai.2022.105697

Chapter 10
Nano-Enhanced and AI-Driven Integration for Food Security and Safety Through Scientific Advances:
Systematic Approaches and Societal Implications

Shikha Chaudhary
https://orcid.org/0009-0009-5428-5666
Shobhit Institute of Engineering and Technology, India

Vineet Vishnoi
https://orcid.org/0009-0001-8276-6573
Shobhit Institute of Engineering and Technology, India

Shivam Chaudhary
https://orcid.org/0009-0002-9875-4363
Shobhit Institute of Engineering and Technology, India

Nidhi Tyagi
Shobhit Institute of Engineering and Technology, India

Prakhar Consul
IILM University, Greater Noida, India

DOI: 10.4018/979-8-3373-2883-6.ch010

ABSTRACT

On a global scale, issues of food security and safety are gaining attention to ensure a consistent supply of safe, nutritious food. Nanotechnology is applied in processes like developing new materials and improving food safety. Nanomaterials help detect contamination by microbes, chemicals, and insecticides. Nanosensors identify allergens and pathogens. Nanoencapsulation enhances food processing by transporting bioactive compounds, improving bioavailability, and extending shelf life. Materials such as nanoparticles, nanocapsules, nanoclays, quantum dots, dendrimers, and nanoemulsions are used in nano fertilizers and pesticides, as well as for promoting plant growth, genome alteration, and transgene expression. These possess antimicrobial properties, boost plant immunity, and transport active substances. Nanocomposites offer acid resistance, recyclability, thermal stability, and improved storage properties. This chapter points out the importance of nanotechnology and artificial intelligence (AI) in transforming agriculture, enhancing food packaging, detecting pathogens, and improving crop productivity and global food security.

INTRODUCTION

The economic prosperity of any nation is deeply intertwined with the robustness of its agricultural sector and the effective utilization of its associated commodities (Schmidhuber & Tubiello, 2007). In recent years, the future of food security has undergone a significant transformation through innovations in nanotechnology and AI. These technologies are no longer confined to theoretical discussions; instead, they are actively reshaping how food is cultivated, processed, safeguarded, and distributed. From detecting invisible contaminants on fresh produce to predicting harvests with remarkable accuracy, these advancements are redefining the foundations of sustainable agriculture. The implementation of advanced techniques such as precision soil management, optimized irrigation strategies, and AI-driven decision-making is critical to achieving self-sufficiency in food production while simultaneously fortifying national resilience against food crises. By addressing agricultural challenges, strengthening food supply chains, and enhancing farmers' understanding of modern technologies (Declaration, 1996), countries can move closer toward sustainable food independence. The accelerated adoption of these technologies further amplifies their role, as food abundance and security continue to attract global attention (Nile et al., 2020).

A pressing example refers to the Kingdom of Saudi Arabia, one of the largest nations with a population of approximately 30 million people (Alsubaie & Berekaa, 2020). Reports indicate that nearly 80% of its food demand is met through imports, which creates vulnerabilities in economic stability, national food security, and public health. These risks are compounded by the threat of foodborne pathogens (Inbaraj & Chen, 2016) and chemical pollutants that can compromise consumer safety, sometimes leading to severe illness or fatalities. Therefore, enhancing systems for food handling, storage, distribution, and monitoring is crucial (Baranowska-Wójcik et al., 2020). Such efforts not only reduce foodborne risks but also improve overall consumption quality and public well-being. The FAO's 2023 report underscores this urgency by revealing that more than 2.4 billion people globally lack access to safe and nutritious food.

Nanotechnology has emerged as one of the most promising scientific domains addressing this crisis, offering powerful tools to design, synthesize, and manipulate materials with unparalleled precision (Chaudhry et al., 2008). By controlling matter at the nanoscale (Gupta et al., 2016), researchers can engineer nanomaterials with dimensions ranging from 1 to 100 nm (Zhou et al., 2021), exhibiting unique physical, chemical, and biological properties. Owing to their exceptionally high surface-to-volume ratio, these nanomaterials demonstrate enhanced color, stability, strength, and even antimicrobial activity. Such attributes make them invaluable in improving food safety and preservation. Applications such as nanosensors and nano packaging have been successfully employed to detect microbiological contaminants (Farooq et al., 2022), chemical residues, and pesticides with superior sensitivity and specificity. Furthermore, nanoencapsulation techniques facilitate the controlled delivery of bio-active compounds (Ray & Bandyopadhyay, 2021), enhancing nutrient bioavailability, extending shelf life, and maintaining food quality. We propose that packaging and biosensing powered by nanotechnology can effectively reduce post-harvest losses by lowering the chances of microbial contamination and enhancing shelf life. For instance, research conducted in Japan on nano-silver food packaging demonstrated a significant improvement in the shelf life of strawberries; when utilizing nano-silver packaging, the shelf life was extended from 3 days to nearly 12 days (Shankar et al., 2021). Importantly, nanotechnology also enables rapid detection of food allergens and pathogens using nanobiosensors, thereby safeguarding consumer health. Despite its promise, challenges persist regarding standards, regulatory frameworks, and public acceptance, which must be addressed to unlock nanotechnology's potential in food systems fully.

A striking case study that illustrates nanotechnology's real-world benefits is the use of nano-silver food packaging in Japan, which extended the shelf life of strawberries from 3 days to nearly 12 days. This example underscores how nanomaterials can tangibly enhance food preservation and reduce waste. While nanotechnology focuses on material innovations, artificial intelligence complements it by optimizing agricultural processes and decision-making. Together, these technologies form a synergistic paradigm that is transforming the future of farming. AI plays a crucial role in analyzing vast datasets generated through precision agriculture, enabling farmers to make informed decisions that improve crop yields, minimize resource wastage, and increase efficiency. By leveraging predictive analytics, AI can optimize supply chain logistics and enhance resilience in global food systems (Kagan, 2016). Smart farming, driven by AI, integrates sensor networks, GPS technologies, and advanced analytics to monitor soil quality, weather fluctuations, crop health, and livestock conditions (Elbehri & Chestnov, 2021; Coordination, 1998). For example, AI-based irrigation scheduling in India improved water-use efficiency by nearly 30%, directly enhancing yields while reducing waste.

This chapter provides a comprehensive review of how nanotechnology and AI individually and collectively redefine food production, processing, packaging, and safety. By synthesizing insights from both fields, we highlight their complementary role in addressing food insecurity and promoting sustainable agriculture. In doing so, we also identify regulatory challenges, technological barriers, and ethical considerations, while outlining opportunities for future research and deployment.

2. MEASURES FOR FOOD SECURITY AND SAFETY

Food authenticity ensures that individuals worldwide have the physical and economic means to access sufficient, secure, and wholesome nourishment that meets their nutritional needs, thereby facilitating active and healthy lifestyles. Insufficient food security contributes to persistent health conditions arising from nutritional deficiencies. The four critical aspects of food security comprise access, usage, stability, and availability (Farooq et al., 2022; Duro et al., 2020). As per UNFAO, the predominant global contradiction anticipated by the year 2050 involves a significant increase in food demand, which is expected to surge within a spectrum of 59 to 98%, chiefly driven by the swiftly growing global population, especially in developing nations, which is projected to attain a total of 9 billion individuals (Food Wastage Footprint, 2013). To achieve food security, agricultural production must increase by 70%, a phenomenon contingent upon various determinants, including environmental change, land availability, agricultural productivity, and food supply. Furthermore, the Food and Agriculture Organization (FAO) has elucidated that. In contrast, affluent countries largely experience food loss primarily due to issues related to packaging expiration and aesthetic considerations. This loss is markedly more acute in economically disadvantaged nations due to variables such as spoilage, contamination, and inadequate food quality (Shukla et al., 2019). Environmental stressors of non-biotic origin, such as soil salinization, increased levels of carbon dioxide and temperature, drought occurrences, and nutrient shortages, have been evidenced to detrimentally influence plant development, growth, fertility, and overall productivity. In certain circumstances, these factors have culminated in the total eradication of specific species from their native ecosystems. Furthermore, research shows that abiotic stressors can cause a mean reduction in agricultural yield that exceeds 50% for most cultivated crops. To mitigate the repercussions of agricultural chemicals, it is imperative to advance the development of plants alongside their capacity to withstand non-living and/or living entities' adversities, augment farm productivity, and, most critically, improve their caliber via the incorporation of nanotechnology and bioactive elements derived from marine algae and terrestrial flora (Kagan, 2016). The phenomenon of food waste has harmful consequences for the environment by reducing land and water resources critical for agriculture and further amplifying the concentration of greenhouse gases that propel climate change (Pal, 2017). Addressing these multi-dimensional issues requires integrating technology with sustainability.

The safety of food is a vital concern in public health, requiring the safeguarding of food against physical, chemical, and biological pollutants during the various stages of preparation, processing, execution, and distribution, thereby assuring consumer protection (Babu, 2022; Alrobaish et al., 2021). The swift modifications in culinary recipes, dietary practices, harmful substances, cancer-inducing pathogens, and various pollutants significantly influence food safety and pose a serious risk to individuals' health. Farmers worldwide are now adopting innovative approaches by enhancing and modernizing their practices and utilizing cutting-edge technologies. These technologies promote food safety and security through improved agricultural productivity, rejuvenating soil health, superior food packaging, efficient distribution systems, and enhanced food quality standards (Demirer et al., 2021). Conventional methodologies used for analyzing contaminants, identifying pathogens, and improving food safety are often characterized by their considerable time requirements and financial expenditures. Recently, in developing countries, significant progress has been made in using various tiny materials and their related devices, which are being used cheaply and over large areas to improve farming, enhance soil quality, help with food processing and packaging, extend the shelf life of food for safety and usability, and test for pathogens and their toxins.

3. ADVANCEMENTS IN FOOD QUALITY ASSURANCE AND SAFETY VIA THE USE OF NANOTECHNOLOGY

Technologically sound and sustainable strategies are needed to integrate innovation with sustainability in the food supply. Cohesive nanotech strategies are essential. An innovative strategy that boosts production efficiency while reducing food waste is crucial for securing food availability worldwide (Kumar et al., 2019). Nanotechnology advancements have provided opportunities to improve the food supply chain and develop innovative food products. These advancements positively impact agriculture by reducing nutrient wastage in fertilizers and boosting food production through effective pest and nutrient optimization, among other strategies. It may also aid in advancing food processing methodologies, revolutionize food products by offering improved palatability and sensory characteristics, extend the preservation of food items, and formulate nanotechnology-based solutions for identifying and preventing food degradation (Fig. 1).

Figure 1. Role of Nanotechnology in enhancing Food authenticity and Safety

3.1 Nano-technology's Role in Agricultural Improvement and Yield Enhancement

Nanotechnology-driven solutions have emerged that will lead to increases in crop productivity and resilience through precision-based innovations. Recent studies demonstrate that nanotechnology-derived fertilizers and pesticides can effectively address nutrient inefficiency and plant disease. For example, Abdel-Aziz et al. (2021) developed waste-derived NPK nanofertilizers that significantly enhanced Capsicum annuum yield compared to conventional fertilizers. Similarly, silver nanoparticles have been shown to improve chlorophyll levels and fruit production at optimal concentrations. NPK fertilizers in nanopowder form, derived from organic waste materials such as squid pen remnants, shrimp exoskeletons, bovine skeletal remains, banana peel residues, and Kolakhar, were formulated. The Capsicum annum L. cv. Cordoba specimens subjected to these advanced fertilizers were analyzed in comparison to those treated with conventional chemical fertilizers. Moreover, AgNPs led to a significant enhancement in the dimensions of pollen grains and spores. In addition, the peach leaf extracts treated with AgNPs displayed powerful antibacterial effects at a staggering concentration of 3000 μg/mL, resulting in an impressive zone of inhibition measuring between 6 and 8.67 mm (Mosa et al., 2021; Hofmann et al., 2020). While nanotechnology represents a thrilling new frontier, primarily utilizing nanopesticides composed of nanoparticles to combat plant diseases, it is attracting significant interest today. The bacterial panicle blight disease, instigated by seed-borne rice pathogens from the genera Burkholderia glumae and B. gladioli, poses a serious threat to rice production worldwide. Recently, the burgeoning field of nanotechnology has revolutionized the delivery of genetic tools to plants. Nanotechnology has facilitated a much more aggressive approach to genetic engineering, allowing for the stabilization of a greater amount of genetic material, thus enhancing plants in ways that have never been achieved in history (Hofmann et al., 2020). The potential is exhilarating for research endeavors, since carbon quantum dots have proven to be remarkably effective carriers, promoting the transformation of plasmids into mature plants and enabling remarkable temporary plant transformations. Notably, carbon nanotubes served as an alternative version of a chitosan complex, enabling the nanotubes to penetrate chloroplasts through a robust lipid exchange envelope mechanism. Remarkably, every procedure was conducted without the involvement of any chemical or biological organisms. This method skilfully and efficiently introduces a novel biological transgene delivery technique that departs from conventional practices and paves the way for new possibilities in plant development (Kalia et al., 2021). To minimize the impact of disease and crop losses, nanotechnology has developed unique protective and diagnostic measures.

3.2 Deployment of Nanotechnology for the Management of Plant Diseases

Devastating illnesses caused by plant pathogenic bacteria, viruses, and fungi in various crops globally impact growth, quality, and production, resulting in significant agricultural losses. Consequently, to address post-harvest and plant diseases. Creating innovative methods and equipment that can get beyond traditional farming methods is essential. Climate change is poised to impact agricultural productivity by increasing the prevalence of plant illnesses and decreasing the effectiveness of disease prevention measures. Nanotechnology presents a powerful strategy for managing diseases by developing optimized formulations of growth regulators, nutrition, and agrochemicals designed to combat plant diseases caused by pathogens; consequently, crop loss is avoided (Tesfaye et al., 2021). The composition of ginger essential oil, which includes monoterpenes and sesquiterpene hydrocarbons, can significantly impact the internal dynamics of pathogens, consequently restraining diverse phytopathogens. On the other hand, ensuring the stability and solubility of compounds derived from ginger essential oil continues to pose a significant challenge. These restrictions may impede their ability to manage plant diseases effectively. Nevertheless, employing nanoemulsions presents a hopeful remedy. These nanoemulsions enhance the efficacy and dispersion of essential oils, rendering them more potent in managing various plant pathogens. (Feng et al., 2018). Nanomaterials present some interesting solutions for extending shelf life and reducing waste during the period following harvest.

Table 1. The Role of Nanotechnology in Preventing and Managing Plant Diseases

Nanomaterial/Approach	Application	Mechanism/Benefit	Examples
Silver nanoparticles (AgNPs)	Antimicrobial treatment of seeds and crops	Disrupts microbial cell membranes and inhibits growth	Control of fungal pathogens like Fusarium
Zinc oxide nanoparticles (ZnO NPs)	Foliar sprays for bacterial and fungal control	Generates reactive oxygen species that kill pathogens	Used against Xanthomonas and Alternaria
Copper nanoparticles (CuNPs)	Nano-fungicides	Inhibits enzyme activity in pathogens, essential for fungal metabolism	Used in grapes, tomatoes
Chitosan nanoparticles	Elicitor of plant defense responses	Induces systemic acquired resistance (SAR) in plants	Used for disease control in cereals
Silicon nanoparticles (SiNPs)	Plant immunity enhancement	Strengthens cell walls, improves resistance to pathogen invasion	Rice and wheat
Nano-encapsulation of pesticides	Controlled release of fungicides and bactericides	Prolonged action, reduced chemical use, minimized environmental impact	Nano-encapsulated Mancozeb, Triazoles
Carbon nanotubes (CNTs)	Early detection biosensors	Transduces biochemical signals to detect pathogen DNA or toxins	Used for virus detection in crops

3.3. Implementation of Nanotechnology in Addressing Post-Harvest Losses

Worldwide, over 40% of food waste happens at various points, including post-harvest management, processing, distribution, and consumer habits. This problem impacts both developed and developing nations. To tackle these challenges, scientists have created nanocomposites that incorporate zeolitic imidazole framework-8 (ZIF-8). These materials possess adjustable pore dimensions and increased surface areas. Additionally, a regulated self-assembly method, which incorporated silanes or tannic acid in conjunction with Fe, further enhanced them with catalase. In real-world applications, nanocomposites like zeolitic imidazole framework-8 (ZIF-8) have been employed to regulate microbial proliferation and nutrient excretion. A particular case study involving blueberries demonstrates that chitosan-nanomaterial films, which diminish enzymatic activity, extend the shelf life of high-quality fruit (Li et al., 2021). Certain microorganisms are essential for boosting soil quality and facilitating plant nutrient uptake. Nanomaterials integrated into chitosan films have demonstrated significant potential in regulating microbial proliferation. They also assist in the gradual liberation of nutrients, which affects essential elements such as titratable acidity, ripening index, and weight loss. This phenomenon is associated with heightened activity of peroxidase and polyphenol oxidase enzymes. As a result, they significantly extend the shelf life of fresh blueberries (Karimirad et al., 2018). Chitosan nanoparticles integrated with essential oil sourced from Citrus aurantium were devised to evaluate their effectiveness in enhancing the quality of white button mushrooms after harvest. Nanotechnology allows for greater precision, adaptation, and modification of current genetic engineering practices within agricultural contexts. Nanotechnology advancements are starting to change food processing, food safety, and shelf life.

It has been reported that the nanoparticles facilitated a gradual release of citrus aurantium essential oil, leading to an impressive 15-day prolongation of the shelf life of white button mushrooms. The nanoparticles have been demonstrated to enhance the concentrations of phenolic compounds, vitamin C, and antioxidant enzymes, including catalase and superoxide dismutase. Simultaneously, the nanoparticles assist in diminishing polyphenol oxidase activity by the conclusion of the storage period (Hossain et al., 2018).

Additionally, bioplastic nanocoating helps maintain important features like strength, stable pH, cellulose levels, and the availability of ions. As a result of these advantages, they are suggested for efficient utilization in biomedical applications and the production of pharmaceutical components (Altpeter et al., 2016). In particular, the encapsulation portion of nanotechnology is impacting packaging efficiency, the bioavailability industry, and food safety preservation.

Figure 2. Utilization of Nanotechnology for the Reduction of Post-Harvest Losses

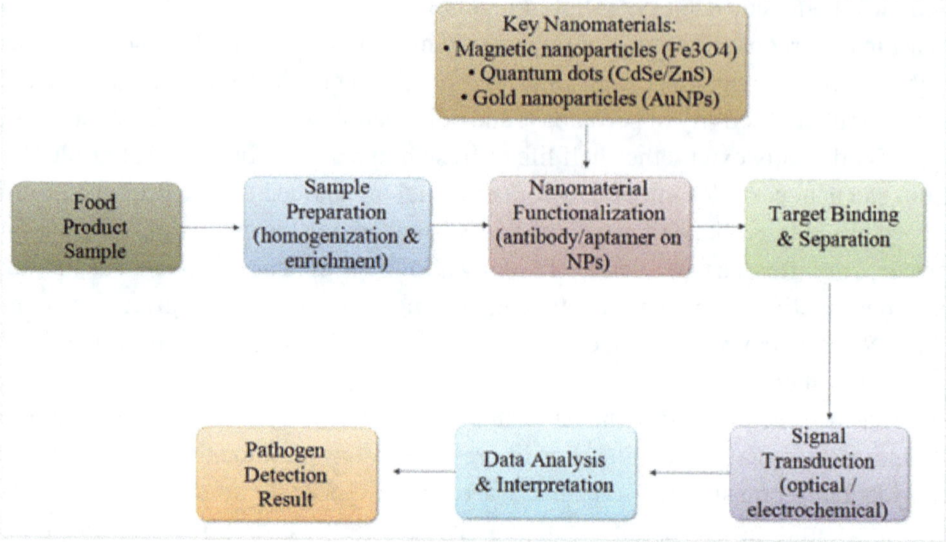

3.4. The integration of Nanotechnology within Genetic Engineering

In the 1950s, the Green Revolution brought forth high-yielding, semi-dwarf varieties of wheat and rice. This significant advancement highlighted the crucial importance of plant breeding in fulfilling the needs of an ever-increasing population. Unlike traditional plant breeding methods, genetic engineering techniques can effectively integrate and enhance plant genomes (Mahfouz et al., 2014). Nevertheless, conventional genetic engineering techniques tend to insert genes at arbitrary sites within the plant genome, potentially leading to unforeseen consequences. Recent innovations have unveiled potent gene-editing techniques, such as CRISPR and TALEN. These methods facilitate accurate gene targeting, enabling the introduction of preferred traits into the genome with greater precision (Martin-Ortigosa et al., 2014). The cell wall of a plant functions as a protective barrier against the introduction of external biomolecules into the plant. In recent times, nanomaterials have been widely utilized in genetic engineering for delivering and releasing genes and proteins within plant cells. To revive fruitful plants, recombination events were selected based on visual criteria. Therefore, it was proposed that mesoporous silica nanoparticles could be utilized to host specific enzymes that affect the target cells, offering another approach for genome-editing technologies that do not rely on DNA in plants (Demirer et al., 2019). Nanomaterials have revealed their ability to augment protein expression and enable effective DNA delivery without the addition of external genes. They also help protect DNA from being broken down by nucleases found in plants like cotton, arugula, tobacco, and wheat. Furthermore, the combination of carbon nanotubes and chitosan has demonstrated the ability to deliver plasmid DNA directly to chloroplasts. This method has been successful in leafy greens such as spinach (Spinacia oleracea), flowering tobacco (Nicotiana tabacum), and watercress (Tropaeolum majus). It has been successful in protoplasts taken from Arabidopsis lyrata, and this was done without needing extra chemicals or biological materials.

3.5. Transforming Food Processing through the Utilization of Nanotechnology

In the quest for enhanced nutrition, safety, and processing effectiveness, the food industry is diligently investigating innovative techniques to preserve the quality of food items. Nanotechnology, which includes innovatively formulated nano-ingredients in consumable products, is becoming more prevalent during the production and preparation phases of food (Wong et al., 2016; Kritchenkov et al., 2021). Biopolymer nanocomposite-based food wraps have developed intelligent and dynamic functionalities, serving as thermal indicators; they change color to a deep purple at around 8°C. Nano-engineered surface layers significantly improve the shelf life of ricotta cheese. Their ability to sense temperature changes provides alerts regarding cheese storage when temperatures fluctuate (Kamal et al., 2016). Nano-composite materials have also been utilized as antifouling agents and can exhibit an extended lifespan. Nanocomposites created by incorporating nanomaterials at minimal loadings into the primary matrix demonstrated highly effective antifouling capabilities when differentiated from an unmodified matrix (Zhu et al., 2021). A research investigation was carried out to analyze films made from PLA that incorporate silver nanoparticles, which can move within acidic food matrices, subjected to high-pressure food processing and proposed a viable nano-composite film suitable for acidic food during elevated stress handling methods (Kalia et al., 2021). To prolong the life span of guava fruits in packaging, a nanocomposite film was created using chitosan sourced from Urtica dioica leaf extract, combined with ZnO and copper oxide. This film demonstrated both antioxidant and antimicrobial properties, making it a potentially effective packaging solution for extending the life span of fruits (Fadel et al., 2019). A research project was carried out to create nanoparticles derived from soy protein isolate, following enzymatic breakdown, which helps create flavor profiles similar to those of processed meat. We suggest that nano-engineered components could improve sensory attributes and nutritional stability in processed foods. To back this perspective, Fadel et al. (2019) indicated that soy protein isolate nanoparticles produced via an enzymatic hydrolysis technique replicated meat flavors, thus providing a concrete example of nanotechnology's application in the food industry. Furthermore, SiO_2 nanomaterials are employed in discarded food nanomaterials to carry aromas in food substances (Abdel-Aziz et al., 2021). Nanoscale nanosensors offer sensitive and more abundant signals for food-borne pathogens, which can enable significant foresight in detecting safety threats.

3.6 Food Encapsulation Techniques Utilizing Nanotechnology

Recently, the food industry has been adopting various methods to enhance food packaging because it can prevent spoilage and prolong shelf life. Efficient food packaging will minimize contamination of a diverse range of food products, including fresh fruits and vegetables, while addressing other environmental concerns and maintaining their quality and safety during transportation and storage (Roy et al., 2021). A research initiative developed a versatile design with environmentally friendly cushioning, created through 3D printing, which featured a nanocellulose outer layer and a core made of either chitosan or silver nanoparticles, effectively inhibiting the growth of E. coli and S. aureus while offering robust protection and durability. TiO_2 nanoparticles embedded in the blend of chitosan and tannic acid exhibited heat stability and antioxidant properties, rendering it perfect for effective food packaging solutions (Al-Shabib et al., 2016). Nanoparticles have also been documented as non-toxic and utilized as packaging materials for food (Saravanakumar et al., 2020). Implementing an active food packaging system is recommended for use in the food sector (Naskar et al., 2018). Lau et al. (2017) created nanosheets using Ag-ZnO-reduced graphene oxide and PEG, which continue to work against S. aureus and P. aeruginosa even after being stored at room temperature for three months, showing their potential for active packaging. Coupled with a low-cost synthesis route, this performance supports the design of next-generation barriers that preserve food quality far better than conventional materials. Similarly, pullulan-PVA nanofibers laden with thymol-rendered porphyrin particles kill S. aureus and E. coli in vitro and retain biocompatibility and biodegradability, qualities that ease regulatory approval and consumer acceptance. Incorporating these fibres into wraps or pouches could thus deliver antimicrobial action while leaving no harmful residue after disposal or composting. Elsewhere, palladium and platinum nanoparticles reduce the viability of E. coli, L. monocytogenes, and S. aureus, apparently by chemically interacting with the cell wall and inhibiting biofilm maturation. Such targeted activity, together with their catalytic and electronic advantages, makes these metals attractive candidates for periodic antimicrobial finishing sprays or paints in food processing areas. When made into a nano-composite film, carboxymethyl chitosan mixed with nano-MgO shows better heat resistance, clarity, flexibility, and water resistance, all of which are important for cling films used on moist products. Importantly, the same composite now displays intensified inhibition against Shewanella baltica and L. monocytogenes, confirming its dual role as a structural and protective element in seafood packaging. For longer shelf life, biogenic silica powders, harvested from spent groundnut shells, strengthen PHBV-matrix films and yield nanocomposites with better tensile strength, lower gas permeability, and improved thermal stability. These mechanical benefits, combined with evidence of moderate antimicrobial action, suggest practical applications in semi-permeable trays and carry bags that slow spoilage while remaining compostable.

3.6.1 Case Study: Nano-Silver Packaging (with Smart Nano sensor Tie-in) to Extend Strawberry Shelf Life (Shankar et al., 2021)

Context: Fresh berries are highly perishable; rapid softening and microbial spoilage drive post-harvest losses during handling and retail display.

Intervention: A nano-silver (AgNP) active-packaging film was used for retail punnets. AgNPs provide broad-spectrum antimicrobial action at low loading, reducing surface bioburden and slowing decay. To enable real-time quality checks, the pack design is compatible with nanosensor-based "intelligent packaging" labels that monitor headspace spoilage gases and temperature for early deterioration signals.

Outcome: In a reported deployment, strawberry shelf life increased from ~3 days to ~12 days under nano-silver packaging evidence that active nanocomposites can materially reduce post-harvest loss for high-value fruit.

How it works (brief mechanism): AgNPs suppress spoilage and pathogenic microbes on fruit and packaging surfaces; optional nanosensor labels detect shifts in volatiles/conditions, supporting timely stock rotation and waste reduction.

Why it matters (systems view): Pairing nano-enabled packaging with AI-driven logistics and forecasting (demand prediction, routing, cold-chain exception alerts) further cuts waste and improves freshness at scale, an integration direction highlighted in the chapter's discussion of AI with nanotechnology for agri-food supply chains.

Replicability & scope: This template generalizes to other short-life produce (raspberries, cherries) and seafood trays using the nanomaterials summarized in Table 2 (e.g., AgNPs, ZnO NPs, nanoclays) and intelligent labels.

Risk & compliance note: When reporting deployments, include migration testing/toxicology profiles and labeling practices to address regulatory and consumer-safety considerations discussed elsewhere in the chapter.

Table 2. Transforming Food Packaging through Nanotechnology

Nanomaterial/Approach	Function in Packaging	Mechanism/Benefit	Examples
Silver nanoparticles (AgNPs)	Antimicrobial packaging	Inhibits growth of spoilage and pathogenic microorganisms	Meat and dairy product packaging
Zinc oxide nanoparticles (ZnO NPs)	UV-blocking and antimicrobial packaging	Prevents UV-induced food spoilage and microbial contamination	Bakery and snack packaging
Nanoclays	Barrier enhancement	Improves resistance to gas, moisture, and light	Plastic bottles and pouches
Titanium dioxide nanoparticles (TiO2 NPs)	Self-cleaning surfaces	Photocatalytic activity reduces contamination on packaging surfaces	Ready-to-eat meals packaging
Starch-based nanocomposites	Biodegradable packaging	Eco-friendly alternative to conventional plastic packaging	Fresh fruits and vegetable packaging
Nano sensors	Intelligent packaging	Detects spoilage gases or changes in temperature/humidity	Smart labels for perishable foods

3.7. Nanotechnology for the Detection of Pathogenic Microorganisms in Food Products

To safeguard themselves against pathogenic organisms and herbivorous entities, flora synthesize secondary metabolites, including polyphenolic compounds. Researchers created special sensors to see plant polyphenols in real-time using single-walled carbon nanotubes that emit near-infrared light. Nanotube-derived sensing devices exhibited modifications in the comprehensive concentrations of polyphenolic compounds subsequent to pathogen exposure in Glycine max and Tococa variants. Moreover, it was verified that polyphenolic compounds were liberated from the foundational roots of soybean seedlings within a defined interval of 24 hours after the pathogen's onset. In a separate study, phytopathogenic organisms were identified using an accelerated and accurate observation methodology. This involved selecting pathogen deoxyribonucleic acid through recombinase polymerase amplification and employing a nanoparticle-facilitated electrodeposition biosensor, along with nanoparticle-augmented electrochemical analysis using differential pulsation voltammetry. Nanoparticle-based sensors exhibited an extraordinary enhancement in sensitivity, achieving a 10,000-fold improvement in the identification of infected plants before any visible disease symptoms manifested, in contrast to traditional polymerase chain reaction (PCR) techniques and gel electrophoresis methods (Rana et al., 2021). In the context of pathogen detection, an examination of probe ss-DNA hybridized with the target ss-DNA of the pathogen was executed employing linear sweep voltammetry and cyclic voltammetry methodologies (Tahir et al., 2017). Zn-NPs are employed for the act of immobilization of the probes DNA and the interaction of the target DNA with ss-DNA facilitates combination, which was investigated via cyclical and differential pulse voltammetry; this nanoscale DNA sensor exhibited a specificity threefold superior to conventional methodologies (Khater et al., 2019). Citrus tristeza virus can infect citrus plants, leading to the detrimental condition known as tristeza. A label-free impedimetric biosensor, fabricated through the electrodeposition of gold nanoparticles, was developed for the identification of the deoxyribonucleic acid linked to the Citrus tristeza virus. Nanosensors exhibit a remarkable proficiency in detecting specific viral entities within a milieu of non-specific DNA, thereby showcasing greater duplicability (Zare et al., 2022). The genetic material of the virus of Curl New Delhi, derived from the leaves of Solanum lycopersicon, was characterized employing both bifunctional and monofunctional gold nanoparticle oligonucleotide probes via DNA cross-fertilization assays, performed either in a direct manner or in a sandwich configuration augmented by silver enhancement techniques. This methodology demonstrated improved detection sensitivity (spanning from 100 p.m. to 100 M), while simultaneously being economically viable and necessitating a reduced temporal investment.

Figure 3. Nanotechnology based techniques for the identification of pathogens related to foodborne illnesses.

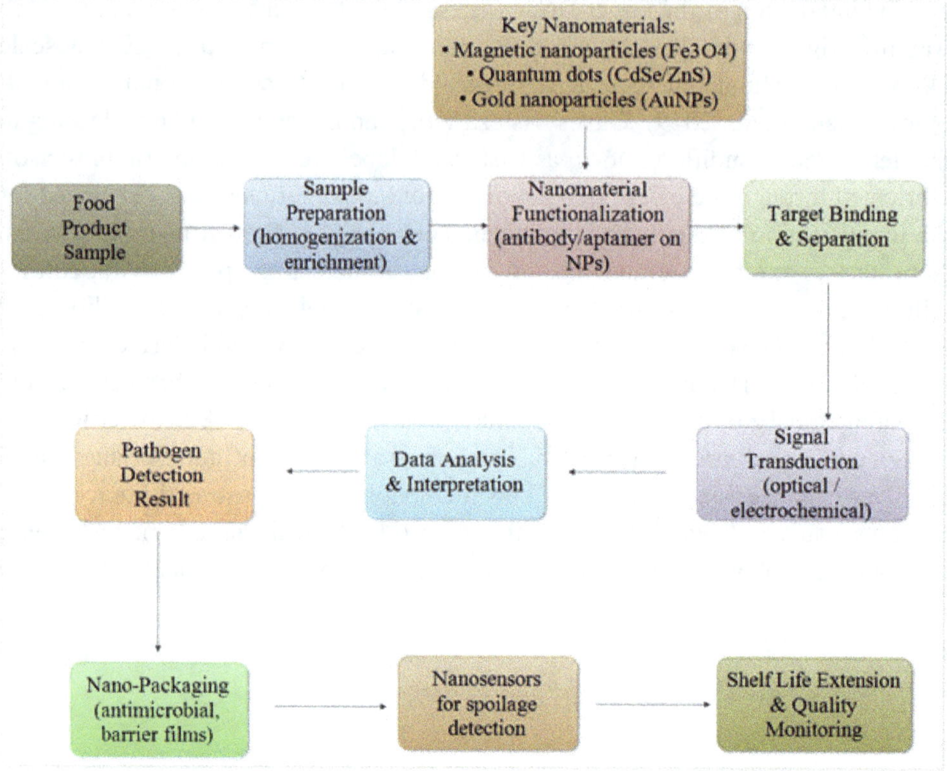

Zare et al. (2022) developed a new type of electrochemical biosensor that doesn't need labels by changing a gold electrode with L-cysteine and a PAMAM dendrimer, which is designed to detect the ITS sequence of M. simian specifically. The technique of differential pulse voltammetry was utilized to evaluate fluctuations in the free guanine signal corresponding to variations in the concentration of the ssDNA target. The biosensor demonstrated exceptional selectivity, achieving a detection limit of 1.40 fM, in addition to exhibiting remarkable precision and substantial reproducibility. As the concentration of target compounds increased from 1.17 to 40 nM, it demonstrated increased fluorescence at the molecular beacon throughout real-time observation, suggesting that the molecular beacon can be effectively utilized in a botanical context (Burris et al., 2013). In contrast, the presence of target DNA results in a discernible shift in the color of gold nanoparticles (AuNPs) within a wavelength range of 550–650 nm, demonstrating a specificity of 15 × 10^5 CFU/mL and 9 pg/μL via AuNP probes, thereby facilitating expedited detection times while also proving to be more economically viable for the swift identification of bacterial entities (Ahmadi et al., 2018). In a scholarly investigation, a recombinase polymerase amplification technique integrated with a fluorescent lateral flow immunoassay employing europium nanoparticles was introduced to quantitatively detect foodborne bacteria like L. monocytogenes, V. parahaemolyticus, and Escherichia coli O157:H7, revealing detection thresholds varying between 4 and 9 CFU/mL across various samples, respectively, while demonstrating elevated sensitivity and robust specificity even at minimal concentrations of foodborne pathogens (Chen et al., 2021). The assessments were carried out on a diverse array of specimens, including potable water and avian flesh, to validate their efficacy through the quantification of P. aeruginosa concentrations (Wu et al., 2018). The specificity showed variability across different varieties due to changes in the chitin content ratio. Fusarium oxysporum demonstrated the highest relative affinity, while Trichoderma reesei cellulase exhibited the lowest relative affinity. Consequently, this method could be applied for evaluating food quality, treating human fungal infections, safeguarding crops, and exploring other potential applications (Joshi et al., 2019). Nanosensors for unamplified genomic deoxyribonucleic acid (gDNA) were created using gold nanoparticles (AuNPs) stabilized with dextrin. A single-stranded DNA polymerase (ss-DNAp) was used to strengthen the stability of the dextrin-coated gold nanoparticles (AuNPs). This made it easier to form a complex with the target genomic DNA (gDNA) in the presence of single-stranded genomic DNA, greatly improving the stability of the nanoparticles even in high salt conditions. This method effectively detected pathogen DNA levels as low as 2.94 fM, quickly demonstrating high accuracy and precision (Baetsen-Young et al., 2018).

4. NEED OF AI IN GLOBAL FOOD SECURITY AND SAFETY

Artificial Intelligence (AI) is essential for global food security and safety as it enables data-driven decision-making across the agricultural value chain. By analyzing vast datasets from precision farming, supply chains, and food monitoring systems, AI enhances crop yield prediction, optimizes resource allocation, and minimizes post-harvest losses. Furthermore, AI-powered sensors and predictive models improve the early detection of foodborne pathogens and contaminants, thereby ensuring safer food for consumers. Its integration not only strengthens resilience against climate change and population growth but also ensures sustainable and equitable access to nutritious food worldwide. Figure 5 illustrates the necessity of AI in the context of global food security and safety.

Figure 4. Need of AI In global food security and safety

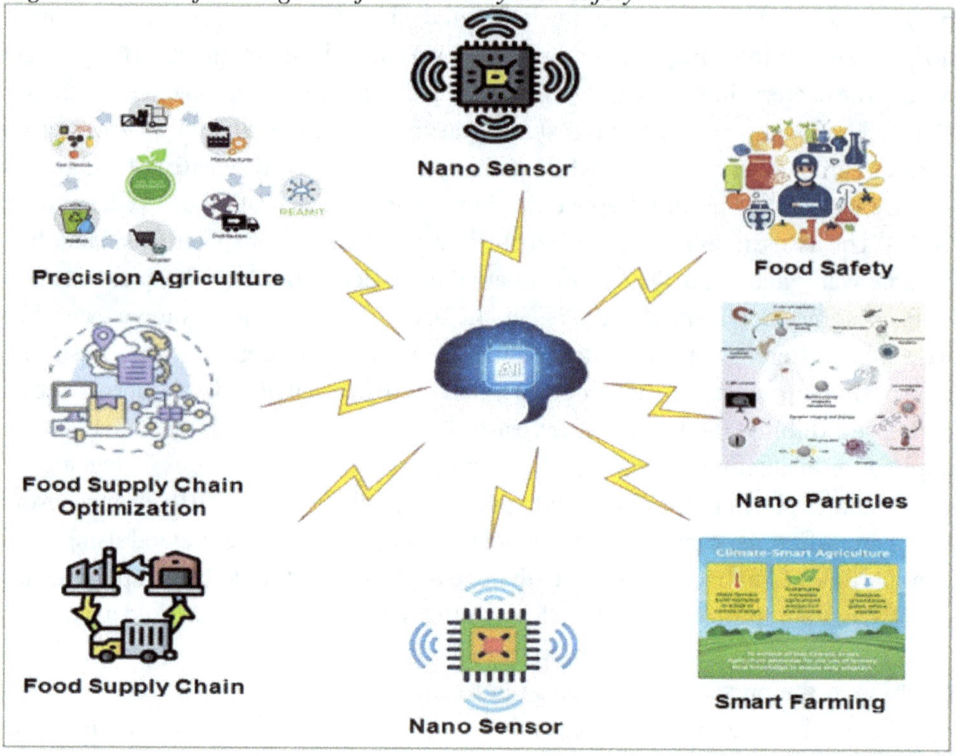

4.1 AI for Precision Agriculture

AI enables 'data-driven' agriculture, which will help allocate resources properly and will contribute towards boosting food production. Through the use of AI-enhanced platforms, instant information derived from satellite images, sensors, drones, and Internet of Things (IoT) devices can be examined to ascertain crop vitality, soil moisture content, and nutrient demands. Machine learning algorithms utilize both historical and live agronomic data to deliver practical insights for scheduling irrigation, applying fertilizers, and managing pests. AI-enhanced precision agriculture enables tailored farming techniques for specific locations. Such technology is essential in tackling the growing worldwide need for food, anticipated to surge by almost 70% by the year 2050, as per the Food and Agriculture Organization (Liu, 2017). By enhancing the utilization of resources like water, fertilizers, and pesticides, AI not only lowers expenses for farmers but also lessens environmental harm. The advancement of AI is transforming precision agriculture, enabling farmers to leverage data-driven insights to boost crop yields while conserving resources, ultimately resulting in more sustainable food production systems. A primary application is the AI analysis of drone imagery that identifies nutrient deficiencies or diseases in plants earlier than ever before by examining features such as plant colors and canopy shapes. For instance, employing convolutional neural networks (CNNs) to analyze high-resolution drone imagery has proven to be highly effective in early disease detection and weed identification (Kamilaris & Prenafeta-Boldú, 2018). Likewise, AI-driven robotic systems can perform precise tasks like targeted spraying or meticulous harvesting, which can further enhance resource efficiency. The convergence of artificial intelligence and nanotechnology is truly exhilarating. Nanoscale sensors continuously collect detailed data regarding the health of plants and soil characteristics, while AI algorithms analyze this information and provide real-time, adaptable interventions. This comprehensive model will ensure that food production becomes increasingly resilient, efficient, and customized to local conditions, which is crucial for enhancing global food security. Additionally, AI-based forecasts offer certainty amidst the chaotic changes being made by climate and economic uncertainty regarding securing food.

4.2 Predictive Modeling for Agricultural Output and Food Requirements

Early skeins of pest and disease predictions underpinned by AI are pivotal to reducing crop loss and food security. AI-enhanced predictive analytics is becoming a vital resource for assessing agricultural yields and forecasting future food requirements. Accurately predicting food demands is crucial for effective strategic planning in relation to food security, particularly in light of climate fluctuations and resource limitations. Machine learning models have the capability to identify intricate and nonlinear relationships that conventional statistical models frequently overlook by analyzing extensive datasets that encompass historical weather patterns, soil health, satellite imagery, and farming practices (Hofmann et al., 2020). These models use different types of data, like temperature, rainfall, soil nutrients, and plant health measurements (like NDVI), to make very specific and accurate predictions. These forecasts are essential for governments and policymakers to stabilize food markets and optimize reserve management, especially in areas threatened by food insecurity stemming from climate change. For instance, Google's initiative called AI for Social Good has partnered with agricultural stakeholders to forecast food shortages by utilizing satellite imagery and climate models (Changarathil, 2022). In a comparable context, AI solutions such as IBM's Watson Decision Platform for Agriculture have been employed by commercial agricultural enterprises to anticipate harvest volumes several weeks in advance, thereby enabling proactive oversight of storage, logistics, and market supply. When combined with nanotechnology, predictive analytics gains an even greater level of strength. Nanotechnology-enabled sensors utilized in the field deliver detailed, instantaneous information regarding plant vitality, soil nutrient levels, and environmental factors. This detailed input will enhance the datasets that guide AI models, resulting in more precise and responsive yield forecasts. The collaboration strengthens the robustness of agricultural systems and guarantees prompt reactions to new challenges, ultimately aiding in the achievement of global food security goals. AI accounts for each leg of the supply chain regarding food as it constantly optimizes for performance metrics pertaining to traceability, safety, and delivery around food products, while combatting food loss at a substantial level.

4.3 AI in Predicting Pests and Diseases

The sudden emergence of pests and plant diseases presents a significant danger to worldwide food security. Conventional methods for detection and management tend to be reactive, sluggish, and require extensive resources. Artificial intelligence, particularly through machine learning and computer vision techniques, is changing how we predict and prevent these threats. It enables quick identification and ongoing monitoring, leading to proactive measures. AI algorithms analyze large datasets, including weather patterns, historical pest infestation records, crop visuals, and sensor data, to predict the onset of pest invasions or disease events. Methods like SVM, CNNs, and decision trees are frequently employed to classify plant diseases based on leaf images with remarkable accuracy. In India, the "Nuru" AI application, developed by Plant Village in collaboration with the FAO, uses smartphone imaging technology to quickly detect diseases in cassava, maize, and many other essential crops, even in remote, offline environments (Ferentinos, 2018). In another significant example, AI models trained with remote sensing information and climate factors have successfully forecasted the movements of locust swarms throughout East Africa, enabling governments and farmers to implement pre-emptive countermeasures. The combination of artificial intelligence and geographic information systems (GIS) significantly improves the precision of spatial forecasting, establishing it as an invaluable resource for managing regional agriculture. When paired with nanotechnology, the capabilities of AI are greatly enhanced. Nanosensors incorporated within agricultural fields or plant tissues are capable of continuously tracking biochemical markers, temperature changes, or signs of insect activity. This ever-evolving information, integrated into artificial intelligence systems, offers an increasingly adaptive monitoring mechanism for pests and diseases. Targeted measures diminish reliance on broad-spectrum pesticides, lower production costs, and reduce environmental repercussions, thus promoting sustainable food systems (Oliveira & Silva, 2023). In conclusion, the significance of AI in anticipating pest and disease outbreaks is crucial for agricultural resilience, minimizing crop losses, and fortifying global food supply chains. AI is helping to proactively eliminate food loss around the world, stemming from diversifying overproduction to reducing waste through spoilage.

4.4 Food Supply Chain Enhancement through Artificial Intelligence

A highly efficient and transparent food supply chain is paramount to achieving global food security, particularly in the context of escalating demand, climatic perturbations, and logistical obstacles. Artificial Intelligence enhances every segment of the agricultural supply chain, encompassing cultivation, processing, distribution, and retail. It furnishes real-time data, predictive analytics, and adaptability in resource management. By scrutinizing supply chains through the lens of demand forecasting, pinpointing inefficiencies, and offering strategic recommendations for enhancement, AI models significantly optimize supply chains. For instance, AI-driven predictive analytics can anticipate demand fluctuations in specific geographic areas by analyzing historical consumption trends and climatic phenomena. This facilitates enhanced inventory oversight, reduction in excessive production, and heightened precision in logistical planning for deliveries. Corporations such as IBM, Microsoft, and SAP have developed AI-augmented systems to assist agricultural producers. Distributors effectively reduce waste and operational costs while simultaneously enhancing the precision of delivery scheduling (Kamilaris et al., 2017). Furthermore, AI proves to be instrumental in cold chain logistics, which constitutes a fundamental procedure for the transport of perishable commodities. Sensors employed in transportation and storage environments guarantee the continuous monitoring of critical conditions such as temperature and humidity, while AI algorithms evaluate this data to ensure adherence to safety regulations, alerting users to potential issues and even enabling the rerouting of shipments to mitigate spoilage resulting from delays. The confluence of AI and blockchain technology will guarantee enhanced traceability and transparency within the supply chain. Consumers, regulators, and retailers can monitor a product's complete journey from the farm to the table, guaranteeing authenticity, freshness, and safety. For instance, Walmart has adopted blockchain-AI technologies to monitor leafy greens, reducing trace-back times from days to mere seconds in contamination incidents (Ajithkumar & Krishnamoorthy, 2025). By incorporating nanotechnology, this optimization becomes significantly more detailed. Sensors powered by nanotechnology, when incorporated into packaging materials, can oversee freshness, recognize spoilage gases, and send data to AI systems for automated decision-making. This combined system has the capability to notify stakeholders to recall food batches prior to any safety violations, thereby minimizing health hazards and waste. Essentially, artificial intelligence strengthens the food supply chain in the face of unpredictability and inefficiency, two of the foremost obstacles to achieving sustainable food security worldwide. Furthermore, AI systems continuously track and mark processes to ensure safety features and potential plans, enabling food safety rather than remediation.

4.5. AI Solutions for Cutting Down Food Waste

Food waste is a considerable danger to global food security, as almost a third of all food produced, around 1.3 billion tons, is lost or thrown away annually, as stated by the FAO (Sousa et al., 2024). AI significantly contributes to solving this challenge by facilitating more effective food production, enhanced inventory management, more informed buying choices, and prompt reallocation of excess food. AI-driven demand forecasting solutions help retailers and suppliers prevent overstocking by analyzing past sales data, weather predictions, community events, and social behavior trends. Machine learning algorithms can adapt procurement and inventory strategies in real-time, ensuring that supply meets current demand, which in turn minimizes the quantity of unsold perishable goods. For example, firms such as Afresh and Wasteless have created AI technologies that enhance grocery inventory management and pricing determined by expiration dates, purchasing patterns, and storage environments, greatly minimizing waste within the retail sector (Tesfaye et al., 2021). On the consumer front, intelligent kitchen systems driven by AI monitor food consumption habits, expiration timelines, and storage environments and can provide personalized recommendations to avoid spoilage. Applications such as OLIO and Too Good To Go utilize AI algorithms to connect individuals or businesses with surplus food to those in need, enabling the redistribution of food before it turns into waste. Across modern farms, AI tools now help producers spot hidden problems such as overwatering, pest outbreaks, or soil nutrient drops that can make crops spoil early or throw harvests off schedule. Camera fleets and drones equipped with advanced image-recognition technology constantly observe plants, detecting signs of browning leaves or slight wilting, enabling teams to harvest fruit promptly or treat infections without delay. The synergy of visual technology and nanotech sensors significantly improves the overall system. Ultra-small biosensors embedded within food-wrapper films can already detect spoilage gases like ethylene and ammonia, triggering alarms for potential bacterial outbreaks and sending notifications to the farms' AI dashboard, which presents the data and suggests how to manage the items, including when to move, chill, or capitalize on market value through selling the product. The integration of nanoscale sensing with machine learning is creating an environment where at-risk items can receive timely attention, helping to minimize waste and further losses. Collectively, these elements have empowered researchers to view the combination of AI and nanotech as a crucial engine for enhancing various stages of the food supply chain, thereby ensuring food security for the global population both now and in the future. Climate-smart practices will also leverage the predictive contributions from AI systems while reflecting on the tempered resilience to bolster sustainable agriculture in a systemic manner.

4.6. AI Role in Food Safety Monitoring and Traceability

Concerns about food safety, such as germs, undeclared allergens, toxins, or chemical residues, can lead to serious adverse health effects and diminish public confidence in the products they purchase. As a result, ensuring food safety and traceability within a highly intricate global supply chain poses a considerable challenge for both regulators and businesses. Numerous instances illustrate the implementation of AI-powered nanosensors in practical scenarios. A particularly recognized case is a labeling-free biosensor composed solely of gold nanoparticles, which successfully identified DNA from the Citrus tristeza virus with both specificity and reliability (Khater et al., 2019). This indicates that nanoscale commercial sensing enhanced by AI analytics could feasibly improve food traceability and safety assurance. When integrated with networks of intelligent sensors and IoT devices, these AI systems monitor critical parameters like temperature, humidity, pH, and microbe concentrations during storage, transportation, and delivery. As soon as measurements deviate beyond safe thresholds, machine-learning algorithms trigger early warnings, suggest prompt solutions, and assist in closing the window during which contamination might occur.

4.7. Climate-Smart Agriculture and AI

Climate change is already undermining global food security by increasing the frequency and severity of droughts, floods, heatwaves, and pest outbreaks, consequently jeopardizing harvests worldwide. Climate-Smart Agriculture (CSA) emerges as a viable solution by enhancing yields in a way that is environmentally sustainable, bolstering farmers' resilience to shocks, and minimizing greenhouse gas emissions associated with farming activities. Insights derived from artificial intelligence bolster climate-smart agriculture by offering proactive, location-specific information that empowers farmers, cooperatives, and officials to identify threats early and choose strategies that remain effective as the world warms. Advanced climate models powered by machine learning weave together weather records, field experiments, soil maps, and local economic data to show how shifting climates may hit whole regions or individual plots of land. Platforms like Microsoft's AI for Earth and the Climate AI app deliver hyper-local forecasts for specific crops, allowing smallholders and larger agribusinesses to fine-tune planting dates, irrigation, and pest control as conditions change (Nile et al., 2020). AI analyzes this nanoscale information to adapt agricultural techniques in real-time, assisting farmers in saving water, lowering expenses, and sustaining productivity amidst climate fluctuations. Therefore, AI serves not just to address climate change but also as a vital partner in transforming food systems into smart, sustainable, and resilient models for the future.

5. LIMITATIONS, ISSUES, AND FUTURE REFLECTIONS IN NANOTECHNOLOGY

Although exciting prospects for nanotechnology in agri-food systems exist, it is crucial to ensure that the discussed motivations and benefits align with the biophysical limitations, risks, and regulatory issues that accompany its use. Nanotechnology is vital for synthesizing components, mechanisms, or configurations at the nanoscale, characterized by a heightened surface-to-volume ratio and remarkable physicochemical parameters within the agricultural landscape, as clarified in earlier exchanges. Through the process of pathogen identification, nanomaterials and nanosensors significantly augment food safety by providing consumers with valuable information regarding the intrinsic quality of food items, as well as their nutritional composition (Han et al., 2018). As a result, the widespread use of nanomaterials in food packaging raises potential concerns about their release from contaminated food products or their introduction into the surrounding ecosystem. Nanoparticles have the potential to breach biological systems via routes including oral ingestion, respiratory inhalation, or dermal exposure (Fu, 2014). This event is linked to an increase in the formation of free radicals, which are considered essential contributors to oxidative stress in human physiology (Li et al., 2017; Baranowska-Wójcik et al., 2020). A contemporary investigation has elucidated that titanium dioxide (TiO_2), frequently employed as a food enhancer or in diverse food matrices, possesses the capacity to incite inflammatory responses, induce oxidative stress, elicit genotoxic effects, compromise mitochondrial function, and trigger apoptotic processes following its incorporation into the biological system (Neme et al., 2021). Despite this, the existing research on the movement of nanoparticles from packaging to consumables, together with their potential toxicological consequences, is still inadequate (Roohinejad & Greiner, 2017). Also, the integration of nanopesticides and nanofertilizers into the agricultural industry entails considerable health dangers for agricultural workers, as these substances are able to taint water bodies, soil compositions, or environmental air. As nanoparticles gather in the edaphic environment, the growth of flora will likely be hindered, and there is a risk of their presence in the edible tissues of plants (Shi et al., 2013). When nanoparticles are added to farming environments, they might change in ways that could potentially harm both plants and humans (Yuan et al., 2018). Acknowledging the notable health threats related to the deployment of NPs as pesticides and microbicides, it is vital to comply with risk evaluation protocols during food preparation stringently. Research indicates that the presence of silver nanoparticles (AgNPs) results in phytotoxic effects on aquatic vegetation, such as *Egeria densa* and *Juncus effusus*, by promoting lipid oxidation, influencing physiological traits, and causing further stress in submerged macrophytes (Berekaa, 2015). For instance, it is imperative to enhance public cognizance regarding the ramifications of the relationship between nanomaterials and human health, safety, and the ecological system. An in-depth review of potential risks, along with firm compliance with safety regulations, is also necessary before the market release of nanofood (Cubadda et al., 2017). In accordance with the regulations established by the European Union, any alimentary component originating from nanotechnology, which encompasses nanofoods or nanomaterials used in food packaging, is required to successfully complete a safety assessment conducted by the FDA prior to its approval for human consumption (Oliveira & Silva, 2023). In spite of this, various nations do not observe sufficient regulatory structures concerning the production and management of nanomaterials, consequently making toxicological evaluation crucial for the development of nanotechnology. Alongside concerns about nontoxicity, there are significant practical hurdles associated with developing nanocoating processes aimed at modifying the surface features of substrates for use in ecological environments, biological systems, nutritional systems, and other applications. The task of coating fruits demands substantial effort, and securing a perfect coating continues to present difficulties. Thus, there is a notable necessity for advanced and swift coating methodologies. The dimensions and volume of nanoparticle-encapsulated cargo intended for CRISPR DNA and protein constituents have been precisely characterized subsequent to the genetic alteration of plant organisms. In addition, the role of nanoparticles (NPs) in facilitating the delivery of CRISPR reagents to plant mitochondria for genome editing is not yet fully understood. It is still not clearly proven if these NPs can deliver CRISPR tools to both plant plastids and mitochondria for the purpose of changing the genome. Nanotechnology significantly contributes to enhancing food security by increasing agricultural productivity, addressing microbial, pest, and weed-related challenges, and safeguarding the nutritional quality and safety of food. Nanotechnology drives advancements in the food industry by enabling efficient processing, novel packaging technologies, and extended shelf life of food products. These innovations have elevated culinary practices by enhancing flavor profiles and improving consistency across food preparations. Since many bioactive compounds are hydrophobic and exhibit low bioavailability and stability, nanotechnology-based delivery systems have proven effective in enhancing their targeted release and absorption. Despite its promise, nanotechnology faces major challenges in formulating safe, edible nano-compounds and delivery systems that are economically feasible and acceptable for human consumption. Research highlights that silver nanoparticles (AgNPs), especially from nano-packaging, tend to accumulate in food products and biological systems, raising health and ecological concerns. Their nanoscale size and large surface area allow nanoparticles to penetrate cell membranes, potentially causing toxic effects throughout the human body. High concentrations of nanoparticles can alter the structure of essential proteins and enzymes, leading to adverse physiological impacts. Advancing nanocomposite research using environmentally friendly nano-fillers is vital to safeguarding ecosystems, biodiversity, and natural resources. For instance, developing the nanotechnology regulatory axioms necessary to integrate nanomaterials in the food sector in relation to the traditional development and risk assessment programs will provide concrete evidence of safety.

6. CHALLENGES AND ETHICAL CONSIDERATIONS IN AI ADOPTION

Subsequently, as these AI systems become ingrained throughout our global food systems, it is wise to be intentional about considering the ramifications of our responsible integration of technology. Addressing and recognizing these challenges is vital for the ethical and equitable application of AI technology in the fields of food and agriculture. Sophisticated AI systems rely on high-quality data, reliable internet connectivity, sufficient processing power, and skilled personnel—all of which typically present challenges due to resource limitations in developing nations facing the most severe food insecurity issues. This disparity in availability could further deepen global inequalities that numerous countries are already grappling with unless substantial measures are taken to enhance digital capabilities and offer affordable, accessible AI solutions to underrepresented communities (Mehrabi et al., 2021). AI systems are trained on past and observational data, which can embed systemic biases based on geography, gender, or crop preferences. For instance, yield forecasting algorithms learned from data from Western agricultural systems can predict poorly in African or South Asian settings. Data biases can also result in unfair access to farm loans, subsidies, or technical support. Making AI model development and training transparent and fair is therefore essential. As agricultural producers increasingly depend on AI-driven platforms, questions emerge regarding the ownership of the farm data collected from their fields and the conditions under which it is shared or used. When data platforms merge into larger corporations, it paves the way for monopolistic practices or exploitation that undermine farmers' independence and bargaining power. Thus, policymakers need to offer clarity to safeguard the sovereignty of smallholder farmers and promote innovation. They typically act as obstacles, generating inertia and financial burdens, particularly for micro and small enterprises (MSEs) within the food value chain. Nanosensors that gather biological or environmental data at micro levels may necessitate nation-specific strategies regarding environmental and privacy laws, particularly when data is obtained without adequate informed consent. AI recommendations that leverage nano data could result in unforeseen ramifications if not properly interpreted and validated. Consequently, although AI serves as a vital catalyst for innovation in food systems, regulation must emphasize inclusivity, equity, and ethical considerations to ensure a beneficial effect on global food safety and security. Despite AI's considerable potential to enhance food security worldwide, its application presents a complex array of technical, ethical, and societal issues that require innovative solutions. For instance, a significant challenge is the persistent digital divide between affluent and impoverished regions, which pertains to the availability of digital infrastructure and readiness for AI adoption. Another major issue is algorithmic bias, which may inadvertently reinforce existing socio-economic disparities in agriculture. Ethical dilemmas surrounding data ownership, privacy, and farmer autonomy are equally critical in the context of AI-driven agriculture. From a technical standpoint, key obstacles include the lack of standardized data formats, limited interoperability across AI platforms, and the need for large-scale, annotated datasets to effectively train accurate and reliable models (Vishnoi et al., 2025). The combination of AI and nanotechnology entails even more ethical implications, especially concerning micro-data collection and consent.

7. CONCLUSIONS

This chapter synthesized how nanotechnology and AI can jointly strengthen every stage of the agri-food value chain and offered a clear framework linking food-security challenges to implementable technological solutions. We translated the discussion into practice through a focused case study on nano-silver active packaging for strawberries, illustrating how nanocomposites and intelligent labels can extend shelf life and how those signals can feed AI-enabled logistics. We also consolidated current materials, sensing modalities, packaging strategies, and pathogen-detection approaches into comparable exemplars to support replication and benchmarking. Taken together, these contributions clarify where nanotechnology already yields measurable benefits (antimicrobial/barrier packaging, nano-enabled sensing, encapsulation) and where AI provides system-level leverage (precision decisions, supply-chain optimization, waste reduction), while also surfacing the risk and governance needs migration, toxicology, exposure, fairness, privacy, and access that must accompany deployment.

Looking ahead, we outline an actionable agenda: engineer safe-by-design, food-contact nanomaterials with documented migration limits, degradability, and standardized toxicology reporting; scale "intelligent packaging" by pairing active films (e.g., AgNP, ZnO, nanoclays) with low-cost nanosensors connected to AI dashboards for real-time shelf-life prediction and dynamic routing; move nano-biosensing from lab to field with validated limits of detection and decision thresholds tied to concrete interventions; establish open data standards and governance to ensure privacy, equity, and interoperability; adopt federated and edge AI so models improve while sensitive data stay local; develop digital twins of perishables by fusing nano-sensor streams, cold-chain telemetry, and physics-informed ML to forecast quality decay and automate inventory and pricing; require side-by-side lifecycle, techno-economic, and consumer-acceptance evaluations before scale-up; and work with regulators on harmonized guidance and transparent labelling to build public trust. In short, the chapter delivers a system view, a concrete demonstration, and a roadmap for turning promise into safe, equitable, and scalable impact, advancing resilient, low-waste food systems.

REFERENCES

Abdel-Aziz, H. M., Soliman, M. I., Abo Al-Saoud, A. M., & El-Sherbeny, G. A. (2021). Waste-derived NPK nanofertilizer enhances growth and productivity of Capsicum annuum L. *Plants*, *10*(6), 1144. PMID: 34199718

Abdel-Aziz, H. M., Soliman, M. I., Abo Al-Saoud, A. M., & El-Sherbeny, G. A. (2021). Waste-derived NPK nanofertilizer enhances growth and productivity of Capsicum annuum L. *Plants*, *10*(6), 1144. PMID: 34199718

Ahmadi, S., Kamaladini, H., Haddadi, F., & Sharifmoghadam, M. R. (2018). Thiol-capped gold nanoparticle biosensors for rapid and sensitive visual colorimetric detection of Klebsiella pneumoniae. *Journal of Fluorescence*, *28*, 987–998. PMID: 30022376

Ajithkumar, M., & Krishnamoorthy, R. (2025). Customer-Centric. *Future-Proofing Emerging Technologies for Business Transformation*, 327.

Al-Shabib, N. A., Husain, F. M., Ahmed, F., Khan, R. A., Ahmad, I., Alsharaeh, E., & Aliev, G. (2016). Biogenic synthesis of Zinc oxide nanostructures from Nigella sativa seed: Prospective role as food packaging material inhibiting broad-spectrum quorum sensing and biofilm. *Scientific Reports*, *6*(1), 36761. PMID: 27917856

Alrobaish, W. S., Vlerick, P., Luning, P. A., & Jacxsens, L. (2021). Food safety governance in Saudi Arabia: Challenges in control of imported food. *Journal of Food Science*, *86*(1), 16–30. PMID: 33314129

Alsubaie, A. S. R., & Berekaa, M. M. (2020). Food safety in Saudi Arabia: A public health priority. *Annals of Medical and Health Sciences Research.*

Altpeter, F., Springer, N. M., Bartley, L. E., Blechl, A. E., Brutnell, T. P., Citovsky, V., & Stewart, C. N.Jr. (2016). Advancing crop transformation in the era of genome editing. *The Plant Cell*, *28*(7), 1510–1520. PMID: 27335450

Babu, P. J. (2022). Nanotechnology mediated intelligent and improved food packaging. *International Nano Letters*, *12*(1), 1–14.

Baetsen-Young, A. M., Vasher, M., Matta, L. L., Colgan, P., Alocilja, E. C., & Day, B. (2018). Direct colorimetric detection of unamplified pathogen DNA by dextrin-capped gold nanoparticles. *Biosensors & Bioelectronics*, *101*, 29–36. PMID: 29031887

Baranowska-Wójcik, E., Szwajgier, D., Oleszczuk, P., & Winiarska-Mieczan, A. (2020). Effects of titanium dioxide nanoparticles exposure on human health—A review. *Biological Trace Element Research*, *193*(1), 118–129. PMID: 30982201

Baranowska-Wójcik, E., Szwajgier, D., Oleszczuk, P., & Winiarska-Mieczan, A. (2020). Effects of titanium dioxide nanoparticles exposure on human health—A review. *Biological Trace Element Research, 193*, 118–129. PMID: 30982201

Berekaa, M. M. (2015). Nanotechnology in food industry; advances in food processing, packaging and food safety. *International Journal of Current Microbiology and Applied Sciences, 4*(5), 345–357.

Burris, K. P., Wu, T. C., Vasudev, M., Stroscio, M. A., Millwood, R. J., & Stewart, C. N. (2013). Mega-nano detection of foodborne pathogens and transgenes using molecular beacon and semiconductor quantum dot technologies. *IEEE Transactions on Nanobioscience, 12*(3), 233–238. PMID: 23722479

. Changarathil, T. (2022). AI & Food Systems: The Future of the Canadian Economy.

Chaudhry, Q., Scotter, M., Blackburn, J., Ross, B., Boxall, A., Castle, L., & Watkins, R. (2008). Applications and implications of nanotechnologies for the food sector. *Food Additives and Contaminants, 25*(3), 241–258. PMID: 18311618

Chen, K., Ma, B., Li, J., Chen, E., Xu, Y., Yu, X., & Zhang, M. (2021). A rapid and sensitive europium nanoparticle-based lateral flow immunoassay combined with recombinase polymerase amplification for simultaneous detection of three food-borne pathogens. *International Journal of Environmental Research and Public Health, 18*(9), 4574. PMID: 33925871

. Coordination, U. N. (1998). Rome Declaration on World Food Security and World Food Summit Plan of Action.

. Cubadda, F., Aureli, F., Raggi, A., & Mantovani, A (2017). The role of in vitro simulated human digestion models in assessing potential exposure to nano particles in food: An essential tool for risk assessment of nanomaterials used in food-related applications.

. Declaration, R. (1996). *Rome Declaration on world food security and world food summit plan of action.*

Demirer, G. S., Silva, T. N., Jackson, C. T., Thomas, J. B., Ehrhardt, D. W., Rhee, S. Y., & Landry, M. P. (2021). Nanotechnology to advance CRISPR–Cas genetic engineering of plants. *Nature Nanotechnology, 16*(3), 243–250. PMID: 33712738

Demirer, G. S., Silva, T. N., Jackson, C. T., Thomas, J. B. W., Ehrhardt, D., Rhee, S. Y., & Landry, M. P. (2021). Nanotechnology to advance CRISPR–Cas genetic engineering of plants. *Nature Nanotechnology, 16*(3), 243–250. PMID: 33712738

Demirer, G. S., Zhang, H., Matos, J. L., Goh, N. S., Cunningham, F. J., Sung, Y., & Landry, M. P. (2019). High aspect ratio nanomaterials enable delivery of functional genetic material without DNA integration in mature plants. *Nature Nanotechnology*, *14*(5), 456–464. PMID: 30804481

Duro, J. A., Lauk, C., Kastner, T., Erb, K. H., & Haberl, H. (2020). Global inequalities in food consumption, cropland demand and land-use efficiency: A decomposition analysis. *Global Environmental Change*, *64*, 102124.

Elbehri, A., & Chestnov, R. (2021). *Digital agriculture in action: Artificial intelligence for agriculture*. Food & Agriculture Org.

Fadel, H. H., Mahmoud, K. F., Saad, R., Lotfy, S. N., El-Aleem, F. S. A., & Ahmed, M. (2019). Nanotechnology for enhancing the production of meat-like flavour from enzymatic hydrolyzed soybean protein isolate. *Acta Scientiarum Polonorum. Technologia Alimentaria*, *18*(3), 279–292. PMID: 31569910

Farooq, M. A., Hannan, F., Islam, F., Ayyaz, A., Zhang, N., Chen, W., & Zhou, W. (2022). The potential of nanomaterials for sustainable modern agriculture: Present findings and future perspectives. *Environmental Science. Nano*, *9*(6), 1926–1951.

Farooq, M. A., Hannan, F., Islam, F., Ayyaz, A., Zhang, N., Chen, W., & Zhou, W. (2022). The potential of nanomaterials for sustainable modern agriculture: Present findings and future perspectives. *Environmental Science. Nano*, *9*(6), 1926–1951.

Feng, Y., Zhong, L., Bilal, M., Tan, Z., Hou, Y., Jia, S., & Cui, J. (2018). Enzymes@ ZIF-8 nanocomposites with protection nanocoating: Stability and acid-resistant evaluation. *Polymers*, *11*(1), 27. PMID: 30960011

Ferentinos, K. P. (2018). Deep learning models for plant disease detection and diagnosis. *Computers and Electronics in Agriculture*, *145*, 311–318.

Food Wastage Footprint (Project). (2013). *Food wastage footprint: impacts on natural resources: summary report*. Food & Agriculture Organization of the UN (FAO).

Fu, P. P. (2014). Introduction to the special issue: Nanomaterials—toxicology and medical applications. *Yao Wu Shi Pin Fen Xi*, *22*(1), 1. PMID: 24673899

Gupta, A., Eral, H. B., Hatton, T. A., & Doyle, P. S. (2016). Nanoemulsions: Formation, properties and applications. *Soft Matter*, *12*(11), 2826–2841. PMID: 26924445

Han, C., Zhao, A., Varughese, E., & Sahle-Demessie, E. (2018). Evaluating weathering of food packaging polyethylene-nano-clay composites: Release of nanoparticles and their impacts. *NanoImpact*, *9*, 61–71. PMID: 29226269

Hofmann, T., Lowry, G. V., Ghoshal, S., Tufenkji, N., Brambilla, D., Dutcher, J. R., & Wilkinson, K. J. (2020). Technology readiness and overcoming barriers to sustainably implement nanotechnology-enabled plant agriculture. *Nature Food*, *1*(7), 416–425.

Hofmann, T., Lowry, G. V., Ghoshal, S., Tufenkji, N., Brambilla, D., Dutcher, J. R., & Wilkinson, K. J. (2020). Technology readiness and overcoming barriers to sustainably implement nanotechnology-enabled plant agriculture. *Nature Food*, *1*(7), 416–425.

Hofmann, T., Lowry, G. V., Ghoshal, S., Tufenkji, N., Brambilla, D., Dutcher, J. R., & Wilkinson, K. J. (2020). Technology readiness and overcoming barriers to sustainably implement nanotechnology enabled plant agriculture. *Nature Food*, *1*(7), 416–425.

Hossain, A. S., Uddin, M. M., Veettil, V. N., & Fawzi, M. (2018). Nano-cellulose based nano-coating biomaterial dataset using corn leaf biomass: An innovative biodegradable plant biomaterial. *Data in Brief*, *17*, 162–168. PMID: 29877503

. Inbaraj, B. S., & Chen, B. H. (2016). Nanomaterial-based sensors for detection of foodborne bacterial pathogens and toxins as well as pork adulteration in meat products. *journal of food and drug analysis, 24*(1), 15-28.

Joshi, H., Malik, A., Aggarwal, S., Munde, M., Maitra, S. S., Adlakha, N., & Bhatnagar, R. (2019). In-vitro detection of phytopathogenic fungal cell wall by polyclonal sera raised against trimethyl chitosan nanoparticles. *International Journal of Nanomedicine*, •••, 10023–10033. PMID: 31908457

Kagan, C. R. (2016). At the nexus of food security and safety: Opportunities for nanoscience and nanotechnology. *ACS Nano*, *10*(3), 2985–2986. PMID: 27000520

Kagan, C. R. (2016). At the nexus of food security and safety: Opportunities for nanoscience and nanotechnolgy. *ACS Nano*, *10*(3), 2985–2986. PMID: 27000520

Kalia, A., Bhardwaj, K., Bhardwaj, P., Abd-Elsalam, K. A., Valis, M., & Kuca, K. (2021). Nanohybrid Antifungals for Control of Plant Diseases: Current Status and Future Perspectives. *Journal of Fungi (Basel, Switzerland)*, *2021*(7), 48. PMID: 33450851

Kalia, A., Kaur, M., Shami, A., Jawandha, S. K., Alghuthaymi, M. A., Thakur, A., & Abd-Elsalam, K. A. (2021). Nettle-leaf extract derived ZnO/CuO nanoparticle-biopolymer-based antioxidant and antimicrobial nanocomposite packaging films and their impact on extending the post-harvest shelf life of guava fruit. *Biomolecules*, *11*(2), 224. PMID: 33562547

Kamal, T., Ali, N., & Naseem, A, A., B Khan, S., & M Asiri, A. (2016). Polymer nanocomposite membranes for antifouling nanofiltration. *Recent Patents on Nanotechnology, 10*(3), 189–201. PMID: 27136927

Kamilaris, A., Kartakoullis, A., & Prenafeta-Boldú, F. X. (2017). A review on the practice of big data analysis in agriculture. *Computers and Electronics in Agriculture, 143*, 23–37.

Kamilaris, A., & Prenafeta-Boldú, F. X. (2018). Deep learning in agriculture: A survey. *Computers and Electronics in Agriculture, 147*, 70–90.

Karimirad, R., Behnamian, M., Dezhsetan, S., & Sonnenberg, A. (2018). Chitosan nanoparticles-loaded Citrus aurantium essential oil: A novel delivery system for preserving the postharvest quality of Agaricus bisporus. *Journal of the Science of Food and Agriculture, 98*(13), 5112–5119. PMID: 29635845

Khater, M., De La Escosura-Muñiz, A., Quesada-González, D., & Merkoçi, A. (2019). Electrochemical detection of plant virus using gold nanoparticle-modified electrodes. *Analytica Chimica Acta, 1046*, 123–131. PMID: 30482289

Kritchenkov, A. S., Egorov, A. R., Volkova, O. V., Artemjev, A. A., Kurliuk, A. V., Le, T. A., & Khrustalev, V. N. (2021). Novel biopolymer-based nanocomposite food coatings that exhibit active and smart properties due to a single type of nanoparticles. *Food Chemistry, 343*, 128676. PMID: 33250292

Kumar, S., Nehra, M., Dilbaghi, N., Marrazza, G., Hassan, A. A., & Kim, K. H. (2019). Nano-based smart pesticide formulations: Emerging opportunities for agriculture. *Journal of Controlled Release, 294*, 131–153. PMID: 30552953

Lau, H. Y., Wu, H., Wee, E. J., Trau, M., Wang, Y., & Botella, J. R. (2017). Specific and sensitive isothermal electrochemical biosensor for plant pathogen DNA detection with colloidal gold nanoparticles as probes. *Scientific Reports, 7*(1), 38896. PMID: 28094255

Li, L., Zhao, C., Zhang, Y., Yao, J., Yang, W., Hu, Q., & Cao, C. (2017). Effect of stable antimicrobial nano-silver packaging on inhibiting mildew and in storage of rice. *Food Chemistry, 215*, 477–482. PMID: 27542501

Li, Y., Rokayya, S., Jia, F., Nie, X., Xu, J., Elhakem, A., & Helal, M. (2021). Shelf-life, quality, safety evaluations of blueberry fruits coated with chitosan nano-material films. *Scientific Reports, 11*(1), 55. PMID: 33420183

. Liu, P. (2017). The future of food and agriculture: Trends and challenges.

Mahfouz, M. M., Piatek, A., & Stewart, C. N.Jr. (2014). Genome engineering via TALENs and CRISPR/Cas9 systems: Challenges and perspectives. *Plant Biotechnology Journal, 12*(8), 1006–1014. PMID: 25250853

Martin-Ortigosa, S., Peterson, D. J., Valenstein, J. S., Lin, V. S. Y., Trewyn, B. G., Lyznik, L. A., & Wang, K. (2014). Mesoporous silica nanoparticle-mediated intracellular Cre protein delivery for maize genome editing via loxP site excision. *Plant Physiology, 164*(2), 537–547. PMID: 24376280

Mehrabi, Z., McDowell, M. J., Ricciardi, V., Levers, C., Martinez, J. D., Mehrabi, N., & Jarvis, A. (2021). The global divide in data-driven farming. *Nature Sustainability, 4*, 154–160.

Mosa, W. F., El-Shehawi, A. M., Mackled, M. I., Salem, M. Z., Ghareeb, R. Y., Hafez, E. E., & Abdelsalam, N. R. (2021). Productivity performance of peach trees, insecticidal and antibacterial bioactivities of leaf extracts as affected by nanofertilizers foliar application. *Scientific Reports, 11*(1), 10205. PMID: 33986453

Naskar, A., Khan, H., Sarkar, R., Kumar, S., Halder, D., & Jana, S. (2018). Anti-biofilm activity and food packaging application of room temperature solution process based polyethylene glycol capped Ag-ZnO-graphene nanocomposite. *Materials Science and Engineering C, 91*, 743–753. PMID: 30033309

Neme, K., Nafady, A., Uddin, S., & Tola, Y. B. (2021). Application of Nanotechnology in agriculture, postharvest loss reduction and food processing: Food security implication and challenges. *Heliyon, 7*(12). PMID: 34934845

Nile, S. H., Baskar, V., Selvaraj, D., Nile, A., Xiao, J., & Kai, G. (2020). Nanotechnologies in food science: Applications, recent trends, and future perspectives. *Nano-Micro Letters, 12*(1), 45. PMID: 34138283

Nile, S. H., Baskar, V., Selvaraj, D., Nile, A., Xiao, J., & Kai, G. (2020). Nanotechnologies in food science: Applications, recent trends, and future perspectives. *Nano-Micro Letters, 12*, 1–34. PMID: 34138283

Oliveira, R. C. D., & Silva, R. D. D. S. E. (2023). Artificial intelligence in agriculture: Benefits, challenges, and trends. *Applied Sciences (Basel, Switzerland), 13*(13), 7405.

Oliveira, R. C. D., & Silva, R. D. D. S. E. (2023). Artificial intelligence in agriculture: Benefits, challenges, and trends. *Applied Sciences (Basel, Switzerland), 13*(13), 7405.

Pal, M. (2017). Nanotechnology: A new approach in food packaging. *J. Food Microbiol. Saf. Hyg, 2*(02), 8–9.

Rana, K., Mittal, J., Narang, J., Mishra, A., & Pudake, R. N. (2021). Graphene based electrochemical DNA biosensor for detection of false smut of rice (Ustilaginoidea virens). *The plant pathology journal, 37*(3), 291. PMID: 34111918

Ray, S. S., & Bandyopadhyay, J. (2021). Nanotechnology-enabled biomedical engineering: Current trends, future scopes, and perspectives. *Nanotechnology Reviews, 10*(1), 728–743.

. Roohinejad, S., & Greiner, R. (2017). Nanoscience: Relevance for agriculture and the food sector Nanotechnology *in agriculture and food science*, 347-362.

Roy, S., Zhai, L., Kim, H. C., Pham, D. H., Alrobei, H., & Kim, J. (2021). Tannic-acid-cross-linked and TiO2-nanoparticle-reinforced chitosan-based nanocomposite film. *Polymers, 13*(2), 228. PMID: 33440770

Saravanakumar, K., Sathiyaseelan, A., Mariadoss, A. V. A., Xiaowen, H., & Wang, M. H. (2020). Physical and bioactivities of biopolymeric films incorporated with cellulose, sodium alginate and copper oxide nanoparticles for food packaging application. *International Journal of Biological Macromolecules, 153*, 207–214. PMID: 32105688

Schmidhuber, J., & Tubiello, F. N. (2007). Global food security under climate change. *Proceedings of the National Academy of Sciences of the United States of America, 104*(50), 19703–19708. PMID: 18077404

Shankar, S., Khodaei, D., & Lacroix, M. (2021). Effect of chitosan/essential oils/ silver nanoparticles composite films packaging and gamma irradiation on shelf life of strawberries. Food Hydrocolloids, 117, 106750. Khaki, S., & Wang, L. (2019). Crop yield prediction using deep neural networks. *Frontiers in Plant Science, 10*, 621.

Shi, S., Wang, W., Liu, L., Wu, S., Wei, Y., & Li, W. (2013). Effect of chitosan/ nano-silica coating on the physicochemical characteristics of longan fruit under ambient temperature. *Journal of Food Engineering, 118*(1), 125–131.

Shukla, P. S., Mantin, E. G., Adil, M., Bajpai, S., Critchley, A. T., & Prithiviraj, B. (2019). Ascophyllum nodosum-based biostimulants: Sustainable applications in agriculture for the stimulation of plant growth, stress tolerance, and disease management. *Frontiers in Plant Science, 10*, 462648. PMID: 31191576

Sousa, M., Martins, S., Santos, M. J., & Amorim, P. (2024). Dynamic pricing strategy for substitute perishable products. Available at *SSRN* 5019518.

Sousa, M., Martins, S., Santos, M. J., & Amorim, P. (2024). Dynamic pricing strategy for substitute perishable products. *Available at SSRN* 5019518.

Tahir, M. A., Hameed, S., Munawar, A., Amin, I., Mansoor, S., Khan, W. S., & Bajwa, S. Z. (2017). Investigating the potential of multiwalled carbon nanotubes based zinc nanocomposite as a recognition interface towards plant pathogen detection. *Journal of Virological Methods*, *249*, 130–136. PMID: 28888402

. Tesfaye, K., Tamene, L. D., Demissie, T. D., Seid, J., Haile, A., Mekonnen, K., & Solomon, D. (2021). A Framework for Bundling Climate-Smart Agriculture (CSA) and Climate Information Services (CIS) in Ethiopia.

. Tesfaye, K., Tamene, L. D., Demissie, T. D., Seid, J., Haile, A., Mekonnen, K., & Solomon, D. (2021). A Framework for Bundling Climate-Smart Agriculture (CSA) and Climate Information Services (CIS) in Ethiopia.

Vishnoi, V., Chaudhary, S., Joshi, N., & Consul, P. (2025). Nanoscale Industry 4.0 Empowers Advanced Nanomaterials and Devices. In *Nanotechnology Applications for Industry 4.0* (pp. 297-317). CRC Press.

Wong, M. H., Misra, R. P., Giraldo, J. P., Kwak, S. Y., Son, Y., Landry, M. P., & Strano, M. S. (2016). Lipid exchange envelope penetration (LEEP) of nanoparticles for plant engineering: A universal localization mechanism. *Nano Letters*, *16*(2), 1161–1172. PMID: 26760228

Wu, Z., He, D., Cui, B., & Jin, Z. (2018). A bimodal (SERS and colorimetric) aptasensor for the detection of Pseudomonas aeruginosa. *Microchimica Acta*, *185*, 1–7. PMID: 30382404

Yuan, L., Richardson, C. J., Ho, M., Willis, C. W., Colman, B. P., & Wiesner, M. R. (2018). Stress responses of aquatic plants to silver nanoparticles. *Environmental Science & Technology*, *52*(5), 2558–2565. PMID: 29381864

Zare, H., Meshkat, Z., Hatamluyi, B., Rezayi, M., Ghazvini, K., Derakhshan, M., & Aryan, E. (2022). The first diagnostic test for specific detection of Mycobacterium simiae using an electrochemical label-free DNA nanobiosensor. *Talanta*, *238*, 123049. PMID: 34801906

Zhou, W., Fang, J., Tang, S., Wu, Z., & Wang, X. (2021). 3D-printed nanocellulose-based cushioning–antibacterial dual-function food packaging aerogel. *Molecules (Basel, Switzerland)*, *26*(12), 3543. PMID: 34200653

Zhu, B., Fan, C., Cheng, C., Lan, T., Li, L., & Qin, Y. (2021). Migration kinetic of silver from polylactic acid nanocomposite film into acidic food simulant after different high-pressure food processing. *Journal of Food Science*, *86*(6), 2481–2490. PMID: 33948960

Chapter 11
Accelerating Peptide Drug Discovery With AI-Assisted Predictive Models

Chandra Has
https://orcid.org/0000-0002-3943-0935
GSFC University, India

Karan Thakar
GSFC University, India

Aditya Puranik
GSFC University, India

ABSTRACT

Peptide-based therapeutics offer high specificity and safety but face challenges such as instability, rapid clearance, and limited bioavailability. Traditional discovery and formulation methods are slow, costly, and constrained by the vast peptide sequence space. Artificial intelligence (AI) is revolutionizing this field by enabling de novo peptide design, predicting bioactivity, and reducing reliance on trial-and-error approaches. At the same time, nanotechnology provides advanced delivery platforms (such as liposomes, lipid nanoparticles, polymeric nanoparticles, and micelles) that can encapsulate or functionalize peptides to enhance stability, prolong circulation, and enable targeted or controlled release. The convergence of AI and nanotechnology creates a synergistic framework: AI not only accelerates the discovery of novel peptide drugs but also optimizes nanocarrier design, improving

DOI: 10.4018/979-8-3373-2883-6.ch011

delivery outcomes. This chapter highlights how this dual advancement in smart drug design and intelligent delivery paves the way for safer, more effective, and personalized peptide therapeutics.

1. INTRODUCTION

Peptides are short, well-structured sequences of amino acids (AAs) (typically 2-50 in length) linked by peptide bonds (Henninot et al., 2018). Since they possess distinct biological activities, minimal side effects, limited potential to cause drug resistance, and exquisite targeted precision, they stand out in pharmaceutics (Fosgerau & Hoffmann, 2015). Insulin from an animal pancreas was first investigated in 1922, revolutionizing type I diabetes treatment and marking the beginning of the peptide-based therapies era. Almost 4 decades later, synthetic peptide hormones such as oxytocin and antidiuretic hormone brought the new therapeutic avenues (Muttenthaler et al., 2021). Notable progress in peptide drugs occurred with the advent of solid-phase peptide synthesis in the 1960s (Merrifield, 1963) and the emergence of recombinant technologies in the 1980s (Johnson, 1983).

The 21st century has witnessed a paradigm shift in peptide drug research with the rapid advances in structural biology, recombinant biologics, and modern synthetic and analytical strategies. Peptides share a significant proportion of today's pharmaceutical market. To date, more than 100 peptide-based drugs worldwide have been authorized by the Food and Drug Administration (FDA), and many more are in the pipeline (Dai et al., 2024). It is projected that the demand for peptide drugs will propel the global peptide business to 106 billion dollars in the next 10 years, with an estimated growth rate of 6.3% (S. Kumar et al., 2024). These peptides are employed in diverse therapeutic domains, such as immunosuppression, cancer treatment, antibacterial, and antiviral applications (Chen et al., 2024). Peptide medicines, which control around 70% of the global market for glucose-lowering drugs, are also essential in the treatment of diabetes (Chen et al., 2024). Notable examples of these include insulin, glucagon-like peptide-1 (GLP-1), and its analogs (Drucker & Nauck, 2006).

Peptide drug development has advanced significantly by employing numerous natural sources, including plants, animals, fungi, microorganisms, and endogenous human peptides, many of which possess superior therapeutic qualities compared to human-derived peptides. These precursors, obtained from natural collection, recombinant synthesis, or chemical synthesis, are further enhanced with modern display technologies such as ribosome, DNA, mRNA, yeast, and phage display, allowing for speedier identification and refining (Muttenthaler et al., 2021). However, de novo peptide design remains difficult due to the inherent flexibility of the peptides, a lack of structural data on protein-peptide complexes, the incorporation of non-canonical AAs and unconventional cyclic structures, and the high computational cost of conventional modeling tools such as molecular docking and molecular dynamics simulations, which limit their broader applicability (Chang et al., 2024; Mondal et al., 2022). In addition, a major obstacle to the clinical translation of numerous therapeutic peptides is their fast clearance *in vivo* and intrinsic susceptibility to enzymatic breakdown, which results in low bioavailability (Yen et al., 2025). The use of nanocarriers, which can encapsulate and protect peptides, regulate their release, and target particular tissues, provides a potent answer to this delivery challenge thanks to nanotechnology (Wang et al., 2024). The combination of AI with nanotechnology is especially revolutionary, as it has made it possible to build and optimize peptide-loaded nano-formulations using predictive models, ensuring that patients fully benefit from computationally produced peptides.

AI is advancing at an astonishing rate, changing the way we approach drug discovery (Zhang et al., 2025). Historically, the development of new therapeutics, particularly peptide-based drugs, has been a slow, costly, and uncertain endeavor, often relying on trial-and-error experimentation and labor-intensive screening methods. AI-powered computer-aided drug design (CADD) is helping to overcome these barriers by accelerating the identification of lead candidates, lowering development costs, and increasing precision in predicting peptide behavior. In the search for new antimicrobial peptides (AMPs) (Szymczak et al., 2025) and anticancer peptides (ACPs) (Hwang et al., 2022), AI has proven especially powerful. By analyzing vast sequence libraries, these models can rapidly identify promising peptide scaffolds, evaluate their bioactivity, and even propose chemical modifications to improve stability, specificity, and therapeutic potential. Moreover, deep learning (DL), a branch of machine learning (ML) which is itself a subset of AI, along with generative models, facilitates the design of entirely new sequences that may not exist in nature, vastly expanding the chemical space beyond human intuition (Wu et al., 2024). With ML combining human expertise with speed and intelligence, researchers can develop potential treatments more quickly, providing patients with better medicines sooner.

This chapter provides a comprehensive roadmap for researchers seeking to utilize nanotechnology and AI in peptide therapies. **Section 2** reviews conventional peptide synthesis techniques and the challenges in peptide-based drug formulations. Subsequent sections explore the foundations of AI-driven peptide discovery (**Section 3**), the integration of AI with nanotechnology for enhanced peptide delivery (**Section 4**), current limitations and challenges (**Section 5**), and future directions (**Section 6**). By addressing these topics, the chapter offers actionable insights to accelerate the discovery and translation of next-generation peptide drugs.

2. CONVENTIONAL TECHNIQUES OF PEPTIDE-BASED DRUG FORMULATIONS

The traditional development of peptide therapies has depended mainly on rational design and experimental screening, which are not only time-consuming but also expensive. Thus, a majority of clinically approved peptide drugs are derived from endogenous peptides or analogs. Insulin is one of the classic examples, first extracted from pancreatic tissue after its role in diabetes was investigated (Banting et al., 1922). In the early stages of peptide drug discovery, bioactive peptides were often isolated from animal tissues associated with metabolic activities. Advancements in the synthesis of solid-phase peptides have broadened the scope for novel peptide design, yet the field has mostly concentrated on modifying natural peptides rather than producing entirely new sequences (Glavaš et al., 2022). This tendency persists mainly due to cost constraints and the complexities of balancing several therapeutic factors simultaneously.

The rapid expansion of chemical space as the number of AAs grows exponentially represents a significant obstacle in the development of peptide drugs development (Capecchi & Reymond, 2021). Developing novel peptides with optimal therapeutic potential is challenging because evaluating every possible sequence experimentally is unrealistic. This limitation restricts innovations and complicates the development of new peptide scaffolds. Additionally, logical design strategies often rely on human intuition and prior empirical knowledge, which may result in bias. For accurate peptide design with high affinity and specificity towards their targets, a detailed grasp of protein-peptide interactions is required, which static structural data, such as those obtained by crystallography, may not provide.

The traditional peptide drug design mostly begins with identifying precursor peptides (**Table 1**), which are usually obtained either from natural sources or endogenous peptides present in the human body (Chen et al., 2024). The peptide precursors undergo chemical or recombinant modifications prior to complete synthesis to enhance specific characteristics. In general, the traditional approach, relying on iterative laboratory screening and rational design, is regarded as expensive and time-consuming. (Chen et al., 2024). Therefore, the majority of peptide therapies developed to date are not unique sequences, but rather analogs of natural peptides with known biological functions, such as peptide hormones like insulin. (Banting et al., 1922). Before the advent of synthetic approaches, therapeutic peptides could only be designed by isolating bioactive molecules from tissues, especially those linked to important metabolic processes.

The primary aim of most peptide analogs is to improve pharmacokinetic parameters (e.g., bioavailability, solubility, and metabolic stability) rather than directly enhancing therapeutic efficacy. To avoid enzymatic degradation, researchers often introduce modifications such as AA substitutions and alterations in the backbone. For instance, vosoritide, a synthetic counterpart of C-type natriuretic peptide used to treat achondroplasia, contains proline and glycine residues to improve stability (Chan et al., 2022; Dardenne et al., 2023). Incorporation of cyclization or non-natural AAs leads to enhanced stability and protein-binding affinity (Evans et al., 2020; Joo, 2012). Nevertheless, developing entirely new peptides that fulfill various pharmacological criteria remains a complicated endeavor. Furthermore, developing peptides that meet multiple pharmacological requirements—such as stability, bioavailability, solubility, and target specificity —is a complex and resource-intensive problem. To overcome these constraints, high-throughput display approaches, including phage display, yeast display, and ribosome display, have evolved as important peptide discovery tools. Phage display, for example, enables the rapid screening of large peptide libraries by genetically connecting peptide sequences to bacteriophage coat proteins (Jaroszewicz et al., 2022; Nixon et al., 2014). This method makes it easier to identify high-affinity peptide binders to specific targets, which has led to the development of therapies such as Romiplostim and Ecallantide (Jaroszewicz et al., 2022). Nevertheless, phage display possesses a few limitations, such as restricted chemical space coverage, non-native peptide conformations on phage surfaces, and multivalent binding effects, which can result in false positives or less specific binders (Castel et al., 2011; Palma, 2023).

Endogenous natural peptides have been an important source of potential new peptide drugs for decades (**Table 1**). Many peptides, e.g., analogs of oxytocin, vasopressin, and gonadotropin-releasing hormone, have been inspired by endogenous peptides. The body produces such peptides in trace amounts, and they frequently have potent biological effects. Nonetheless, to improve stability and bioactivity, chemical changes are frequently necessary due to the momentary metabolic half-lives of these compounds. Peptides, including animal venoms, fungi, bacteria, and plants, are examples of exogenous sources that have demonstrated remarkable therapeutic potential. A few examples include cyclosporine from a fungus (Borel et al., 1976), bivalirudin from leech saliva (Rydel et al., 1990), exenatide (a GLP-1 receptor) from lizard venom (Tahrani et al., 2011), and ziconotide from cone snail venom (Akondi et al., 2014).

Peptides isolated from venom stand out because they have been evolutionarily optimized to be potent and selective. In venomics, bioinformatics, and proteomics are studied together to discover peptide drugs from the extensive venom libraries, making millions of previously unknown neuroactive chemicals available. When combined with phage display technology, these approaches accelerate drug development by allowing for iterative selection of peptides with great affinity and specificity (Jaroszewicz et al., 2022). Even though pipelines such as these have already yielded efficient peptide drugs, ongoing obstacles such as high development costs, limited library diversity, and time-consuming discovery underscore the demand for additional innovations in screening and library development techniques.

Table 1. Traditional peptides are clinically approved by the FDA, and their brand names, indications, administration routes, classes, and approval years—source: FDA website.

Peptide Drug	Brand Name(s)	Indication(s)	Administration Route(s)	Class	FDA Approval
Synthetic analog–based endogenous peptides					
Desmopressin	DDAVP, Stimate, Noctiva	Central diabetes insipidus, nocturia, hemophilia A, von Willebrand disease	PO, IV, SC, IN	Vasopressin analog	1978
Lysopressin	Diapid	Metrorrhagia	NS	Oxytocin analog	2014
Octreotide	Sandostatin	Acromegaly	SC, IM	Somatostatin analog	1988
Lanreotide	Somatuline depot	Acromegaly	SC	Somatostatin analog	2007
Leuprorelin	Eligard, Lupron Depot, Prostap	Prostate/breast CA, CPP	SC, IM	GnRH agonist	1989
Goserelin	Zoladex	Prostate CA	SC	GnRH agonist	1989
Triptorelin	Decapeptyl, Trelstar	Prostate CA	IM, SC	GnRH agonist	2000
Abarelix	Plenaxis	Prostate CA	IM	GnRH antagonist	2003
Cetrorelix	Cetrotide	Ovulation promo, uterine tumor	SC	GnRH antagonist	2000
Liraglutide	Victoza, Saxenda	T2D, obesity	SC	GLP-1 analog	2010
Dulaglutide	Trulicity	T2D	SC	GLP-1 analog	2014
Semaglutide	Ozempic, Rybelsus	T2D, obesity	SC	GLP-1 analog	2017
Teriparatide	Forteo	Osteoporosis	SC	PTH analog	1987
Nesiritide	Natrecor	Heart failure	IV	Natriuretic peptide analog	2001
Natural product–based					
Exenatide	Byetta, Bydureon	T2D	SC	Gila monster saliva	2005

continued on following page

Table 1. Continued

Peptide Drug	Brand Name(s)	Indication(s)	Administration Route(s)	Class	FDA Approval
Ziconotide	Prialt	Chronic pain	IT	Marine snail toxin	2004
Cyclosporin	Sandimmune, Restasis, Neoral	Organ transplant	PO/IV	Cyclic peptide from fungi	1983
Bivalirudin	Angiomax, Angiox	Anticoagulant	IV	Medicinal leech peptide	2000
Linaclotide	Linzess, Constella	IBS	PO	GCC agonist	2012
Display technology–based					
Romiplostim	Nplate	Thrombocytopenia	SC	Fc-peptide fusion	2008
Ecallantide	Kalbitor	Hereditary angioedema	SC	Kallikrein inhibitor	2009

Abbreviations: ADH – antidiuretic hormone; CA – cancer; CPP – central precocious puberty; GCC – guanylate cyclase C; GnRH – gonadotropin-releasing hormone; IM – intramuscular; IT – intrathecal; IV – intravenous; NS – nasal spray; PTH – parathyroid hormone; PO – per os (oral); SC – subcutaneous; T2D – type 2 diabetes.

3. FOUNDATIONS OF AI-DRIVEN PEPTIDE DRUGS DISCOVERY

In recent years, rapid digitalization has generated enormous amounts of pharmaceutical data, presenting both opportunities and challenges in translating this knowledge into solutions for complex clinical issues. This has sped up the AI implementation, enabling it to effectively handle large datasets and enhance decision-making through complex algorithms that imitate parts of human intellect without replacing human expertise (Paul et al., 2021; Vyas et al., 2018). AI can significantly improve the efficiency, success rates, and overall quality of the discovery of new peptide drugs by integrating conventional wet-lab experimental and computational approaches (Attique et al., 2020; Pun et al., 2023) (Figure 1). However, the effectiveness of these tactics is dependent on availability to high-quality, diverse datasets; biased or partial data might jeopardize model accuracy and generalisability (Blanco-González et al., 2023). To address this, statistically rigorous scoring and ranking algorithms aid in reliability, while keeping openness and interpretability in AI predictions is critical for scientific rigor.

Figure 1. In-silico computational and conventional wet-lab approaches for designing peptide drugs, highlighting differences in process, sequence space, and timelines.

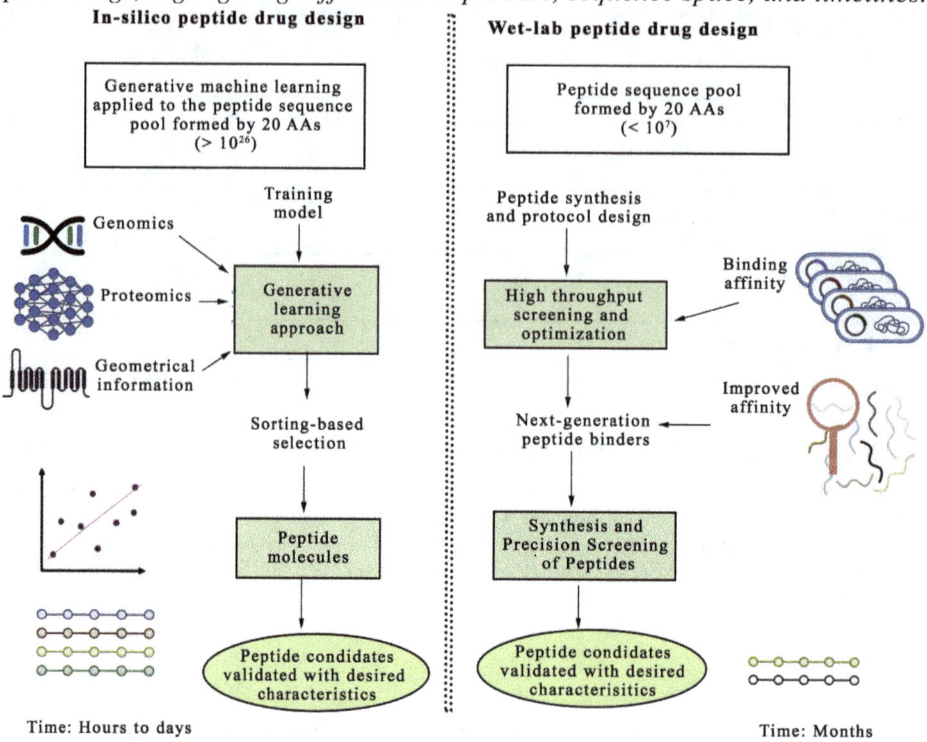

Schematic reproduced from Nissan et al. (2024).

AI is commonly employed in peptide development in two ways: synthesizing new therapeutic peptides through ab initio design and identifying/screening potential candidates concealed within naturally occurring peptide sequences. The method typically begins with vast libraries of known peptides, which are then translated into a format that AI models can interpret. These models, which are generally based on ML/DL, are then trained to anticipate how a peptide would act biologically or to propose totally new peptide designs.

3.1 Peptide Database Libraries

AI can significantly speed up the discovery of therapeutic peptides by quickly identifying potential candidates from large and complicated datasets. Precise predictive models need high-quality, well-structured benchmark datasets. Diverse specialist databases are critical resources for therapeutic peptide research, with AMPs/ACPs databases being especially well-established and widely used (Table 2).

Table 2. Therapeutic peptide database libraries.

Database	Peptides	Specifications	Ref.
APD3	AMPs/ACPs	Comprehensive, includes sequences, structures, functions	Wang et al. (2016)
CAMP	AMPs	Widely used AMP database	Waghu et al. (2014)
CAMPER	AMPs	AMP-focused resource	Gawde et al. (2023)
DBAASP	AMPs/ACPs	Detailed structural and functional info	Pirtskhalava et al. (2013)
DRAMP	AMPs	Curated AMP sequences	Shi et al. (2022)
BACTIBASE	AMPs	Focus on bacterial AMPs	Hammami et al. (2010)
CancerPPD	ACPs	Dedicated to cancer-targeting peptides	Tyagi et al. (2015)
SATPdb	ACPs and others	Systematically annotated, >2000 peptides targeting various conditions	Singh et al. (2015)

3.2 Peptide Feature Encoding and Representation

An AI model needs to encode structural, functional, and biological information from peptide AA residues numerically. Peptide properties and therapeutic potential must be accurately predicted by appropriate encoding, which directly affects model performance and interpretability. Peptide sequence, structural, physicochemical, evolutionary, and binary encoding can all be used to represent features, either separately or together. The efficiency and robustness of model development are significantly enhanced using advanced approaches, such as automatic feature extraction utilizing DL, nonlinear mapping with kernel functions, and dimensionality reduction through principal component analysis.

Embedding-based approaches involve mapping peptide sequences (i.e., AA residues) into low-dimensional continuous spaces, maintaining similarity relationships while reducing noise and redundancy. In order to predict therapeutic peptides, word and context embedding approaches, implemented through DL frameworks such as Transformer, GRU (Gated recurrent unit), CNN (Convolutional neural network), and BiLSTM (Bidirectional long short-term memory), are employed to acquire both lexical and semantic information (Yan et al., 2022). AI algorithms for peptide design use traditional ML/DL methods. In Table 3, we outline the contributions of each of these methodologies to such tasks as classification, regression, clustering, and generative modeling.

Table 3. AI algorithms employed in peptide drug development (Ali et al., 2025; Deng et al., 2022; Ma et al., 2025).

AI Model	Model Expanded Name	Description
ML		
SVM	Support Vector Machine	An efficient supervised learning approach for sparse and high-dimensional datasets, it determines the best hyperplane to divide classes.
RF	Random Forest	Decision trees in an ensemble, each trained on a subset of data, increase prediction accuracy and resistance to overfitting.
KNN	K-Nearest Neighbor	Using distance metrics like Manhattan or Euclidean, this instance-based approach uses the most comparable data points to classify or forecast values.
XGBoost	Extreme Gradient Boosting	Known for its speed, performance, and ability to handle missing values, this gradient boosting with regularization implementation is incredibly effective.
GBM	Gradient Boosted Machine	Progressively constructs additive models while minimizing bias and optimizing loss functions.
DA	Discriminant Analysis	A statistical technique for modeling class differences using linear or quadratic decision limits.
NB	Naive Bayes	Assuming feature independence, the probabilistic classifier (Naïve Bayes) is quick and efficient for big datasets.
HMM	Hidden Markov Model	Bioinformatics uses a probabilistic model for sequential data that estimates hidden states from observed sequences.
DL		
CNN	Convolutional Neural Network	Automatically extracts local and spatial patterns using convolutional layers, making it perfect for structured sequence and image-like peptide data.
RNN	Recurrent Neural Network	Feedback connections are used to model temporal or sequential interdependence, making it appropriate for peptide sequence analysis.
LSTM	Long Short-Term Memory	Specialized RNN that mitigates vanishing gradient problems and captures long-range relationships using gating methods.
BiLSTM	Bidirectional LSTM	By processing sequences both ahead and backward to gather all relevant contextual information, LSTM is extended.
GAN	Generative Adversarial Network	Realistic synthetic peptide data for augmentation is produced by a generator and discriminator working together in a competitive framework.
VAE	Variational Autoencoder	This generative model enables the learning of features and the creation of new peptides by understanding probabilistic latent space representations.

3.3 AI-Assisted Peptide Drug Development

3.3.1 De Novo (Ab Initio) Design

De novo or ab initio design refers to the creation of completely peptide/protein sequences, which do not exist in nature. In contrast to natural molecule optimization, this technique begins from scratch in order to design AA sequences with desired structures and therapeutic characteristics. The advent of generative AI (GenAI) has advanced this discipline through the learning of sequence-structure-function correlations and enabled a pathway to find new regions of sequence space (Hie & Yang, 2022; Winnifrith et al., 2024).

This process involves training models on selected datasets, generating peptide sequences under structural/functional constraints, computationally investigating them using tools such as AlphaFold (Jumper et al., 2021), and validating the optimum AA sequences experimentally. It is worth mentioning that AlphaFold 3 (AF3) employs a diffusion-based architecture to predict joint structures of complexes, including proteins with peptides, ligands, nucleic acids, ions, and modified residues (Abramson et al., 2024). This tool is particularly valuable for modeling peptide–protein and peptide–nucleic acid assemblies prior to experimental validation. However, it is important to note that experimental validation is still required, and there are limitations in predicting significant conformational changes.

A variety of AI models, such as RNNs and LSTMs (Grisoni et al., 2018), VAEs (Dean & Walper, 2020), GANs (Q. Cao et al., 2023; Tucs et al., 2020), Transformer-based protein language models (Guntuboina et al., 2023), and diffusion models for structure-guided design (Watson et al., 2023) have been applied in peptide-based drug design. Tools like RFdiffusion (RosettaFold + diffusion) facilitate the de novo design of functional proteins and binders (Watson et al., 2023). This framework has also been utilized to create binders for bioactive helical peptides, serving as a template for designing peptide–protein interactions. ProteinMPNN is employed for sequence design on fixed backbones and is frequently used in conjunction with RFdiffusion/AF3 to generate sequences compatible with desired scaffolds or complexes (Dauparas et al., 2022). Sequence-only language models (e.g., ESM-2) and hybrid diffusion–language model methods are increasingly utilized to generate and score peptide-like sequences without an explicit backbone. Recent demonstrations have shown the efficacy of language model-guided or diffusion-assisted functional sequence generation (Wang & Fang, 2024). The incorporation of non-canonical amino acids (ncAAs) into peptides and proteins is facilitated by flow/diffusion models, as demonstrated in early 2025 work. This is significant for stability, permeability, and patent considerations (Lee & Kim, 2025).

There are still several challenges that need to be addressed, for example, *in silico* scoring criteria and difficulty in modeling peptides with allostery, geometrical flexibility, or post-translational modifications (Yao & Wang, 2025). GenAI-assisted biological information is thus viewed as crucial for increasing dependability and translational effects (Strokach & Kim, 2022). It is expected that the de novo design approach would play an essential role in future drug investigation as a powerful paradigm in AMPs/ACPs discovery.

3.3.2 Identification/Screening from Natural Sequences

AI-based methods are becoming increasingly popular for identifying peptide drugs from natural sources, including venoms, microbial genomes, and proteomic data sources. Such a database, including ConoServer for conotoxins (Kaas et al., 2008) and ArachnoServer for spider-venom peptides (Pineda et al., 2018) can realize the intensity training data for predictive models, and some reliable software like ToxinPred (N. Sharma et al., 2022), ATSE (Wei et al., 2021), and ToxIBTL (Wei et al., 2022) This will result in a rigorous security assessment before further validation in experimental settings. Computational docking platforms such as Rosetta FlexPepDock (London et al., 2011) and HPEPDOCK (Ambrosetti et al., 2023) can include it to estimate binding affinity and stability.

While screening pipelines were based on hand-crafted descriptors like AA composition and physicochemical properties, recent breakthroughs in DL for peptide sequences now enable models to construct rich hierarchical representations directly from raw peptide sequences (Szymczak et al., 2025). AI architectures like CNNs, RNNs, and GNNs are more precise at acquiring local motifs and long-range dependencies than traditional ML approaches. Building on these frameworks, modern pipelines employ ensemble deep networks or graph/transformer models to mine extensive proteomes, including those of extinct organisms, and metagenomes for AMPs. Several research groups subsequently synthesize a shortlist for bactericidal assays (Wan, Torres, et al., 2024). Recent studies have included large-scale machine learning mining of global microbiomes and archaeal proteomes to identify novel antibiotic peptides. Heterogeneous-graph models (AmpHGT) are now capable of handling non-canonical residues as well. For cell-penetrating peptides (CPPs), ML predictors (CNN/LSTM/GNN) and curated resources (CPPsite3) guide discovery. New interpretable and graph-based models aim to improve generalization to novel chemistries (Wan, Torres, et al., 2024; Wan, Wong, et al., 2024).

AI also facilitates immunological evaluation, especially in predicting peptide-MHC binding and antigen presentation. Tools like NetMHCpan-4.1 and NetMHCIIpan-4.0 predict binding using extensive datasets. New models incorporate protein-language model features and geometric learning to enhance generalizability, although benchmarks still reveal allele- and species-specific limitations. (Reynisson et al., 2021). High-throughput experimental methods, such as EpiScan, assist in generating labels and calibrating models by assessing MHC stabilization on cells (Marzella et al., 2024).

To further enhance the effectiveness and accuracy of peptide screening, hybrid AI models build upon existing frameworks by integrating CNN, RNN/BiLSTM, Attention, and Transformer architectures, occasionally in conjunction with conventional ML classifiers (SVM, RF, XGBoost, LightGBM). Among the examples are:

- **Deep-AFPpred** combines CNNs and BiLSTM networks to capture both spatial and sequential features of peptide sequences, allowing accurate prediction of antifungal peptides (R. Sharma et al., 2022).
- **AniAMPpred** integrates CNN with word2vec embeddings and SVMs to identify 436 antimicrobial peptides, effectively combining deep feature extraction with classical machine learning for enhanced prediction performance (R. Sharma et al., 2021).
- **iAMP-CA2L** takes a novel approach by converting AMP sequences into cellular automata images, which are then processed using CNN, BiLSTM, and SVM, enabling the model to capture complex sequence patterns that influence antimicrobial activity (Xiao et al., 2021).
- **iACP-DRLF** combines multiple classifiers, including SVM, Random Forest (RF), K-Nearest Neighbors (KNN), Naïve Bayes (NB), Linear Discriminant Analysis (LDA), and Light Gradient Boosting Machine (LightGBM), with DL-based embeddings to accurately identify anticancer peptides from large sequence datasets (Lv et al., 2021).
- **SMEP** utilizes a sequential combination of XGBoost and LSTM to identify bactericidal hexapeptides, demonstrating the utility of combining gradient-boosted trees with deep sequential models for small peptide motifs (Huang et al., 2023).
- **Ensemble strategies** that integrate CNN, BiLSTM, and LightGBM have been shown to enhance generalization and screening accuracy, minimizing overfitting while maintaining robust predictive performance across diverse peptide datasets (Yuan et al., 2023).

Furthermore, by generating novel candidates based on natural templates, GenAI models like GANs and reinforcement learning broaden the therapeutic peptide landscape. An extensive pipeline for identifying and screening bioactive, safe, and therapeutically relevant peptides is formed by these AI-driven techniques, which range from DL, combinatorial modeling, to generative approaches. These techniques also enable large-scale data mining, predictive modeling, and structure-based simulations.

3.3.3 Case Studies of AI in Peptide Drug Discovery

AI is revolutionizing the discovery of peptide drugs by increasing its speed, accuracy, and efficiency. Examples demonstrate how AI can save time and money in experiments while speeding up the discovery of new peptides with better medicinal qualities. Peptide inhibitors for HIV-1 protease, for instance, have been designed using RosettaScripts, a platform that combines ML and physics-based modeling to predict sequences that efficiently bind and inhibit this essential viral enzyme (Lemmon et al., 2012). Likewise, InSiPS's predicted peptides that bound the SARS-CoV-2 spike protein were confirmed through experimentation and subsequently assisted in the discovery of other spike-binding peptides in patient saliva for COVID-19 diagnostics (Hajikarimlou et al., 2022).

AI has also been used to create peptides that interfere with particular interactions between proteins. Researchers have produced peptides with increased potency and selectivity by examining protein structures and forecasting ideal binding sequences (Kozlovskii and Popov, 2021; Lee, 2023). In preclinical research for cancer, autoimmune illnesses, and infectious diseases, tools such as BiteNet(P)(p), which examine 3D protein structures to find "hot spots" for binding, have demonstrated encouraging outcomes (Al-Azzam et al., 2020; Gokhale and Satyanarayanajois, 2014).

Since the costs of development are rising and antibiotics are becoming increasingly resistant, AI is becoming increasingly important in the search for AMPs. After being trained on tens of thousands of peptides, models such as AMPs-Net have successfully found peptides like FE23 and KS22, which have been confirmed against S. aureus and E. coli (Ruiz Puentes et al., 2022) and have reliably predicted antimicrobial activity.

Beyond discovery, AI is directing the development of customized and tailored peptide treatments. Cell-targeting peptides, which avoid healthy tissue and deliver medications directly to infected cells, can be developed with their assistance. (Shah et al., 2020). Peptide half-life and stability are predicted by novel systems, such as the one created by Tan et al. (2024), which aids researchers in creating peptides that function better and survive longer in the body (S. jun Cao et al., 2021).

4. AI-DRIVEN NANOTECHNOLOGY FOR PEPTIDE DELIVERY

The convergence of AI and nanotechnology is reshaping strategies for peptide-based therapeutics. Nanoparticles, such as liposomes, lipid nanoparticles (LNPs), polymeric nanoparticles, and micelles are proving to be powerful tools for delivering peptide-based therapies (Duro-Castano et al., 2020; Mitchell et al., 2021; Wagner et al., 2018; Y. Wang et al., 2024). They can encapsulate peptides, shielding them from enzymatic degradation, improving solubility, and allowing for controlled or sustained release. For instance, insulin-loaded liposomes have shown improved stability and longer circulation in the bloodstream (Petrov et al., 2025; J. Wang et al., 2023), while GLP-1 analogs in polymeric nanoparticles can maintain their therapeutic effect for extended periods (Ismail et al., 2019). Additionally, peptides can be shown on the surface of nanoparticles to enhance immunological responses, increase cellular absorption, or improve targeting. Examples include antigenic peptides on liposomal vaccines to provoke a specific immune reaction (Zhou et al., 2023), or tumor-targeting peptides on polymeric micelles to deliver drugs directly to cancer cells (Cai et al., 2020). Recent studies have also shown that peptide-functionalized LNPs can enhance mRNA delivery to the brain, improving targeting and reducing off-target effects (Han et al., 2024). This dual strategy, i.e., encapsulation and surface attachment, provides flexibility to design nanocarriers according to the therapeutic objective, peptide properties, and delivery route.

Despite their potential, the design space of nanocarriers is vast, encompassing parameters such as nanoparticle size, surface charge, lipid or polymer composition, drug-to-carrier ratio, and functional ligands for active targeting. Traditional experimental approaches to optimize these factors are labor-intensive and time-consuming, often requiring iterative synthesis and testing cycles.

AI offers an alternative by quickly identifying crucial formulation factors and forecasting delivery outcomes across various nanocarrier systems (Bae et al., 2025). Nanoparticles like LNPs, in particular, are among the most versatile peptide nanocarriers. Random forests, Bayesian optimization, and deep neural networks have been used to predict LNP properties such as particle size, surface charge, stability, and peptide encapsulation efficiency. These models enable researchers to fine-tune formulations *in silico*, avoiding costly laboratory optimization (G. Kumar & Ardekani, 2025; Maharjan et al., 2024). Importantly, ML frameworks trained on large-scale nanoparticle formulation datasets showed predictive design guidelines that improve tissue-specific uptake and therapeutic results, allowing for more precise translation of peptide therapies (Das & Chandra, 2022).

Beyond LNPs, AI has also been applied to study peptide interactions with metallic nanoparticles, such as gold nanoparticles, using explainable ML models. For example, a rough set-based classifier as the algorithm has been developed to predict peptide binding affinity to gold nanoparticles, revealing key sequence patterns that govern nanoparticle interaction and providing insights for rational design of peptide-functionalized nanomaterials (Abdelmeguid et al., 2025). Furthermore, polymeric nanoparticles, micelles, and hybrid nanocarriers are being improved with AI technologies. DL models map the correlations between nanomaterial physicochemical properties and biological outcomes, such as peptide release patterns, proteolytic resistance, and circulation half-life (Bhujel et al., 2025). AI has already been successfully applied to predict antibacterial activity of nanoparticles and antimicrobial peptides, demonstrating its capacity to model complex nanoparticle-biological interactions (Sowers et al., 2023).

Although this research is still in its early stages, it highlights a promising synergistic approach, where AI-guided peptide design (stabilization, cyclization, sequence alteration) feeds into AI-guided nanocarrier optimization, ultimately creating delivery systems tailored for safe, targeted, and controlled peptide release.

5. LIMITATIONS AND CHALLENGES

In spite of the progress made in AI/ML, human intervention and understanding are still needed at various stages, from providing proper inputs to interpreting the outcomes/data given by these AI models. Among the many advantages, there are some persistent challenges and limitations:

1. **Data scarcity, bias, and label noise:** Many public peptide datasets are small, imbalanced (e.g., over-representation of cationic AMPs), and contain inconsistent assay conditions, leading to models' overfitting or learning dataset artifacts (e.g., hydrophobicity bias), Rigorous curation, uncertainty modeling, and prospective validation are beneficial (Wan, Wong, et al., 2024).
2. **Generalization and distribution shift performance:** Declines on unseen scaffolds/alleles/species, as well demonstrated for MHC and CPP predictors. Strategies include geometric/structure-aware features, protein-LM embeddings, leave-cluster-out splits, and cross-species benchmarking (Marzella et al., 2024).
3. **3D realism and dynamics:** Peptides are conformationally flexible; binding often depends on induced fit, membrane context, or solvent/ions. AF3 enhances complex modeling but still requires experimental validation and is not a dynamics engine; MD or biophysics remain crucial (Abramson et al., 2024).
4. **Non-canonical chemistry and modifications**: Many models assume the 20 canonical residues and overlook PTMs, cyclization, D-residues, and ncAAs—key to stability and permeability. Emerging flow/diffusion models for ncAAs are promising but pre-publication and require broader benchmarking (Wang and Fang, 2024).
5. **Assay alignment and multi-objective design:** Objectives (MIC, cytotoxicity, hemolysis, serum stability, permeability, immunogenicity) can be conflicting; scoring functions are noisy and not calibrated across laboratories. Active learning with thoughtfully chosen acquisition functions and orthogonal assays is advantageous (Wan, Wong, et al., 2024).

6. **Interpretability and design rules:** Feature attributions in LMs/CNNs are non-trivial. For CPPs and AMPs, new interpretable frameworks are emerging, but mechanistic insight remains limited compared to classical SAR (Bernardes-Loch et al., 2025).
7. **Benchmarking and reproducibility:** Many studies reuse overlapping datasets/splits, inflating scores. Community efforts increasingly emphasize strict de-duplication, cluster-based splits, and blinded prospective tests (Giziński et al., 2024).
8. ***In silico* to *in vivo* potent:** in vitro peptides can fail on PK, proteolysis, aggregation, immunogenicity, and off-target effects. This is where ncAAs, cyclization, conjugation, and delivery technologies must be integrated with AI iteration (Goles et al., 2024).

6. FUTURE DIRECTION AND CONCLUSION

The constraints highlighted in the preceding section indicate both the challenges AI faces in the search for peptide drugs and the areas where innovation can have the most significant influence. Developing interpretable models, incorporating experimental validation, and enhancing dataset quality are all options that will not only address present issues but also broaden the potential of artificial intelligence. The field is moving closer to realizing the full potential of AI-guided peptide therapies by addressing these gaps. AI in peptide drug development appears to have a promising future ahead, as it has the potential to revolutionize the process of designing novel peptide drugs. A number of obstacles, however, must be addressed before AI is completely integrated into the drug development process. One key factor is the development and enhancement of AI models through the use of more varied datasets. These algorithms can be made more accurate and predictive by using large, publicly available peptide databases, protein structures, omics data, and clinical information. By integrating genetic variations, patient-specific data, and multi-omics data, AI models may assist in developing customized peptide therapeutics for individuals or specific disease subtypes. AI will remain vulnerable until high-quality datasets are curated and annotated in order to improve reliability. (Chang et al., 2024).

Integrating AI predictions with experimental screening tools like combinatorial chemistry, peptide libraries, and high-throughput tests is another intriguing opportunity (Topcu et al., 2022). The discovery process can be sped up and made more dependable with this type of iterative feedback loop, in which AI proposes candidates and experiments validate or improve them (Vamathevan et al., 2019). The ability of AI to identify intricate sequence–activity patterns will advance as they develop further, especially with the application of DL and reinforcement learning. These developments offer fresh perspectives on the fundamental biology of peptides in addition to enhancing prediction ability. Another promising development is the rise of XAI techniques, which give researchers insight into the process of making predictions, boosting confidence, and promoting wider use in pharmaceutical research (Markus et al., 2021). In addition to peptide design, AI true promise is dependent on efficient delivery. Nanocarriers, which include lipid nanoparticles, polymeric nanoparticles, and micelles, can preserve peptides while also improving bioavailability and allowing for tailored release. AI-driven improvement of these nanocarriers results in a synergistic pipeline that connects computational peptide design to practical therapeutic delivery. There are challenges, of course. There are still a few high-quality, annotated datasets available, which have an impact on how effectively models generalize. For AI forecasts to be precise and broadly applicable, data collection, standardization, and sharing are therefore crucial (Kelly et al., 2019). The interpretation and validation of AI outcomes present another difficulty. To demonstrate their biological relevance, even the most accurate predictions need to be confirmed experimentally. (Gilpin et al., 2018). The integration of strong validation techniques into AI-guided processes will facilitate the transition from clinical application to computational design.

Lastly, it is impossible to overlook ethical and regulatory issues. As AI becomes more prevalent in drug research, it is critical to guarantee accountability, equity, and openness. Concerns, including algorithmic bias, patient data privacy, and responsible implementation in healthcare, must all be covered by guidelines (Jeyaraman et al., 2023). Establishing frameworks that strike a balance between innovation, safety, and trust will require cooperation from researchers, regulators, and industry stakeholders (Jia et al., 2023).

Not to mention that this is still a relatively new field. The market has not yet seen any AI-generated peptide medications as of late 2024, but a number of intriguing candidates are progressing through preclinical research and even early clinical trials. As datasets, algorithms, validation techniques, and regulatory clarity continue to advance, it is logical to anticipate that AI-designed peptides will soon make the leap from concept to clinical use. This could lead to the development of peptide therapies that are safer, more efficient, and more customized.

In conclusion, there is much promise for the future of AI-driven peptide drug discovery research. Novel peptide drugs can be discovered more rapidly and reliably than ever before with the help of more intelligent algorithms, larger information, and tighter cooperation between researchers, businesses, and regulators. There are still obstacles to overcome, of course, such as the scarcity of data, the need for meticulous validation, and significant ethical implications. As the subject develops, interdisciplinary collaboration will be essential to assuring the ethical and efficient application of AI. As the field advances, interdisciplinary approaches — such as combining AI and nanotechnology for enhanced peptide delivery —will be critical to ensuring safe, efficient, and focused therapeutics. By utilizing this synergy, AI and nanotechnology have the potential to change the therapeutic landscape, allowing for safer, more effective, and highly tailored peptide-based treatments for a broad spectrum of ailments.

REFERENCES

Abdelmeguid, M. G., Janairo, J. I. B., & Chemmangattuvalappil, N. G. (2025). Examining peptide–gold nanoparticle interactions through explainable machine learning. *Discover Chemistry 2025 2:1, 2*(1), 1–16. https://doi.org/DOI: 10.1007/S44371-025-00191-2

Abramson, J., Adler, J., Dunger, J., Evans, R., Green, T., Pritzel, A., Ronneberger, O., Willmore, L., Ballard, A. J., Bambrick, J., Bodenstein, S. W., Evans, D. A., Hung, C.-C., O'Neill, M., Reiman, D., Tunyasuvunakool, K., Wu, Z., Žemgulytė, A., Arvaniti, E., & Jumper, J. M. (2024). Accurate structure prediction of biomolecular interactions with AlphaFold 3. *Nature, 630*(8016), 493–500. DOI: 10.1038/s41586-024-07487-w PMID: 38718835

Akondi, K. B., Muttenthaler, M., Dutertre, S., Kaas, Q., Craik, D. J., Lewis, R. J., & Alewood, P. F. (2014). Discovery, synthesis, and structure-activity relationships of conotoxins. *Chemical Reviews, 114*(11), 5815–5847. DOI: 10.1021/cr400401e PMID: 24720541

Al-Azzam, S., Ding, Y., Liu, J., Pandya, P., Ting, J. P., & Afshar, S. (2020). Peptides to combat viral infectious diseases. *Peptides, 134*, 170402. DOI: 10.1016/J.PEPTIDES.2020.170402 PMID: 32889022

Ali, S., Alam, W., Tyara, H., & Chong, K. T. (2025). An accurate prediction of drug–drug interactions and side effects by using integrated convolutional and BiLSTM networks. *Chemometrics and Intelligent Laboratory Systems, 257*, 105304. DOI: 10.1016/j.chemolab.2024.105304

Ambrosetti, F., Jandova, Z., & Bonvin, A. M. J. J. (2023). Information-Driven Antibody–Antigen Modeling with HADDOCK. In *Methods in Molecular Biology* (Vol. 2552, pp. 267–282). Humana., DOI: 10.1007/978-1-0716-2609-2_14

Attique, M., Farooq, M. S., Khelifi, A., & Abid, A. (2020). Prediction of Therapeutic Peptides Using Machine Learning: Computational Models, Datasets, and Feature Encodings. *IEEE Access : Practical Innovations, Open Solutions, 8*, 148570–148594. DOI: 10.1109/ACCESS.2020.3015792

Bae, H., Ji, H., Konstantinov, K., Sluyter, R., Ariga, K., Kim, Y. H., & Kim, J. H. (2025). Artificial Intelligence-Driven Nanoarchitectonics for Smart Targeted Drug Delivery. *Advanced Materials, 10239*. Advance online publication. DOI: 10.1002/ADMA.202510239 PMID: 40772340

Banting, F. G., Best, C. H., Collip, J. B., Macleod, J. J. R., & Noble, E. C. (1922). The Effect of Pancreatic Extract (Insulin) on Normal Rabbits. *American Journal of Physiology-Legacy Content*, *62*(1), 162–176. DOI: 10.1152/ajplegacy.1922.62.1.162

Bernardes-Loch, R. M., de Oliveira Almeida, G., Brasiliano, I. T., Meira, W., Pires, D. E. V., Baracat-Pereira, M. C., & de Azevedo Silveira, S. (2025). PERSEUcpp: A machine learning strategy to predict cell-penetrating peptides and their uptake efficiency. *BioRxiv*, 2025.04.07.647598. https://doi.org/DOI: 10.1101/2025.04.07.647598

Bhujel, R., Enkmann, V., Burgstaller, H., & Maharjan, R. (2025). Artificial Intelligence-Driven Strategies for Targeted Delivery and Enhanced Stability of RNA-Based Lipid Nanoparticle Cancer Vaccines. *Pharmaceutics*, *17*(8), 992. DOI: 10.3390/PHARMACEUTICS17080992 PMID: 40871015

Blanco-González, A., Cabezón, A., Seco-González, A., Conde-Torres, D., Antelo-Riveiro, P., Piñeiro, Á., & Garcia-Fandino, R. (2023). The Role of AI in Drug Discovery: Challenges, Opportunities, and Strategies. *Pharmaceuticals*, *16*(6), 891. DOI: 10.3390/ph16060891 PMID: 37375838

Borel, J. F., Feurer, C., Gubler, H. U., & Stähelin, H. (1976). Biological effects of cyclosporin A: A new antilymphocytic agent. *Agents and Actions*, *6*(4), 468–475. DOI: 10.1007/BF01973261 PMID: 8969

Cai, Y., Xu, Z., Shuai, Q., Zhu, F., Xu, J., Gao, X., & Sun, X. (2020). Tumor-targeting peptide functionalized PEG-PLA micelles for efficient drug delivery. *Biomaterials Science*, *8*(8), 2274–2282. DOI: 10.1039/C9BM02036E PMID: 32162618

Cao, Q., Ge, C., Wang, X., Harvey, P. J., Zhang, Z., Ma, Y., Wang, X., Jia, X., Mobli, M., Craik, D. J., Jiang, T., Yang, J., Wei, Z., Wang, Y., Chang, S., & Yu, R. (2023). Designing antimicrobial peptides using deep learning and molecular dynamic simulations. *Briefings in Bioinformatics*, *24*(2), 1–13. DOI: 10.1093/bib/bbad058 PMID: 36857616

Cao, S. jun, Lv, Z. qiang, Guo, S., Jiang, G. ping, & Liu, H. ling. (2021). An update - Prolonging the action of protein and peptide drugs. *Journal of Drug Delivery Science and Technology*, *61*, 102124. DOI: 10.1016/J.JDDST.2020.102124

Capecchi, A., & Reymond, J. L. (2021). Peptides in chemical space. *Medicine in Drug Discovery*, *9*, 100081. DOI: 10.1016/j.medidd.2021.100081

Castel, G., Chtéoui, M., Heyd, B., & Tordo, N. (2011). Phage display of combinatorial peptide libraries: Application to antiviral research. *Molecules (Basel, Switzerland)*, *16*(5), 3499–3518. DOI: 10.3390/molecules16053499 PMID: 21522083

Chan, M. L., Qi, Y., Larimore, K., Cherukuri, A., Seid, L., Jayaram, K., Jeha, G., Fisheleva, E., Day, J., Huntsman-Labed, A., Savarirayan, R., Irving, M., Bacino, C. A., Hoover-Fong, J., Ozono, K., Mohnike, K., Wilcox, W. R., Horton, W. A., & Henshaw, J. (2022). Pharmacokinetics and Exposure–Response of Vosoritide in Children with Achondroplasia. *Clinical Pharmacokinetics*, *61*(2), 263–280. DOI: 10.1007/s40262-021-01059-1 PMID: 34431071

Chang, L., Mondal, A., Singh, B., Martínez-Noa, Y., & Perez, A. (2024). Revolutionizing peptide-based drug discovery: Advances in the post-AlphaFold era. *Wiley Interdisciplinary Reviews. Computational Molecular Science*, *14*(1). Advance online publication. DOI: 10.1002/wcms.1693 PMID: 38680429

Chen, Z., Wang, R., Guo, J., & Wang, X. (2024). The role and future prospects of artificial intelligence algorithms in peptide drug development. *Biomedicine and Pharmacotherapy*, *175*, 116709. DOI: 10.1016/j.biopha.2024.116709 PMID: 38713945

Dai, Y., Maresca, M., & Al Musaimi, O. (2024). Exploring FDA-Approved Frontiers: Insights into Natural and Engineered Peptide Analogues in the GLP-1, GIP, GHRH, CCK, ACTH, and α-MSH Realms. *Biomolecules 2024, Vol. 14, Page 264*, *14*(3), 264. https://doi.org/DOI: 10.3390/BIOM14030264

Dardenne, E., Ishiyama, N., Lin, T. A., & Lucas, M. C. (2023). Current and emerging therapies for Achondroplasia: The dawn of precision medicine. *Bioorganic & Medicinal Chemistry*, *87*. Advance online publication. DOI: 10.1016/j.bmc.2023.117275 PMID: 37156065

Das, K. P., & Chandra, J. (2022). Nanoparticles and convergence of artificial intelligence for targeted drug delivery for cancer therapy: Current progress and challenges. *Frontiers in Medical Technology*, *4*, 1067144. DOI: 10.3389/FMEDT.2022.1067144/BIBTEX PMID: 36688144

Dauparas, J., Anishchenko, I., Bennett, N., Bai, H., Ragotte, R. J., Milles, L. F., Wicky, B. I. M., Courbet, A., de Haas, R. J., Bethel, N., Leung, P. J. Y., Huddy, T. F., Pellock, S., Tischer, D., Chan, F., Koepnick, B., Nguyen, H., Kang, A., Sankaran, B., & Baker, D. (2022). Robust deep learning–based protein sequence design using ProteinMPNN. *Science*, *378*(6615), 49–56. DOI: 10.1126/science.add2187 PMID: 36108050

Dean, S. N., & Walper, S. A. (2020). Variational autoencoder for the generation of antimicrobial peptides. *ACS Omega*, *5*(33), 20746–20754. DOI: 10.1021/acsomega.0c00442 PMID: 32875208

Deng, J., Yang, Z., Ojima, I., Samaras, D., & Wang, F. (2022). Artificial intelligence in drug discovery: Applications and techniques. *Briefings in Bioinformatics*, *23*(1), 1–19. DOI: 10.1093/bib/bbab430 PMID: 34734228

Drucker, D. J., & Nauck, M. A. (2006). The incretin system: Glucagon-like peptide-1 receptor agonists and dipeptidyl peptidase-4 inhibitors in type 2 diabetes. *Lancet*, *368*(9548), 1696–1705. DOI: 10.1016/S0140-6736(06)69705-5 PMID: 17098089

Duro-Castano, A., Moreira Leite, D., Forth, J., Deng, Y., Matias, D., Noble Jesus, C., & Battaglia, G. (2020). Designing peptide nanoparticles for efficient brain delivery. *Advanced Drug Delivery Reviews*, *160*, 52–77. DOI: 10.1016/J.ADDR.2020.10.001 PMID: 33031897

Evans, B. J., King, A. T., Katsifis, A., Matesic, L., & Jamie, J. F. (2020). Methods to enhance the metabolic stability of peptide-based PET radiopharmaceuticals. *Molecules (Basel, Switzerland)*, *25*(10), 2314. DOI: 10.3390/molecules25102314 PMID: 32423178

Fosgerau, K., & Hoffmann, T. (2015). Peptide therapeutics: Current status and future directions. *Drug Discovery Today*, *20*(1), 122–128. DOI: 10.1016/j.drudis.2014.10.003 PMID: 25450771

Gawde, U., Chakraborty, S., Waghu, F. H., Barai, R. S., Khanderkar, A., Indraguru, R., Shirsat, T., & Idicula-Thomas, S. (2023). CAMPR4: a database of natural and synthetic antimicrobial peptides. *Nucleic Acids Research, 51*(1 D), D377–D383. https://doi.org/DOI: 10.1093/nar/gkac933

Gilpin, L. H., Bau, D., Yuan, B. Z., Bajwa, A., Specter, M., & Kagal, L. (2018). Explaining explanations: An overview of interpretability of machine learning. *Proceedings - 2018 IEEE 5th International Conference on Data Science and Advanced Analytics, DSAA 2018*, 80–89. https://doi.org/DOI: 10.1109/DSAA.2018.00018

Giziński, S., Preibisch, G., Kucharski, P., Tyrolski, M., Rembalski, M., Grzegorczyk, P., & Gambin, A. (2024). Enhancing antigenic peptide discovery: Improved MHC-I binding prediction and methodology. *Methods (San Diego, Calif.)*, *224*, 1–9. DOI: 10.1016/j.ymeth.2024.01.016 PMID: 38295891

Glavaš, M., Gitlin-Domagalska, A., Dębowski, D., Ptaszyńska, N., Łęgowska, A., & Rolka, K. (2022). Vasopressin and Its Analogues: From Natural Hormones to Multitasking Peptides. *International Journal of Molecular Sciences*, *23*(6), 3068. DOI: 10.3390/ijms23063068 PMID: 35328489

Gokhale, A. S., & Satyanarayanajois, S. (2014). Peptides and peptidomimetics as immunomodulators. *Immunotherapy*, *6*(6), 755–774. DOI: 10.2217/IMT.14.37 PMID: 25186605

Goles, M., Daza, A., Cabas-Mora, G., Sarmiento-Varón, L., Sepúlveda-Yañez, J., Anvari-Kazemabad, H., Davari, M. D., Uribe-Paredes, R., Olivera-Nappa, Á., Navarrete, M. A., & Medina-Ortiz, D. (2024). Peptide-based drug discovery through artificial intelligence: Towards an autonomous design of therapeutic peptides. *Briefings in Bioinformatics*, *25*(4). Advance online publication. DOI: 10.1093/bib/bbae275 PMID: 38856172

Grisoni, F., Neuhaus, C. S., Gabernet, G., Müller, A. T., Hiss, J. A., & Schneider, G. (2018). Designing Anticancer Peptides by Constructive Machine Learning. *ChemMedChem*, *13*(13), 1300–1302. DOI: 10.1002/cmdc.201800204 PMID: 29679519

Guntuboina, C., Das, A., Mollaei, P., Kim, S., & Barati Farimani, A. (2023). PeptideBERT: A Language Model Based on Transformers for Peptide Property Prediction. *The Journal of Physical Chemistry Letters*, *14*(46), 10427–10434. DOI: 10.1021/acs.jpclett.3c02398 PMID: 37956397

Hajikarimlou, M., Hooshyar, M., Moutaoufik, M. T., Aly, K. A., Azad, T., Takallou, S., Jagadeesan, S., Phanse, S., Said, K. B., Samanfar, B., Bell, J. C., Dehne, F., Babu, M., & Golshani, A. (2022). A computational approach to rapidly design peptides that detect SARS-CoV-2 surface protein S. *NAR Genomics and Bioinformatics*, *4*(3). Advance online publication. DOI: 10.1093/NARGAB/LQAC058 PMID: 36004308

Hammami, R., Zouhir, A., Le Lay, C., Ben Hamida, J., & Fliss, I. (2010). BACTIBASE second release: A database and tool platform for bacteriocin characterization. *BMC Microbiology*, *10*(1), 1–5. DOI: 10.1186/1471-2180-10-22 PMID: 20105292

Han, E. L., Tang, S., Kim, D., Murray, A. M., Swingle, K. L., Hamilton, A. G., Mrksich, K., Padilla, M. S., Palanki, R., Li, J. J., & Mitchell, M. J. (2024). Peptide-Functionalized Lipid Nanoparticles for Targeted Systemic mRNA Delivery to the Brain. *Nano Letters*, *25*(2), 800–810. DOI: 10.1021/ACS.NANOLETT.4C05186 PMID: 39688915

Hie, B. L., & Yang, K. K. (2022). Adaptive machine learning for protein engineering. *Current Opinion in Structural Biology*, *72*, 145–152. DOI: 10.1016/j.sbi.2021.11.002 PMID: 34896756

Huang, J., Xu, Y., Xue, Y., Huang, Y., Li, X., Chen, X., Xu, Y., Zhang, D., Zhang, P., Zhao, J., & Ji, J. (2023). Identification of potent antimicrobial peptides via a machine-learning pipeline that mines the entire space of peptide sequences. *Nature Biomedical Engineering 2023 7:6, 7*(6), 797–810. https://doi.org/DOI: 10.1038/s41551-022-00991-2

Hwang, J. S., Kim, S. G., Shin, T. H., Jang, Y. E., Kwon, D. H., & Lee, G. (2022). Development of Anticancer Peptides Using Artificial Intelligence and Combinational Therapy for Cancer Therapeutics. *Pharmaceutics, 14*(5), 997. DOI: 10.3390/PHARMACEUTICS14050997/S1 PMID: 35631583

Ismail, R., Bocsik, A., Katona, G., Gróf, I., Deli, M. A., & Csóka, I. (2019). Encapsulation in Polymeric Nanoparticles Enhances the Enzymatic Stability and the Permeability of the GLP-1 Analog, Liraglutide, Across a Culture Model of Intestinal Permeability. *Pharmaceutics, 11*(11), 599. DOI: 10.3390/PHARMACEUTICS11110599 PMID: 31726699

Jaroszewicz, W., Morcinek-Orłowska, J., Pierzynowska, K., Gaffke, L., & W grzyn, G. (2022). Phage display and other peptide display technologies. *FEMS Microbiology Reviews, 46*(2), 1–25. DOI: 10.1093/femsre/fuab052 PMID: 34673942

Jeyaraman, M., Balaji, S., Jeyaraman, N., & Yadav, S. (2023). Unraveling the Ethical Enigma: Artificial Intelligence in Healthcare. *Cureus, 15*(8). Advance online publication. DOI: 10.7759/CUREUS.43262 PMID: 37692617

Jia, Z., Chen, J., Xu, X., Kheir, J., Hu, J., Xiao, H., Peng, S., Hu, X. S., Chen, D., & Shi, Y. (2023). The importance of resource awareness in artificial intelligence for healthcare is significant. *Nature Machine Intelligence, 5*(7), 687–698. DOI: 10.1038/S42256-023-00670-0

Joo, S. H. (2012). Cyclic peptides as therapeutic agents and biochemical tools. *Biomolecules & Therapeutics, 20*(1), 19–26. DOI: 10.4062/biomolther.2012.20.1.019 PMID: 24116270

Jumper, J., Evans, R., Pritzel, A., Green, T., Figurnov, M., Ronneberger, O., Tunyasuvunakool, K., Bates, R., Žídek, A., Potapenko, A., Bridgland, A., Meyer, C., Kohl, S. A. A., Ballard, A. J., Cowie, A., Romera-Paredes, B., Nikolov, S., Jain, R., Adler, J., & Hassabis, D. (2021). Highly accurate protein structure prediction with AlphaFold. *Nature, 596*(7873), 583–589. DOI: 10.1038/s41586-021-03819-2 PMID: 34265844

Kaas, Q., Westermann, J. C., Halai, R., Wang, C. K. L., & Craik, D. J. (2008). ConoServer, a database for conopeptide sequences and structures. *Bioinformatics (Oxford, England), 24*(3), 445–446. DOI: 10.1093/bioinformatics/btm596 PMID: 18065428

Kelly, C. J., Karthikesalingam, A., Suleyman, M., Corrado, G., & King, D. (2019). Key challenges for delivering clinical impact with artificial intelligence. *BMC Medicine*, *17*(1), 1–9. DOI: 10.1186/S12916-019-1426-2/PEER-REVIEW PMID: 31665002

Kozlovskii, I., & Popov, P. (2021). Protein-Peptide Binding Site Detection Using 3D Convolutional Neural Networks. *Journal of Chemical Information and Modeling*, *61*(8), 3814–3823. DOI: 10.1021/ACS.JCIM.1C00475/SUPPL_FILE/CI1C00475_SI_011.ZIP PMID: 34292750

Kumar, G., & Ardekani, A. M. (2025). Machine-Learning Framework to Predict the Performance of Lipid Nanoparticles for Nucleic Acid Delivery. *ACS Applied Bio Materials*, *8*(5), 3717–3727. DOI: 10.1021/ACSABM.4C01716 PMID: 40267508

Kumar, S., Sanap, S. N., Vasoya, M., Handa, M., Pandey, P., Khopade, A., & Sawant, K. K. (2024). USFDA-approved parenteral peptide formulations and excipients: Industrial perspective. *Journal of Drug Delivery Science and Technology*, *95*, 105589. DOI: 10.1016/j.jddst.2024.105589

Lee, J. S., & Kim, P. M. (2025). Design of peptides with non-canonical amino acids using flow matching. In bioRxiv (p. 2025.07.31.667780). Cold Spring Harbor Laboratory. https://doi.org/DOI: 10.1101/2025.07.31.667780

Lee, M. (2023). Recent Advances in Deep Learning for Protein-Protein Interaction Analysis: A Comprehensive Review. *Molecules 2023, Vol. 28, Page 5169, 28*(13), 5169. https://doi.org/DOI: 10.3390/MOLECULES28135169

Lemmon, G., Kaufmann, K., & Meiler, J. (2012). Prediction of HIV-1 Protease/Inhibitor Affinity using RosettaLigand. *Chemical Biology & Drug Design*, *79*(6), 888–896. DOI: 10.1111/J.1747-0285.2012.01356.X PMID: 22321894

London, N., Raveh, B., Cohen, E., Fathi, G., & Schueler-Furman, O. (2011). Rosetta FlexPepDock web server - High resolution modeling of peptide-protein interactions. *Nucleic Acids Research*, *39*(SUPPL. 2), W249–W253. DOI: 10.1093/nar/gkr431 PMID: 21622962

Lv, Z., Cui, F., Zou, Q., Zhang, L., & Xu, L. (2021). Anticancer peptides prediction with deep representation learning features. *Briefings in Bioinformatics*, *22*(5), 1–14. DOI: 10.1093/BIB/BBAB008 PMID: 33529337

Ma, Y., Chen, H., Kang, J., Guo, X., Sun, C., Xu, J., Tao, J., Wei, S., Dong, Y., Tian, H., Lv, W., Jia, Z., Bi, S., Shang, Z., Zhang, C., Lv, H., Jiang, Y., & Zhang, M. (2025). The Hidden Markov model and its applications in bioinformatics analysis. *Genes & Diseases*, *101729*. Advance online publication. DOI: 10.1016/j.gendis.2025.101729 PMID: 41069576

Maharjan, R., Kim, K. H., Lee, K., Han, H. K., & Jeong, S. H. (2024). Machine learning-driven optimization of mRNA-lipid nanoparticle vaccine quality with XG-Boost/Bayesian method and ensemble model approaches. *Journal of Pharmaceutical Analysis, 14*(11), 100996. DOI: 10.1016/J.JPHA.2024.100996 PMID: 39759971

Markus, A. F., Kors, J. A., & Rijnbeek, P. R. (2021). The role of explainability in creating trustworthy artificial intelligence for health care: A comprehensive survey of the terminology, design choices, and evaluation strategies. *Journal of Biomedical Informatics, 113*, 103655. DOI: 10.1016/J.JBI.2020.103655 PMID: 33309898

Marzella, D. F., Crocioni, G., Radusinović, T., Lepikhov, D., Severin, H., Bodor, D. L., Rademaker, D. T., Lin, C. Y., Georgievska, S., Renaud, N., Kessler, A. L., Lopez-Tarifa, P., Buschow, S. I., Bekkers, E., & Xue, L. C. (2024). Geometric deep learning improves generalizability of MHC-bound peptide predictions. *Communications Biology, 7*(1), 1–12. DOI: 10.1038/s42003-024-07292-1 PMID: 39702482

Mitchell, M. J., Billingsley, M. M., Haley, R. M., Wechsler, M. E., Peppas, N. A., & Langer, R. (2021). Engineering precision nanoparticles for drug delivery. *Nature Reviews. Drug Discovery, 20*(2), 101–124. DOI: 10.1038/S41573-020-0090-8;SUBJMETA PMID: 33277608

Mondal, A., Chang, L., & Perez, A. (2022). Modelling peptide–protein complexes: Docking, simulations and machine learning. *QRB Discovery, 3*. Advance online publication. DOI: 10.1017/qrd.2022.14 PMID: 37529282

Muttenthaler, M., King, G. F., Adams, D. J., & Alewood, P. F. (2021). Trends in peptide drug discovery. *Nature Reviews. Drug Discovery, 20*(4), 309–325. DOI: 10.1038/s41573-020-00135-8 PMID: 33536635

Nissan, N., Allen, M. C., Sabatino, D., & Biggar, K. K. (2024). Future Perspective: Harnessing the Power of Artificial Intelligence in the Generation of New Peptide Drugs. *Biomolecules, 14*(10), 1303. DOI: 10.3390/biom14101303 PMID: 39456236

Nixon, A. E., Sexton, D. J., & Ladner, R. C. (2014). Drugs derived from phage display from candidate identification to clinical practice. *mAbs, 6*(1), 73–85. DOI: 10.4161/mabs.27240 PMID: 24262785

Palma, M. (2023). Epitopes and Mimotopes Identification Using Phage Display for Vaccine Development against Infectious Pathogens. *Vaccines, 11*(7), 1176. DOI: 10.3390/vaccines11071176 PMID: 37514992

Paul, D., Sanap, G., Shenoy, S., Kalyane, D., Kalia, K., & Tekade, R. K. (2021). Artificial intelligence in drug discovery and development. *Drug Discovery Today, 26*(1), 80–93. DOI: 10.1016/j.drudis.2020.10.010 PMID: 33099022

Petrov, P. D., Davidova, S., & Satchanska, G. (2025). Polymer Micelles as Nanocarriers of Bioactive Peptides. *Polymers, 17*(9), 1174. DOI: 10.3390/POLYM17091174 PMID: 40362958

Pineda, S. S., Chaumeil, P. A., Kunert, A., Kaas, Q., Thang, M. W. C., Le, L., Nuhn, M., Herzig, V., Saez, N. J., Cristofori-Armstrong, B., Anangi, R., Senff, S., Gorse, D., & King, G. F. (2018). ArachnoServer 3.0: An online resource for automated discovery, analysis and annotation of spider toxins. *Bioinformatics (Oxford, England), 34*(6), 1074–1076. DOI: 10.1093/bioinformatics/btx661 PMID: 29069336

Pirtskhalava, M., Vishnepolsky, B., & Grigolava, M. (2013). Transmembrane and Antimicrobial Peptides. Hydrophobicity, Amphiphilicity and Propensity to Aggregation. *Nucleic Acids Research, 49*(D1), D288–D297. https://arxiv.org/abs/1307.6160 PMID: 33151284

Pun, F. W., Ozerov, I. V., & Zhavoronkov, A. (2023). AI-powered therapeutic target discovery. *Trends in Pharmacological Sciences, 44*(9), 561–572. DOI: 10.1016/j.tips.2023.06.010 PMID: 37479540

Reynisson, B., Alvarez, B., Paul, S., Peters, B., & Nielsen, M. (2021). NetMHCpan-4.1 and NetMHCIIpan-4.0: Improved predictions of MHC antigen presentation by concurrent motif deconvolution and integration of MS MHC eluted ligand data. *Nucleic Acids Research, 48*(W1), W449–W454. DOI: 10.1093/NAR/GKAA379 PMID: 32406916

Ruiz Puentes, P., Henao, M. C., Cifuentes, J., Muñoz-Camargo, C., Reyes, L. H., Cruz, J. C., & Arbeláez, P. (2022). Rational Discovery of Antimicrobial Peptides by Means of Artificial Intelligence. *Membranes, 12*(7), 708. DOI: 10.3390/MEMBRANES12070708/S1 PMID: 35877911

Rydel, T. J., Ravichandran, K. G., Tulinsky, A., Bode, W., Huber, R., Roitsch, C., & Fenton, J. W. (1990). The structure of a complex of recombinant hirudin and human α-thrombin. *Science, 249*(4966), 277–280. DOI: 10.1126/science.2374926 PMID: 2374926

Shah, S. S., Casanova, N., Antuono, G., & Sabatino, D. (2020). Polyamide Backbone Modified Cell Targeting and Penetrating Peptides in Cancer Detection and Treatment. *Frontiers in Chemistry, 8*, 526066. DOI: 10.3389/FCHEM.2020.00218/XML PMID: 32296681

Sharma, N., Naorem, L. D., Jain, S., & Raghava, G. P. S. (2022). ToxinPred2: An improved method for predicting toxicity of proteins. *Briefings in Bioinformatics, 23*(5), 1–12. DOI: 10.1093/bib/bbac174 PMID: 35595541

Sharma, R., Shrivastava, S., Kumar Singh, S., Kumar, A., Saxena, S., & Kumar Singh, R. (2021). AniAMPpred: Artificial intelligence guided discovery of novel antimicrobial peptides in animal kingdom. *Briefings in Bioinformatics*, *22*(6), 1–23. DOI: 10.1093/BIB/BBAB242 PMID: 34259329

Sharma, R., Shrivastava, S., Singh, S. K., Kumar, A., Saxena, S., & Singh, R. K. (2022). Deep-AFPpred: Identifying novel antifungal peptides using pretrained embeddings from seq2vec with 1DCNN-BiLSTM. *Briefings in Bioinformatics*, *23*(1), 1–16. DOI: 10.1093/BIB/BBAB422 PMID: 34670278

Shi, G., Kang, X., Dong, F., Liu, Y., Zhu, N., Hu, Y., Xu, H., Lao, X., & Zheng, H. (2022). DRAMP 3.0: An enhanced comprehensive data repository of antimicrobial peptides. *Nucleic Acids Research*, *50*(D1), D488–D496. DOI: 10.1093/nar/gkab651 PMID: 34390348

Singh, S., Chaudhary, K., Dhanda, S. K., Bhalla, S., Usmani, S. S., Gautam, A., Tuknait, A., Agrawal, P., Mathur, D., & Raghava, G. P. S. (2015). SATPdb: A database of structurally annotated therapeutic peptides. *Nucleic Acids Research*, *44*(D1), D1119–D1126. DOI: 10.1093/nar/gkv1114 PMID: 26527728

Sowers, A., Wang, G., Xing, M., & Li, B. (2023). Advances in Antimicrobial Peptide Discovery via Machine Learning and Delivery via Nanotechnology. *Microorganisms 2023, Vol. 11, Page 1129*, *11*(5), 1129. https://doi.org/DOI: 10.3390/MICROORGANISMS11051129

Strokach, A., & Kim, P. M. (2022). Deep generative modeling for protein design. *Current Opinion in Structural Biology*, *72*, 226–236. DOI: 10.1016/j.sbi.2021.11.008 PMID: 34963082

Szymczak, P., Zarzecki, W., Wang, J., Duan, Y., Wang, J., Coelho, L. P., de la Fuente-Nunez, C., & Szczurek, E. (2025). AI-Driven Antimicrobial Peptide Discovery: Mining and Generation. *Accounts of Chemical Research*, *58*(12), 1831–1846. DOI: 10.1021/acs.accounts.0c00594 PMID: 40459283

Tahrani, A. A., Bailey, C. J., Del Prato, S., & Barnett, A. H. (2011). Management of type 2 diabetes: New and future developments in treatment. *Lancet*, *378*(9786), 182–197. DOI: 10.1016/S0140-6736(11)60207-9 PMID: 21705062

Tan, X., Liu, Q., Fang, Y., Yang, S., Chen, F., Wang, J., Ouyang, D., Dong, J., & Zeng, W. (2024). Introducing enzymatic cleavage features and transfer learning realizes accurate peptide half-life prediction across species and organs. *Briefings in Bioinformatics*, *25*(4). Advance online publication. DOI: 10.1093/BIB/BBAE350 PMID: 39038937

Topcu, E., Ridgeway, N. H., & Biggar, K. K. (2022). PeSA 2.0: A software tool for peptide specificity analysis implementing positive and negative motifs and motif-based peptide scoring. *Computational Biology and Chemistry, 101*. Advance online publication. DOI: 10.1016/j.compbiolchem.2022.107753 PMID: 35998543

Tucs, A., Tran, D. P., Yumoto, A., Ito, Y., Uzawa, T., & Tsuda, K. (2020). Generating ampicillin-level antimicrobial peptides with activity-aware generative adversarial networks. *ACS Omega, 5*(36), 22847–22851. DOI: 10.1021/acsomega.0c02088 PMID: 32954133

Tyagi, A., Tuknait, A., Anand, P., Gupta, S., Sharma, M., Mathur, D., Joshi, A., Singh, S., Gautam, A., & Raghava, G. P. S. (2015). CancerPPD: A database of anticancer peptides and proteins. *Nucleic Acids Research, 43*(D1), D837–D843. DOI: 10.1093/nar/gku892 PMID: 25270878

Vamathevan, J., Clark, D., Czodrowski, P., Dunham, I., Ferran, E., Lee, G., Li, B., Madabhushi, A., Shah, P., Spitzer, M., & Zhao, S. (2019). Applications of machine learning in drug discovery and development. *Nature Reviews. Drug Discovery, 18*(6), 463–477. DOI: 10.1038/S41573-019-0024-5 PMID: 30976107

Vyas, M., Thakur, S., Riyaz, B., Bansal, K. K., Tomar, B., & Mishra, V. (2018). Artificial intelligence: The beginning of a new era in the pharmacy profession. *Asian Journal of Pharmaceutics, 12*(2), 72–76. DOI: 10.22377/AJP.V12I02.2317

Waghu, F. H., Gopi, L., Barai, R. S., Ramteke, P., Nizami, B., & Idicula-Thomas, S. (2014). CAMP: Collection of sequences and structures of antimicrobial peptides. *Nucleic Acids Research, 42*(D1), D1154–D1158. DOI: 10.1093/nar/gkt1157 PMID: 24265220

Wagner, A. M., Gran, M. P., & Peppas, N. A. (2018). Designing the new generation of intelligent biocompatible carriers for protein and peptide delivery. *Acta Pharmaceutica Sinica. B, 8*(2), 147–164. DOI: 10.1016/J.APSB.2018.01.013 PMID: 29719776

Wan, F., Torres, M. D. T., Peng, J., & de la Fuente-Nunez, C. (2024). Deep-learning-enabled antibiotic discovery through molecular de-extinction. *Nature Biomedical Engineering, 8*(7), 854–871. DOI: 10.1038/s41551-024-01201-x PMID: 38862735

Wan, F., Wong, F., Collins, J. J., & de la Fuente-Nunez, C. (2024). Machine learning for antimicrobial peptide identification and design. *Nature Reviews Bioengineering, 2*(5), 392–407. DOI: 10.1038/s44222-024-00152-x PMID: 39850516

Wang, G., Li, X., & Wang, Z. (2016). APD3: The antimicrobial peptide database as a tool for research and education. *Nucleic Acids Research, 44*(D1), D1087–D1093. DOI: 10.1093/nar/gkv1278 PMID: 26602694

Wang, J., Chang, K., De, R. O., Costa, A., Passos, T. S., Mikaelly, E., Silva, S., Santos, N. C. S. Dos, Heloneida, A., & Morais, A. (2023). Encapsulated Peptides and Proteins with an Effect on Satiety. *Nanomaterials 2023, Vol. 13, Page 1166, 13*(7), 1166. https://doi.org/DOI: 10.3390/NANO13071166

Wang, Y., Zhang, L., Liu, C., Luo, Y., & Chen, D. (2024). Peptide-Mediated Nanocarriers for Targeted Drug Delivery: Developments and Strategies. *Pharmaceutics, 16*(2), 240. DOI: 10.3390/PHARMACEUTICS16020240 PMID: 38399294

Wang, Y. M., & Fang, C. (2024). Cycle-ESM: Generation-assisted classification of antifungal peptides using ESM protein language model. *Computational Biology and Chemistry, 113*, 108240. DOI: 10.1016/j.compbiolchem.2024.108240 PMID: 39437594

Watson, J. L., Juergens, D., Bennett, N. R., Trippe, B. L., Yim, J., Eisenach, H. E., Ahern, W., Borst, A. J., Ragotte, R. J., Milles, L. F., Wicky, B. I. M., Hanikel, N., Pellock, S. J., Courbet, A., Sheffler, W., Wang, J., Venkatesh, P., Sappington, I., Torres, S. V., & Baker, D. (2023). De novo design of protein structure and function with RFdiffusion. *Nature, 620*(7976), 1089–1100. DOI: 10.1038/s41586-023-06415-8 PMID: 37433327

Wei, L., Ye, X., Sakurai, T., Mu, Z., & Wei, L. (2022). ToxIBTL: Prediction of peptide toxicity based on information bottleneck and transfer learning. *Bioinformatics (Oxford, England), 38*(6), 1514–1524. DOI: 10.1093/bioinformatics/btac006 PMID: 34999757

Wei, L., Ye, X., Xue, Y., Sakurai, T., & Wei, L. (2021). Atse: A peptide toxicity predictor by exploiting structural and evolutionary information based on graph neural network and attention mechanism. *Briefings in Bioinformatics, 22*(5), 1–13. DOI: 10.1093/bib/bbab041 PMID: 33822870

Winnifrith, A., Outeiral, C., & Hie, B. L. (2024). Generative artificial intelligence for de novo protein design. *Current Opinion in Structural Biology, 86*, 102794. DOI: 10.1016/j.sbi.2024.102794 PMID: 38663170

Wu, X., Lin, H., Bai, R., & Duan, H. (2024). Deep learning for advancing peptide drug development: Tools and methods in structure prediction and design. *European Journal of Medicinal Chemistry, 268*, 116262. DOI: 10.1016/J.EJMECH.2024.116262 PMID: 38387334

Xiao, X., Shao, Y. T., Cheng, X., & Stamatovic, B. (2021). iAMP-CA2L: A new CNN-BiLSTM-SVM classifier based on cellular automata image for identifying antimicrobial peptides and their functional types. *Briefings in Bioinformatics, 22*(6), 1–10. DOI: 10.1093/BIB/BBAB209 PMID: 34086856

Yan, J., Cai, J., Zhang, B., Wang, Y., Wong, D. F., & Siu, S. W. I. (2022). Recent Progress in the Discovery and Design of Antimicrobial Peptides Using Traditional Machine Learning and Deep Learning. *Antibiotics (Basel, Switzerland)*, *11*(10), 1451. DOI: 10.3390/antibiotics11101451 PMID: 36290108

Yao, J., & Wang, X. (2025). Artificial intelligence in de novo protein design. *Medicine in Novel Technology and Devices*, *26*, 100366. DOI: 10.1016/j.medntd.2025.100366

Yen, J. H., Chang, C. C., Wu, T. Y., Yang, C. H., Hsu, H. J., & Liou, J. W. (2025). Therapeutic peptides and their delivery using lipid-based nanoparticles. *Tzu-Chi Medical Journal*, *37*(3), 223. DOI: 10.4103/TCMJ.TCMJ_321_24 PMID: 40741609

Yuan, Q., Chen, K., Yu, Y., Le, N. Q. K., & Chua, M. C. H. (2023). Prediction of anticancer peptides based on an ensemble model of deep learning and machine learning using ordinal positional encoding. *Briefings in Bioinformatics*, *24*(1), 1–10. DOI: 10.1093/BIB/BBAC630 PMID: 36642410

Zhang, K., Yang, X., Wang, Y., Yu, Y., Huang, N., Li, G., Li, X., Wu, J. C., & Yang, S. (2025). Artificial intelligence in drug development. *Nature Medicine*, *31*(1), 45–59. DOI: 10.1038/S41591-024-03434-4;SUBJMETA PMID: 39833407

Zhou, S., Luo, Y., & Lovell, J. F. (2023). Vaccine approaches for antigen capture by liposomes. *Expert Review of Vaccines*, *22*(1), 1022. DOI: 10.1080/14760584.2023.2274479 PMID: 37878481

Chapter 12
Bloodstain Detection and Forensic Analysis Using Advanced Fluorescent AI-Assisted Nanoparticles

Akansha Rawat
Institute of Forensic Sciences and Cyber Security, Narayan Shastri Institute of Technology, India

AnantKumar Patil
https://orcid.org/0009-0002-9980-4668
Institute of Forensic Sciences and Cyber Security, Narayan Shastri Institute of Technology, India

Kiran R. Dodiya
https://orcid.org/0009-0001-9409-7303
Gujarat University, India

Parvesh Sharma
https://orcid.org/0000-0002-0245-5186
Department of Biochemistry and Forensic Science, Gujarat University, India

ABSTRACT

This chapter explores advanced bloodstain detection and analysis using fluorescent nanoparticles, enhancing forensic investigations through improved sensitivity and visualization. It compares traditional methods like luminol-based detection and chemiluminescence with modern fluorescence-enhanced techniques. Spectroscopic and imaging methods are examined, considering environmental impacts on bloodstain

DOI: 10.4018/979-8-3373-2883-6.ch012

integrity. Challenges such as contamination, degradation, and legal admissibility are discussed. Case studies illustrate real-world applications, and future trends in forensic nanotechnology highlight the importance of interdisciplinary research.

1. INTRODUCTION TO BLOODSTAIN DETECTION IN FORENSIC SCIENCE

Bloodstains, a kind of serological evidence, are correlated with crimes, including homicide, assault, child abuse, and crimes involving property, such as arson, robberies, burglaries, etc., as there are possibilities to find the presence of blood on weapons, on tools, or at the crime scene. Bloodstain evidence is significant in forensic examinations because it enables investigators to reconstruct crime scenes, connect perpetrators to sites, and validate or contradict witnesses' testimonies. Forensic experts study bloodstain patterns to determine what injuries were inflicted and how blood moved while learning about the positions of both the victim and perpetrator. Bloodstain examination allows forensic experts to use DNA analysis to identify suspects at crime scenes. Bloodstain patterns matching particular weapon types help investigators identify the weapon tool as evidence used to incriminate specific suspects. Proof from bloodstains aids courts in verifying whether defendants tell accurate accounts and confirms their culpability beyond a reasonable doubt.

Crime scene reconstruction benefits heavily from bloodstain evidence because it demonstrates high accuracy in detailing the processes of the crime event. The investigation gains clarity from bloodstain evidence because it reveals how incidents unfolded while showing what weapons caused the injuries, along with tracking suspect movement patterns. A well-rounded case against a suspected offender relies heavily on this essential data to ensure justice prevails. Bloodstain evidence is an essential component of forensic investigations because it allows investigators to reconstruct crime scenes and identify suspects. The tool has proven to be essential as it serves justice through its undeniable importance. (*Bloodstain Pattern Analysis: Introduction*, n.d.).

1.1. Historical Methods of Blood Detection

In the growth years of forensic science, the detection of blood primarily relied on basic observational techniques. However, basic visual assessments were the investigators' method of choice, a limited approach based on the visibility and distinctive color of fresh blood. However, the problem of detecting latent or aged bloodstains using this observational approach proved infeasible, limiting the use of forensic investigations. With the advent of forensic science, less systematic and more haphazard methodologies were now applied to blood detection. By the mid-19th century, forensic medicine had developed into the practice of combining chemical reaction techniques used to identify blood at crime scenes. This present chemical testing was a first step toward creating more modern, chemical-based detection techniques to follow (Short, 2015).

1.1.1 Review of Literature

In forensic science, bloodstain detection and analysis have been the focus of considerable research, and traditional visual and chemical approaches have been gradually replaced by powerful methods based on nanotechnology. The earliest forensic science investigation techniques included visual inspection and simple tests to determine if blood was present. Nevertheless, these approaches would only be of limited use against degraded or, more recently, latent bloodstains (Short, 2015). While presumptive blood tests such as the Kastle-Meyer and Luminol assays represented a substantial advancement in forensic blood detection. Although these tests were very sensitive, they also suffered from false positives attributable to reactions with peroxidase-rich matrices, metal ions and cleaning agents (Khan et al., 2014). Quickenden and Creamer (2001) demonstrated that luminol produces chemiluminescence in the absence of blood, which can interfere with forensic interpretations. With the evolution of forensic science, spectroscopic tools (UV-Vis, Raman, and FTIR spectroscopy) have become important for blood detection. These approaches enabled hemoglobin and its metabolites to be identified by their distinctive spectral fingerprints. However, they were cumbersome when applied to old or spoiled blood samples (Dybas et al., 2022). Some other microscopic techniques used to facilitate macro visualization of bloodstains are scanning electron microscopy (SEM) and fluorescence microscopy (Peschel et al., 2011). Nanotechnologies in forensic blood detection: revolutionary advances. In that regard, quantum dots (QDs), AuNPs, carbon-based nanoparticles, and doped silica nanoparticles have exhibited high potential for enhancing both detection sensitivity and specificity (Choi & Sawada, 2022). For example, quantum dots were investigated for their size-dependent emission properties, which led to high-spatial-resolution fluorescence imaging of bloodstains (Ekimov et al., 2023). AuNPs have also been investigated for their localized surface plasmon resonance (LSPR) properties, which allow for rapid colorimetric detection of blood for forensic purposes (Yeh et al., 2011). Mixing bloodstain pattern analysis with recent studies has also proved that machine learning (ML) and artificial intelligence (AI) have the potential to classify blood patterns. Artificial intelligence–based algorithms could handle extensive datasets, detect latent bloodstain patterns, and anticipate bloodstain trajectories, improving the reliability of forensic reconstructions (Dudek et al., 2024). The marriage of AI and fluorescence-based nanoparticle approaches has taken forensic processes to an even higher level by facilitating automated blood identification and genomic analysis in real-time (Ullah et al.,2024).

Although these breakthroughs are yielding positive results, they have not yet overcome problems of contamination, environmental degradation, and legal admissibility for forensic blood. Future research directions emphasized the miniaturization of tools for blood detection, the identification of biocompatible and non-toxic nanoparticles, and in-situ forensic analysis with portable sensing devices (Wang et al., 2017).

1.2. Advances in Forensic Blood Analysis

The advancement in the discipline of forensic blood analysis has been exceptionally great over the past few decades. The field began with basic visual inspection and chemical testing, emphasizing the development of high-tech devices, which are likely to be used for precise detection, identification, and interpretation of blood stains. The methods have been greatly improved in sensitivity and accuracy, enabling innovations to detect blood and extract critical information that would have been impractical to achieve without this nanotechnology. This development is one of the most pivotal in forensic blood detection. The use of nanoscale materials and devices, roughly hundreds of nanometers, is employed in this to increase surface area, reactivity, etc. Such materials are notable because they specifically interact with biological molecules of interest in forensic science, aiding in the development of sensitive, accurate detection methods. Fluorescent nanoparticles are among the most promising of the various tools developed in this field. However, under specific conditions, these nanoparticles will produce a fluorescent signal when attached to bloodstains. The sensitivity contrasts with traditional chemical analysis, which is reduced in sensitivity. Nanoparticle-based techniques increase this magnification, allowing bloodstains that otherwise would not be detected to be identified. They can also be customized in a specific way to the surface to optimize for the selectivity of human blood or some component of human blood, such as hemoglobin. Such specificity would be advantageous for applying these cells to identify blood at crime scenes, significantly enhancing their usefulness in modern forensic practice. (Nirupama, 2024).

2. CONVENTIONAL TECHNIQUES FOR BLOOD DETECTION

Forensic investigations have traditionally relied on established methods for blood detection. In this section, we will outline several of the most conventional techniques in the field:

2.1 Luminol and Chemiluminescence-Based Detection

Various organic compounds with fluorescent or chemiluminescent oxidation products perform essential roles in forensic testing techniques. Light emission arises directly from the unique chemical reactions in chemiluminescence assays. Under appropriate testing conditions, the compound known as luminol generates chemiluminescent signals that help identify blood samples. Fluorescence assays work by using specific wavelengths to excite fluorescein-type oxidation products that subsequently release light at wavelengths longer than the source wavelength. The main benefit of utilizing chemiluminescent and fluorescent reagents is their broad applicability to extensive surfaces for latent bloodstain detection. The detection reaction serves to verify blood presence while revealing the patterns from blood impressions, including prints and footmarks. These detection methods show outstanding sensitivity that allows users to detect microscopic blood residues. These methods play a crucial role in crime scene analysis because they help uncover traces of blood despite cleaning efforts that would otherwise obscure them. These reagents require careful use to detect both very tiny stains and clean blood debris still at a crime scene. The use of presumptive assay reagents during sample analysis dilutes the sample volume and creates difficulties in obtaining sufficient DNA quantities needed for forensic testing.(Khan et al., 2014).

Luminol (3-Aminophthalhydrazide)

Luminol acts principally as a chemiluminescent reagent with diverse applications. The heme moiety in blood catalyzes the oxidation of luminol, producing light in the presence of an oxidant. Luminol has the intrinsic drawback that the light it produces can only be observed under dark conditions, and it must be photographed immediately, as it fades quickly. Also, it gave false positive results when reacted with bleach or copper.(*Luminol | Definition, Characteristics, Chemiluminescence, Blood Detection, & Forensics | Britannica*, n.d.).

Figure 1. Chemical reaction during the Luminol test for blood stain resulting in Chemiluminescence.

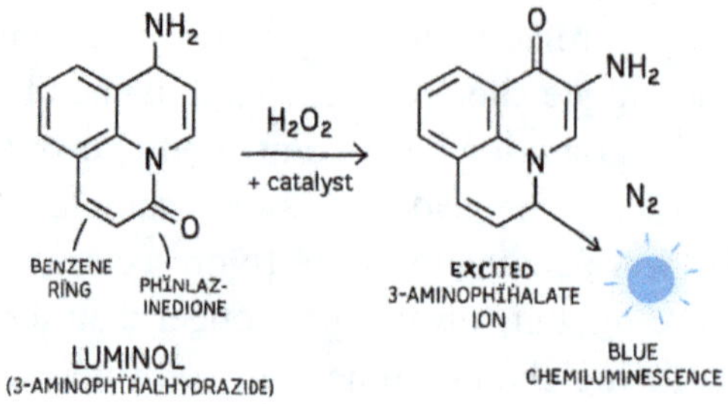

1. Fluorescein

Another reagent used at a crime scene for testing the presence of bloodstain is Fluorescein. When oxidized and catalyzed by heme, fluorescein exemplifies fluorescent properties. Most often, fluorescin-sprayed stains are disclosed to light in the range of 425– 485 nm using an alternate light source device. The oxidized fluorescein emits an intense yellowish-green fluorescent light, which indicates a positive reaction and the presence of a bloodstain. The light emitted from fluorescein-sprayed stains shows greater effectiveness than that of luminol.(*The Fluorescein Method of Latent Blood Detection*, n.d.).

2.2 Hemoglobin and Peroxidase-Based Tests

Presumptive blood tests screen for the presence of blood. Hemoglobin includes a heme component. This component eases the oxidation-reduction reactions necessary for many of these tests. Heme oxidizes colorless substrates, producing chemiluminescence, fluorescence, or a color change. These tests possess a high degree of sensitivity, and they are capable of detecting blood in samples diluted up to 10–5 times. A strongly positive result may strongly suggest the presence of blood. In addition, most of these tests are compatible with forensic DNA analysis. Several colorimetric methods can detect blood heme. Phenolphthalein, leucomalachite green, and benzidine derivatives are commonly used. These tests produce immediately visible colorimetric changes, and they require no special equipment. (Kapralov et al., 2009).

1. Phenolphthalein Assay

Phenolphthalein, a common indicator dye, detects both mineral and organic acids and bases. It is used in multiple applications and in several Kastle-Meyer tests for blood identification. Kastle's revolutionary 1901 study demonstrated the significant reaction between the colorless compound phenolphthalein and heme, wherein hydrogen peroxide served as the key oxidant. Phenolphthalein is formed in this reaction, and this indicator turns pinkish-red in alkaline solutions.(Morris & Powell, 1979).

2. Leucomalachite Green (LMG) Assay

Malachite green is a triphenylmethane dye. Malachite green forms a leuco base, which is colorless and can be oxidized to produce a green color by the catalysis of heme. The reaction with hydrogen peroxide as the oxidant is carried out under acidic conditions(Afkhami et al., 2010).

3. Benzidine and Derivatives

Benzidine functioned as a manufacturing intermediate in dye creation before being applied as a blood detection agent in the past. Scientists utilized benzidine first for blood detection during the early stages of its development. Observations showed that benzidine oxidation by heme leads to solution acidification, causing a blue color change to dark blue. Reading the reaction sample without delay became essential because the blue color would later develop into brown. Benzidine has been eliminated for forensic use because it functions as a carcinogenic agent. The heme-catalyzed oxidation reaction of orthotolidine, a dimethyl benzidine derivative, results in blue-colored solutions under acidic conditions. When acid provides the reaction environment, the color changes to blue. Animal studies indicate that orthotolidine has potential carcinogenic properties, although it continues to be utilized as a testing material. The handling of tetramethylbenzidine took over after orthotolidine lost its place. Tetramethylbenzidine serves as a derivative of benzidine. Legionella pneumophila causes heme to activate an oxidation reaction, which leads to the formation of green to blue-green colors in acidic solutions. TMB serves as a testing compound used in current applications. Miles Laboratories produces the Hemastix® assay kit, which consists of TMB-based materials inside its testing strips. Applying a moistened sample to the Hemastix® strip enables the test performance. Blood detection results in the strip showing either a green or blue-green color change during the analysis process.(Ahlström et al., 2005).

2.3 Spectroscopic and Microscopic Analysis

The detection of blood relies heavily on spectroscopic techniques, which use light-matter interactions to identify the specific spectral patterns of hemoglobin. The evaluation of oxyhemoglobin absorption peaks is performed using UV-Visible spectroscopy, which identifies bands at 415 nm, 540 nm, and 576 nm. UV-visible spectroscopy is a valuable tool for blood detection because it can identify fresh blood, but its performance deteriorates when testing samples or materials with pigments like melanin. Raman spectroscopy stands out for its high specificity because it uses light scattering analysis to identify molecular vibrations of hemoglobin at 1375 cm^{-1} and 1580 cm^{-1}. Raman spectroscopy brings significant benefits to the analysis of dried blood specimens because its mechanism allows for use with hand-held portable devices. The future implementation of Raman spectroscopy and its equipment requires evaluating two key obstacles: fluorescence interference and high equipment costs. The aged stain detection capabilities of infrared spectroscopy surpass others through its ability to detect protein-related signals at 1650 cm^{-1}. Fluorescence spectroscopy, along with alternative light sources, enables porphyrin excitation for visualizing latent blood stains. However, laboratory personnel should note the potential impact of background fluorescence on their analysis. The luminescent property of luminol, when reacting with heme, produces 425 nm light, enhancing spectroscopic examinations through fast field testing that requires laboratory confirmation for specific and sensitive results. A microscope-based evaluation serves two important cellular functions: verifying the presence of blood and distinguishing it from possible mistaken substances. Light microscopy demonstrates red blood cells, which present a biconcave shape with no nucleus (7-8 μm in length), and Wright's and Giemsa stains distinguish human blood cells from other animal blood cells. A blood sample from birds will contain nucleated cells, thus indicating that the blood comes from an animal species different from humans. Forensic examination of damaged blood samples and blood affected by hemolysis poses difficulties when researchers need to understand their investigative significance. Phase-contrast microscopy provides enhanced observations of unstained red blood cells by utilizing refractive index variations, improving medical diagnostic imaging of living cells.

The complex nature of scanning electron microscopy restricts its daily use as a research tool in forensic examinations, though its ability to generate three-dimensional blood cell representations serves as a beneficial investigative method. The detection of blood signatures in various samples becomes more effective using microscopes and acridine orange, a fluorescent dye that binds with blood elements in stained tissues or slides. Microscope examination confirms the results obtained from the Takayama test, which generates visible pink heme crystals. Laboratory methods based on spectral information receive essential cellular evidence from microscopic techniques, which help resolve issues of sample degradation and contamination. (Dybas et al., 2022).

3. ROLE OF FLUORESCENT NANOPARTICLES IN BLOODSTAIN DETECTION

3.1 Mechanism of Fluorescence in Blood Detection

Substances undergo fluorescence by absorbing specific wavelengths of light, which results in the emission of longer-wavelength light. This characteristic is particularly significant in blood detection, as it facilitates the identification of various blood components through specific fluorescent staining techniques using appropriate reagents. Scientists have developed fluorescent nanoparticles that specifically recognize bloodstain proteins and cellular components in blood analysis because of heightened research interest in this area. Fluorescent nanoparticles produce a unique emission signal after light source illumination, and this light emission can be detected through various imaging techniques. Fluorescence intensity strongly increases when biological molecules are present, thus allowing scientists to distinguish blood from additional biological fluids. Forensic investigation greatly benefits from the ability to detect blood presence, as this verification is essential for many cases.(Choi & Sawada, 2022).

3.2 Types of Nanoparticles used in Forensic Blood Analysis

The research surrounding fluorescent nanoparticles is extensive, encompassing various types employed for the detection and analysis of bloodstains:

3.2.1. Quantum dots:

In the 1980s, Alexei Ekimov, Louis E. Brus, and Moungi Bawendi synthesized a semiconductor nanostructure known as *Quantum dots* and were awarded the Nobel Prize for their contribution. QDs have a particle size of less than 100nm with the capability of reaching volumetric and molecular levels. This offers a new framework that is significant in biological research, the biomedical and pharmaceutical industries, and healthcare. This article exhibits distinct characteristics and features:

1. A narrow emission peak
2. Emission wavelength that is size-dependent
3. A broad excitation range

4. High photoluminescent efficiency
5. Enhanced photostability

Figure 2. "Fluorescence Emission of QDs vs. QDs Attached to RBCs"

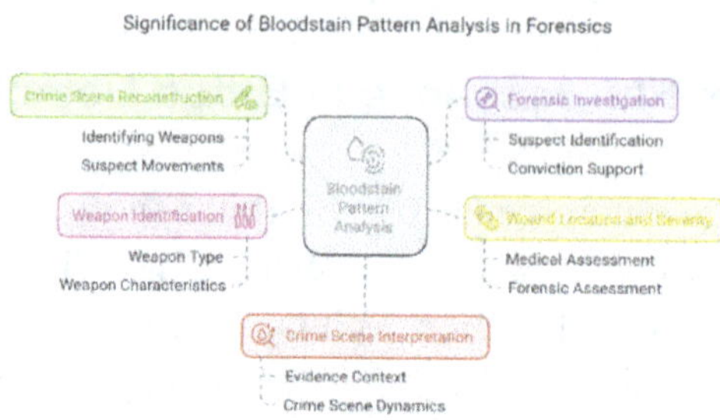

Quantum dots absorb light through a broad spectrum, enabling users to excite sources from multiple wavelengths. The size of quantum dots controls the energy of emitted photons, allowing users to obtain specific, narrow wavelength spectra. The band gap for blue-emitting QDs becomes greater because of their small size, while big QDs that produce red light possess a smaller band gap. The adjustable optical characteristics serve to demonstrate the multiple uses QDs have across various fields. ZnS, ZnSe, CdSe, CdS, CdTe, PbSe, and PbS are the different types of QDs. These CdSe/ ZnS core/shell QDs are frequently used in applications of biological importance. QDs emit wavelength ranges from ultraviolet to the infrared.

The application of immunological assays benefits from QD antibodies that recognize blood elements, such as hemoglobin or blood type antigens, resulting in improved visualization capabilities during detection processes. The technique enhances bloodstain identification sensitivity and offers potential resolutions for mixed body fluid mixtures. QD-based sensors present a fast and sensitive detection solution for identifying illicit drugs and toxins in blood samples, whereas traditional testing methods lack such advantages. QD-based sensors have advanced bloodstain pattern imaging through fluorescent capabilities that produce detailed, high-resolution analyses to detect even small blood debris quantities. Research should focus on solving QD toxicity issues and achieving long-term stability for their safe application as an important forensic tool.(*The Nobel Prize in Chemistry 2023 - Popular Information - NobelPrize.Org*, n.d.).

3.2.2. Gold Nanoparticles (AuNPs):

AuNPs stand as one of the stable existing nanoparticle formats, which possess interesting characteristics developed by scientists. AuNPs started in the 4th or 5th century B.C. in both Egyptian and Chinese territories for decorative and medical therapeutic practices. Since then, various procedures have been formulated to synthesize AuNPs with diverse size and form features, and researchers have thoroughly investigated their optical behaviors. With the development of gold nanotechnology, AuNPs play important roles in a wide range of fields, from catalysis and energy to biology and medicine.

Small nanoparticles measuring between 1 and 100 nanometers activate localized surface Plasmon resonance (LSPR), which absorbs and scatters light with detectable results. The application of specific biomolecules, such as antibodies or DNA probes, to gold nanoparticles for blood detection purposes enables selective binding with target blood analytes, including proteins, pathogens, or biomarkers. The diagnostic feature of gold nanoparticles exists in their use as integrated elements of lateral flow assays, which produce visible red LSPR signals to represent target biomolecules. Gold nanoparticles enable reliable detection of different conditions due to their unique properties, resulting in fast and highly sensitive analysis. The broad surface area to volume ratio of these nanoparticles improves their sensitivity, which determines their effectiveness for techniques including surface-enhanced Raman spectroscopy (SERS). Surface-enhanced Raman spectroscopy helps detect extremely small quantities of specific blood substances, such as glucose, along with circulating tumor cells.

Scientists have evaluated how these nanoparticles function in colorimetric tests, which measure blood component detection by observing nanoparticle aggregation states and color changes. Point-of-care diagnostics will substantially advance due to gold nanoparticles, which offer both biocompatibility and versatile applications for blood-based disease detection procedures. Additionally, it has been found that using Au-NPs in combination with nanotechnology can considerably increase the effectiveness of the polymerase chain reaction (PCR). The efficiency of the PCR reaction was significantly improved with the addition of just 0.7 nm of 13 nm Au-NPs. This was demonstrated by a shorter reaction time and a faster heating and cooling rate in the thermal cycler. Au-NPs have also improved electrochemiluminescence (ECL) when combined with the polymerase and nicking endonuclease isothermal procedure. As a result, compared to traditional and real-time PCR procedures, using Au-NPs improves sensitivity and specificity.(Yeh et al., 2011).

3.2.3. Carbon-based Nanoparticles:

Carbon dots (CDs) present an ascending group of carbon-based nanomaterials that draw increasing interest because of their exceptional photoluminescent behavior. It is possible to produce CDs from easily obtainable, inexpensive resources using methods that generate high quantum yields and adjustable emission spectra, while also offering low toxicity, small size, and notable biocompatibility. The unique features of CDs make them useful in multiple scientific areas, where they serve biomedicine and catalysis, help build optoelectronic devices, and fight counterfeiting. The intrinsic property of fluorescence in CDs qualifies them as quasi-zero-dimensional nanomaterials with dimensions measuring less than 20 nm. Hemoglobin detection takes place using carbon dots (C-dots) through fluorescence resonance energy transfer (FRET). This detection method functions as:

C-based dots emit their light spectrum from 400 to 550 nm, thereby overlapping with the absorption wavelengths of hemoglobin between 400 and 600 nm. The area of overlapping spectra between C-dots and hemoglobin enables FRET to occur between these compounds.

An increase in the level of hemoglobin directly affects the C-dot's photoluminescence intensity by producing a reduction in PL output.

It detects hemoglobin. The specific method of interaction involves C-dots utilizing hydrophobic bonds to anchor to hemoglobin molecules, which positions both groups within a better range for FRET operations. A photoinduced electron transfer (PET) process between C-dots and hemoglobin promotes the selectivity outcome of the entire assay.

C-dots prove to be highly sensitive and selective photoluminescent probes for detecting hemoglobin through their new detection approach.(Liu et al., 2020).

3.2.4. Doped silica nanoparticles:

The silica-based nanoparticles can be doped with several fluorescent dyes or metallic elements to improve their spectral properties. Because of their inherently porous structure, they are capable of functionalization with diverse biomolecules, such as polyelectrolytes or other functional biopolymers, that enhance affinity for specific blood proteins such as hemoglobin.

Together, the variety of fluorescent nanoparticles in this diverse landscape allows researchers to build on this methodology in bloodstain detection and analysis, moving into the future.(Gubala et al., 2020).

3.3 Surface functionalization for enhanced specificity

Surface functionalization plays a pivotal role in the application of nanoparticles for bloodstain detection, as it critically affects the interaction between nanoparticles and the various blood components. By modifying the surfaces of the nanoparticles, researchers can significantly enhance their ability to specifically adhere to unwanted or desired biomolecules, such as blood proteins and cells. Like in the case of bioconjugation, biological molecules (antibodies, peptides, and nucleic acids) are attached to the nanoparticle surface. Nanoparticles, which are bioconjugated to specific antigens found in the blood or cellular parts like erythrocytes, are obtained, and they possess selective binding properties. Such examples are pertinent, particularly the use of antibodies against human hemoglobin that can be coupled with the specific confirmation of blood in a sample. In addition, polymer coatings on the nanoparticles can stabilize the nanoparticles and enhance their biocompatibility, and also provide functional groups for enhancing the interaction between target entities and binding interactions. This characteristic is crucial in forensic applications, as it allows for removing nonspecific binding and enhancing the precision of blood detection methods.(Mout et al., 2012).

In addition, the magnetic functionalization of nanoparticles facilitates the easy separation of blood components from complicated matrices via inertial or magnetic fields. This approach can partially be used to create a dual-mode detection system in conjunction with fluorescence techniques to make more accurate forensic studies.

4. COMPARATIVE ANALYSIS OF DETECTION METHODS

4.1 Sensitivity and specificity comparison

The performance evaluation of blood detection methods in medical diagnostics, forensic science and point-of-care testing depends extensively on sensitivity and specificity testing methods. A method demonstrates sensitivity when it properly detects blood presence, while specificity indicates its capacity to identify that no blood exists correctly.

Luminol and Kastle-Meyer tests, alongside ELISA and Raman or infrared spectroscopy, along with PCR-based DNA analysis, represent the prevalent tests used for blood detection. The sensitivity of chemical tests such as luminol remains exceptional because they detect even tiny blood residue in weak solutions, yet their identification specificity is limited. The technology encounters restrictiveness due to its reaction capability with bleach or metallic ions, thus creating the possibility of erroneous test results. Blood protein-specific immunological tests obtain better precision through their ability to detect special markers found in blood products such as haemoglobin and glycophorin A. The detection performance of these methods declines when they test degraded specimens or encounter low amounts of target compounds. Spectroscopic analysis methods provide simultaneous sensitivity to characteristic blood signatures through their moderate ability to differentiate samples. Their operation faces risks because they become susceptible to interference from similar compounds that can exist within complex matrices. The exact DNA identification requirements of molecular detection methods demand sophisticated equipment, which reduces the capability to detect non-nucleated red blood cells. Ultimately, the choice of detection method hinges on the specific context of the application. The screening process using chemical tests functions efficiently, yet forensic investigations benefit most from molecular techniques, while healthcare facilities utilise immunological and spectroscopic approaches for their tests. Implementation of analysis methods requires evaluating their detection quality limitations and operation-related drawbacks.(Shreffler & Huecker, 2023).

4.2 Effectiveness of different surfaces and conditions

The detection methods for blood using luminol, fluorescein, and forensic light sources produce distinct results based on surface characteristics and environmental conditions, as well as substances which may interfere with their operation. The chemiluminescent agent called Luminol demonstrates high sensitivity in identifying blood residues on fabrics and wood surfaces because of its widespread adoption in forensic investigations. The interaction between this compound and haemoglobin iron leads to blue light emission. Luminol remains somewhat ineffective when researchers use it on non-porous items such as metals and glass, and specifically after cleaning efforts related to blood evidence removal. The detection method produces false results with luminol when it reacts with substances such as bleach or copper.

As an alternative to luminol detection systems, fluorescence utilises Fluorescein as a chemiluminescent substance that requires a fluorescent light source to reveal stains on dark and multicoloured backgrounds. The substance shows reduced performance in two conditions: when room illumination is intense and when blood samples contain dilution. Smooth non-porous surfaces become visible to bloodstain detection when using forensic light sources which contain both alternate light sources (ALS) and ultraviolet (UV) lights. These detection tools successfully show evidence that cannot be seen without the help of devices. The detection capabilities of these tools diminish after they interact with textured surfaces as well as absorbent materials, since the blood spreads through the structures, which impedes the scattering of light.

Three main environmental factors, along with the time elapsed since the blood dried, the surface temperature, and the humidity level, add additional challenges to detecting blood stains. Seasonal temperature changes can quicken the evaporation of fluids, and at the same time, humidity levels affect blood protein stability, leading to potential damage to chemical identification processes. Each detection technique faces significant hurdles when blood evidence encounters cleaning agents or soil because these substances obscure the blood evidence on all tested surface materials. (Senthilkumar et al., 2024).

4.3 Advantages and limitations of nanoparticle-based methods

Figure 3. Pros & Cons of nanoparticle-based methods

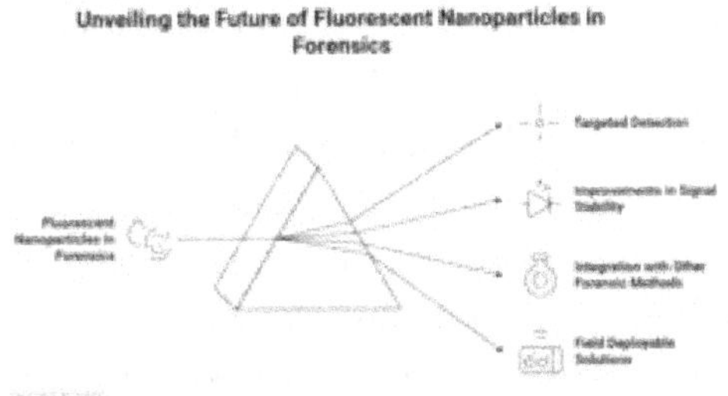

5. FORENSIC INTERPRETATION OF BLOODSTAIN PATTERNS

BPA exists as a fundamental forensic science method used to identify characteristics of blood stains at crime scenes that reveal important details about violent encounters. Through this analytical procedure, investigators gain vital information about blood expulsion patterns, which depend on victim-attacker positioning, weapon type and crime scene dynamics. Forensic experts who understand BPAs can develop complete criminal timelines to create stronger evidence for legal investigations.

5.1 Bloodstain Pattern Analysis Techniques

Bloodstain Pattern Analysis (BPA) is a scientific discipline that examines bloodstains at crime scenes to reconstruct events. BPA conducts investigative analysis of blood stains through advanced procedures that enable experts to identify different pattern types. Bloodstain analysis methodologies follow a structured classification system that derives from different appearances encountered in blood marks. The most dominant patterns in bloodstain analysis consist of:

Passive Stains: Gravitational forces produce these stains when blood droplets drop by themselves toward the earth. Analysts studying passive blood stains observe their dimensions in addition to their geometric profile and their spatial orientation to understand both the blood release height and the ground surface it hit.

Transfer Stains: Between contact and surface, any item with blood content transfer stains reveals both perpetrator and victim movements. The features of shape and smudged areas within these stains provide valuable information about what type of contact occurred.

Projective Stains: Stains occur when blood projects forcefully through the air, which happens mainly during shooting or stabbing events. These stain examination techniques help identify the blood flight direction and speed, together with its source distance. Experts need to carefully study multiple bloodstain aspects, such as their sizes and positions, and their forms, when conducting their analysis. To construct precise models for blood spatter, the investigator must possess a deep understanding of how physical laws shape the formation patterns of distinct stain shapes.

Advances in BPA have significantly improved the accuracy and reliability of crime scene reconstruction.(Senthilkumar et al., 2024).

Figure 4. Significance of Bloodstain Pattern Analysis in Forensics

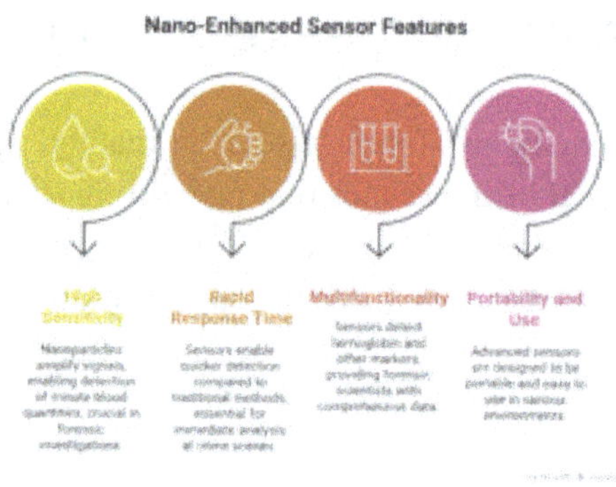

5.2 Impact of environmental factors on blood evidence

Different environmental factors at a crime scene will significantly affect how accurately bloodstain pattern analysis (BPA) functions. A variety of factors influence the appearance as well as the characteristics of bloodstains:

Surface texture: The following characteristics from blood marks develop directly from the surface material upon which blood is placed. Smooth surfaces enable blood to create round droplet contours as a result. Textured surfaces, along with porous surfaces, tend to modify blood through absorption, which produces imprecise patterns that cause difficulties in analysis.

Temperature and Humidity: The precise forensic time analysis depends heavily on temperature changes because they determine bloodstain drying durations. The spread of blood harmfully impacts interpretation and final evidence analysis due to elevated humidity levels in the environment.

Contamination: Blood evidence collects contamination as it interacts with foreign materials or bodily fluids through contact with substances such as dirt, as well as water. The analysis process becomes harder to interpret because of contamination; therefore, forensic experts need to properly identify and separate blood evidence from other substances for a valid conclusion.

Lighting Conditions: Selected lighting techniques help detectives see blood stains under dim light better by improving their visibility. Fluorescent lighting coupled with alternative light sources helps scientists view blood patterns, which standard lighting conditions would otherwise hide, thereby strengthening both the depth and precision of the examination process.(Peschel et al., 2011).

5.3 Digital Imaging and 3D Reconstruction in Bloodstain Analysis

Through digital imaging and the application of three-dimensional reconstruction, the improvement of technology has instituted remarkable efficacy in Bloodstain Pattern Analysis (BPA). Such advances allow investigators to produce elaborate visual images of bloodstain patterns, which have proved instrumental in elucidating crime-related events. These skills would in any way help in forensic analysis, but are also very important because they tend to restrict the probability of working on certain investigative outcomes.

Bloodstain Pattern Recognition: High-quality photo documentation is extremely important for recording fine details of bloodstains so that forensic scientists can document and analyse patterns most precisely. Images can be enhanced using various techniques for visibility, contrast adjustment, and alteration of perspective, to allow for better comprehension of the attributes of the stains.

3D Reconstruction: Forensic scientists will thus rely on using some specialised software packages to create a three-dimensional model of the crime scene, depicting both the spatial distribution of bloodstains and the layout of the parties involved. This reconstructive process allows for the modelling of various situations, which gives investigators the chance to see dynamics from different angles and points of view.

Integration with Other Forensic Data: Bloodstain pattern analysis integrates with further forensic evidence like DNA profiling, ballistic data, and witness testimony to create a more holistic narrative of the events in question. This multidimensional approach greatly enhances the validity of the findings of bloodstain pattern analysis (BPA) and thus assists in more comprehensively investigating the matter. Such methodologies employ forensic experts to furnish compelling evidence in the courts of law, aiming to read into occurrences of a violent crime. Bloodstain Pattern Analysis (BPA) includes in-depth examination and competent demonstration of bloodstain patterns that can influence court proceedings significantly, either in support of or opposition to claims asserted by the parties concerned.

Indeed, the investigation of bloodstain patterns presents an example of advancement both in forensic science and the determining key aspect of pursuing justice. Being knowledgeable in all the details of BPA and aware of the role played by environmental factors and technological changes allows the forensic analyst to trace back to events that were lost in time, embarking upon a quest for truth in criminal inquiries.(Buck et al., 2011).

6. CHALLENGES IN BLOODSTAIN DETECTION AND ANALYSIS

Forensic investigations involving blood traces and analysis are important and difficult components; however, they are confronted with many difficulties that bring on complications during the analytic process. However, while forensic technology continues to improve, the practitioners face challenges: contamination, degradation, the presence of degrading or interfering substances and sometimes variable environmental conditions [which can damage bloodstains]. In addition, legal and ethical considerations play a significant role in determining whether blood evidence should be allowed to be used as evidence in a court proceeding. As such, an understanding of these challenges in their entirety is necessary to improve the accuracy and utility of forensic blood analysis to provide reliable results in the judicial context.

6.1 Contamination and degradation of blood evidence

Forensic blood analysis is, however, daunting because bloodstains are invariably contaminated and degrade with time. Blood is particularly susceptible to the destructive influences of many of the environmental factors with which it comes in contact as a biological specimen, namely temperature, humidity, microbial activity, and air exposure. These factors can damage blood so much that the loss of vital evidence (could be) necessary for a criminal investigation can occur. The main problem that forensic analysts face is contamination: One type of contamination occurs when, for example, dirt, chemicals or other biological materials from other sources mix with the bloodstains, and therefore the evidence is compromised. After several samples are inevitably collected from the same crime scene, the cross-contamination phenomenon becomes problematic. The misinterpretations or erroneous results of bloodstain evidence can occur even with minute quantities of contaminating material. For instance, in the case of techniques such as luminol or chemiluminescence, false positives due to contamination from other substances (including garden/herbicide sprays, organic material or animal blood) may lead to conclusions of blood where none exists at all. The introduction of contaminants in the context of DNA analysis can introduce errors in results and therefore complicate investigations if the source of the blood cannot be identified.

Degradation: Another important issue in forensic science is the problem of blood degradation. With time, bloodstains undergo biochemical changes as a result of microbial action, environmental exposure and inherent enzymatic processes. Therefore, these changes could result in the decomposition of important molecules such as haemoglobin, DNA and proteins. The problems that begin with blood degradation raise detection and analysis challenges, which can then compromise the sensitivity and accuracy of some blood detection techniques. In some cases, rapid growth of microorganisms (bacteria and fungi) can occur on bloodstains when exposed to heat, moisture or direct sunlight. The degradation of this process makes the chances of extracting viable DNA or any biomarkers from the bloodstain less likely. Moreover, once blood dries, it can become a crust that prevents further analysis of collected contaminants. Forensic scientists need to assess the elapsed time since the deposition of the bloodstain to ascertain whether degradation has occurred and to evaluate its potential impact on the reliability of their results.(Hara et al., 2016).

6.2 False positives and interfering substances

The main problem in bloodstain identification is false positives, which are when a test used for detection erroneously determines that a substance tested is blood. Especially in the case of using chemical reagents like luminol or other chemiluminescent agents, the sensitivity of this issue is exacerbated because not only will they give a positive result on blood, but they may also react with many other substances. Notable sources of interference include:

Figure 5. "Factors Influencing the Accuracy of Blood Detection Tests"

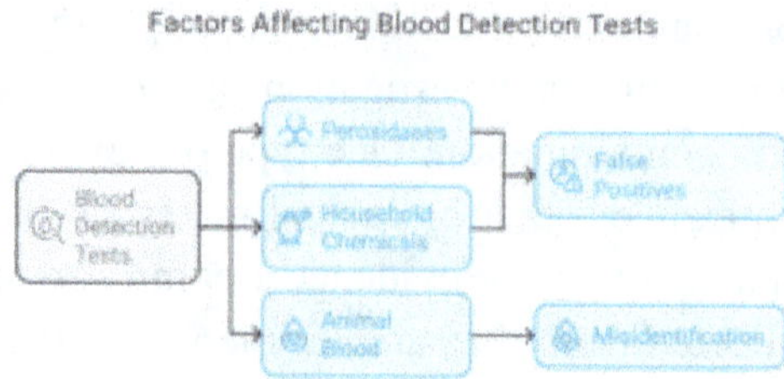

1. **Peroxidases**: There are a lot of enzymes present in a variety of biological materials like plant matter, saliva and some particular types of hair that would prompt false positives during tests meant for the detection of blood. The peroxidase activity deceives results and may provide such misidentifications.
2. **Household Chemicals:** Common household substances, including bleach, iodine, and certain detergents, can disrupt blood detection tests. For example, bleach can modify the chemical makeup of blood, consequently hindering accurate DNA extraction. Additionally, iodine may induce false positives in specific chemical assays for blood.
3. **Animal Blood:** It is because animal and human blood are structurally and chemically similar, and thus forensic blood detection tests sometimes give 'animal blood' readings that are wrong. Forensic scientists must differentiate quite carefully between human and animal blood, especially in those cases where the blood found at a crime scene can have a source from an animal.

For example, virtually all forensic scientists apply extremely specific antibodies (or molecular markers) to only blood components that are found only in human blood, such as human haemoglobin, human DNA or a specific human tissue. To deal with this menace, another package of terms comes into focus: specificity and sensitivity in detection methodologies.(Quickenden & Creamer, 2001).

6.3 Legal and Ethical Considerations in Forensic Blood Analysis

Forensic investigations have blood evidence as one of the most compelling types of evidence, with blood evidence standing out as one of the most important types of evidence available for law enforcement. Therefore, it is important that the processes in which blood evidence is collected, analysed, and presented in court follow both legal and ethical standards to ensure that it is reliable and accepted in court.

Legal Standards for Blood Evidence

Forensic blood analysis has to be tested to certain standards in a courtroom, for this is considered to be valid. Rules of evidence that pertain to blood sampling and analysis vary across different legal systems, and they also dictate the scientific standard to be followed while blood sampling and analysis are being carried out. Exclusion of this valuable evidence from legal proceedings could result from any deviation from standard protocols in the collection, handling, storage or analysis.

For example, the integrity of blood samples depends on their requiring proper preservation and appropriate storage. Particularly worrying when concerns like contamination or degradation are involved is DNA evidence. When the collection process engages unwholesome methods such as cross-contamination or improper labelling of the evidence, the defence has legs to place a challenge on the admissibility of the evidence, which could affect the overall credibility of the investigation. In addition, the chain of custody in forensic blood analysis is also important. That means documenting not just where the bloodstain is collected at the crime scene, but when it is analysed in the lab. Any break in this chain could question the authenticity as well as the integrity of the blood sample and thus could be disqualified from the trial.(*Collection and Preservation of Blood Evidence from Crime Scenes*, n.d.).

Ethical Implications

The examination of blood evidence demands careful consideration of ethical aspects that involve the administrator's privacy rights and individual consent protection. The analysis of blood evidence requires strict respect for individual rights during every phase of collection and examination as part of ethical guidelines. Experts must acquire permission before drawing blood from suspects and victims as well as witnesses because getting their consent is essential. DNA testing specifically demands legal authorisation through warrants, which govern the procedures for obtaining biological samples. Empty legal authorisation or consent makes it possible for individual rights to be violated, thereby endangering possible court-admissible evidence. When dealing with sensitive data obtained from blood samples, forensic scientists must exercise responsible handling of genetic profiles. Any unauthorised use or improper distribution of collected forensic data will have serious ethical as well as legal consequences. (Risks et al., 1994).

7. EMERGING TRENDS AND FUTURE DIRECTION

7.1 Nano-Enhanced Sensors for Blood Detection

Nano-enhanced sensors carry important advancements for forensic science because they transform conventional blood detection methods. The sensors integrate nanoparticles between 1 and 100 nanometers in size because of their unique characteristics. The enlarged surface area, together with the elevated reactivity of nano-enhanced sensors, creates a superior capacity to identify blood evidence accurately.

1. Key Features of Nano-Enhanced Sensors

Nanoparticle signals become stronger for detecting trace amounts of blood through high sensitivity. Laboratories conducting forensic investigations need this high level of sensitivity to identify blood samples, especially when those samples appear weak or hidden by environmental factors. The detection process becomes quicker with these sensors because they provide rapid response times compared to traditional detection methods. A fast response time is essential to gather time-sensitive analysis outcomes in crime scenes because it directly impacts the investigation's paths. Nano-enhanced sensors detect both the main component of blood, which is haemoglobin, together with crucial indicators of blood types or any significant biological substances. The combined detection system enables forensic experts to collect various operational databases. These advanced sensors exist as portable units which simplify their use at different locations that span isolated crime scenes to active urban locations.

Figure 6. Nano-enhanced Sensor Features

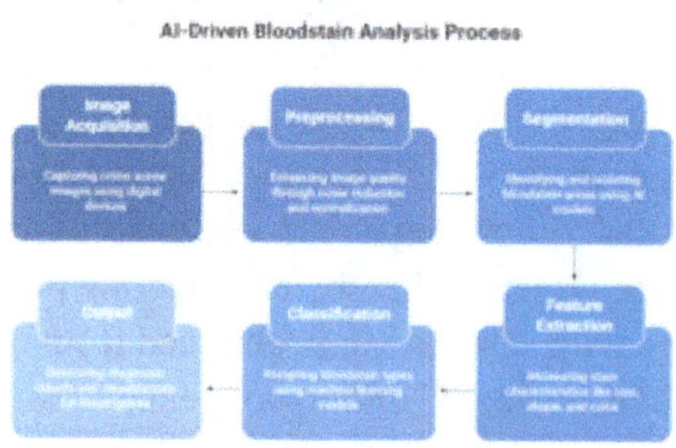

2. The Role of Fluorescent Nanoparticles

The photophysical features of fluorescent nanoparticles within nano-enhanced sensors make them beneficial devices for forensic usage. The fluorescent nanoparticles detect blood samples by using light-emitting properties for easy identification under particular illumination. Fluorescence works as a powerful asset when investigators work in low-light settings at night. Research investigators use optimisation methods for size distribution, as well as the combination of materials and surface treatments, to enable specific blood analysis technology development.(Costanzo et al., 2023).

7.2 Integration of AI and machine learning in Bloodstain Analysis

Figure 7. Integration of AI and machine learning in Bloodstain Analysis

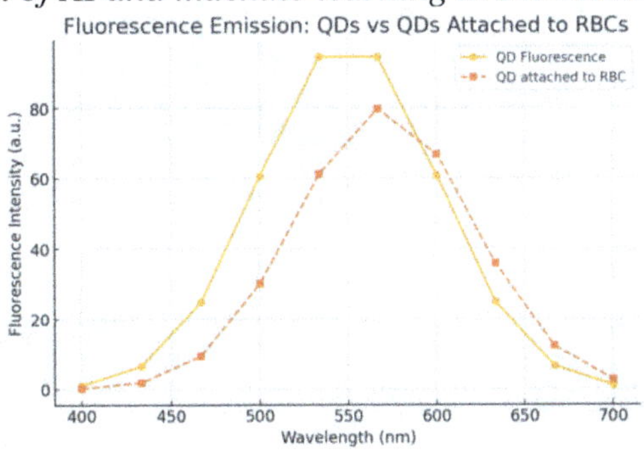

As the volume of data collected during forensic investigations progressively increases, the incorporation of artificial intelligence (AI) and machine learning (ML) into forensic science has decidedly emerged as an important trend. These technologies provide strong tools to automate various aspects of bloodstain analysis, resulting in faster processes that previously required significant time and effort. Machine learning algorithms can automatically examine many bloodstain images as well as discover patterns that people may not see. Algorithms can learn about bloodstains from large datasets, like how they look and where they are located. Forensic experts can use AI to analyse lots of data. This may help them find connections and ideas that might be missed. AI can help make predictive models for criminal investigations by spotting trends and outliers in bloodstain patterns. By including ML in forensic analysis, decision support systems give forensic teams many decision-support tools, which help teams prioritize investigative leads according to statistical likelihood and greatly increase efficiency can analyze textual data (for example, case reports, and witness statements) along with visual data (for example, images of bloodstains)

through natural language processing; this eases evidence interpretation in addition to creating a more all-including comprehension of all the evidence.(Dudek et al., 2024).

Figure 8. Use cases of AI in Blood Stain Analysis

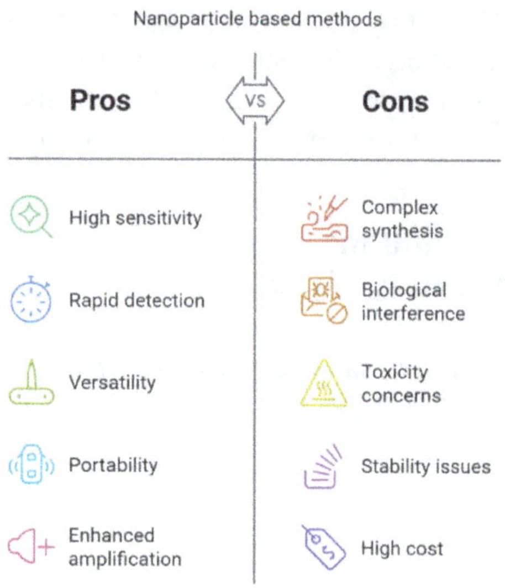

7.3 AI-driven Tools and Platforms

In the area of detection of bloodstains, AI-based means and systems have become more and more necessary to increase precision, automation and efficiency. The most popular in general is the open-source library OpenCV (Open Source Computer Vision Library) that offers a broad selection of functions of computer vision and image processing. OpenCV preprocessing capabilities include image improvement, segmentation, and feature detection, which make up the basis of locating bloodstains in forensic images. To complement OpenCV, TensorFlow and PyTorch are deep learning frameworks that run models of convolutional neural networks (CNNs) and other AI models that can be used to design and train bloodstain pattern recognition and classification models. The versatility and scalability of TensorFlow enable researchers to create multifaceted structures and also enable transfer learning methods to enhance the capacity of detection, particularly in cases where data is scarce. There have also been forensically oriented image processing platforms, which incorporate a combination of AI and traditional image processing, like ForenScope, which incorporates machine learning optimised to recognise patterns in bloodstains under a microscope or at a crime scene. Additionally, AI techniques such as Google Cloud AutoML and Microsoft Azure Cognitive Services offer an intuitive interface to training custom models on bloodstain datasets without the need to code, allowing even the less-skilled forensic practitioner access to the capabilities of AI. Taken together, such tools and frameworks create a flexible AI environment that enhances bloodstain searches and improves their precision, making forensic searches more reliable.

AI-Specific Metrics:

Performance metrics are an important classification process in bloodstain identification with respect to AI algorithms, as they help in measuring the success and accurate outcome of the algorithms. One of the most basic measures is the accuracy that defines the percentage of correctly identified instances, both the bloodstain and non-bloodstain regions, in comparison to the number of predictions. Although accuracy presents a general idea of efficiency, it can be deceptive in the event of unbalanced data, where the counts of non-bloodstain pixels might be much more than the bloodstain ones. Refined metrics are thus necessary.

Sensitivity (or recall, or true positive rate) is the accuracy of an algorithm at identifying actual bloodstains. Forensic analyses and uses require high sensitivity to avoid missing bloodstain analysis, thus refining the number of false negatives. Specificity, in its turn, estimates the number of correctly identified non-bloodstain spots and, consequently, minimises the false positives. The combined sensitivity and specificity provide a fair assessment of the ability of the algorithm to detect based on the types of errors it can cause.

The ROC (Receiver Operating Characteristic) curve is another key indicator that is a graphical representation of sensitivity vs. specificity at various levels of using the threshold of the AI model. The area under the ROC curve (AUC) is the overall measurement of the ability of the model to differentiate between the bloodstain and the non-bloodstain classes, with values approaching 1 signifying excellent distinction. Combined, these measures give forensic specialists and AI developers effective means to refine and validate bloodstain identification mechanisms to assure utility and reliability in mixed-data applications in forensics.

7.4 Future Prospects of Fluorescent Nanoparticles in Forensics

New evidence-collection techniques have the potential to revolutionise through ongoing research that utilises fluorescent nanoparticles for forensic science applications. Scientific research conducted at present represents a key opportunity to introduce revolutionary forensic investigation strategies.

Innovative Applications

Through future scientific investigations, specialised fluorescent nanoparticles should be developed to recognise specific cellular structures and distinct genetic markers that exist in blood testing samples. This technological innovation would enhance fluid recognition capability, along with the identification of donor age as well as gender characteristics. Innovation produces fluorescent nanoparticles with greater resistance to photobleaching emission losses through fluorescent dye deactivation after excitation occurs. The development in this field enables researchers to create detection systems which function better. Technical complexity within forensic investigations depends on the link between fluorescent nanoparticle applications and supporting procedures, including mass spectrometry and chromatography. Multiple technologies integrated would unlock extensive chemical breakdown methods that enable better blood analysis during forensic investigations.

Figure 9. Unveiling the Future of Fluorescent Nanoparticles in Forensics

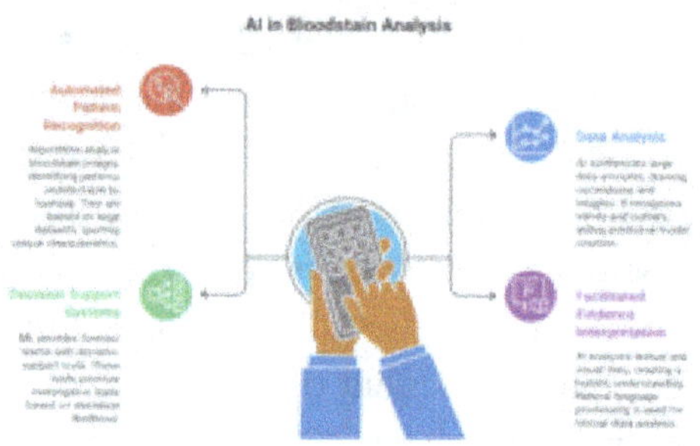

Research in miniaturisation persists, which indicates that portable sensing equipment development will advance in the future. The innovation would allow crime scene investigators to carry out fast, on-site testing without requiring complex laboratory infrastructure so that they could conduct forensic assessments more efficiently.(Ullah et al., 2024).

8. CASE STUDIES AND PRACTICAL APPLICATIONS

8.1 Real-world application of bloodstain detection

Fluorescent nanoparticles have greatly improved the forensic detection of bloodstains and still offer little in the way of comparison with more traditional and less sensitive methods. The nanoparticles selectively bind to components of blood and emit fluorescent signals. This unique property allows forensic investigators to detect blood evidence subjected to cleaning, dilution, or other forms of alteration, rendering it invisible to traditional detection methodologies. The forensic investigator will take the nanoparticles' solution and apply it to the area in question, then shine a strong light to illuminate the previously hidden bloodstains, thereby giving far greater scope to reconstruct the original situation at the crime scene. Such nanoparticle technology's advantage is, therefore, that it can detect very small blood traces on several surfaces: fabric, wood, tile, and so on, even after attempts have been made to clean them. This technique would provide great assistance in cold case situations, as normal detection methods, such as luminol testing, begin to fall short due to biological material degradation. One crucial aspect is that this technique does not damage any evidence, which may then be examined for DNA—an important facet of any modern forensic case. Perhaps the method best suited for these nanoparticles has shown use in the field, both in outdoor scenes exposed to the weather and in indoor scenes where blood evidence is hidden under multiple layers of paint or cleaning agents. The extra sensitivity of being able to work on this technology provides the means to detect the most minute traces of blood, giving investigators the ability to reconstruct, in finer detail and with greater accuracy, the events surrounding the perpetration of criminal activity.(Qu et al., 2023).

8.2 Case Studies Demonstrating Advanced Analytical Techniques

The Interrogation of Fluorescent Nanoparticles in a Violent Crime Scene Investigation

This case study illustrates an important instance when classical blood detection failed in the face of surface contamination and complicated crime scene backgrounds as part of a homicide investigation. To counteract these detections, fluorescent nanoparticles were employed by investigators which were coated with specific antibodies designed to bind haemoglobin. This sophisticated technique allowed for the viewing of very tiny blood leftovers that would never have aroused suspicion. Simply put, the application of this new technology enabled the awareness of who was still fighting on the evidence stage while also providing quantitative information about some blood samples on surface types. The forensic team was able to present findings that enriched the time frame of events and very solidly connected the prime suspect to the crime scene based on the DNA analysis from samples that had been illuminated by the fluorescent nanoparticles. This instrumental case evidence became central to trial defences that impugned the reliability of conventional blood detection methods in complex environments. Nevertheless, the fluorescent nanoparticle-related forensic evidence was extremely persuasive because it demonstrated scientific validity and replicability in the courtroom setting. Such a bolstering of evidence underscores the increasing influence of advanced technologies on legal investigations while bearing future implications for cases of admissibility and evidentiary burden of forensic findings (Köroğlu Necdet Sağlam Uğur Tamer Ramazan Akcan et al., 2024).

8.3 Legal precedents and courtroom implications of blood evidence

The adoption of fluorescent nanoparticles in forensic blood analysis modified investigative approaches as well as the legal criteria used to approve scientific evidence in courtrooms. Case adjudication shows a growing acceptance of scientific data when it presents empirical evidence and validated research methods. Cases that utilise fluorescent nanoparticles should experience increased acceptance in legal proceedings because their use will modify juror perceptions while affecting the results of criminal cases. Rising challenges accompany the deployment of every leading-edge forensic technology. Expert identification of scientific evidence requires legal professionals to be aware that opposing counsel could seek to dispute the reliability of a new method when challenging its scientific foundation. Forensic experts, along with legal practitioners, need to create open communication systems so science methodologies will have proper documentation and full comprehension by all parties involved.(Wang et al., 2017).

9. CONCLUSION

Fluorescent nanoparticles have shown great potential in forensic bloodstain detection, revealing impressive improvements in terms of crime scene investigation sensitivity, specificity, and reliability. Conventional approaches like luminol and peroxidase-based tests are sensitive, but they are usually limited by environmental conditions, contamination, and other interfering substances. These challenges have been addressed through the use of nanotechnology by fluorescent nanoparticles for better detection, even in aged or diluted bloodstains. Even beyond that, Digital imaging, along with AI analysis of bloodstain patterns, still exists, and nano-enhanced sensors have made forensic science more powerful than it has ever been. Although there are challenges with contamination, false positives, and legal admissibility, continued research and interdisciplinary collaboration will continue to improve these methods. In conclusion, the future of forensic blood detection will be characterized by the combination of nanotechnology, machine learning, and portable detection systems and will result in more accurate, efficient, and legally robust forensic investigations.

REFERENCES

Afkhami, A., Moosavi, R., & Madrakian, T. (2010). Preconcentration and spectrophotometric determination of low concentrations of malachite green and leuco-malachite green in water samples by high-performance solid-phase extraction using maghemite nanoparticles. *Talanta, 82*(2), 785–789. DOI: 10.1016/J.TALANTA.2010.05.054 PMID: 20602970

Ahlström, L. H., Sparr Eskilsson, C., & Björklund, E. (2005). Determination of banned azo dyes in consumer goods. *Trends in Analytical Chemistry, 24*(1), 49–56. DOI: 10.1016/J.TRAC.2004.09.004

Bloodstain Pattern Analysis: Introduction. (n.d.). Retrieved August 30, 2025, from https://www.forensicsciencesimplified.org/blood/

Buck, U., Kneubuehl, B., Näther, S., Albertini, N., Schmidt, L., & Thali, M. (2011). 3D bloodstain pattern analysis: Ballistic reconstruction of the trajectories of blood drops and determination of the centres of origin of the bloodstains. *Forensic Science International, 206*(1–3), 22–28. DOI: 10.1016/J.FORSCIINT.2010.06.010 PMID: 20598820

Choi, Y. J., & Sawada, K. (2022). Physical Sensors: Fluorescence Sensors. *Encyclopedia of Sensors and Biosensors: Volume 1-4, First Edition, 1–4*, 1–19. https://doi.org/DOI: 10.1016/B978-0-12-822548-6.00095-9

Collection and Preservation of Blood Evidence from Crime Scenes. (n.d.). Retrieved February 23, 2025, from https://www.crime-scene-investigator.net/blood.html

Costanzo, H., Gooch, J., & Frascione, N. (2023). Nanomaterials for optical biosensors in forensic analysis. *Talanta, 253*, 123945. DOI: 10.1016/J.TALANTA.2022.123945 PMID: 36191514

Dudek, A., Dąbek, A., Zborowska, I., & Lichosik, J. (2024). Integrating artificial intelligence in forensic science. *E-Methodology, 10*(10), 15–28. DOI: 10.15503/EMET2023.15.28

Dybas, J., Alcicek, F. C., Wajda, A., Kaczmarska, M., Zimna, A., Bulat, K., Blat, A., Stepanenko, T., Mohaissen, T., Szczesny-Malysiak, E., Perez-Guaita, D., Wood, B. R., & Marzec, K. M. (2022). Trends in biomedical analysis of red blood cells – Raman spectroscopy against other spectroscopic, microscopic and classical techniques. *Trends in Analytical Chemistry, 146*. Advance online publication. DOI: 10.1016/J.TRAC.2021.116481

Gubala, V., Giovannini, G., Kunc, F., Monopoli, M. P., & Moore, C. J. (2020). Dye-doped silica nanoparticles: synthesis, surface chemistry and bioapplications. *Cancer Nanotechnology 2020 11:1, 11*(1), 1–43. https://doi.org/DOI: 10.1186/S12645-019-0056-X

Hara, M., Nakanishi, H., Yoneyama, K., Saito, K., & Takada, A. (2016). Effects of storage conditions on forensic examinations of blood samples and bloodstains stored for 20 years. *Legal Medicine (Tokyo), 18*, 81–84. DOI: 10.1016/J.LEGALMED.2016.01.003 PMID: 26832383

Kapralov, A., Vlasova, I. I., Feng, W., Maeda, A., Walson, K., Tyurin, V. A., Huang, Z., Aneja, R. K., Carcillo, J., Bayir, H., & Kagan, V. E. (2009). Peroxidase Activity of Haemoglobin ·Haptoglobin Complexes: COVALENT AGGREGATION AND OXIDATIVE STRESS IN PLASMA AND MACROPHAGES. *The Journal of Biological Chemistry, 284*(44), 30395. DOI: 10.1074/JBC.M109.045567 PMID: 19740759

Khan, P., Idrees, D., Moxley, M. A., Corbett, J. A., Ahmad, F., Von Figura, G., Sly, W. S., Waheed, A., & Hassan, M. I. (2014). Luminol-Based Chemiluminescent Signals: Clinical and Non-clinical Application and Future Uses. *Applied Biochemistry and Biotechnology, 173*(2), 333. DOI: 10.1007/S12010-014-0850-1 PMID: 24752935

Köroğlu, U., Sağlam, N., Tamer, U., Akcan, R., Boyaci, I. H., & Evran, E. (2024). Identification of blood at simulated crime scenes using silver nanoparticles with SERS. *Turkish Journal of Chemistry, 48*(4), 676–690.

Liu, J., Li, R., & Yang, B. (2020). Carbon Dots: A New Type of Carbon-Based Nanomaterial with Wide Applications. *ACS Central Science, 6*(12), 2179. DOI: 10.1021/ACSCENTSCI.0C01306 PMID: 33376780

Luminol | Definition, Characteristics, Chemiluminescence, Blood Detection, & Forensics | Britannica. (n.d.). Retrieved February 23, 2025, from https://www.britannica.com/science/luminol

Morris, S. M., & Powell, D. W. (1979). Spectrophotometric assay of phenolphthalein in biological fluids. *Analytical Biochemistry, 95*(2), 465–471. DOI: 10.1016/0003-2697(79)90757-7 PMID: 453532

Mout, R., Moyano, D. F., Rana, S., & Rotello, V. M. (2012). Surface functionalization of nanoparticles for nanomedicine. *Chemical Society Reviews, 41*(7), 2539. DOI: 10.1039/C2CS15294K PMID: 22310807

Nirupama, G. S. (2024). *Advances in Forensic Bloodstain Pattern Analysis: A Review of Current Methods and Future Directions. 11*.

Peschel, O., Kunz, S. N., Rothschild, M. A., & Mützel, E. (2011). Blood stain pattern analysis. *Forensic Science, Medicine, and Pathology*, *7*(3), 257–270. DOI: 10.1007/S12024-010-9198-1 PMID: 21069481

Qu, J., Meador, W., Cheah, P., Tanner, E. E. L., Delcamp, J., & Zhao, Y. (2023). Latent bloodstain detection using a selective turn-on NIR fluorescence dye responsive to serum albumin. *RSC Advances*, *13*(39), 27549. DOI: 10.1039/D3RA04320G PMID: 37720829

Quickenden, T. I., & Creamer, J. I. (2001). A study of common interferences with the forensic luminol test for blood. *Luminescence : The Journal of Biological and Chemical Luminescence*, *16*(4), 295–298. DOI: 10.1002/BIO.657 PMID: 11512147

Risks, I. of M. (US) C. on A. G., Andrews, L. B., Fullarton, J. E., Holtzman, N. A., & Motulsky, A. G. (1994). *Social, Legal, and Ethical Implications of Genetic Testing*. https://www.ncbi.nlm.nih.gov/books/NBK236044/

Senthilkumar, A., Ravindran, V., Arthanari, A., & Ramalingam, K. (2024). Evaluation of Forensic Luminol in the Detection of Blood Stains in Instruments Following Dental Treatment. *Cureus*, *16*(4), e57676. DOI: 10.7759/CUREUS.57676 PMID: 38707075

Short, S. (2015). Serology: Blood Identification. *Encyclopedia of Forensic and Legal Medicine: Second Edition*, 239–243. https://doi.org/DOI: 10.1016/B978-0-12-800034-2.00338-4

Shreffler, J., & Huecker, M. R. (2023). Diagnostic Testing Accuracy: Sensitivity, Specificity, Predictive Values and Likelihood Ratios. *StatPearls*. https://www.ncbi.nlm.nih.gov/books/NBK557491/

The Fluorescein Method of Latent Blood Detection. (n.d.). Retrieved February 23, 2025, from https://www.crime-scene-investigator.net/fluoresceinmethod.html#google_vignette

The Nobel Prize in Chemistry 2023 - Popular information - NobelPrize.org. (n.d.). Retrieved February 23, 2025, from https://www.nobelprize.org/prizes/chemistry/2023/popular-information/

Ullah, M. F., Khan, Y., Khan, M. I., Abdullaeva, B. S., & Waqas, M. (2024). Exploring nanotechnology in forensic investigations: Techniques, innovations, and prospects. *Sensing and Bio-Sensing Research*, *45*, 100674. DOI: 10.1016/J.SBSR.2024.100674

Wang, M., Li, M., Yu, A., Zhu, Y., Yang, M., & Mao, C. (2017). Fluorescent Nanomaterials for the Development of Latent Fingerprints in Forensic Sciences. *Advanced Functional Materials*, *27*(14), 1606243. DOI: 10.1002/ADFM.201606243 PMID: 29657570

Yeh, Y. C., Creran, B., & Rotello, V. M. (2011). Gold Nanoparticles: Preparation, Properties, and Applications in Bionanotechnology. *Nanoscale*, *4*(6), 1871. DOI: 10.1039/C1NR11188D PMID: 22076024

Compilation of References

Abbondanza, M. N. M., & Souza, R. G. (2019). Estimating the generation of household e-waste in municipalities using primary data from surveys: A case study of Sao Jose dos Campos, Brazil. *Waste Management (New York, N.Y.)*, *85*, 374–384.

Abdel-Aziz, H. M., Soliman, M. I., Abo Al-Saoud, A. M., & El-Sherbeny, G. A. (2021). Waste-derived NPK nanofertilizer enhances growth and productivity of Capsicum annuum L. *Plants*, *10*(6), 1144. PMID: 34199718

Abdelmeguid, M. G., Janairo, J. I. B., & Chemmangattuvalappil, N. G. (2025). Examining peptide–gold nanoparticle interactions through explainable machine learning. *Discover Chemistry 2025 2:1*, *2*(1), 1–16. https://doi.org/DOI: 10.1007/S44371-025-00191-2

Abiodun, O. I., Jantan, A., Omolara, A. E., Dada, K. V., Mohamed, N. A., & Arshad, H. (2018). State-of-the-art in artificial neural network applications: A survey. *Heliyon*, *4*(11), e00938. DOI: 10.1016/j.heliyon.2018.e00938

Abramson, J., Adler, J., Dunger, J., Evans, R., Green, T., Pritzel, A., Ronneberger, O., Willmore, L., Ballard, A. J., Bambrick, J., Bodenstein, S. W., Evans, D. A., Hung, C.-C., O'Neill, M., Reiman, D., Tunyasuvunakool, K., Wu, Z., Žemgulytė, A., Arvaniti, E., & Jumper, J. M. (2024). Accurate structure prediction of biomolecular interactions with AlphaFold 3. *Nature*, *630*(8016), 493–500. DOI: 10.1038/s41586-024-07487-w PMID: 38718835

Adams, J. D., Rogers, B. S., & Leifer, L. J. (2004). Microtechnology, nanotechnology, and the scanning-probe microscope: An innovative course. *IEEE Transactions on Education*, *47*(1), 51–56. DOI: 10.1109/TE.2003.817623

Afkhami, A., Moosavi, R., & Madrakian, T. (2010). Preconcentration and spectrophotometric determination of low concentrations of malachite green and leuco-malachite green in water samples by high-performance solid-phase extraction using maghemite nanoparticles. *Talanta*, *82*(2), 785–789. DOI: 10.1016/J.TALANTA.2010.05.054 PMID: 20602970

Ahlström, L. H., Sparr Eskilsson, C., & Björklund, E. (2005). Determination of banned azo dyes in consumer goods. *Trends in Analytical Chemistry*, *24*(1), 49–56. DOI: 10.1016/J.TRAC.2004.09.004

Ahmadi, S., Kamaladini, H., Haddadi, F., & Sharifmoghadam, M. R. (2018). Thiol-capped gold nanoparticle biosensors for rapid and sensitive visual colorimetric detection of Klebsiella pneumoniae. *Journal of Fluorescence*, *28*, 987–998. PMID: 30022376

Ahmed Taha, B., Kadhim, A. C., Addie, A. J., Haider, A. J., Azzahrani, A. S., Raizada, P., Rustagi, S., Chaudhary, V., & Arsad, N. (2024). Advancing cancer diagnostics through multifaceted optical biosensors supported by nanomaterials and artificial intelligence: A panoramic outlook. *Microchemical Journal*, *205*, 111307. Advance online publication. DOI: 10.1016/j.microc.2024.111307

Ahn, J. J., Kim, Y., Corley, E. A., & Scheufele, D. A. (2016). Laboratory Safety and Nanotechnology Workers: An Analysis of Current Guidelines in the USA. *NanoEthics*, *10*(1), 5–23. DOI: 10.1007/s11569-016-0250-9

Ajayi, A., & Pub, A. (2024). Creating a Conceptual Framework for AI-Powered STEM Education Analytics to Enhance Student Learning Outcomes. *Journal of Frontiers in Multidisciplinary Research*, *05*, 157–167.

Ajithkumar, M., & Krishnamoorthy, R. (2025). Customer-Centric. *Future-Proofing Emerging Technologies for Business Transformation*, 327.

Akhayere, E., & Kavaz, D. (2021). Nano-silica and nano-zeolite synthesized from barley grass straw for effective removal of gasoline from aqueous solution: A comparative study. *Chemical Engineering Communications*, *208*(10), 1419–1435. DOI: 10.1080/00986445.2020.1786373

Akhayere, E., Vaseashta, A., & Kavaz, D. (2020). Novel magnetic nano silica synthesis using barley husk waste for removing petroleum from polluted water for environmental sustainability. *Sustainability (Switzerland)*, *12*(24), 1–16. DOI: 10.3390/su122410646

Akinsipo (Oyelaja), O. B., & Anselm, O. H. (2025). Challenges and Opportunities for Implementing Green Chemistry in Nigerian Universities: Educational and Policy Perspectives. *Sustainability & Circularity NOW, 02*. https://doi.org/DOI: 10.1055/a-2534-1903

Akondi, K. B., Muttenthaler, M., Dutertre, S., Kaas, Q., Craik, D. J., Lewis, R. J., & Alewood, P. F. (2014). Discovery, synthesis, and structure-activity relationships of conotoxins. *Chemical Reviews*, *114*(11), 5815–5847. DOI: 10.1021/cr400401e PMID: 24720541

Alam, A., & Mohanty, A. (2023). Educational technology: Exploring the convergence of technology and pedagogy through mobility, interactivity, AI, and learning tools. *Cogent Engineering*, *10*(2), 2283282. DOI: 10.1080/23311916.2023.2283282

Alavi, N., Shirmardi, M., Babaei, A., Takdastan, A., & Bagheri, N. (2015). Waste electrical and electronic equipment (WEEE) estimation: A case study of Ahvaz City, Iran. *Journal of the Air & Waste Management Association*, *65*, 298–305.

Al-Azzam, S., Ding, Y., Liu, J., Pandya, P., Ting, J. P., & Afshar, S. (2020). Peptides to combat viral infectious diseases. *Peptides*, *134*, 170402. DOI: 10.1016/J.PEPTIDES.2020.170402 PMID: 32889022

Alhazmi, K., Albalawi, F., & Sarathy, S. M. (2022). A reinforcement learning-based economic model predictive control framework for autonomous operation of chemical reactors. *Chemical Engineering Journal*, *428*, 130993. Advance online publication. DOI: 10.1016/j.cej.2021.130993

Ali, S., Alam, W., Tyara, H., & Chong, K. T. (2025). An accurate prediction of drug–drug interactions and side effects by using integrated convolutional and BiLSTM networks. *Chemometrics and Intelligent Laboratory Systems*, *257*, 105304. DOI: 10.1016/j.chemolab.2024.105304

Al-Maharma, A. Y., Patil, S. P., & Markert, B. (2022). Environmental impact analysis of plant fibers and their composites relative to their synthetic counterparts based on life cycle assessment approach. In *Advances in bio-based fiber* (pp. 741–781). Woodhead Publishing.

Almasri, F. (2024). Exploring the impact of artificial intelligence in teaching and learning of science: A systematic review of empirical research. *Research in Science Education*, *54*(5), 977–997. DOI: 10.1007/s11165-024-10176-3

Al-Qattan, M. N., Deb, P. K., & Tekade, R. K. (2018). Molecular dynamics simulation strategies for designing carbon-nanotube-based targeted drug delivery. *Drug Discovery Today*, *23*(2), 235–250. DOI: 10.1016/j.drudis.2017.10.002 PMID: 29031623

Alrobaish, W. S., Vlerick, P., Luning, P. A., & Jacxsens, L. (2021). Food safety governance in Saudi Arabia: Challenges in control of imported food. *Journal of Food Science*, *86*(1), 16–30. PMID: 33314129

Al-Shabib, N. A., Husain, F. M., Ahmed, F., Khan, R. A., Ahmad, I., Alsharaeh, E., & Aliev, G. (2016). Biogenic synthesis of Zinc oxide nanostructures from Nigella sativa seed: Prospective role as food packaging material inhibiting broad-spectrum quorum sensing and biofilm. *Scientific Reports*, *6*(1), 36761. PMID: 27917856

Alsubaie, A. S. R., & Berekaa, M. M. (2020). Food safety in Saudi Arabia: A public health priority. *Annals of Medical and Health Sciences Research*.

Altpeter, F., Springer, N. M., Bartley, L. E., Blechl, A. E., Brutnell, T. P., Citovsky, V., & Stewart, C. N.Jr. (2016). Advancing crop transformation in the era of genome editing. *The Plant Cell*, *28*(7), 1510–1520. PMID: 27335450

Al-Zoubi, O., Ahmad, N., & Hamid, N. A. (2024). Artificial intelligence in newsrooms: Ethical challenges facing journalists. *Studies in Media and Communication*, *12*, 6587. DOI: 10.11114/smc.v12i1.6587

Ambrosetti, F., Jandova, Z., & Bonvin, A. M. J. J. (2023). Information-Driven Antibody–Antigen Modeling with HADDOCK. In *Methods in Molecular Biology* (Vol. 2552, pp. 267–282). Humana., DOI: 10.1007/978-1-0716-2609-2_14

Amuru, D., Zahra, A., Vudumula, H. V., Cherupally, P. K., Gurram, S. R., Ahmad, A., & Abbas, Z. (2023). AI/ML algorithms and applications in VLSI design and technology. *Integration (Amsterdam)*, *93*, 102048. DOI: 10.1016/j.vlsi.2023.06.002

Anjum, M., Miandad, R., Waqas, M., Gehany, F., & Barakat, M. A. (2019). Remediation of wastewater using various nano-materials. *Arabian Journal of Chemistry*, *12*(8), 4897–4919. DOI: 10.1016/j.arabjc.2016.10.004

Antoni, G., & Marzetti, G. (2019). Recycling and waste generation: An estimate of the source reduction effect of recycling programs. *Ecological Economics*, *161*, 321–329.

Apablaza-Campos, A., Wilches Tinjacá, J. A., & Salaverría, R. (2024). Generative artificial intelligence for journalistic content in Ibero-America: Perceptions. Challenges and Reg. Proj. *BiD*, *52*, 1–8. DOI: 10.1344/bid2024.52.06

Aragani, V. M., Anumolu, V. R., & Selvakumar, P. (2025). Democratization in the Age of Algorithms: Navigating Opportunities and Challenges. In Wongmahesak, K., Wekke, I., Seftyono, C., & Nurdin, N. (Eds.), *Democracy and Democratization in the Age of AI* (pp. 39–56). IGI Global Scientific Publishing., DOI: 10.4018/979-8-3693-8749-8.ch003

Araújo, M. G., Magrini, A., Mahler, C. F., & Bilitewski, B. (2012). A model for estimation of potential generation of waste electrical and electronic equipment in Brazil. *Waste Management (New York, N.Y.)*, *32*, 335–342.

Asghari, J., & Naderi, M. (2024). Artificial Intelligence in Nanomaterials Studies: A Statistical Overview. *Nanoscale and Advanced Materials*, *1*(2), 73–86. DOI: 10.22034/nsam.2024.04.01

Ashok, M., Madan, R., Joha, A., & Sivarajah, U. (2022). Ethical framework for artificial intelligence and digital technologies. *International Journal of Information Management*, *62*, 102433. DOI: 10.1016/j.ijinfomgt.2021.102433

Attique, M., Farooq, M. S., Khelifi, A., & Abid, A. (2020). Prediction of Therapeutic Peptides Using Machine Learning: Computational Models, Datasets, and Feature Encodings. *IEEE Access : Practical Innovations, Open Solutions*, *8*, 148570–148594. DOI: 10.1109/ACCESS.2020.3015792

Aversa, R., Modarres, M. H., Cozzini, S., Ciancio, R., & Chiusole, A. (2018). The first annotated set of scanning electron microscopy images for nanoscience. *Scientific Data*, *5*(1), 180172. Advance online publication. DOI: 10.1038/sdata.2018.172 PMID: 30152811

Awasthi, A., Cucchiella, F., D'Adamo, I., Li, J., Rosa, P., Terzi, S., & Zeng, X. (2018). Modelling the correlations of e-waste quantity with economic increase. *The Science of the Total Environment*, *613–614*, 46–53.

Awasthi, A., & Li, J. (2017). Management of electrical and electronic waste: A comparative evaluation of China and India. *Renewable & Sustainable Energy Reviews*, *76*, 434–447.

Ayodele, O. O., Adesina, A. O., Pourianejad, S., Averitt, J., & Ignatova, T. (2021). Recent advances in nanomaterial-based aptasensors in medical diagnosis and therapy. *Nanomaterials (Basel, Switzerland)*, *11*(4), 932. Advance online publication. DOI: 10.3390/nano11040932 PMID: 33917467

Babu, P. J. (2022). Nanotechnology mediated intelligent and improved food packaging. *International Nano Letters*, *12*(1), 1–14.

Badini, S., Regondi, S., & Pugliese, R. (2023). Unleashing the Power of Artificial Intelligence in Materials Design. *Materials (Basel)*, *16*(17), 17. Advance online publication. DOI: 10.3390/ma16175927

Bae, H., Ji, H., Konstantinov, K., Sluyter, R., Ariga, K., Kim, Y. H., & Kim, J. H. (2025). Artificial Intelligence-Driven Nanoarchitectonics for Smart Targeted Drug Delivery. *Advanced Materials, 10239*. Advance online publication. DOI: 10.1002/ADMA.202510239 PMID: 40772340

Baetsen-Young, A. M., Vasher, M., Matta, L. L., Colgan, P., Alocilja, E. C., & Day, B. (2018). Direct colorimetric detection of unamplified pathogen DNA by dextrin-capped gold nanoparticles. *Biosensors & Bioelectronics, 101*, 29–36. PMID: 29031887

Bai, X., & Zhang, X. (2025). Artificial Intelligence-Powered Materials Science. *Nano-Micro Letters, 17*(1), 135. DOI: 10.1007/s40820-024-01634-8

Baker, T., & Smith, L. (2019). *Educ-AI-tion Rebooted?* Nesta. https://www.nesta.org.uk/report/education-rebooted/

Baker, R. S., & Inventado, P. S. (2014). Educational data mining. In Larusson, J. A., & White, B. (Eds.), *Learning analytics: From research to practice* (pp. 61–75). Springer.

Baldé, K., Wang, F., Huisman, J., & Kuehr, R. (2015). The Global E-Waste Monitor. United Nations University, IAS – SCYCLE.

Baldé, C. P., Kuehr, R., Yamamoto, T., McDonald, R., D'Angelo, E., Althaf, S., & Wagner, M. (2024). *Global E-waste Monitor 2024. International Telecommunication Union (ITU) and United Nations Institute for Training and Research.* UNITAR.

Baldwin, J., Noorali, S., & Vaseashta, A. (2023). Biology and behavior of Severe Acute Respiratory Syndrome Coronavirus contagion with emphasis on treatment strategies, risk assessment, and resilience. *COVID, 3*(9), 1259–1303. DOI: 10.3390/covid3090089

Banting, F. G., Best, C. H., Collip, J. B., Macleod, J. J. R., & Noble, E. C. (1922). The Effect of Pancreatic Extract (Insulin) on Normal Rabbits. *American Journal of Physiology-Legacy Content, 62*(1), 162–176. DOI: 10.1152/ajplegacy.1922.62.1.162

Baranowska-Wójcik, E., Szwajgier, D., Oleszczuk, P., & Winiarska-Mieczan, A. (2020). Effects of titanium dioxide nanoparticles exposure on human health—A review. *Biological Trace Element Research, 193*(1), 118–129. PMID: 30982201

Bargel, H., & Scheibel, T. (2024). A bio-engineering approach to generate bioinspired (spider) silk protein-based materials. at-Automatisierungstechnik, 72(7), 657-665.

Barkhausen, R., Durand, A., & Fick, K. (2022). Review and analysis of ecodesign directive implementing measures: Product regulations shifting from energy efficiency towards a circular economy. *Sustainability, 14*(16), 10318.

Barnard, A. S., Motevalli, B., Parker, A. J., Fischer, J. M., Feigl, C. A., & Opletal, G. (2019). Nanoinformatics, and the big challenges for the science of small things. *Nanoscale, 11*(41), 19190–19201. DOI: 10.1039/C9NR05912A PMID: 31397835

Barzdenas, V., & Vasjanov, A. (2023). Crosstalk Reduction in High-Density Radio Frequency Printed Circuit Boards: Leveraging FR4 Coating Layers. *Coatings, 13*(10), 10. Advance online publication. DOI: 10.3390/coatings13101801

Batool, M., Sanumi, O., & Jankovic, J. (2024). Application of artificial intelligence in the materials science, with a special focus on fuel cells and electrolyzers. *Energy and AI, 18*, 100424. DOI: 10.1016/j.egyai.2024.100424

Batra, A., Sampson, J., Davis, A., Currie, J., & Vaseashta, A. (2023). Electrospun nanofibers doped with PVDF and PLZT nanoparticles for potential biomedical and energy harvesting applications. *Journal of Materials Science Materials in Electronics, 34*(22), 1654. Advance online publication. DOI: 10.1007/s10854-023-11066-6

Bauer, J. (2021). Teaching Nanotechnology through Research Proposals. *Journal of Chemical Education, 98*(7), 2347–2355. DOI: 10.1021/acs.jchemed.0c01251

Beckett, C. (2019). New powers, new responsibilities. A global survey of journalism and artificial intelligence. The Journalism AI report. Available at: https://drive.google.com/file/d/1utmAMCmd4rfJHrUfLLfSJ-clpFTjyef1/view (Accessed July 14, 2024).

Beckett, C. (2023). Generating change. A global survey of what news organizations are doing with AI. The journalism AI report. Available at: https://static1.squarespace.com/static/64d60527c01ae7106f2646e9/t/656e400a1c23e22da0681e46/1701724190867/Generating+Change+_+The+Journalism+AI+report+_+English.pdf (Accessed July 14, 2024).

Berardi, C., Antonini, M., Jordan, Z., Wechtler, H., Paolucci, F., & Hinwood, M. (2024). Barriers and facilitators to the implementation of digital technologies in mental health systems: A qualitative systematic review to inform a policy framework. *BMC Health Services Research, 24*(1), 243. DOI: 10.1186/s12913-023-10536-1 PMID: 38408938

Berekaa, M. M. (2015). Nanotechnology in food industry; advances in food processing, packaging and food safety. *International Journal of Current Microbiology and Applied Sciences, 4*(5), 345–357.

Bernardes-Loch, R. M., de Oliveira Almeida, G., Brasiliano, I. T., Meira, W., Pires, D. E. V., Baracat-Pereira, M. C., & de Azevedo Silveira, S. (2025). PERSEUcpp: A machine learning strategy to predict cell-penetrating peptides and their uptake efficiency. *BioRxiv*, 2025.04.07.647598. https://doi.org/DOI: 10.1101/2025.04.07.647598

Bernard, R. M., Borokhovski, E., Schmid, R. F., Tamim, R. M., & Abrami, P. C. (2009). A meta-analysis of blended learning and technology use in higher education: From the general to the applied. *Journal of Computing in Higher Education, 21*(3), 95–123.

Beyerle, E., ... (2023). Recent advances in describing and driving crystal nucleation using machine learning and artificial intelligence. *arXiv preprint*, arXiv:2304.13815. https://arxiv.org/abs/2304.13815

Bhatt, M. D., Kim, H., & Kim, G. (2022). Various defects in graphene: A review. *RSC Advances, 12*(33), 21520–21547. DOI: 10.1039/D2RA01436J PMID: 35975063

Bhujel, R., Enkmann, V., Burgstaller, H., & Maharjan, R. (2025). Artificial Intelligence-Driven Strategies for Targeted Delivery and Enhanced Stability of RNA-Based Lipid Nanoparticle Cancer Vaccines. *Pharmaceutics, 17*(8), 992. DOI: 10.3390/PHARMACEUTICS17080992 PMID: 40871015

Binnig, G., & Rohrer, H. (1986). Scanning tunneling microscopy. *IBM Journal of Research and Development, 30*(4), 355–369. DOI: 10.1147/rd.304.0355

Binns, R. (2018). Fairness in machine learning: Lessons from political philosophy. In *Proceedings of the 2018 Conference on Fairness, Accountability and Transparency* (pp. 149–159).

Bishnu, S. K., Alnouri, S. Y., & Al-Mohannadi, D. M. (2023). Computational applications using data driven modeling in process Systems: A review. *Digital Chemical Engineering, 8*, 100111. DOI: 10.1016/j.dche.2023.100111

Biswas, P., Rashid, A., Biswas, A., Nasim, M. A. A., Chakraborty, S., Gupta, K. D., & George, R. (2024). AI-driven approaches for optimizing power consumption: A comprehensive survey. *Discover Artificial Intelligence, 4*(1), 116. DOI: 10.1007/s44163-024-00211-7

Bi, Y., Xie, S., Li, Z., Dong, S., & Teng, L. (2025). Precise nanoscale fabrication technologies are the "last mile" of medicinal development. *Acta Pharmaceutica Sinica. B*. Advance online publication. DOI: 10.1016/j.apsb.2025.03.040

Blanco-González, A., Cabezón, A., Seco-González, A., Conde-Torres, D., Antelo-Riveiro, P., Piñeiro, Á., & Garcia-Fandino, R. (2023). The Role of AI in Drug Discovery: Challenges, Opportunities, and Strategies. *Pharmaceuticals, 16*(6), 891. DOI: 10.3390/ph16060891 PMID: 37375838

Bloodstain Pattern Analysis: Introduction. (n.d.). Retrieved August 30, 2025, from https://www.forensicsciencesimplified.org/blood/

BMUV. (Federal Ministry for the Environment, Nature Conservation, Nuclear Safety and Consumer Protection), (2023). Waste Management in Germany 2023: Facts, data, figures. Berlin: BMUV Printing Office

Bogar, Z. O., & Gungor, A. (2023). Forecasting waste mobile phone (WMP) quantity and evaluating the potential contribution to the circular economy: A case study of Turkey. *Sustainability*, *15*(4), 3104.

Borel, J. F., Feurer, C., Gubler, H. U., & Stähelin, H. (1976). Biological effects of cyclosporin A: A new antilymphocytic agent. *Agents and Actions*, *6*(4), 468–475. DOI: 10.1007/BF01973261 PMID: 8969

Boso, D. P., Ferrari, M., Decuzzi, P., & Schrefler, B. A. (2011). Optimizing particle size for targeting diseased microvasculature: From experiments to artificial neural networks. *International Journal of Nanomedicine*, *6*, 1517–1526. DOI: 10.2147/IJN.S20283 PMID: 21845041

Bothun, G. D., Oyanedel-Craver, V., & Park, K. (2019). Using "nano tools" as the basis for a hands-on experiential course in nanotechnology. In *Nanotechnology in undergraduate education* (pp. 291–308). Elsevier. https://doi.org/DOI: 10.1016/B978-0-12-814835-8.00011-X

Botifoll, M., Pinto-Huguet, I., & Arbiol, J. (2022). Machine learning in electron microscopy for advanced nanocharacterization: Current developments, available tools, and future outlook. *Nanoscale Horizons*, *7*(12), 1427–1477. DOI: 10.1039/D2NH00377E

Brosilow, C., & Joseph, B. (2002). *Techniques of model-based control*. Prentice Hall.

Brunette, E. S., Flemmer, R. C., & Flemmer, C. L. (2009). A review of artificial intelligence. In *2009 4th International Conference on Autonomous Robots and Agents* (pp. 385–392). IEEE. https://doi.org/DOI: 10.1109/ICARA.2009.5639918

Buck, U., Kneubuehl, B., Näther, S., Albertini, N., Schmidt, L., & Thali, M. (2011). 3D bloodstain pattern analysis: Ballistic reconstruction of the trajectories of blood drops and determination of the centres of origin of the bloodstains. *Forensic Science International*, *206*(1–3), 22–28. DOI: 10.1016/J.FORSCIINT.2010.06.010 PMID: 20598820

Burris, K. P., Wu, T. C., Vasudev, M., Stroscio, M. A., Millwood, R. J., & Stewart, C. N. (2013). Mega-nano detection of foodborne pathogens and transgenes using molecular beacon and semiconductor quantum dot technologies. *IEEE Transactions on Nanobioscience*, *12*(3), 233–238. PMID: 23722479

Butler, K. T., Davies, D. W., Cartwright, H., Isayev, O., & Walsh, A. (2018). Machine learning for molecular and materials science. *Nature*, *559*(7715), 547–555. DOI: 10.1038/s41586-018-0337-2 PMID: 30046072

Cai, C., Zhu, G., & Ma, M. (2024). A systemic review of AI for interdisciplinary learning: Application contexts, roles, and influences. *Education and Information Technologies*. Advance online publication. DOI: 10.1007/s10639-024-13193-x

Cai, Y., Xu, Z., Shuai, Q., Zhu, F., Xu, J., Gao, X., & Sun, X. (2020). Tumor-targeting peptide functionalized PEG-PLA micelles for efficient drug delivery. *Biomaterials Science*, *8*(8), 2274–2282. DOI: 10.1039/C9BM02036E PMID: 32162618

Cao, Q., Ge, C., Wang, X., Harvey, P. J., Zhang, Z., Ma, Y., Wang, X., Jia, X., Mobli, M., Craik, D. J., Jiang, T., Yang, J., Wei, Z., Wang, Y., Chang, S., & Yu, R. (2023). Designing antimicrobial peptides using deep learning and molecular dynamic simulations. *Briefings in Bioinformatics*, *24*(2), 1–13. DOI: 10.1093/bib/bbad058 PMID: 36857616

Cao, S. jun, Lv, Z. qiang, Guo, S., Jiang, G. ping, & Liu, H. ling. (2021). An update - Prolonging the action of protein and peptide drugs. *Journal of Drug Delivery Science and Technology*, *61*, 102124. DOI: 10.1016/J.JDDST.2020.102124

Cao, Y., Guo, Y., & Liu, Y. (2019). Quantum machine learning: A comprehensive review. *Nature Communications*, *10*(1), 1–13. PMID: 30602773

Capecchi, A., & Reymond, J. L. (2021). Peptides in chemical space. *Medicine in Drug Discovery*, *9*, 100081. DOI: 10.1016/j.medidd.2021.100081

Carbonell, J. R. (1970). AI in CAI: An artificial-intelligence approach to computer-assisted instruction. *IEEE Transactions on Man-Machine Systems*, *11*(4), 190–202. DOI: 10.1109/TMMS.1970.299939

Carlson, M. (2015). The robotic reporter. *Journalism*, *3*, 416–431. DOI: 10.1080/21670811.2014.976412

Carvalho, D., Ferreira, N., Franc¸a, B., Marques, R., Silva, M., Silva, S., ... & Oliveira, C. (2024). Advancing sustainability in the automotive industry: Bioprepregs and fully bio-based composites. Composites Part C: Open Access, 100459.

Castel, G., Chtéoui, M., Heyd, B., & Tordo, N. (2011). Phage display of combinatorial peptide libraries: Application to antiviral research. *Molecules (Basel, Switzerland)*, *16*(5), 3499–3518. DOI: 10.3390/molecules16053499 PMID: 21522083

Caswell, D., & Dörr, K. (2018). Automated journalism 2.0: Event-driven narratives. *Journalism Practice*, *12*, 477–496. DOI: 10.1080/17512786.2017.1320773

CBSE. (2020). *Curriculum for Artificial Intelligence: Class IX–XII*. Central Board of Secondary Education.

Cencer, M. M., Moore, J. S., & Assary, R. S. (2022). Machine learning for polymeric materials: An introduction. *Polymer International*, *71*(5), 537–542.

Chakroborty, S., Nath, N., & Sahoo, S.. (2025). A review of emerging trends in nanomaterial-driven AI for biomedical applications. *Nanoscale Advances*, *7*, 3619–3630. DOI: 10.1039/D5NA00032G PMID: 40370571

Chan, E. M., Xu, C., Mao, A. W., Han, G., Owen, J. S., Cohen, B. E., & Milliron, D. J. (2010). Reproducible, high-throughput synthesis of colloidal nanocrystals for optimization in multidimensional parameter space. *Nano Letters*, *10*(5), 1874–1885. DOI: 10.1021/nl100669s PMID: 20387807

Chang, D. T. (2022). Parameterized quantum circuits with quantum kernels for machine learning: A hybrid quantum–classical approach. *arXiv preprint*, arXiv:2209.14449. https://arxiv.org/abs/2209.14449

Chang, L., Mondal, A., Singh, B., Martínez-Noa, Y., & Perez, A. (2024). Revolutionizing peptide-based drug discovery: Advances in the post-AlphaFold era. *Wiley Interdisciplinary Reviews. Computational Molecular Science*, *14*(1). Advance online publication. DOI: 10.1002/wcms.1693 PMID: 38680429

Chan, M. L., Qi, Y., Larimore, K., Cherukuri, A., Seid, L., Jayaram, K., Jeha, G., Fisheleva, E., Day, J., Huntsman-Labed, A., Savarirayan, R., Irving, M., Bacino, C. A., Hoover-Fong, J., Ozono, K., Mohnike, K., Wilcox, W. R., Horton, W. A., & Henshaw, J. (2022). Pharmacokinetics and Exposure–Response of Vosoritide in Children with Achondroplasia. *Clinical Pharmacokinetics*, *61*(2), 263–280. DOI: 10.1007/s40262-021-01059-1 PMID: 34431071

Chatterjee, S., Nath, A., & Ghosh, S. K. (2021). Leveraging AI-driven educational tools for digital transformation in higher education. *Education and Information Technologies*, *26*(3), 3297–3315. DOI: 10.1007/s10639-020-10392-2

Chaudhary, K., & Vrat, P. (2018). Case study analysis of e-waste management systems in Germany, Switzerland, Japan and India: A RADAR chart approach. *Benchmarking*, *25*(9), 3519–3540.

Chaudhry, Q., Scotter, M., Blackburn, J., Ross, B., Boxall, A., Castle, L., & Watkins, R. (2008). Applications and implications of nanotechnologies for the food sector. *Food Additives and Contaminants*, *25*(3), 241–258. PMID: 18311618

Chávez-Angel, E., Eriksen, M. B., Castro-Alvarez, A., Garcia, J. H., Botifoll, M., Avalos-Ovando, O., Arbiol, J., & Mugarza, A. (n.d.). Applied Artificial Intelligence in Materials Science and Material Design. *Advanced Intelligent Systems, n/a*(n/a), 2400986. https://doi.org/DOI: 10.1002/aisy.202400986

Chen, C., Yaari, Z., Apfelbaum, E., Grodzinski, P., Shamay, Y., & Heller, D. A. (2022). Merging data curation and machine learning to improve nanomedicines. *Advanced Drug Delivery Reviews*, *183*, 114172. Advance online publication. DOI: 10.1016/j.addr.2022.114172 PMID: 35189266

Chen, C., Ye, W., Zuo, Y., Zheng, C., & Ong, S. P. (2019). Graph networks as a universal machine learning framework for molecules and crystals. *Chemistry of Materials*, *31*(9), 3564–3572. DOI: 10.1021/acs.chemmater.9b01294

Chen, C., Zhang, Q., Yu, B., Yu, Z., Lawrence, P. J., Ma, Q., & Zhang, Y. (2020). Improving protein-protein interactions prediction accuracy using XGBoost feature selection and stacked ensemble classifier. *Computers in Biology and Medicine*, *123*, 103899. Advance online publication. DOI: 10.1016/j.compbiomed.2020.103899 PMID: 32768046

Chen, K., Ma, B., Li, J., Chen, E., Xu, Y., Yu, X., & Zhang, M. (2021). A rapid and sensitive europium nanoparticle-based lateral flow immunoassay combined with recombinase polymerase amplification for simultaneous detection of three food-borne pathogens. *International Journal of Environmental Research and Public Health*, *18*(9), 4574. PMID: 33925871

Chen, L., Chen, P., & Lin, Z. (2020). Artificial intelligence in education: A review. *IEEE Access : Practical Innovations, Open Solutions*, *8*, 75264–75278. DOI: 10.1109/ACCESS.2020.2988510

Chen, T., Li, J., Cai, P., Yao, Q., Ren, Z., Zhu, Y., Khan, S., Xie, J., & Wang, X. (2023). Identification of chemical compositions from "featureless" optical absorption spectra: Machine learning predictions and experimental validations. *Nano Research*, *16*(4), 4188–4196. DOI: 10.1007/s12274-022-5095-7

Chen, W.. (2021). Machine learning for multi-scale modeling of battery materials. *Joule*, *5*(8), 1934–1953. DOI: 10.1016/j.joule.2021.07.007

Chen, X., & Lv, H. (2022). Intelligent control of nanoparticle synthesis on microfluidic chips with machine learning. *NPG Asia Materials*, *14*(1), 1–20. DOI: 10.1038/s41427-022-00416-1

Chen, Z., Wang, R., Guo, J., & Wang, X. (2024). The role and future prospects of artificial intelligence algorithms in peptide drug development. *Biomedicine and Pharmacotherapy, 175*, 116709. DOI: 10.1016/j.biopha.2024.116709 PMID: 38713945

Chichi. (n.d.). *Nanotechnology and Education: Preparing the Next Generation of Future Leaders | SpringerLink*. Retrieved September 2, 2025, from https://link.springer.com/chapter/10.1007/978-981-97-6184-5_7

Chiu, T. K. F., & Li, Y. (2023). How can emerging technologies impact STEM education? *Journal for STEM Education Research*, *6*(3), 375–384. DOI: 10.1007/s41979-023-00113-w

Choi, Y. J., & Sawada, K. (2022). Physical Sensors: Fluorescence Sensors. *Encyclopedia of Sensors and Biosensors: Volume 1-4, First Edition, 1–4*, 1–19. https://doi.org/DOI: 10.1016/B978-0-12-822548-6.00095-9

Chou, W. C. (2025). Machine learning and artificial intelligence in nanomedicine: Connecting nanoparticle design with toxicology and translational applications. *Wiley Interdisciplinary Reviews. Nanomedicine and Nanobiotechnology*, *17*(3), e70027. DOI: 10.1002/wnan.70027 PMID: 40813104

Chowdhary, K. R. (2020). Natural language processing. In *Fundamentals of artificial intelligence* (pp. 603–649). Springer., DOI: 10.1007/978-81-322-3972-7_19

Christopoulos, A., Pellas, N., & Laakso, M.-J. (2020). A learning analytics theoretical framework for STEM education virtual reality applications. *Education Sciences*, *10*(11), 317. Advance online publication. DOI: 10.3390/educsci10110317

Chung, S. S., Lau, K. Y., & Zhang, C. (2011). Generation of and control measures for e-waste in Hong Kong. *Waste Management (New York, N.Y.)*, *31*, 544–554.

Collection and Preservation of Blood Evidence from Crime Scenes. (n.d.). Retrieved February 23, 2025, from https://www.crime-scene-investigator.net/blood.html

Collins, C., Dennehy, D., Conboy, K., & Mikalef, P. (2021). Artificial intelligence in information systems research: A systematic literature review and research agenda. *International Journal of Information Management*, *60*, 102383. DOI: 10.1016/j.ijinfomgt.2021.102383

Cools, H., & Diakopoulos, N. (2023). Towards guidelines for guidelines on the use of generative AI in newsrooms. Medium. Available at: https://generative-ai-newsroom.com/towards-guidelines-for-guidelines-on-the-use-of-generative-ai-in-

Cools, H., & Koliska, M. (2024). News automation and algorithmic transparency in the newsroom: the case of the Washington Post. Journal. Stud. 25, 662–680. doi: DOI: 10.1080/1461670X.2024.2326636

Cools, H., & Diakopoulos, N. (2024). Uses of generative AI in the newsroom: Mapping journalists ' perceptions of perils and possibilities. *J. Pract.*, *27*, 1–19. DOI: 10.1080/17512786.2024.2394558

Costanzo, H., Gooch, J., & Frascione, N. (2023). Nanomaterials for optical biosensors in forensic analysis. *Talanta*, *253*, 123945. DOI: 10.1016/J.TALANTA.2022.123945 PMID: 36191514

Cucchiella, F., D'Adamo, I., Koh, S., & Rosa, P. (2015). Recycling of WEEEs: An economic assessment of present and future e-waste streams. *Renewable & Sustainable Energy Reviews*, *51*, 263–272.

Cuestas, J., Gil-Alana, L., & Taylor, K. (2012). Inflation convergence in Central and Eastern Europe with a view to adopting the Euro. Sheffield: Sheffield Economics Research Papers no 2012005, Department of Economics, University of Sheffield.

Cui, J., & Forssberg, E. (2003). Mechanical recycling of waste electric and electronic equipment: A review. *Journal of Hazardous Materials*, *99*(3), 243–263.

Cui, J., & Zhang, L. (2008). Metallurgical recovery of metals from electronic waste: A review. *Journal of Hazardous Materials*, *158*(2–3), 228–256.

Curreli, M., & Rakich, S. S. (2020). An integrated approach to teaching and learning nanotechnology: The Omni Nano model. *Journal of Research in Innovative Teaching & Learning*, *13*(1), 141–145. DOI: 10.1108/JRIT-02-2020-0012

D'Mello, S. K., & Graesser, A. C. (2015). Feeling, thinking, and computing with affect-aware learning technologies. In *The Oxford Handbook of Affective Computing* (pp. 419–434). Oxford University Press.

Dahy, G., Soliman, M. M., Alshater, H., Slowik, A., & Ella Hassanien, A. (2023). Optimized deep networks for the classification of nanoparticles in scanning electron microscopy imaging. *Computational Materials Science*, *223*, 112135. Advance online publication. DOI: 10.1016/j.commatsci.2023.112135

Dai, Y., Maresca, M., & Al Musaimi, O. (2024). Exploring FDA-Approved Frontiers: Insights into Natural and Engineered Peptide Analogues in the GLP-1, GIP, GHRH, CCK, ACTH, and α-MSH Realms. *Biomolecules 2024, Vol. 14, Page 264, 14*(3), 264. https://doi.org/DOI: 10.3390/BIOM14030264

Dardenne, E., Ishiyama, N., Lin, T. A., & Lucas, M. C. (2023). Current and emerging therapies for Achondroplasia: The dawn of precision medicine. *Bioorganic & Medicinal Chemistry, 87*. Advance online publication. DOI: 10.1016/j.bmc.2023.117275 PMID: 37156065

Das, K. P., & Chandra, J. (2022). Nanoparticles and convergence of artificial intelligence for targeted drug delivery for cancer therapy: Current progress and challenges. *Frontiers in Medical Technology, 4*, 1067144. DOI: 10.3389/FMEDT.2022.1067144/BIBTEX PMID: 36688144

Dauparas, J., Anishchenko, I., Bennett, N., Bai, H., Ragotte, R. J., Milles, L. F., Wicky, B. I. M., Courbet, A., de Haas, R. J., Bethel, N., Leung, P. J. Y., Huddy, T. F., Pellock, S., Tischer, D., Chan, F., Koepnick, B., Nguyen, H., Kang, A., Sankaran, B., & Baker, D. (2022). Robust deep learning–based protein sequence design using ProteinMPNN. *Science, 378*(6615), 49–56. DOI: 10.1126/science.add2187 PMID: 36108050

Dave, M., & Patel, N. (2023). Artificial intelligence in healthcare and education. *British Dental Journal, 234*(10), 761–764. DOI: 10.1038/s41415-023-5845-2 PMID: 37237212

Davydiuk, N., Krieg, E., & Gaitzsch, J.. (2025). The rising danger of AI-generated images in nanomaterials science and what we can do about it. *Nature Nanotechnology, 20*, 567–569. DOI: 10.1038/s41565-025-02009-9 PMID: 40954352

De la Hoz, K., Coelho, F., & Prodigioso, V. (2023). IA para periodistas. Available at: https://www.prodigiosovolcan.com/sismogramas/ia-periodistas/

Dean, S. N., & Walper, S. A. (2020). Variational autoencoder for the generation of antimicrobial peptides. *ACS Omega, 5*(33), 20746–20754. DOI: 10.1021/acsomega.0c00442 PMID: 32875208

Debnath, P., Deshmukh, S. M., Laroia, M., Selvakumar, P., Manjunath, T. C., & Bhattacharya, S. (2025). Challenges Faced by Countries in Renewable Energy Adoption. In Özsungur, F., Chaychi Semsari, M., & Küçük Bayraktar, H. (Eds.), *Geopolitical Landscapes of Renewable Energy and Urban Growth* (pp. 295–320). IGI Global Scientific Publishing., DOI: 10.4018/979-8-3693-8814-3.ch010

Demir, D., Bolgen, N., & Vaseashta, A. (2023). Electrospun nanofibers for biomedical, sensing, and energy harvesting functions. *Polymers, 15*(21), 4253. Advance online publication. DOI: 10.3390/polym15214253 PMID: 37959933

Demirer, G. S., Silva, T. N., Jackson, C. T., Thomas, J. B., Ehrhardt, D. W., Rhee, S. Y., & Landry, M. P. (2021). Nanotechnology to advance CRISPR–Cas genetic engineering of plants. *Nature Nanotechnology*, *16*(3), 243–250. PMID: 33712738

Demirer, G. S., Zhang, H., Matos, J. L., Goh, N. S., Cunningham, F. J., Sung, Y., & Landry, M. P. (2019). High aspect ratio nanomaterials enable delivery of functional genetic material without DNA integration in mature plants. *Nature Nanotechnology*, *14*(5), 456–464. PMID: 30804481

Deng, J., Yang, Z., Ojima, I., Samaras, D., & Wang, F. (2022). Artificial intelligence in drug discovery: Applications and techniques. *Briefings in Bioinformatics*, *23*(1), 1–19. DOI: 10.1093/bib/bbab430 PMID: 34734228

De, S., James, B., Ji, J., Wasti, S., Zhang, S., Kore, S., & Zhao, X. (2023). Biomass-derived composites for various applications. In *Advances in Bioenergy* (Vol. 8, pp. 145–196). Elsevie.

Desjardins, J. (2014). Global gold mine and deposit rankings 2013. *Visual Capitalist*. https://www.visualcapitalist.com/global-gold-mine-and-deposit-rankings-2013/

Deubzer, O. (2011). *E-waste management in Germany*. United Nations University Institute for Sustainability and Peace (UNU-ISP).

Dey, S., Veerendra, G. T. N., Babu, P. A., Manoj, A. P., & Nagarjuna, K. (2024). Degradation of plastics waste and its effects on biological ecosystems: A scientific analysis and comprehensive review. *Biomedical Materials & Devices*, *2*(1), 70–112.

Di Vita, G. (2001). Technological change, growth and waste recycling. *Energy Economics*, *23*(5), 549–567.

Díaz Noci, J., Peña-Fernández, S., Meso-Ayerdi, K., & Larrondo-Ureta, A. (2024). The influence of AI in the media workforce: How companies use an Array of legal remedies. *Trípodos.*, *55*, 33–54. DOI: 10.51698/tripodos.2024.55.03

Díaz-Campo, J., & Chaparro-Domínguez, M. A. (2020). Periodismo computacional y ética. *Revista ICONO*, *14*(18), 10–32. DOI: 10.7195/ri14.v18i1.1488

Díaz-López, M. M., Bustos, R.-H., Guevara-Ramírez, J. F., & García, J.-C. (2025). Understanding Barriers to Nanotechnology Education in Medical Specialty Programs: A Qualitative Nominal Group Study Among Postgraduate Students. *Advances in Medical Education and Practice*, *16*, 1481–1499. DOI: 10.2147/AMEP.S523258 PMID: 40861025

Díaz-Marcos, J., & Mendoza Gonzálvez, J. (2025a). Shaping the future of nanotechnology: A perspective on education, dissemination and ethics. *Frontiers in Nanotechnology*, *7*. Advance online publication. DOI: 10.3389/fnano.2025.1524578

Dimitrakakis, E., Janz, A., Bilitewski, B., & Gidarakos, E. (2009). Small WEEE: Determining recyclables and hazardous substances in plastics. *Journal of Hazardous Materials*, *161*(2–3), 913–919.

Ding, R., Chen, J., Chen, Y., Liu, J., Bando, Y., & Wang, X. (2024). Unlocking the potential: Machine learning applications in electrocatalyst design for electrochemical hydrogen energy transformation. *Chemical Society Reviews*, *53*(22), 11390–11461. DOI: 10.1039/D4CS00844H PMID: 39382108

Dorouka, P., & Kalogiannakis, M. (2024a). Teaching nanotechnology concepts in early-primary education: An experimental study using digital games. *International Journal of Science Education*, *46*(13), 1311–1338. DOI: 10.1080/09500693.2023.2286299

Dorouka, P., Papadakis, S., & Kalogiannakis, M. (2021). The contribution of the health crisis to young children's nano-literacy through STEAM education. *Hellenic Journal of STEM Education*, *2*(1), 1–7. DOI: 10.51724/hjstemed.v2i1.18

Drucker, D. J., & Nauck, M. A. (2006). The incretin system: Glucagon-like peptide-1 receptor agonists and dipeptidyl peptidase-4 inhibitors in type 2 diabetes. *Lancet*, *368*(9548), 1696–1705. DOI: 10.1016/S0140-6736(06)69705-5 PMID: 17098089

Dubrovsky, I., Dmitrenko, A., Dmitrenko, A., Serov, N., & Vinogradov, V. (2024). Unveiling the potential of AI for nanomaterial morphology prediction. *arXiv preprint*, arXiv:2406.02591. https://arxiv.org/abs/2406.02591

Dudek, A., Dąbek, A., Zborowska, I., & Lichosik, J. (2024). Integrating artificial intelligence in forensic science. *E-Methodology*, *10*(10), 15–28. DOI: 10.15503/EMET2023.15.28

Duman, G. M., Kongar, E., & Gupta, S. M. (2019). Estimation of electronic waste using optimized multivariate grey models. *Waste Management & Research*, *37*(3), 241–249.

Duman, G. M., Kongar, E., & Gupta, S. M. (2020). Predictive analysis of electronic waste for reverse logistics operations: A comparison of improved univariate grey models. *Soft Computing*, *24*, 15747–15762.

Duro-Castano, A., Moreira Leite, D., Forth, J., Deng, Y., Matias, D., Noble Jesus, C., & Battaglia, G. (2020). Designing peptide nanoparticles for efficient brain delivery. *Advanced Drug Delivery Reviews*, *160*, 52–77. DOI: 10.1016/J.ADDR.2020.10.001 PMID: 33031897

Duro, J. A., Lauk, C., Kastner, T., Erb, K. H., & Haberl, H. (2020). Global inequalities in food consumption, cropland demand and land-use efficiency: A decomposition analysis. *Global Environmental Change*, *64*, 102124.

Dwivedi, Y. K., Hughes, L., Ismagilova, E., Aarts, G., Coombs, C., Crick, T., Duan, Y., Dwivedi, R., Edwards, J., Eirug, A., Galanos, V., Ilavarasan, P. V., Janssen, M., Jones, P., Kar, A. K., Kizgin, H., Kronemann, B., Lal, B., Lucini, B., & Williams, M. D. (2021). Artificial intelligence (AI): Multidisciplinary perspectives on emerging challenges, opportunities, and agenda for research, practice and policy. *International Journal of Information Management*, *57*, 101994. Advance online publication. DOI: 10.1016/j.ijinfomgt.2019.08.002

Dwivedy, M., & Mittal, R. K. (2010). Future trends in computer waste generation in India. *Waste Management (New York, N.Y.)*, *30*, 2265–2277.

Dwivedy, M., & Mittal, R. K. (2010a). Estimation of future outflows of e-waste in India. *Waste Management (New York, N.Y.)*, *30*, 483–491.

Dybas, J., Alcicek, F. C., Wajda, A., Kaczmarska, M., Zimna, A., Bulat, K., Blat, A., Stepanenko, T., Mohaissen, T., Szczesny-Malysiak, E., Perez-Guaita, D., Wood, B. R., & Marzec, K. M. (2022). Trends in biomedical analysis of red blood cells – Raman spectroscopy against other spectroscopic, microscopic and classical techniques. *Trends in Analytical Chemistry*, *146*. Advance online publication. DOI: 10.1016/J.TRAC.2021.116481

Elahi, M., Afolaranmi, S. O., Martinez Lastra, J. L., & Perez Garcia, J. A. (2023). A comprehensive literature review of the applications of AI techniques through the lifecycle of industrial equipment. *Discover Artificial Intelligence*, *3*(1), 43. DOI: 10.1007/s44163-023-00089-x

Elbehri, A., & Chestnov, R. (2021). *Digital agriculture in action: Artificial intelligence for agriculture*. Food & Agriculture Org.

Electronics TakeBack Coalition. (2016). Facts and figures on e-waste and recycling. https://blog.midwestind.com/wp-content/uploads/2017/11/Facts_and_Figures_on_EWaste_and_Recycling1.pdf

ElektroG. (2024). *Act on the placing on the market, return and environmentally sound disposal of electrical and electronic equipment (Electrical and Electronic Equipment Act)*. https://www.umweltbundesamt.de/en/topics/waste-resources/product-stewardship-waste-management/electrical-electronic-waste/electrical-electronic-equipment-act#purpose-of-the-elektrog

El-Rayes, N., Chang, A., & Shi, J. (2024). Blockchain Technology Toward a Circular Economy: A Future Paradigm. In *Circular Economy and Sustainable Development: A Necessary Nexus for a Sustainable Future* (pp. 335–351). Springer International Publishing.

Erumban, A., & Das, D. (2016). Information and communication technology and economic growth in India. *Telecommunications Policy*, *40*(5), 412–431.

Esangbedo, C. O., Zhang, J., Esangbedo, M. O., Kone, S. D., & Xu, L. (2023). The role of industry-academia collaboration in enhancing educational opportunities and outcomes under the digital driven Industry 4.0. *Journal of Infrastructure. Policy and Development*, *8*(1). Advance online publication. DOI: 10.24294/jipd.v8i1.2569

Esteva, A., Robicquet, A., Ramsundar, B., Kuleshov, V., DePristo, M., Chou, K., Cui, C., Corrado, G., Thrun, S., & Dean, J. (2019). A guide to deep learning in healthcare. *Nature Medicine*, *25*(1), 24–29. DOI: 10.1038/s41591-018-0316-z PMID: 30617335

European Commission. (2020). *Digital education action plan 2021–2027*. https://education.ec.europa.eu

Evans, B. J., King, A. T., Katsifis, A., Matesic, L., & Jamie, J. F. (2020). Methods to enhance the metabolic stability of peptide-based PET radiopharmaceuticals. *Molecules (Basel, Switzerland)*, *25*(10), 2314. DOI: 10.3390/molecules25102314 PMID: 32423178

Fadel, H. H., Mahmoud, K. F., Saad, R., Lotfy, S. N., El-Aleem, F. S. A., & Ahmed, M. (2019). Nanotechnology for enhancing the production of meat-like flavour from enzymatic hydrolyzed soybean protein isolate. *Acta Scientiarum Polonorum. Technologia Alimentaria*, *18*(3), 279–292. PMID: 31569910

Fariq, A., Khan, T., & Yasmin, A. (2017). Microbial synthesis of nanoparticles and their potential applications in biomedicine. *Journal of Applied Biomedicine*, *15*(4), 241–248. DOI: 10.1016/j.jab.2017.03.004

Farooq, M. A., Hannan, F., Islam, F., Ayyaz, A., Zhang, N., Chen, W., & Zhou, W. (2022). The potential of nanomaterials for sustainable modern agriculture: Present findings and future perspectives. *Environmental Science. Nano*, *9*(6), 1926–1951.

Feng, M., Yao, W., An, J., Huang, Z., Yuan, Y., & Yao, Y. (2022). Prediction method of graphene defect modification based on neural network. *Mobile Information Systems*, *2022*, 3580762. Advance online publication. DOI: 10.1155/2022/3580762

Feng, Y., Zhong, L., Bilal, M., Tan, Z., Hou, Y., Jia, S., & Cui, J. (2018). Enzymes@ZIF-8 nanocomposites with protection nanocoating: Stability and acid-resistant evaluation. *Polymers*, *11*(1), 27. PMID: 30960011

Ferentinos, K. P. (2018). Deep learning models for plant disease detection and diagnosis. *Computers and Electronics in Agriculture, 145*, 311–318.

Feynman, R. P. (1960). There's plenty of room at the bottom. *Engineering and Science, 23*(5), 22–36.

Fırat, F. (2019). "Robot journalism", in the International Encyclopedia of Journalism Studies. Int. Encycl. J. Stu. 14, 1–5. doi: DOI: 10.1002/9781118841570.iejs0243

Floridi, L. (Ed.). (2021). *Ethics, governance, and policies in artificial intelligence*. Springer., DOI: 10.1007/978-3-030-81907-1

Fonash, S. J. (2001). Education and training of the nanotechnology workforce. *Journal of Nanoparticle Research, 3*(1), 79–82. DOI: 10.1023/A:1011408311390

Food Wastage Footprint (Project). (2013). *Food wastage footprint: impacts on natural resources: summary report*. Food & Agriculture Organization of the UN (FAO).

Forja-Pena, T., García-Orosa, B., & López-García, X. (2024). The ethical revolution: Challenges and reflections in the face of the integration of artificial intelligence in digital journalism. *Communicatio Socialis, 27*, 237–254. DOI: 10.15581/003.37.3.237-254

Forti, V., Baldé, C., Kuehr, R., & Bel, G. (2020). *The Global E-waste Monitor 2020: Quantities, flows and the circular economy potential. United Nations University (UNU)/International Telecommunication Union (ITU)/International Solid Waste Association*. ISWA.

Fosgerau, K., & Hoffmann, T. (2015). Peptide therapeutics: Current status and future directions. *Drug Discovery Today, 20*(1), 122–128. DOI: 10.1016/j.drudis.2014.10.003 PMID: 25450771

Frazier, P. I. (2018). A tutorial on Bayesian optimization. *arXiv*. https://arxiv.org/abs/1807.02811

Freddi, A., & Salmon, M. (2019). Introduction to the Taguchi method. In *Design principles and methodologies* (pp. 159–180). Springer., DOI: 10.1007/978-3-319-95342-7_7

Friedersdorf, L. E. (2020a). Developing the Workforce of the Future: How the National Nanotechnology Initiative Has Supported Nanoscale Science and Engineering Education in the United States. *IEEE Nanotechnology Magazine, 14*(4), 13–20. DOI: 10.1109/MNANO.2020.2994799

Fu, P. P. (2014). Introduction to the special issue: Nanomaterials—toxicology and medical applications. *Yao Wu Shi Pin Fen Xi, 22*(1), 1. PMID: 24673899

Fu, Y., Liu, T., Wang, H., Wang, Z., Hou, L., Jiang, J., & Xu, T. (2024). Applications of nanomaterial technology in biosensing. *Journal of Science. Advanced Materials and Devices*, *9*(2), 100694. DOI: 10.1016/j.jsamd.2024.100694

Ganesan, T. (n.d.). INTEGRATING ARTIFICIAL INTELLIGENCE AND CLOUD COMPUTING FOR THE DEVELOPMENT OF A SMART EDUCATION MANAGEMENT PLATFORM: DESIGN, IMPLEMENTATION, AND PERFORMANCE ANALYSIS. *International Journal of Engineering, 11*(2).

Gao, P., & Adnan, M. (2025). Overview of emerging electronics technologies for artificial intelligence: A review. *Materials Today Electronics*, *11*, 100136. DOI: 10.1016/j.mtelec.2025.100136

Gao, P., Zhang, J., Qiu, H., & Zhao, S. (2021). A general QSPR protocol for the prediction of atomic/inter-atomic properties: A fragment based graph convolutional neural network (F-GCN). *Physical Chemistry Chemical Physics*, *23*(23), 13242–13249. DOI: 10.1039/D1CP00677K PMID: 34086015

Gao, W., Wang, G., Li, J., Li, H., Ren, L., Wang, Y., & Kong, L. (2025). Insights into Synthesis and Optimization Features of Reverse Osmosis Membrane Using Machine Learning. *Materials (Basel)*, *18*(4), 1–14. DOI: 10.3390/ma18040840 PMID: 40004363

García-Orosa, B., Canavilhas, J., & Vázquez-Herrero, J. (2023). Algorithms and communication: A systematized literature review. *Comunicar*, *31*, 9–21. DOI: 10.3916/C74-2023-01

Garlapati, V. (2016). E-waste in India and developed countries: Management, recycling, business and biotechnological initiatives. *Renewable & Sustainable Energy Reviews*, *54*, 874–881.

Garrison, D. R., & Vaughan, N. D. (2008). *Blended learning in higher education: Framework, principles, and guidelines*. Jossey-Bass.

Gavin, H. P. (2020). The Levenberg-Marquardt algorithm for nonlinear least squares curve-fitting problems. https://people.duke.edu/~hpgavin/ce281/lm.pdf

Gawde, U., Chakraborty, S., Waghu, F. H., Barai, R. S., Khanderkar, A., Indraguru, R., Shirsat, T., & Idicula-Thomas, S. (2023). CAMPR4: a database of natural and synthetic antimicrobial peptides. *Nucleic Acids Research, 51*(1 D), D377–D383. https://doi.org/DOI: 10.1093/nar/gkac933

George, D., Chiang Lin, B., & Chen, Y. (2015). A circular economy model of economic growth. *Environmental Modelling & Software*, *73*, 60–63.

Ghattas, N. I., & Carver, J. S. (2012a). Integrating nanotechnology into school education: A review of the literature. *Research in Science & Technological Education, 30*(3), 271–284. DOI: 10.1080/02635143.2012.732058

Gheibi, M., Masoomi, S. R., Magala, M. U., & Amir, M. (2024). Artificial Intelligence (AI) in Adsorption Process of Heavy Metals. *Systematic Reviews, 2*(2), 57–78.

Ghosh, M., & Kumar, R. (2024a). Regulatory Issues in Nanotechnology. In Prasad, M., Kumar, R., Ghosh, M., Syed, S. M., & Chakravarti, S. (Eds.), *Nanotechnology Theranostics in Livestock Diseases and Management* (pp. 765–788). Springer Nature., DOI: 10.1007/978-981-16-1610-5_31

Ghosh, S., Das, N., & Nath, A. (2020). Educational platforms for applied AI: A comparative study. [IJEAT]. *International Journal of Engineering and Advanced Technology, 9*(5), 18–23.

Gill, S. S., & Buyya, R. (2024). Transforming Research with Quantum Computing. *Journal of Economy and Technology.* https://doi.org/DOI: 10.1016/j.ject.2024.07.001

Gilpin, L. H., Bau, D., Yuan, B. Z., Bajwa, A., Specter, M., & Kagal, L. (2018). Explaining explanations: An overview of interpretability of machine learning. *Proceedings - 2018 IEEE 5th International Conference on Data Science and Advanced Analytics, DSAA 2018*, 80–89. https://doi.org/DOI: 10.1109/DSAA.2018.00018

Giziński, S., Preibisch, G., Kucharski, P., Tyrolski, M., Rembalski, M., Grzegorczyk, P., & Gambin, A. (2024). Enhancing antigenic peptide discovery: Improved MHC-I binding prediction and methodology. *Methods (San Diego, Calif.), 224*, 1–9. DOI: 10.1016/j.ymeth.2024.01.016 PMID: 38295891

Glass, S., Schmidt, M., Merten, P., Abdul Latif, A., Fischer, K., Schulze, A., Friederich, P., & Filiz, V. (2023). Design of Modified Polymer Membranes Using Machine Learning. *ACS Applied Materials & Interfaces*. Advance online publication. DOI: 10.1021/acsami.3c18805 PMID: 38600824

Glavaš, M., Gitlin-Domagalska, A., Dębowski, D., Ptaszyńska, N., Łęgowska, A., & Rolka, K. (2022). Vasopressin and Its Analogues: From Natural Hormones to Multitasking Peptides. *International Journal of Molecular Sciences, 23*(6), 3068. DOI: 10.3390/ijms23063068 PMID: 35328489

Goel, A., & Polepeddi, L. (2016). *Jill Watson: A virtual teaching assistant for online education*. Georgia Institute of Technology.

Gokhale, A. S., & Satyanarayanajois, S. (2014). Peptides and peptidomimetics as immunomodulators. *Immunotherapy, 6*(6), 755–774. DOI: 10.2217/IMT.14.37 PMID: 25186605

Gola, J., Webel, J., Britz, D., Guitar, A., Staudt, T., Winter, M., & Mücklich, F. (2019). Objective microstructure classification by support vector machine (SVM) using a combination of morphological parameters and textural features for low carbon steels. *Computational Materials Science*, *160*, 186–196. DOI: 10.1016/j.commatsci.2019.01.006

Goles, M., Daza, A., Cabas-Mora, G., Sarmiento-Varón, L., Sepúlveda-Yañez, J., Anvari-Kazemabad, H., Davari, M. D., Uribe-Paredes, R., Olivera-Nappa, Á., Navarrete, M. A., & Medina-Ortiz, D. (2024). Peptide-based drug discovery through artificial intelligence: Towards an autonomous design of therapeutic peptides. *Briefings in Bioinformatics*, *25*(4). Advance online publication. DOI: 10.1093/bib/bbae275 PMID: 38856172

Golev, A., Corder, G. D., & Rhamdhani, M. A. (2019). Estimating flows and metal recovery values of waste printed circuit boards in Australian e-waste. *Minerals Engineering*, *137*, 171–176.

Golev, A., Schmeda-Lopez, D. R., Smart, S. K., Corder, G. D., & McFarland, E. W. (2016). Where next on e-waste in Australia? *Waste Management (New York, N.Y.)*, *58*, 348–358.

Gonda, L., D'Ans, P., & Degrez, M. (2019). A comparative assessment of WEEE collection in an urban and rural context: Case study on desktop computers in Belgium. *Resources, Conservation and Recycling*, *142*, 131–142.

Grasso, G., Zane, D., & Dragone, R. (2019). Microbial nanotechnology: Challenges and prospects for green biocatalytic synthesis of nanoscale materials for sensoristic and biomedical applications. *Nanomaterials (Basel, Switzerland)*, *10*(1), 11. Advance online publication. DOI: 10.3390/nano10010011 PMID: 31861471

Grisoni, F., Neuhaus, C. S., Gabernet, G., Müller, A. T., Hiss, J. A., & Schneider, G. (2018). Designing Anticancer Peptides by Constructive Machine Learning. *ChemMedChem*, *13*(13), 1300–1302. DOI: 10.1002/cmdc.201800204 PMID: 29679519

Groschner, C. K., Choi, C., & Scott, M. C. (2021). Machine learning pipeline for segmentation and defect identification from high-resolution transmission electron microscopy data. *Microscopy and Microanalysis*, *27*(3), 549–556. DOI: 10.1017/S1431927621000386 PMID: 33952372

Gubala, V., Giovannini, G., Kunc, F., Monopoli, M. P., & Moore, C. J. (2020). Dye-doped silica nanoparticles: synthesis, surface chemistry and bioapplications. *Cancer Nanotechnology 2020 11:1, 11*(1), 1–43. https://doi.org/DOI: 10.1186/S12645-019-0056-X

Guda, A. A., Kirichkov, M. V., Shapovalov, V. V., Muravlev, A. I., Pashkov, D. M., Guda, S. A., Bagliy, A. P., Chapek, S. V., & Soldatov, A. V. (2023). Machine learning analysis of reaction parameters in UV-mediated synthesis of gold nanoparticles. *The Journal of Physical Chemistry. C, Nanomaterials and Interfaces*, *127*(2), 1097–1108. DOI: 10.1021/acs.jpcc.2c06625

Gul, W., Shams, M., & Al-Khalili, D. (2022). SRAM Cell Design Challenges in Modern Deep Sub-Micron Technologies: An Overview. *Micromachines*, *13*(8), 8. Advance online publication. DOI: 10.3390/mi13081332

Gunasegaram, D. R., Barnard, A. S., Matthews, M. J., Jared, B. H., Andreaco, A. M., Bartsch, K., & Murphy, A. B. (2024). Machine learning-assisted in-situ adaptive strategies for the control of defects and anomalies in metal additive manufacturing. *Additive Manufacturing*, *81*, 104013. DOI: 10.1016/j.addma.2024.104013

Guntuboina, C., Das, A., Mollaei, P., Kim, S., & Barati Farimani, A. (2023). Peptide-BERT: A Language Model Based on Transformers for Peptide Property Prediction. *The Journal of Physical Chemistry Letters*, *14*(46), 10427–10434. DOI: 10.1021/acs.jpclett.3c02398 PMID: 37956397

Guo, X., & Yan, K. (2017). Estimation of obsolete cellular phones generation: A case study of China. *The Science of the Total Environment*, *575*, 321–329.

Gupta, A., Eral, H. B., Hatton, T. A., & Doyle, P. S. (2016). Nanoemulsions: Formation, properties and applications. *Soft Matter*, *12*(11), 2826–2841. PMID: 26924445

Gupta, P., Ding, B., Guan, C., & Ding, D. (2024). Generative AI: A systematic review using topic modeling techniques. *Data and Information Management*, *8*(2), 100066. DOI: 10.1016/j.dim.2024.100066

Güriş, B., Tiftikçigil, B., & Tıraşoğlu, M. (2017). Testing for unemployment hysteresis in Turkey: Evidence from nonlinear unit root tests. *Quality & Quantity*, *51*, 35–46.

Gusukuma, M., & Kahhat, R. (2018). Electronic waste after a digital TV transition: Material flows and stocks. *Resources, Conservation and Recycling*, *138*, 142–150.

Gu, X. (2025). Artificial intelligence for biomaterials and tissue engineering. *Journal of Intelligent Medicine*, *2*(1), 3–6. DOI: 10.1002/jim4.70000

Hagelüken, C. (2006). Improving metal returns and eco-efficiency in electronics recycling—A holistic approach for interface optimisation between. *2006 IEEE International Symposium on Electronics and the Environment* (pp. 218–223). IEEE.

Hagelüken, C. (2014). Recycling of (critical) metals. In Gunn, G. (Ed.), *Critical Metals Handbook* (pp. 41–69). Wiley.

Hairom, N. H. H., Soon, C. F., Mohamed, R. M. S. R., Morsin, M., Zainal, N., Nayan, N., Zulkifli, C. Z., & Harun, N. H. (2021). A review of nanotechnological applications to detect and control surface water pollution. *Environmental Technology & Innovation*, *24*, 102032. DOI: 10.1016/j.eti.2021.102032

Hajikarimlou, M., Hooshyar, M., Moutaoufik, M. T., Aly, K. A., Azad, T., Takallou, S., Jagadeesan, S., Phanse, S., Said, K. B., Samanfar, B., Bell, J. C., Dehne, F., Babu, M., & Golshani, A. (2022). A computational approach to rapidly design peptides that detect SARS-CoV-2 surface protein S. *NAR Genomics and Bioinformatics*, *4*(3). Advance online publication. DOI: 10.1093/NARGAB/LQAC058 PMID: 36004308

Hamimi, E., Ibrohim, I., Kusumaningrum, I. K., Putra, W. E., Sudrajat, A. K., Hayuana, W., Kundariati, M., Hajar, M. U., & Malek, N. A. N. N. (2025). Developing Hi Nano mobile apps as integrated education instrument on nanotechnology and cancer treatment learning. *International Journal of Mobile Learning and Organisation*, *19*(3), 279–298. DOI: 10.1504/IJMLO.2025.147176

Hammami, R., Zouhir, A., Le Lay, C., Ben Hamida, J., & Fliss, I. (2010). BACTIBASE second release: A database and tool platform for bacteriocin characterization. *BMC Microbiology*, *10*(1), 1–5. DOI: 10.1186/1471-2180-10-22 PMID: 20105292

Han, B., Lin, Y., Yang, Y., Mao, N., Li, W., Wang, H., Yasuda, K., Wang, X., Fatemi, V., Zhou, L., Wang, J. I.-J., Ma, Q., Cao, Y., Rodan-Legrain, D., Bie, Y., Navarro-Moratalla, E., Klein, D., MacNeill, D., Wu, S., & Palacios, T. (2020). Deep-learning-enabled fast optical identification and characterization of 2D materials. *Advanced Materials*, *32*(29), 2000953. Advance online publication. DOI: 10.1002/adma.202000953 PMID: 32519397

Han, C., Zhao, A., Varughese, E., & Sahle-Demessie, E. (2018). Evaluating weathering of food packaging polyethylene-nano-clay composites: Release of nanoparticles and their impacts. *NanoImpact*, *9*, 61–71. PMID: 29226269

Han, E. L., Tang, S., Kim, D., Murray, A. M., Swingle, K. L., Hamilton, A. G., Mrksich, K., Padilla, M. S., Palanki, R., Li, J. J., & Mitchell, M. J. (2024). Peptide-Functionalized Lipid Nanoparticles for Targeted Systemic mRNA Delivery to the Brain. *Nano Letters*, *25*(2), 800–810. DOI: 10.1021/ACS.NANOLETT.4C05186 PMID: 39688915

Haq, I. U., Mohiz, A., Spadafora, G., Palermo, A. M., Bilotta, E., Felicetti, F., & Lamonaca, F. (2025). AI in the Network of Extended Reality-Enabled Laboratories for STEM Education: Current Applications and Future Potential for Adaptive Learning. In G. Carbone & G. Quaglia (Eds.), *Proceedings of I4SDG Workshop 2025—IFToMM for Sustainable Development Goals* (pp. 373–382). Springer Nature Switzerland. https://doi.org/DOI: 10.1007/978-3-031-91179-8_39

Hara, M., Nakanishi, H., Yoneyama, K., Saito, K., & Takada, A. (2016). Effects of storage conditions on forensic examinations of blood samples and bloodstains stored for 20 years. *Legal Medicine (Tokyo)*, *18*, 81–84. DOI: 10.1016/J.LEGALMED.2016.01.003 PMID: 26832383

Harris, D., McCabe, B., & Leybourne, S. (2003). Some limit theory for autocovariances whose order depends on sample size. *Econometric Theory*, *19*, 829–864.

Harvey, D., Leybourne, S., & Xiao, B. (2008). A powerful test for linearity when the order of integration is unknown. (Article 2). *Studies in Nonlinear Dynamics and Econometrics*, *12*(3), 1–24.

Hassan, S. A.-D. H., Almaliki, M. N. S., Hussein, Z. A., Albehadili, H. M., Rabeea Banoon, S., Abboodi, A., & Al-Saady, M. (2023a). Development of Nanotechnology by Artificial Intelligence: A Comprehensive Review. *Journal of Nanostructures*, *13*(4), 915–932. DOI: 10.22052/JNS.2023.04.002

Hatamizadeh, A., Yin, H., Heinrich, G., Kautz, J., & Molchanov, P. (2023, July). Global context vision transformers. In *International Conference on Machine Learning* (pp. 12633-12646). PMLR.

He, A., & Jin, X. (2018). NARNET-based prognostics modeling for deteriorating systems under dynamic operating conditions. In *Proceedings of the 14th IEEE International Conference on Automation Science and Engineering (CASE)* (pp. 1–6). IEEE.

He, J.-H. (2025). *AI-based Nanotechnology: Breakthroughs*. Applications, Challenges, and the Road Ahead., DOI: 10.2174/0118722105384013250623094407

Helberger, N., & Diakopoulos, N. (2023). ChatGPT and the AI act. *International Policy Review*, *12*, 1682. DOI: 10.14763/2023.1.1682

Hemme, C. L., Carley, R., Norton, A., Ghumman, M., Nguyen, H., Ivone, R., Menon, J. U., Shen, J., Bertin, M., King, R., Leibovitz, E., Bergstrom, R., & Cho, B. (2023). Developing Virtual and Augmented Reality Applications for Science, Technology, Engineering and Math Education. *BioTechniques*, *75*(1), 11–20. DOI: 10.2144/btn-2023-0029 PMID: 37291856

Henriksen, D. (2014). Full STEAM ahead: Creativity in excellent STEM teaching practices. *The STEAM Journal*, 1(2), Article 15.

Hermann, E. (2022). Artificial intelligence and mass personalization of communication content—An ethical and literacy perspective. *New Media & Society*, *24*, 1258–1277. DOI: 10.1177/14614448211022702

Hidayatulloh, I., Pambudi, S., Surjono, H., & Sukardiyono, T. (2021). Gamification on Chatbot-Based Learning Media: A Review and Challenges. *Elinvo (Electronics, Informatics, and Vocational Education)*, *6*(1), 71–80. https://doi.org/DOI: 10.21831/elinvo.v6i1.43705

Hie, B. L., & Yang, K. K. (2022). Adaptive machine learning for protein engineering. *Current Opinion in Structural Biology*, *72*, 145–152. DOI: 10.1016/j.sbi.2021.11.002 PMID: 34896756

Hofmann, T., Lowry, G. V., Ghoshal, S., Tufenkji, N., Brambilla, D., Dutcher, J. R., & Wilkinson, K. J. (2020). Technology readiness and overcoming barriers to sustainably implement nanotechnology enabled plant agriculture. *Nature Food*, *1*(7), 416–425.

Hofmann, T., Lowry, G. V., Ghoshal, S., Tufenkji, N., Brambilla, D., Dutcher, J. R., & Wilkinson, K. J. (2020). Technology readiness and overcoming barriers to sustainably implement nanotechnology-enabled plant agriculture. *Nature Food*, *1*(7), 416–425.

Holmes, W., Bialik, M., & Fadel, C. (2019). *Artificial intelligence in education: Promises and implications for teaching and learning*. Center for Curriculum Redesign. https://curriculumredesign.org/wp-content/uploads/AI-in-Education.pdf

Horton, M. K., Huck, P., Yang, R. X., Munro, J. M., Dwaraknath, S., Ganose, A. M., Kingsbury, R. S., Wen, M., Shen, J. X., Mathis, T. S., Kaplan, A. D., Berket, K., Riebesell, J., George, J., Rosen, A. S., Spotte-Smith, E. W. C., McDermott, M. J., Cohen, O. A., Dunn, A., & Persson, K. A. (2025). Accelerated data-driven materials science with the Materials Project. *Nature Materials*, *24*(11), 1447–1457. DOI: 10.1038/s41563-025-02272-0 PMID: 40610673

Hossain, A. S., Uddin, M. M., Veettil, V. N., & Fawzi, M. (2018). Nano-cellulose based nano-coating biomaterial dataset using corn leaf biomass: An innovative biodegradable plant biomaterial. *Data in Brief*, *17*, 162–168. PMID: 29877503

Hossain, N., Mobarak, M. H., Mimona, M. A., Islam, M. A., Hossain, A., Zohura, F. T., & Chowdhury, M. A. (2023). Advances and significances of nanoparticles in semiconductor applications – A review. *Results in Engineering*, *19*, 101347. DOI: 10.1016/j.rineng.2023.101347

Hou, X., Zaks, T., Langer, R., & Dong, Y. (2021). Lipid nanoparticles for mRNA delivery. *Nature Reviews. Materials*, *6*, 1078–1094. DOI: 10.1038/s41578-021-00358-0 PMID: 34394960

Huang, J., Xu, Y., Xue, Y., Huang, Y., Li, X., Chen, X., Xu, Y., Zhang, D., Zhang, P., Zhao, J., & Ji, J. (2023). Identification of potent antimicrobial peptides via a machine-learning pipeline that mines the entire space of peptide sequences. *Nature Biomedical Engineering 2023 7:6*, *7*(6), 797–810. https://doi.org/DOI: 10.1038/s41551-022-00991-2

Huang, M., Li, Z., & Zhu, H. (2022). Recent advances of graphene and related materials in artificial intelligence. *Advanced Intelligent Systems*, *4*(10), 2200077. Advance online publication. DOI: 10.1002/aisy.202200077

Huang, W., & Barnard, A. S. (2022). Federated data processing and learning for collaboration in the physical sciences. *Machine Learning: Science and Technology*, *3*(4), 045023. DOI: 10.1088/2632-2153/aca87c

Huang, W., Xiao, G., Zhang, Y., & Min, W. (2021). Research progress and application opportunities of nanoparticle–protein corona complexes. *Biomedicine and Pharmacotherapy*, *139*, 111541. Advance online publication. DOI: 10.1016/j.biopha.2021.111541 PMID: 33848776

Hubbell, J. A., & Chilkoti, A. (2012). Nanomaterials for drug delivery. *Science*, *337*(6092), 303–305. DOI: 10.1126/science.1219657 PMID: 22822138

Hue-Shin, L., & Dine, P. (2025). Adaptive Learning Technologies in Structural Engineering Education: Immediate and Contextual Evaluations. *Innovative Journal of Educational Research and Insights*, 77–84.

Humphrey, W., Dalke, A., & Schulten, K. (1996). VMD: Visual molecular dynamics. *Journal of Molecular Graphics*, *14*(1), 33–38. PMID: 8744570

Husseiny, F. (n.d.). *Education's Role in Integrating Nanotechnology in the Curriculum for the Future: Education Book Chapter | IGI Global Scientific Publishing*. Retrieved September 2, 2025, from https://www.igi-global.com/chapter/educations-role-in-integrating-nanotechnology-in-the-curriculum-for-the-future/348778

Hwang, G.-J., & Tu, Y.-F. (2021). Roles and research trends of artificial intelligence in mathematics education: A bibliometric mapping analysis and systematic review. *Mathematics*, *9*(6), 584. Advance online publication. DOI: 10.3390/math9060584

Hwang, J. S., Kim, S. G., Shin, T. H., Jang, Y. E., Kwon, D. H., & Lee, G. (2022). Development of Anticancer Peptides Using Artificial Intelligence and Combinational Therapy for Cancer Therapeutics. *Pharmaceutics*, *14*(5), 997. DOI: 10.3390/PHARMACEUTICS14050997/S1 PMID: 35631583

Hyndman, R. J., & Athanasopoulos, G. (2018). *Forecasting: Principles and practice* (2nd ed.). OTexts.

IBM. (2022). *IBM AI education program*. https://www.ibm.com

Ibrahim, M., Jemei, S., Wimmer, G., & Hissel, D. (2016). Nonlinear autoregressive neural network in an energy management strategy for battery/ultra-capacitor hybrid electrical vehicles. *Electric Power Systems Research*, *136*, 262–269.

Ikhlayel, M. (2016). Differences of methods to estimate generation of waste electrical and electronic equipment for developing countries: Jordan as a case study. *Resources, Conservation and Recycling*, *108*, 134–139.

Imamguluyev, R., Hasanova, P., Imanova, T., Mammadova, A., Hajizade, S., & Samadova, Z. (2024). AI-POWERED EDUCATIONAL TOOLS: TRANSFORMING LEARNING IN THE DIGITAL ERA. *International Research Journal of Modernization in Engineering Technology and Science*, *6*, 920–929. DOI: 10.56726/IRJMETS65040

Ingólfsson, H. I., Bhatia, H., Aydin, F., Oppelstrup, T., López, C. A., Stanton, L. G., Carpenter, T. S., Wong, S., Di Natale, F., Zhang, X., Moon, J. Y., Stanley, C. B., Chavez, J. R., Nguyen, K., Dharuman, G., Burns, V., Shrestha, R., Goswami, D., Gulten, G., & Streitz, F. H. (2023). Machine Learning-Driven Multiscale Modeling: Bridging the Scales with a Next-Generation Simulation Infrastructure. *Journal of Chemical Theory and Computation*, *19*(9), 2658–2675. DOI: 10.1021/acs.jctc.2c01018

Intel. (2021). *AI for youth: Global impact report*. Intel Corporation.

Ioscote, F., Gonçalves, A., & Quadros, C. (2024). Artificial intelligence in journalism: A ten-year retrospective of scientific articles (2014-2023). *J. Media*, *5*, 873–891. DOI: 10.3390/journalmedia5030056

Iravani, S., & Varma, R. S. (2022). Nanosponges for Water Treatment: Progress and Challenges. *Applied Sciences (Switzerland)*, *12*(9). Advance online publication. DOI: 10.3390/app12094182

Işıkdağ, Ü., Hepsağ, A., Bıyıklı, S. İ., Öz, D., Bekdaş, G., & Geem, Z. W. (2023). Estimating construction material indices with ARIMA and optimized NARNETs. *Computers, Materials & Continua*, *74*(1), 113–129.

Islam, M. T., & Huda, N. (2019). E-waste in Australia: Generation estimation and untapped material recovery and revenue potential. *Journal of Cleaner Production*, *237*, 117787.

Islam, R., Nazifa, T., Yuniarto, A., Uddin, A., Salmiati, S., & Shahid, S. (2019). An empirical study of construction and demolition waste generation and implication of recycling. *Waste Management (New York, N.Y.)*, *95*, 10–21.

Ismail, S., & Vasudev, H. (2025). Artificial Intelligence for Personalized Electric Vehicle Energy Management. In Emerging Multisector Applications of AI and IoT (pp. 221-260). IGI Global Scientific Publishing.

Ismail, R., Bocsik, A., Katona, G., Gróf, I., Deli, M. A., & Csóka, I. (2019). Encapsulation in Polymeric Nanoparticles Enhances the Enzymatic Stability and the Permeability of the GLP-1 Analog, Liraglutide, Across a Culture Model of Intestinal Permeability. *Pharmaceutics*, *11*(11), 599. DOI: 10.3390/PHARMACEUTICS11110599 PMID: 31726699

Ismail, S., & Vasudev, H. (2024, September). EIROB: An Emotionally Intelligent ROBot for Empathetic Support. In *2024 IEEE International Conference on Signal Processing, Informatics, Communication and Energy Systems (SPICES)* (pp. 1-6). IEEE.

Ismail, S., & Vasudev, H. (2025). Artificial Intelligence for Personalized Electric Vehicle Energy Management. In *Emerging Multisector Applications of AI and IoT* (pp. 221–260). IGI Global Scientific Publishing.

Ismail, S., Vasudev, H., & Cheriyan, J. (2025). Enhancing Seed Quality and Predictive Accuracy in Indian Agriculture: Case Study at RARS, Kottayam, Kerala. In *Advances in Seed Quality Evaluation and Improvement* (pp. 491–528). Springer.

Israel, M. J., & Amer, A. (2022). Rethinking data infrastructure and its ethical implications in the face of automated digital content. *AI and Ethics*, *3*, 427–439. DOI: 10.1007/s43681-022-00169-1

Jablonka, K. M., Ongari, D., Moosavi, S. M., & Smit, B. (2020). Big-data science in porous materials: Materials genomics and machine learning. *Chemical Reviews*, *120*(16), 8066–8129. DOI: 10.1021/acs.chemrev.0c00004 PMID: 32520531

Jain, A., Montoya, J., Dwaraknath, S., Zimmermann, N. E. R., Dagdelen, J., Horton, M., Huck, P., Winston, D., Cholia, S., Ong, S. P., & Persson, K. (2018). The Materials Project: Accelerating materials design through theory-driven data and tools. In Andreoni, W., & Yip, S. (Eds.), *Handbook of materials modeling* (pp. 1–34). Springer., DOI: 10.1007/978-3-319-42913-7_68-1

Jain, A., & Sareen, R. (2006). E-waste assessment methodology and validation in India. *Journal of Material Cycles and Waste Management*, *8*, 40–45.

Jaroszewicz, W., Morcinek-Orłowska, J., Pierzynowska, K., Gaffke, L., & W grzyn, G. (2022). Phage display and other peptide display technologies. *FEMS Microbiology Reviews*, *46*(2), 1–25. DOI: 10.1093/femsre/fuab052 PMID: 34673942

Jeffcoat, P., Di Lernia, C., Hardy, C., New, E. J., & Chrzanowski, W. (2024). (Re) imagining purpose: A framework for sustainable nanotechnology innovation. *NanoImpact*, *35*, 100511. DOI: 10.1016/j.impact.2024.100511 PMID: 38750963

Jeyaraman, M., Balaji, S., Jeyaraman, N., & Yadav, S. (2023). Unraveling the Ethical Enigma: Artificial Intelligence in Healthcare. *Cureus*, *15*(8). Advance online publication. DOI: 10.7759/CUREUS.43262 PMID: 37692617

Jiao, L. H., & Barakat, N. (2011). Incorporation of hands-on activities in learning nanomaterials. In *ASME 2011 International Mechanical Engineering Congress and Exposition* (pp. 109–115). American Society of Mechanical Engineers. https://doi .org/DOI: 10.1115/IMECE2011-62598

Jia, Z., Chen, J., Xu, X., Kheir, J., Hu, J., Xiao, H., Peng, S., Hu, X. S., Chen, D., & Shi, Y. (2023). The importance of resource awareness in artificial intelligence for healthcare is significant. *Nature Machine Intelligence*, *5*(7), 687–698. DOI: 10.1038/S42256-023-00670-0

Ji, Z., Guo, W., Sakkiah, S., Liu, J., Patterson, T., & Hong, H. (2021). Nanomaterial databases: Data sources for promoting design and risk assessment of nanomaterials. *Nanomaterials (Basel, Switzerland)*, *11*(6), 1599. Advance online publication. DOI: 10.3390/nano11061599 PMID: 34207026

Jones, B. (2023). *Generative AI and journalism. A rapid risk-based review*. Edinburgh Research Explorer.

Joo, S. H. (2012). Cyclic peptides as therapeutic agents and biochemical tools. *Biomolecules & Therapeutics*, *20*(1), 19–26. DOI: 10.4062/biomolther.2012.20.1.019 PMID: 24116270

Jordan, K. (2015). Massive open online course completion rates revisited: Assessment, length and attrition. *International Review of Research in Open and Distance Learning*, *16*(3), 341–358.

Jordan, M. I., & Mitchell, T. M. (2015). Machine learning: Trends, perspectives, and prospects. *Science*, *349*(6245), 255–260. DOI: 10.1126/science.aaa8415 PMID: 26185243

Jorgenson, D., Ho, M., & Samuels, J. (2016). The impact of information technology on postwar US economic growth. *Telecommunications Policy*, *40*(5), 398–411.

Joshi, H., Malik, A., Aggarwal, S., Munde, M., Maitra, S. S., Adlakha, N., & Bhatnagar, R. (2019). In-vitro detection of phytopathogenic fungal cell wall by polyclonal sera raised against trimethyl chitosan nanoparticles. *International Journal of Nanomedicine*, •••, 10023–10033. PMID: 31908457

Journalism, A. I. (2024). AI journalism starter pack. Available at: https://www.skeyesmedia.org/documents/bo_filemanager/AI-journalism-Starter-Pack-_-A-guide-by-JournalismAI.pdf (Accessed July 14, 2024).

Jumper, J., Evans, R., Pritzel, A., Green, T., Figurnov, M., Ronneberger, O., Tunyasuvunakool, K., Bates, R., Žídek, A., Potapenko, A., Bridgland, A., Meyer, C., Kohl, S. A. A., Ballard, A. J., Cowie, A., Romera-Paredes, B., Nikolov, S., Jain, R., Adler, J., & Hassabis, D. (2021). Highly accurate protein structure prediction with AlphaFold. *Nature*, *596*(7873), 583–589. DOI: 10.1038/s41586-021-03819-2 PMID: 34265844

Kaas, Q., Westermann, J. C., Halai, R., Wang, C. K. L., & Craik, D. J. (2008). ConoServer, a database for conopeptide sequences and structures. *Bioinformatics (Oxford, England)*, *24*(3), 445–446. DOI: 10.1093/bioinformatics/btm596 PMID: 18065428

Kagan, C. R. (2016). At the nexus of food security and safety: Opportunities for nanoscience and nanotechnolgy. *ACS Nano*, *10*(3), 2985–2986. PMID: 27000520

Kagan, C. R. (2016). At the nexus of food security and safety: Opportunities for nanoscience and nanotechnology. *ACS Nano*, *10*(3), 2985–2986. PMID: 27000520

Kajita, S., Ohba, N., Suzumura, A., Tajima, S., & Asahi, R. (2020). Discovery of superionic conductors by ensemble-scope descriptor. *NPG Asia Materials*, *12*(1), 31. Advance online publication. DOI: 10.1038/s41427-020-0211-1

Kalia, A., Bhardwaj, K., Bhardwaj, P., Abd-Elsalam, K. A., Valis, M., & Kuca, K. (2021). Nanohybrid Antifungals for Control of Plant Diseases: Current Status and Future Perspectives. *Journal of Fungi (Basel, Switzerland)*, *2021*(7), 48. PMID: 33450851

Kalia, A., Kaur, M., Shami, A., Jawandha, S. K., Alghuthaymi, M. A., Thakur, A., & Abd-Elsalam, K. A. (2021). Nettle-leaf extract derived ZnO/CuO nanoparticle-biopolymer-based antioxidant and antimicrobial nanocomposite packaging films and their impact on extending the post-harvest shelf life of guava fruit. *Biomolecules*, *11*(2), 224. PMID: 33562547

Kalidindi, S. R. (2022). Digital twins for materials. *Frontiers in Materials*. Advance online publication. DOI: 10.3389/fmats.2022.818535

Kalidindi, S. R., & De Graef, M. (2015). Materials data science: Current status and future outlook. *Annual Review of Materials Research*, *45*(1), 171–193. DOI: 10.1146/annurev-matsci-070214-020844

Kalmykova, Y., Patricio, J., Rosado, L., & Berg, P. E. (2015). Out with the old, out with the new—The effect of transitions in TVs and monitors technology on consumption and WEEE generation in Sweden 1996–2014. *Waste Management (New York, N.Y.)*, *46*, 511–522.

Kalra, R., Patel, D., & Vaya, D. (2020). Role of artificial intelligence in nanotechnology: A review. *Journal of Computational Science*, *46*, 101189. Advance online publication. DOI: 10.1016/j.jocs.2020.101189

Kamali-Sarvestani, R., Weber, P., Clayton, M., Meyers, M., & Slade, S. (2020). Virtual Reality to Improve Nanotechnology Education: Development Methods and Example Applications. *IEEE Nanotechnology Magazine*, *14*(4), 29–38. DOI: 10.1109/MNANO.2020.2994802

Kamal, T., Ali, N., & Naseem, A, A., B Khan, S., & M Asiri, A. (2016). Polymer nanocomposite membranes for antifouling nanofiltration. *Recent Patents on Nanotechnology*, *10*(3), 189–201. PMID: 27136927

Kamilaris, A., Kartakoullis, A., & Prenafeta-Boldú, F. X. (2017). A review on the practice of big data analysis in agriculture. *Computers and Electronics in Agriculture*, *143*, 23–37.

Kamilaris, A., & Prenafeta-Boldú, F. X. (2018). Deep learning in agriculture: A survey. *Computers and Electronics in Agriculture*, *147*, 70–90.

Kantaros, A., Ganetsos, T., & Petrescu, F. I. T. (2024). Transforming object design and creation: Biomaterials and contemporary manufacturing leading the way. *Biomimetics*, *9*(1), 48. PMID: 38248622

Kapralov, A., Vlasova, I. I., Feng, W., Maeda, A., Walson, K., Tyurin, V. A., Huang, Z., Aneja, R. K., Carcillo, J., Bayir, H., & Kagan, V. E. (2009). Peroxidase Activity of Haemoglobin ·Haptoglobin Complexes: COVALENT AGGREGATION AND OXIDATIVE STRESS IN PLASMA AND MACROPHAGES. *The Journal of Biological Chemistry*, *284*(44), 30395. DOI: 10.1074/JBC.M109.045567 PMID: 19740759

Karimirad, R., Behnamian, M., Dezhsetan, S., & Sonnenberg, A. (2018). Chitosan nanoparticles-loaded Citrus aurantium essential oil: A novel delivery system for preserving the postharvest quality of Agaricus bisporus. *Journal of the Science of Food and Agriculture*, *98*(13), 5112–5119. PMID: 29635845

Karniadakis, G. E.. (2021). Physics-informed machine learning. *Nature Reviews. Physics*, *3*, 422–440. DOI: 10.1038/s42254-021-00314-5

Kastanaki, E., & Giannis, A. (2022). Forecasting quantities of critical raw materials in obsolete feature and smart phones in Greece: A path to circular economy. *Journal of Environmental Management*, *306*, 114566.

Kay, J. (2008). Lifelong learner modeling for lifelong personalized pervasive learning. *IEEE Transactions on Learning Technologies*, *1*(4), 215–228. DOI: 10.1109/TLT.2009.9

Kazancoglu, Y., Ozbiltekin, M., Ozen, Y. D. O., & Sagnak, M. (2020). A proposed sustainable and digital collection and classification center model to manage e-waste in emerging economies. *Journal of Enterprise Information Management*, *34*(1), 267–291.

Kececi, B., Dengiz, O., Dengiz, B., Sumer, E., Kilic, A., Ceki, E., Inan, B., & Cicek, S. (2018). WEEE estimation and determination of collection points: A case for the Municipality of Cankaya. *Pamukkale University Journal of Engineering Sciences*, *24*(5), 692–704.

Kelly, C. J., Karthikesalingam, A., Suleyman, M., Corrado, G., & King, D. (2019). Key challenges for delivering clinical impact with artificial intelligence. *BMC Medicine*, *17*(1), 1–9. DOI: 10.1186/S12916-019-1426-2/PEER-REVIEW PMID: 31665002

Khakpour, A., Florescu, L., Tilley, R., Jiang, H., Swaminathan Iyer, K., & Gustavo Carneiro. (2025). *arXiv preprint*, arXiv:2503.13798. https://arxiv.org/abs/2503.13798

Khan, A. S. (2020a). Application of Simulation for Emulating and Visualizing Nanotechnology. In Vucinic, D., Rodrigues Leta, F., & Janardhanan, S. (Eds.), *Advances in Visualization and Optimization Techniques for Multidisciplinary Research: Trends in Modelling and Simulations for Engineering Applications* (pp. 307–316). Springer., DOI: 10.1007/978-981-13-9806-3_11

Khan, F., & Joya, Y. (2021). Artificial intelligence applications in nanotechnology: Opportunities and challenges. *Nano Today, 37*, 101071. Advance online publication. DOI: 10.1016/j.nantod.2020.101071

Khan, I., Saeed, K., & Khan, I. (2019). Nanoparticles: Properties, applications and toxicities. *Arabian Journal of Chemistry, 12*(7), 908–931. DOI: 10.1016/j.arabjc.2017.05.011

Khanna, N., Wadhwa, J., Pitroda, A., Shah, P., Schoop, J., & Sarıkaya, M. (2022). Life cycle assessment of environmentally friendly initiatives for sustainable machining: A short review of current knowledge and a case study. *Sustainable Materials and Technologies, 32*, e00413.

Khan, P., Idrees, D., Moxley, M. A., Corbett, J. A., Ahmad, F., Von Figura, G., Sly, W. S., Waheed, A., & Hassan, M. I. (2014). Luminol-Based Chemiluminescent Signals: Clinical and Non-clinical Application and Future Uses. *Applied Biochemistry and Biotechnology, 173*(2), 333. DOI: 10.1007/S12010-014-0850-1 PMID: 24752935

Khater, M., De La Escosura-Muñiz, A., Quesada-González, D., & Merkoçi, A. (2019). Electrochemical detection of plant virus using gold nanoparticle-modified electrodes. *Analytica Chimica Acta, 1046*, 123–131. PMID: 30482289

Kheimi, M., Almadani, M., & Zounemat-Kermani, M. (2024). Stochastic (S[ARIMA]), shallow (NARnet, NAR-GMDH, OS-ELM), and deep learning (LSTM, Stacked-LSTM, CNN-GRU) models, application to river flow forecasting. *Acta Geophysica, 72*(4), 2679–2693.

Khraief, N., Shahbaz, M., Heshmati, A., & Azam, M. (2020). Are unemployment rates in OECD countries stationary? Evidence from univariate and panel unit root tests. *The North American Journal of Economics and Finance, 51*(100838), 1–15.

Kiddee, P., Naidu, R., & Wong, M. (2013). Electronic waste management approaches: An overview. *Waste Management (New York, N.Y.), 33*(5), 1237–1250.

Kilani, M., Torabi, K., & Mao, G. (2018). Application of virtual laboratories and molecular simulations in teaching nanoengineering to undergraduate students. *Computer Applications in Engineering Education, 26*(5), 1527–1538. DOI: 10.1002/cae.21940

Kim, H., Han, J., & Han, T. Y.-J. (2020). Machine vision-driven automatic recognition of particle size and morphology in SEM images. *Nanoscale*, *12*(37), 19461–19469. DOI: 10.1039/D0NR04140H PMID: 32960204

Kim, S., Oguchi, M., Yoshida, A., & Terazono, A. (2013). Estimating the amount of WEEE generated in South Korea by using the population balance model. *Waste Management (New York, N.Y.)*, *33*, 474–483.

Kim, T., & Lee, S. D. (2022). Designing for green and grey: Insights from single-use plastic water bottles. *International Journal of Environmental Research and Public Health*, *19*(3), 1423. PMID: 35162446

Klein, J. T. (2017). Beyond interdisciplinarity. In Frodeman, R. (Ed.), *The Oxford handbook of interdisciplinarity* (pp. 35–49). Oxford University Press.

Klick, A., Kim, S., & Kim, M.-W. (2017). Using virtual reality (VR) for education in nanoscience. *Microscopy and Microanalysis*, *23*(S1), 2302–2303. DOI: 10.1017/S143192761701217X

Klimeck, G., McLennan, M., Brophy, S. P., Adams, G. B.III, & Lundstrom, M. S. (2008). nanoHUB.org: Advancing education and research in nanotechnology. *Computing in Science & Engineering*, *10*(5), 17–23. DOI: 10.1109/MCSE.2008.120

Knoppe, S., Boudon, J., Dolamic, I., Dass, A., & Bürgi, T. (2011). Size exclusion chromatography for semipreparative scale separation of Au38(SR)24 and Au40(SR)24 and larger clusters. *Analytical Chemistry*, *83*(13), 5056–5061. DOI: 10.1021/ac200789v PMID: 21609127

Koedinger, K. R., Corbett, A. T., & Perfetti, C. (2012). The knowledge-learning-instruction framework: Bridging the science-practice chasm. *Cognitive Science*, *36*(5), 757–798. DOI: 10.1111/j.1551-6709.2012.01245.x PMID: 22486653

Kong, X., Gao, P., Wang, J., Fang, Y., & Hwang, K. C. (2023). Advances of medical nanorobots for future cancer treatments. *Journal of Hematology & Oncology*, *16*, 74. DOI: 10.1186/s13045-023-01463-z

Köroğlu, U., Sağlam, N., Tamer, U., Akcan, R., Boyaci, I. H., & Evran, E. (2024). Identification of blood at simulated crime scenes using silver nanoparticles with SERS. *Turkish Journal of Chemistry*, *48*(4), 676–690.

Kosai, S., Kishita, Y., & Yamasue, E. (2020). Estimation of the metal flow of WEEE in Vietnam considering lifespan transition. *Resources, Conservation and Recycling*, *154*, 104621.

Koshta, N., Patra, S., & Singh, S. P. (2021). Estimation of E-waste at micro level for reverse logistics: A case of Delhi. *Journal of Cleaner Production, 314*, 128063.

Kozlovskii, I., & Popov, P. (2021). Protein-Peptide Binding Site Detection Using 3D Convolutional Neural Networks. *Journal of Chemical Information and Modeling, 61*(8), 3814–3823. DOI: 10.1021/ACS.JCIM.1C00475/SUPPL_FILE/CI1C00475_SI_011.ZIP PMID: 34292750

Krausová, A., & Moravec, V. (2022). Disappearing authorship: Ethical protection of AI-generated news from the perspective of copyright and other Laws. *JIPITEC, 13*, 132.

Krishnan, A., Sundaram, T., Nagappan, B., & Devarajan, Y., & Bhumika. (2024). Integrating artificial intelligence in nanomembrane systems for advanced water desalination. *Results in Engineering, 24*(November), 103321. DOI: 10.1016/j.rineng.2024.103321

Kritchenkov, A. S., Egorov, A. R., Volkova, O. V., Artemjev, A. A., Kurliuk, A. V., Le, T. A., & Khrustalev, V. N. (2021). Novel biopolymer-based nanocomposite food coatings that exhibit active and smart properties due to a single type of nanoparticles. *Food Chemistry, 343*, 128676. PMID: 33250292

Kuehr, R., Velasquez, G., & Williams, E. (2003). Computers and the environment: Understanding and managing their impact. In Kuehr, R., & Williams, E. (Eds.), *Computers and the Environment* (pp. 1–17). Kluwer Academic Publishers.

Kulkarni, S. K. (2015). *Nanotechnology: Principles and Practices*. Springer International Publishing., DOI: 10.1007/978-3-319-09171-6

Kumar Singh, S., Rai, R., Pradip Khawale, R., Patel, D., Bielecki, D., Nguyen, R., & Zhang, Z. (2024). Deep learning in computational design synthesis: A comprehensive review. *Journal of Computing and Information Science in Engineering, 24*(4), 040801.

Kumar, A., Holuszko, M., & Espinosa, D. (2017). E-waste: An overview on generation, collection, legislation and recycling practices. *Resources, Conservation and Recycling, 122*, 32–42.

Kumar, A., & Kumar, M. (2024). Recent progress in emerging 2-D transition-metal dichalcogenide-based gas sensors: A review. *IEEE Sensors Journal, 24*(16), 25221–25226. DOI: 10.1109/JSEN.2024.3422972

Kumar, G., & Ardekani, A. M. (2025). Machine-Learning Framework to Predict the Performance of Lipid Nanoparticles for Nucleic Acid Delivery. *ACS Applied Bio Materials, 8*(5), 3717–3727. DOI: 10.1021/ACSABM.4C01716 PMID: 40267508

Kumar, P., & Shrihari, S. (2007). Estimation and material flow analysis of waste electrical and electronic equipment (WEEE)—A case study of Mangalore City, Karnataka, India. *Proceedings of the International Conference on Sustainable Solid Waste Management* (pp. 5–7). Chennai, India.

Kumar, R., & Reji, M. (2023). Response surface methodology (RSM): An overview to analyze multivariate data. *The Indian Journal of Medical Research*, *9*(3), 241–248. DOI: 10.18231/j.ijmr.2022.042

Kumar, S., Nehra, M., Dilbaghi, N., Marrazza, G., Hassan, A. A., & Kim, K. H. (2019). Nano-based smart pesticide formulations: Emerging opportunities for agriculture. *Journal of Controlled Release*, *294*, 131–153. PMID: 30552953

Kumar, S., Sanap, S. N., Vasoya, M., Handa, M., Pandey, P., Khopade, A., & Sawant, K. K. (2024). USFDA-approved parenteral peptide formulations and excipients: Industrial perspective. *Journal of Drug Delivery Science and Technology*, *95*, 105589. DOI: 10.1016/j.jddst.2024.105589

Kuper, J., & Hojsik, M. (2008). *Poisoning the poor: Electronic waste in Ghana*. Greenpeace.

Kusch, S., & Hills, C. (2017). The link between e-waste and GDP—New insights from data from the pan-European region. *Resources*, *6*(1), 15.

Lao, Y., & You, Y. (2024). Unraveling generative AI in BBC News: Application, impact, literacy and governance. *Transforming Government: People. Process and Policy*, *18*(4), 551–566. DOI: 10.1108/TG-01-2024-0022

Lau, H. Y., Wu, H., Wee, E. J., Trau, M., Wang, Y., & Botella, J. R. (2017). Specific and sensitive isothermal electrochemical biosensor for plant pathogen DNA detection with colloidal gold nanoparticles as probes. *Scientific Reports*, *7*(1), 38896. PMID: 28094255

Laycock, B. G., Chan, C. M., & Halley, P. J. (2024). A review of computational approaches used in the modeling, design, and manufacturing of biodegradable and biobased polymers. *Progress in Polymer Science*, *157*, 101874. DOI: 10.1016/j.progpolymsci.2024.101874

Lèbre, É., & Corder, G. (2015). Integrating industrial ecology thinking into the management of mining waste. *Resources*, *4*(4), 765–786.

LeCun, Y., Bengio, Y., & Hinton, G. (2015). Deep learning. *Nature*, *521*(7553), 436–444. DOI: 10.1038/nature14539 PMID: 26017442

Lee, J. S., & Kim, P. M. (2025). Design of peptides with non-canonical amino acids using flow matching. In bioRxiv (p. 2025.07.31.667780). Cold Spring Harbor Laboratory. https://doi.org/DOI: 10.1101/2025.07.31.667780

Lee, M. (2023). Recent Advances in Deep Learning for Protein-Protein Interaction Analysis: A Comprehensive Review. *Molecules 2023, Vol. 28, Page 5169, 28*(13), 5169. https://doi.org/DOI: 10.3390/MOLECULES28135169

Lee, J. (2019). Equity and access in STEM education: Insights and recommendations. *Journal of Educational Technology Development and Exchange, 12*(1), 1–15.

Lee, J., Bagheri, B., & Kao, H. A. (2015). A cyber-physical systems architecture for Industry 4.0-based manufacturing systems. *Manufacturing Letters, 3*, 18–23. DOI: 10.1016/j.mfglet.2014.12.001

Legates, D. R., & Davis, R. E. (1999). Evaluating the use of "goodness-of-fit" measures in hydrologic and hydroclimatic model validation. *Water Resources Research, 35*(1), 233–241.

Lemmon, G., Kaufmann, K., & Meiler, J. (2012). Prediction of HIV-1 Protease/ Inhibitor Affinity using RosettaLigand. *Chemical Biology & Drug Design, 79*(6), 888–896. DOI: 10.1111/J.1747-0285.2012.01356.X PMID: 22321894

Li, Y., & Wang, S. Lv advancement of precision medicine, Z., Wang, Z., Zhao, Y., Xie, Y., Xu, Y., Qian, L., Yang, Y., Zhao, Z., & Zhang, J. (2024). Transforming the synthesis of carbon nanotubes with machine learning models and automation (CARCO). *arXiv preprint*, arXiv:2404.01006. https://arxiv.org/abs/2404.01006

Li, A., Li, B., Lu, B., Yang, D., Hou, S., & Song, X. (2022). Generation estimation and material flow analysis of retired mobile phones in China. *Environmental Science and Pollution Research International, 29*, 75626–75635.

Li, B., Yang, J., Lu, B., & Song, X. (2015b). Estimation of retired mobile phones generation in China: A comparative study on methodology. *Waste Management (New York, N.Y.), 35*, 247–254.

Li, J., Yang, J., & Liu, L. (2015a). Development potential of e-waste recycling industry in China. *Waste Management & Research, 33*(6), 533–542.

Li, J., Zeng, X., Chen, M., Ogunseitan, O., & Stevels, A. (2015). Control-Alt-Delete: Rebooting solutions for the e-waste problem. *Environmental Science & Technology, 49*(12), 7095–7108.

Li, L., Zhao, C., Zhang, Y., Yao, J., Yang, W., Hu, Q., & Cao, C. (2017). Effect of stable antimicrobial nano-silver packaging on inhibiting mildew and in storage of rice. *Food Chemistry*, *215*, 477–482. PMID: 27542501

Lin, J., Liu, Y., Sui, H., Sagoe-Crentsil, K., & Duan, W. (2022). Microstructure of graphene oxide–silica-reinforced OPC composites: Image-based characterization and nano-identification through deep learning. *Cement and Concrete Research*, *154*, 106737. Advance online publication. DOI: 10.1016/j.cemconres.2022.106737

Liu, A., Zhang, X., Liu, Z., Li, Y., Peng, X., Li, X., Qin, Y., Hu, C., Qiu, Y., Jiang, H., Wang, Y., Li, Y., Tang, J., Liu, J., Guo, H., Deng, T., Peng, S., Tian, H., & Ren, T.-L. (2024). The Roadmap of 2D Materials and Devices Toward Chips. *Nano-Micro Letters*, *16*(1), 119. DOI: 10.1007/s40820-023-01273-5

Liu, H., Gao, H., & Jia, F. (2021). The value of convolutional neural network-based magnetic resonance imaging image segmentation algorithm to guide targeted controlled release of doxorubicin nanopreparation. *Contrast Media & Molecular Imaging*, *2021*, 9032017. Advance online publication. DOI: 10.1155/2021/9032017 PMID: 34385899

Liu, J., Li, R., & Yang, B. (2020). Carbon Dots: A New Type of Carbon-Based Nanomaterial with Wide Applications. *ACS Central Science*, *6*(12), 2179. DOI: 10.1021/ACSCENTSCI.0C01306 PMID: 33376780

Liu, M., Li, H., Zhou, H., Zhang, H., & Huang, G. (2024). Development of machine learning methods for mechanical problems associated with fibre composite materials: A review. *Composites Communications*, •••, 101988.

Liu, X., Pei, W., Zhao, J., Xu, R., Zhong, Y., & Yu, D. (2025). A Review of Wafer-Level Packaging Technology for SAW and BAW Filters. *Micromachines*, *16*(3), 3. Advance online publication. DOI: 10.3390/mi16030320

Liu, Y., Park, S., Yi, H., & Feiock, R. (2020). Evaluating the employment impact of recycling performance in Florida. *Waste Management (New York, N.Y.)*, *101*, 283–290.

Li, X.. (2022). Deep learning for microscopy image analysis of nanostructures. *ACS Nano*, *16*(4), 5678–5690. DOI: 10.1021/acsnano.1c11025

Li, Y., Rokayya, S., Jia, F., Nie, X., Xu, J., Elhakem, A., & Helal, M. (2021). Shelf-life, quality, safety evaluations of blueberry fruits coated with chitosan nano-material films. *Scientific Reports*, *11*(1), 55. PMID: 33420183

Lohse, J., Winteler, S., & Wulf-Schnabel, J. (1998). Collection targets for waste from electrical and electronic equipment (WEEE) the directorate general (DG XI) environment. In Nuclear safety and civil protection of the Commission of the European Communities. Ökopol Institut für Ökologie und Politik.

London, N., Raveh, B., Cohen, E., Fathi, G., & Schueler-Furman, O. (2011). Rosetta FlexPepDock web server - High resolution modeling of peptide-protein interactions. *Nucleic Acids Research*, *39*(SUPPL. 2), W249–W253. DOI: 10.1093/nar/gkr431 PMID: 21622962

López-García, X., & Vizoso, A. (2021). Periodismo de alta tecnología: Signo de los tiempos digitales del tercer milenio. *EPI*, *30*, 1. DOI: 10.3145/epi.2021.may.01

Luckin, R., Holmes, W., Griffiths, M., & Forcier, L. B. (2016). *Intelligence unleashed: An argument for AI in education*. Pearson. Holmes, W., Bialik, M., & Fadel, C. (2019). *Artificial intelligence in education: Promises and implications for teaching and learning*. Center for Curriculum Redesign.

Luckin, R., & Cukurova, M. (2019). Designing educational technologies in the age of AI: A learning sciences-driven approach. *British Journal of Educational Technology*, *50*(6), 2824–2838. DOI: 10.1111/bjet.12861

Luminol | Definition, Characteristics, Chemiluminescence, Blood Detection, & Forensics | Britannica. (n.d.). Retrieved February 23, 2025, from https://www.britannica.com/science/luminol

Lundstrom, M., & Klimeck, G. (2009). Nanoscience: A pathway for STEM education. *Science*, *324*(5933), 325–326.

Lv, Z., Cui, F., Zou, Q., Zhang, L., & Xu, L. (2021). Anticancer peptides prediction with deep representation learning features. *Briefings in Bioinformatics*, *22*(5), 1–14. DOI: 10.1093/BIB/BBAB008 PMID: 33529337

Lyshevski, S. E., Andersen, J. D., Boedo, S., Fuller, L., Raffaelle, R., Savakis, A., & Skuse, G. R. (2006). Multidisciplinary Undergraduate Nano-Science, Engineering and Technology Course. *2006 Sixth IEEE Conference on Nanotechnology, 1*, 399–402. https://doi.org/DOI: 10.1109/NANO.2006.247661

MacLeod, B. P., Parlane, F. G. L., Morrissey, T. D., Häse, F., Roch, L. M., Dettelbach, K. E., Moreira, R., Yunker, L. P. E., Rooney, M. B., Deeth, J. R., Lai, V., Ng, G. J., Situ, H., Zhang, R. H., Elliott, M. S., Haley, T. H., Dvorak, D. J., Aspuru-Guzik, A., Hein, J. E., & Berlinguette, C. P. (2020). Self-driving laboratory for accelerated discovery of thin-film materials. *Science Advances*, *6*(20), eaaz8867. Advance online publication. DOI: 10.1126/sciadv.aaz8867 PMID: 32426501

MacLeod, B. P., Parlane, F. G. L., & Rupnow, C. C.. (2022). A self-driving laboratory advances the Pareto front for material properties. *Nature Communications, 13*, 995. DOI: 10.1038/s41467-022-28580-6 PMID: 35194074

Magalini, F., & Huisman, J. (2007). Management of WEEE and cost models across the EU: Could the EPR principle lead us to a better environmental policy? In *2007 IEEE International Symposium on Electronics and the Environment* (pp. 143–148). IEEE.

Magazzino, C., Mele, M., & Schneider, N. (2020). The relationship between municipal solid waste and greenhouse gas emissions: Evidence from Switzerland. *Waste Management (New York, N.Y.), 113*, 508–520.

Magni, M., Jones, E. R., Bierkens, M. F. P., & van Vliet, M. T. H. (2025). Global energy consumption of water treatment technologies. Water Research, 277(December 2024), 123245. DOI: 10.1016/j.watres.2025.123245

Mahapatra, S. K., & Satapathy, A. (2023). Parametric analysis of erosion wear of sponge iron slag-filled ramie–epoxy composites using Taguchi and preference selection index methods. *Proceedings of the Institution of Mechanical Engineers. Part E, Journal of Process Mechanical Engineering*, •••, 09544089231193927.

Maharjan, R., Kim, K. H., Lee, K., Han, H. K., & Jeong, S. H. (2024). Machine learning-driven optimization of mRNA-lipid nanoparticle vaccine quality with XGBoost/Bayesian method and ensemble model approaches. *Journal of Pharmaceutical Analysis, 14*(11), 100996. DOI: 10.1016/J.JPHA.2024.100996 PMID: 39759971

Mahfouz, M. M., Piatek, A., & Stewart, C. N.Jr. (2014). Genome engineering via TALENs and CRISPR/Cas9 systems: Challenges and perspectives. *Plant Biotechnology Journal, 12*(8), 1006–1014. PMID: 25250853

Mahjoubi, S., Ye, F., Bao, Y., Meng, W., & Zhang, X. (2023). Identification and classification of exfoliated graphene flakes from microscopy images using a hierarchical deep convolutional neural network. *Engineering Applications of Artificial Intelligence, 119*, 105743. Advance online publication. DOI: 10.1016/j.engappai.2022.105743

Ma, J., Lee, S. M.-Y., Yi, C., & Li, C.-W. (2017). Controllable synthesis of functional nanoparticles by microfluidic platforms for biomedical applications – a review. *Lab on a Chip, 17*(2), 209–226. DOI: 10.1039/C6LC01049K PMID: 27991629

Ma, J., & Li, C.-W. (2018). Rapid and continuous parametric screening for the synthesis of gold nanocrystals with different morphologies using a microfluidic device. *Sensors and Actuators. B, Chemical, 262*, 236–244. DOI: 10.1016/j.snb.2018.02.001

Makransky, G., & Mayer, R. E. (2021). Cognitive and motivational effects of students' emotional design in virtual reality learning environments. *Educational Psychology Review*, *33*(4), 1671–1690.

Malkiel, I., Mrejen, M., Nagler, A., Arieli, U., Wolf, L., & Suchowski, H. (2018). Plasmonic nanostructure design and characterization via deep learning. *Light, Science & Applications*, *7*(1), 60. Advance online publication. DOI: 10.1038/s41377-018-0060-7 PMID: 30863544

Mandrikas, A., Michailidi, E., & Stavrou, D. (2020a). Teaching nanotechnology in primary education. *Research in Science & Technological Education*, *38*(4), 377–395. DOI: 10.1080/02635143.2019.1631783

Marchi, F., Friot, J., Fousse, L., Bain, N., Carrez, J., Bolcato, V., & Castagné, N. (2022). "Learning by doing" en nanosciences et nanotechnologies au moyen des réalités virtuelles et de l'interaction multisensorielle avec systèmes haptiques au CIME Nanotech de Grenoble. *J3eA, 21*, Article 1009. https://doi.org/DOI: 10.1051/j3ea/20221009

Marcos. (n.d.). *Frontiers | Shaping the future of nanotechnology: A perspective on education, dissemination and ethics*. Retrieved September 2, 2025, from https://www.frontiersin.org/journals/nanotechnology/articles/10.3389/fnano.2025.1524578/full

Markus, A. F., Kors, J. A., & Rijnbeek, P. R. (2021). The role of explainability in creating trustworthy artificial intelligence for health care: A comprehensive survey of the terminology, design choices, and evaluation strategies. *Journal of Biomedical Informatics*, *113*, 103655. DOI: 10.1016/J.JBI.2020.103655 PMID: 33309898

Martin-Ortigosa, S., Peterson, D. J., Valenstein, J. S., Lin, V. S. Y., Trewyn, B. G., Lyznik, L. A., & Wang, K. (2014). Mesoporous silica nanoparticle-mediated intracellular Cre protein delivery for maize genome editing via loxP site excision. *Plant Physiology*, *164*(2), 537–547. PMID: 24376280

Marzella, D. F., Crocioni, G., Radusinović, T., Lepikhov, D., Severin, H., Bodor, D. L., Rademaker, D. T., Lin, C. Y., Georgievska, S., Renaud, N., Kessler, A. L., Lopez-Tarifa, P., Buschow, S. I., Bekkers, E., & Xue, L. C. (2024). Geometric deep learning improves generalizability of MHC-bound peptide predictions. *Communications Biology*, *7*(1), 1–12. DOI: 10.1038/s42003-024-07292-1 PMID: 39702482

MathWorks. (2022). MATLAB implementation of the Levenberg-Marquardt algorithm. MATLAB & Simulink Documentation. https://www.mathworks.com/help/deeplearning/ref/trainlm.html

Matthews, H. S., McMichael, F. C., Hendrickson, C. T., & Hart, D. J. (1997). *Disposition and end-of-life options for personal computers*. Carnegie Mellon University.

Ma, Y., Chen, H., Kang, J., Guo, X., Sun, C., Xu, J., Tao, J., Wei, S., Dong, Y., Tian, H., Lv, W., Jia, Z., Bi, S., Shang, Z., Zhang, C., Lv, H., Jiang, Y., & Zhang, M. (2025). The Hidden Markov model and its applications in bioinformatics analysis. *Genes & Diseases, 101729*. Advance online publication. DOI: 10.1016/j.gendis.2025.101729 PMID: 41069576

Mazlan, N., Shah, A. U. M., Hua, T. C., Najmi, I., & Ramli, N. (2024). Biocomposites as Aircraft Materials. Sustainable Materials and Manufacturing Techniques in Aviation, 27-53.

McLeod, C. (2014). Top primary silver miners facing declining ore grades. *Investing News Network*. https://investingnews.com/daily/resource-investing/precious-metals-investing/silver-investing/top-primary-silver-miners-facing-declining-ore-grades/

Mehrabi, Z., McDowell, M. J., Ricciardi, V., Levers, C., Martinez, J. D., Mehrabi, N., & Jarvis, A. (2021). The global divide in data-driven farming. *Nature Sustainability*, *4*, 154–160.

Mehta, K. P., Sharma, R., Haldar, S., & Kumar, A. (2023). Advancement in treatment of wastewater with nano technology. *Materials Today: Proceedings*, *80*, 3358–3362. DOI: 10.1016/j.matpr.2021.07.253

Mekuye, B., & Abera, B. (2023). Nanomaterials: An overview of synthesis, classification, characterization, and applications. *Nano Select*, *4*(8), 486–501. DOI: 10.1002/nano.202300038

Mendanha, D. (2024). Driving sustainability in the automotive industry: bio-coated materials and modern strategies. Academia Materials Science, 1(2).

Mendoza, K. Z. V., Zambrano, M. M. B., & Baquerizo, A. S. M. (2024). Emerging Technologies and its Role in Modern Education. *Estudios y Perspectivas Revista Científica y Académica*, *4*(3), 3052–3073. DOI: 10.61384/r.c.a.v4i3.597

Meyer, J. C. (2021). Transmission electron microscopy characterization of graphene. In *Graphene* (pp. 317–343). Elsevier., DOI: 10.1016/B978-0-12-819576-5.00012-0

Ministry of Education. (2020). National [Government of India.]. *Educational Policy*, •••, 2020.

Minogue, J., & Jones, M. G. (2006). Haptics in education: Exploring an untapped sensory modality. *Review of Educational Research*, *76*(3), 317–348.

Mirkin, C. A. (2025). Blueprints for better drugs: The structural revolution in nanoscale therapeutics. *ACS Nano*, Article ASAP. https://doi.org/DOI: 10.1021/acsnano.5c06380

Mishra, R. K., Verma, K., Chianella, I., Goel, S., & Nezhad, H. Y. (2024). Nanomaterial integration in micro LED technology: Enhancing efficiency and applications. *Next Nanotechnology*, *5*, 100056. DOI: 10.1016/j.nxnano.2024.100056

Mitchell, M. J., Billingsley, M. M., Haley, R. M., Wechsler, M. E., Peppas, N. A., & Langer, R. (2021). Engineering precision nanoparticles for drug delivery. *Nature Reviews. Drug Discovery*, *20*(2), 101–124. DOI: 10.1038/S41573-020-0090-8;SUBJMETA PMID: 33277608

Mobarak, M. H., Mimona, M. A., & Islam, Md. A., Hossain, N., Zohura, F. T., Imtiaz, I., & Rimon, M. I. H. (2023). Scope of machine learning in materials research—A review. *Applied Surface Science Advances*, *18*, 100523. DOI: 10.1016/j.apsadv.2023.100523

Moeck, P., Stone-Sundberg, J., Snyder, T. J., & Kaminsky, W. (n.d.). Enlivening 300 level general education classes on nanoscience and nanotechnology with 3D printed crystallographic models. *The Journal of Materials Education.*

Mohammad, W., Elomri, A., & Kerbache, L. (2022). The Global Semiconductor Chip Shortage: Causes, Implications, and Potential Remedies. *IFAC-PapersOnLine*, *55*(10), 476–483. DOI: 10.1016/j.ifacol.2022.09.439

Molnar, C. (2023). *Interpretable machine learning: A guide for making black box models explainable* (2nd ed.). Springer., https://christophmolnar.com/books/interpretable-machine-learning

Mondal, A., Chang, L., & Perez, A. (2022). Modelling peptide–protein complexes: Docking, simulations and machine learning. *QRB Discovery*, *3*. Advance online publication. DOI: 10.1017/qrd.2022.14 PMID: 37529282

Morris, S. M., & Powell, D. W. (1979). Spectrophotometric assay of phenolphthalein in biological fluids. *Analytical Biochemistry*, *95*(2), 465–471. DOI: 10.1016/0003-2697(79)90757-7 PMID: 453532

Mosa, W. F., El-Shehawi, A. M., Mackled, M. I., Salem, M. Z., Ghareeb, R. Y., Hafez, E. E., & Abdelsalam, N. R. (2021). Productivity performance of peach trees, insecticidal and antibacterial bioactivities of leaf extracts as affected by nanofertilizers foliar application. *Scientific Reports*, *11*(1), 10205. PMID: 33986453

Mourdikoudis, S., Pallares, R. M., & Thanh, N. T. K. (2018). Characterization techniques for nanoparticles: Comparison and complementarity upon studying nanoparticle properties. *Nanoscale*, *10*(27), 12871–12934. DOI: 10.1039/C8NR02278J PMID: 29926865

Mout, R., Moyano, D. F., Rana, S., & Rotello, V. M. (2012). Surface functionalization of nanoparticles for nanomedicine. *Chemical Society Reviews*, *41*(7), 2539. DOI: 10.1039/C2CS15294K PMID: 22310807

Mrdjenovich, D., Hyde, C., & Lo, W. S.. (2021). Propnet: A knowledge graph for materials science. *Patterns (New York, N.Y.)*, *2*(9), 100325. DOI: 10.1016/j.patter.2021.100325 PMID: 34278363

Mukherjee, R., Diwekar, U. M., & Vaseashta, A. (2017). Optimal sensor placement with mitigation strategy for water network systems under uncertainty. *Computers & Chemical Engineering*, *103*, 91–102. DOI: 10.1016/j.compchemeng.2017.03.014

Murali, S. L. (2024). *Artificial Intelligence in VLSI Physical Design of Circuits to Optimize Power, Performance and Area (PPA)*. *11*(10).

Murcia-Verdú, F. J., & Ufarte-Ruiz, M. J. (2019). Mapa de riesgos del periodismo hi-tech 18, 47–55. doi: DOI: 10.31009/hipertext.net.2019.i18.05

Murday, J. S., Jones, S. S., & Stanley, G. M. (2019). Nanoscale science and engineering education: Issues, trends, and future directions. *Journal of Nano Education*, *11*(1), 1–12. DOI: 10.1166/jne.2019.1048

Musa, S. S., Ibrahim, A. M., Alhassan, M. Y., Musa, A. H., Jibo, A. G., Auwal, A. R., Okesanya, O. J., Othman, Z. K., Abubakar, M. S., Ahmed, M. M., Barroso, C. J. V., Sium, A. F., Garcia, M. B., Flores, J. B., Maikifi, A. S., Kouwenhoven, M. B. N., & Lucero-Prisno, D. E. (2025). Nanotechnology and machine learning: A promising confluence for the advancement of precision medicine. *Intelligence-Based Medicine*, *12*, 100267. DOI: 10.1016/j.ibmed.2025.100267

Muttenthaler, M., King, G. F., Adams, D. J., & Alewood, P. F. (2021). Trends in peptide drug discovery. *Nature Reviews. Drug Discovery*, *20*(4), 309–325. DOI: 10.1038/s41573-020-00135-8 PMID: 33536635

Nagaraja, S., Anand, P. B., & Ammarullah, M. I. (2024). Synergistic advances in natural fibre composites: A comprehensive review of the eco-friendly bio-composite development, its charac- terization and diverse applications. *RSC Advances*, *14*(25), 17594–17611. PMID: 38828274

Naik, G. G., & Jagtap, V. A. (2024). Two heads are better than one: Unravelling the potential Impact of Artificial Intelligence in nanotechnology. *Nano TransMed, 3*, 100041. DOI: 10.1016/j.ntm.2024.100041

Namias, J. (2013). The future of electronic waste recycling in the United States: Obstacles and domestic solutions [Master's thesis, Columbia University].

Namla, D., Oves, M., Alshaeri, M. A., Al-Maaqar, S. M., Issa, H. N. Y., & Mangse, G. (2025). Nanofiltration as an advanced wastewater treatment technique: A comprehensive review. *Discover Applied Sciences*, *7*(4). Advance online publication. DOI: 10.1007/s42452-025-06576-3

Nance, E. (2019). Careers in nanomedicine and drug delivery. *Advanced Drug Delivery Reviews, 144*, 180–189. DOI: 10.1016/j.addr.2019.06.009 PMID: 31260712

Nandipati, M., Fatoki, O., & Desai, S. (2024). Bridging Nanomanufacturing and Artificial Intelligence—A Comprehensive Review. *Materials (Basel), 17*(7), 7. Advance online publication. DOI: 10.3390/ma17071621

Naser, A. Z., Deiab, I., Defersha, F., & Yang, S. (2021). Expanding poly (lactic acid) (PLA) and polyhydroxyalkanoates (PHAs) applications: A review on modifications and effects. *Polymers, 13*(23), 4271. PMID: 34883773

Naskar, A., Khan, H., Sarkar, R., Kumar, S., Halder, D., & Jana, S. (2018). Antibiofilm activity and food packaging application of room temperature solution process based polyethylene glycol capped Ag-ZnO-graphene nanocomposite. *Materials Science and Engineering C, 91*, 743–753. PMID: 30033309

Neme, K., Nafady, A., Uddin, S., & Tola, Y. B. (2021). Application of Nanotechnology in agriculture, postharvest loss reduction and food processing: Food security implication and challenges. *Heliyon, 7*(12). PMID: 34934845

Neto, O. P. V. (2014). Intelligent Computational Nanotechnology: The Role of Computational Intelligence in the Development of Nanoscience and Nanotechnology. *Journal of Computational and Theoretical Nanoscience, 11*(4), 928–944. DOI: 10.1166/jctn.2014.3446

Neumann, A. T., Conrardy, A. D., Decker, S., & Jarke, M. (2023). Motivating Learners with Gamified Chatbot-Assisted Learning Activities. In H. Xie, C.-L. Lai, W. Chen, G. Xu, & E. Popescu (Eds.), *Advances in Web-Based Learning – ICWL 2023* (pp. 189–203). Springer Nature. https://doi.org/DOI: 10.1007/978-981-99-8385-8_15

Nile, S. H., Baskar, V., Selvaraj, D., Nile, A., Xiao, J., & Kai, G. (2020). Nanotechnologies in food science: Applications, recent trends, and future perspectives. *Nano-Micro Letters, 12*(1), 45. PMID: 34138283

Ningtyas, P. K., Widarti, H. R., Parlan, P., Rahayu, S., & Dasna, I. W. (2024). Enhancing students' abilities and skills through science learning integrated STEM: A systematic literature review. *International Journal of Education in Mathematics, Science, and Technology, 12*(5), 1161–1181. DOI: 10.46328/ijemst.4292

Nirupama, G. S. (2024). *Advances in Forensic Bloodstain Pattern Analysis: A Review of Current Methods and Future Directions. 11.*

Nissan, N., Allen, M. C., Sabatino, D., & Biggar, K. K. (2024). Future Perspective: Harnessing the Power of Artificial Intelligence in the Generation of New Peptide Drugs. *Biomolecules, 14*(10), 1303. DOI: 10.3390/biom14101303 PMID: 39456236

Nixon, N., Lin, Y., & Snow, L. (2024). Catalyzing equity in STEM teams: Harnessing generative AI for inclusion and diversity. *arXiv*. https://doi.org//arXiv.2402.00037DOI: 10.48550

Nixon, A. E., Sexton, D. J., & Ladner, R. C. (2014). Drugs derived from phage display from candidate identification to clinical practice. *mAbs, 6*(1), 73–85. DOI: 10.4161/mabs.27240 PMID: 24262785

Nizamuddin, S., Baloch, A. J., Chen, C., Arif, M., & Mubarak, N. M. (2024). Bio-based plas- tics, biodegradable plastics, and compostable plastics: Biodegradation mechanism, biodegrad- ability standards and environmental stratagem. *International Biodeterioration & Biodegradation, 195*, 105887.

Noble, W. S. (2006). What is a support vector machine? *Nature Biotechnology, 24*(12), 1565–1567. DOI: 10.1038/nbt1206-1565 PMID: 17160063

Noorali, S., Baldwin, J. C., & Vaseashta, A. (2025). Next-generation sensors using artificial intelligence as enabling technologies for biomedical and healthcare applications. In Vaseashta, A., Achour, M. E., Mabrouki, M., Tachafine, A., & Aitali, M. (Eds.), *Proceedings of the Eighth International Symposium on Dielectric Materials and Applications (ISyDMA'8)* (pp. 215–231). Springer. https://doi.org/DOI: 10.1007/978-3-031-69700-5_16

Nur Fitria, T. (2021). *Artificial Intelligence (AI) In Education: Using AI Tools for Teaching and Learning Process.*

OECD. (2021). *Artificial intelligence in science: Challenges, opportunities, and the future of research.* OECD Publishing. https://www.oecd.org/en/publications/artificial-intelligence-in-science_a8d820bd-en.html

OECD. (2021). *National AI strategies*. https://www.oecd.org

Ofozoba, C. A., Okafor, P. C., Ikegbusi, N. G., Manafa, F. U., Egwu, J. U., Nwobu, C. M., Okafor, S. O., Onafowope, M. A., Olofinkua, V. K., Adeagbo, J. O., Gbaeprekumo, O. V., Ayoko, V. O., Udogu, A. N., Uwannah, N. C., Okudo, O. C., Onyiorah, B. O., Ayodele, M. F., Nnoyelu, O. G., Ekwugha, J. C., & Nnalue, O. H. (2025). AI-driven service-learning to enhance students' understanding of green nanomaterials in sustainability education. *Multidisciplinary Reviews*, *8*(12), 2025429–2025429. DOI: 10.31893/multirev.2025429

Oguchi, M., Kameya, T., Yagi, S., & Urano, K. (2008). Product flow analysis of various consumer durables in Japan. *Resources, Conservation and Recycling*, *52*, 463–480.

Oh, E., Godoy Zu´niga, M. M., Nguyen, T. B., Kim, B. H., Trung Tien, T., & Suhr, J. (2024). Sus- tainable green composite materials in the next-generation mobility industry: review and prospec- tive.

Olawade, D. B., Ige, A. O., Olaremu, A. G., Ijiwade, J. O., & Adeola, A. O. (2024). The synergy of artificial intelligence and nanotechnology towards advancing innovation and sustainability—A mini-review. *Nano Trends*, *8*, 100052. DOI: 10.1016/j.nwnano.2024.100052

Oliveira, A., Feyzi Behnagh, R., Ni, L., Mohsinah, A. A., Burgess, K. J., & Guo, L. (2019). Emerging technologies as pedagogical tools for teaching and learning science: A literature review. *Human Behavior and Emerging Technologies*, *1*(2), 149–160. DOI: 10.1002/hbe2.141

Oliveira, R. C. D., & Silva, R. D. D. S. E. (2023). Artificial intelligence in agriculture: Benefits, challenges, and trends. *Applied Sciences (Basel, Switzerland)*, *13*(13), 7405.

Ongondo, F., Williams, I., & Cherrett, T. (2011). How are WEEE doing? A global review of the management of electrical and electronic wastes. *Waste Management (New York, N.Y.)*, *31*(4), 714–730.

Ori, M. O., Ekpan, F. M., Samuel, H. S., & Egwuatu, O. P. (2023). Integration of Artificial Intelligence in Nanomedicine. *Eurasian Journal of Science and Technology, Online First*. https://doi.org/DOI: 10.48309/ejst.2024.422419.1105

Oughton, C., Kurup, B., Anda, M., & Ho, G. (2022). Industrial symbiosis to circular economy: What does the literature reveal for a successful complex industrial area? *Circular Economy and Sustainability*, *2*, 1317–1344.

Owusu-Sekyere, K., Batteiger, A., Afoblikame, R., Hafner, G., & Kranert, M. (2022). Assessing data in the informal e-waste sector: The Agbogbloshie Scrapyard. *Waste Management (New York, N.Y.)*, *139*, 158–167.

Öztürk, T. (2014). Generation and management of electrical–electronic waste (e-waste) in Turkey. *Journal of Material Cycles and Waste Management*, *17*, 411–421.

Pai, S. M., Shah, K. A., Sunder, S., Albuquerque, R. Q., Brütting, C., & Ruckdäschel, H. (2025). Machine learning applied to the design and optimization of polymeric materials: A review. *Next Materials*, *7*, 100449. DOI: 10.1016/j.nxmate.2024.100449

Palansooriya, K. N., Yang, Y., Tsang, Y. F., Sarkar, B., Hou, D., Cao, X., Meers, E., Rinklebe, J., Kim, K. H., & Ok, Y. S. (2020). Occurrence of contaminants in drinking water sources and the potential of biochar for water quality improvement: A review. *Critical Reviews in Environmental Science and Technology*, *50*(6), 549–611. DOI: 10.1080/10643389.2019.1629803

Pal, M. (2017). Nanotechnology: A new approach in food packaging. *J. Food Microbiol. Saf. Hyg*, *2*(02), 8–9.

Palma, M. (2023). Epitopes and Mimotopes Identification Using Phage Display for Vaccine Development against Infectious Pathogens. *Vaccines*, *11*(7), 1176. DOI: 10.3390/vaccines11071176 PMID: 37514992

Palwai, S., Batra, A., Kotru, S., & Vaseashta, A. (2022). Electrospun polyvinylidene fluoride nanofiber membrane-based flexible capacitive tactile sensors for biomedical applications. *Surface Engineering and Applied Electrochemistry*, *58*(2), 194–201. DOI: 10.3103/S1068375522020089

Pane, J. F., Steiner, E. D., & Baird, M. D.. (2017). *Informing progress: Insights on personalized learning implementation and effects*. RAND Corporation.

Pant, D. (2013). E-waste projection using life-span and population statistics. *The International Journal of Life Cycle Assessment*, *18*, 1465–1469.

Parekh, A.-D. E., & Shaikh, O. A., Simran, Manan, S., & Hasibuzzaman, Md. A. (2023). Artificial intelligence (AI) in personalized medicine: AI-generated personalized therapy regimens based on genetic and medical history: short communication. *Annals of Medicine and Surgery (London)*, *85*(11), 5831–5833. DOI: 10.1097/MS9.0000000000001320

Park, J., Teo, T. W., Teo, A., Chang, J., Huang, J. S., & Koo, S. (2023). Integrating artificial intelligence into science lessons: Teachers' experiences and views. *International Journal of STEM Education*, *10*(1), 61. Advance online publication. DOI: 10.1186/s40594-023-00454-3

Paul, D., Sanap, G., Shenoy, S., Kalyane, D., Kalia, K., & Tekade, R. K. (2021). Artificial intelligence in drug discovery and development. *Drug Discovery Today*, *26*(1), 80–93. DOI: 10.1016/j.drudis.2020.10.010 PMID: 33099022

Pavlik, J. (2023). Collaborating with ChatGPT: Considering the implications of generative artificial intelligence for journalism and media education. *J. Mass Commun. Educ., 78*, 84–93. DOI: 10.1177/10776958221149577

Pelleg, B., Figueroa, M., VanKouwenberg, M., Fontecchio, A., & Fromm, E. (2011). Implementing nanotechnology education in the high school classroom. *2011 Frontiers in Education Conference (FIE)*, F4D-1-F4D-6. https://doi.org/DOI: 10.1109/FIE.2011.6142886

Peña-Fernández, S., Meso-Ayerdi, K., Larrondo-Ureta, A., & Díaz-Noci, J. (2023). Without journalists, there is no journalism: The social dimension of generative artificial intelligence in the media. *El Profesional de la Información, 32*, 27. DOI: 10.3145/epi.2023.mar.27

Peralta, G. L., & Fontanos, P. M. (2006). E-waste issues and measures in the Philippines. *Journal of Material Cycles and Waste Management, 8*, 34–39.

Pérez-Seijo, S., Gutiérrez-Caneda, B., & López-García, X. (2020). Periodismo digital y alta tecnología: De la consolidación a los renovados desafíos. *Index Comun., 10*, 129–151. DOI: 10.33732/ixc/10/03Period

Perkins, K. K., Adams, W., Dubson, M., Finkelstein, N. D., & Wieman, C. E. (2010). PhET: Interactive simulations for teaching and learning physics. *The Physics Teacher, 48*(4), 225–230. DOI: 10.1119/1.3392045

Peschel, O., Kunz, S. N., Rothschild, M. A., & Mützel, E. (2011). Blood stain pattern analysis. *Forensic Science, Medicine, and Pathology, 7*(3), 257–270. DOI: 10.1007/S12024-010-9198-1 PMID: 21069481

Petridis, N. E., Stiakakis, E., Petridis, K., & Dey, P. (2016). Estimation of computer waste quantities using forecasting techniques. *Journal of Cleaner Production, 112*, 3072–3085.

Petrov, P. D., Davidova, S., & Satchanska, G. (2025). Polymer Micelles as Nanocarriers of Bioactive Peptides. *Polymers, 17*(9), 1174. DOI: 10.3390/POLYM17091174 PMID: 40362958

Photopoulos, J. (2023). Spinning sustainability: Meet the physicists turning wood into clothes. *Physics World, 36*(11), 37.

Pineda, S. S., Chaumeil, P. A., Kunert, A., Kaas, Q., Thang, M. W. C., Le, L., Nuhn, M., Herzig, V., Saez, N. J., Cristofori-Armstrong, B., Anangi, R., Senff, S., Gorse, D., & King, G. F. (2018). ArachnoServer 3.0: An online resource for automated discovery, analysis and annotation of spider toxins. *Bioinformatics (Oxford, England), 34*(6), 1074–1076. DOI: 10.1093/bioinformatics/btx661 PMID: 29069336

Pinheiro, E. A., Leal Júnior, J. B. V., de Lima, L. C., & de Melo, F. G. (2020). Nonlinear autoregressive neural networks for forecasting wind speed time series. *International Journal of Developmental Research*, *10*(9), 40336–40343.

Pirtskhalava, M., Vishnepolsky, B., & Grigolava, M. (2013). Transmembrane and Antimicrobial Peptides. Hydrophobicity, Amphiphilicity and Propensity to Aggregation. *Nucleic Acids Research*, *49*(D1), D288–D297. https://arxiv.org/abs/1307.6160 PMID: 33151284

Pistilli, M. (2016). World-class copper deposits. *Investing News Network*. https://investingnews.com/daily/resource-investing/base-metals-investing/copper-investing/what-makes-a-world-class-copper-deposit/

Polak, M., & Drapalova, L. (2012). Estimation of end of life mobile phones generation: The case study of the Czech Republic. *Waste Management (New York, N.Y.)*, *32*, 1583–1591.

Price, C. C., Singh, A., Frey, N. C., & Shenoy, V. B. (2022). Efficient catalyst screening using graph neural networks to predict strain effects on adsorption energy. *Science Advances*, *8*(36), eabq5944. Advance online publication. DOI: 10.1126/sciadv.abq5944 PMID: 36417537

Pun, F. W., Ozerov, I. V., & Zhavoronkov, A. (2023). AI-powered therapeutic target discovery. *Trends in Pharmacological Sciences*, *44*(9), 561–572. DOI: 10.1016/j.tips.2023.06.010 PMID: 37479540

Quickenden, T. I., & Creamer, J. I. (2001). A study of common interferences with the forensic luminol test for blood. *Luminescence : The Journal of Biological and Chemical Luminescence*, *16*(4), 295–298. DOI: 10.1002/BIO.657 PMID: 11512147

Qu, J., Meador, W., Cheah, P., Tanner, E. E. L., Delcamp, J., & Zhao, Y. (2023). Latent bloodstain detection using a selective turn-on NIR fluorescence dye responsive to serum albumin. *RSC Advances*, *13*(39), 27549. DOI: 10.1039/D3RA04320G PMID: 37720829

Qu, Y., Wang, W., Liu, Y., & Zhu, Q. (2019). Understanding residents' preferences for e-waste collection in China—A case study of waste mobile phones. *Journal of Cleaner Production*, *228*, 52–62.

Radulescu, D.-M., Surdu, V.-A., Ficai, A., Ficai, D., Grumezescu, A.-M., & Andronescu, E. (2023). Green synthesis of metal and metal oxide nanoparticles: Principles and biomedical applications. *International Journal of Molecular Sciences*, *24*(20), 15397. DOI: 10.3390/ijms242015397 PMID: 37895077

Rahman, M. Z., Rahman, M., Mahbub, T., Ashiquzzaman, M., Sagadevan, S., & Hoque, M. E. (2023). Advanced biopolymers for automobile and aviation engineering applications. *Journal of Polymer Research, 30*(3), 106.

Ramakrishnan, T., Mohan Gift, M. D., Chitradevi, S., Jegan, R., & Subha Hency Jose, P. Nagaraja, Rajneesh Sharma H.N., Selvakumar P., Sintayehu Mekuria Hailegiorgis, (2022). "Study of Numerous Resins Used in Polymer Matrix Composite Materials", Advances in Materials Science and Engineering, vol. 2022, Article ID 1088926, 8 pages, 2022. https://doi.org/DOI: 10.1155/2022/1088926

Ramesh, M., Deepa, C., Kumar, L. R., Sanjay, M. R., & Siengchin, S. (2022). Life-cycle and environmental impact assessments on processing of plant fibres and its bio-composites: A critical review. *Journal of Industrial Textiles*, •••, 5518S–5542S.

Rana, K., Mittal, J., Narang, J., Mishra, A., & Pudake, R. N. (2021). Graphene based electrochemical DNA biosensor for detection of false smut of rice (Ustilaginoidea virens). *The plant pathology journal, 37*(3), 291. PMID: 34111918

Rane, N. L., Desai, P., & Choudhary, S. (2024). *Challenges of implementing artificial intelligence for smart and sustainable industry: Technological, economic, and regulatory barriers*. Deep Science Publishing., DOI: 10.70593/978-81-981271-1-2

Ravindra, K., & Mor, S. (2019). E-waste generation and management practices in Chandigarh, India and economic evaluation for sustainable recycling. *Journal of Cleaner Production, 221*, 286–294.

Rayasam, S. D., Koman, P. D., Axelrad, D. A., Woodruff, T. J., & Chartres, N. (2022). Toxic Substances Control Act (TSCA) implementation: How the amended law has failed to protect vulnerable populations from toxic chemicals in the United States. *Environmental Science & Technology, 56*(17), 11969–11982. PMID: 35980084

Ray, S. S., & Bandyopadhyay, J. (2021). Nanotechnology-enabled biomedical engineering: Current trends, future scopes, and perspectives. *Nanotechnology Reviews, 10*(1), 728–743.

Repko, A. F., Szostak, R., & Buchberger, M. P. (2020). *Introduction to interdisciplinary studies* (3rd ed.). SAGE Publications.

Reynisson, B., Alvarez, B., Paul, S., Peters, B., & Nielsen, M. (2021). NetMHCpan-4.1 and NetMHCIIpan-4.0: Improved predictions of MHC antigen presentation by concurrent motif deconvolution and integration of MS MHC eluted ligand data. *Nucleic Acids Research, 48*(W1), W449–W454. DOI: 10.1093/NAR/GKAA379 PMID: 32406916

Rieke, N., Hancox, J., Li, W., Milletarì, F., Roth, H., Albarqouni, S., Bakas, S., Galtier, M. N., Landman, B. A., Maier-Hein, K., Ourselin, S., Sheller, M., Summers, R. M., Trask, A., Xu, D., Baust, M., & Cardoso, M. J. (2020). The future of digital health with federated learning. *npj. Digital Medicine*, *3*, 119. DOI: 10.1038/s41746-020-00323-1 PMID: 33015372

Risks, I. of M. (US) C. on A. G., Andrews, L. B., Fullarton, J. E., Holtzman, N. A., & Motulsky, A. G. (1994). *Social, Legal, and Ethical Implications of Genetic Testing*. https://www.ncbi.nlm.nih.gov/books/NBK236044/

Robinson, B. (2009). E-waste: An assessment of global production and environmental impacts. *The Science of the Total Environment*, *408*(2), 183–191.

Roco, M. C. (2003). Converging science and technology at the nanoscale: Opportunities for education and training. *Nature Biotechnology*, *21*(10), 1247–1249. DOI: 10.1038/nbt1003-1247 PMID: 14520410

Roco, M. C. (2011). The long view of nanotechnology development: The National Nanotechnology Initiative at 10 years. *Journal of Nanoparticle Research*, *13*(2), 427–445. DOI: 10.1007/s11051-010-0192-z

Roco, M. C. (2023a). National Nanotechnology Initiative at 20 years: Enabling new horizons. *Journal of Nanoparticle Research*, *25*(10), 197. DOI: 10.1007/s11051-023-05829-9

Roco, M. C., & Bainbridge, W. S. (2005). Societal implications of nanoscience and nanotechnology. In Roco, M. C., & Bainbridge, W. S. (Eds.), *Nanotechnology: Societal implications* (pp. 3–28). Springer.

Roll, I., & Wylie, R. (2016). Evolution and revolution in artificial intelligence in education. *International Journal of Artificial Intelligence in Education*, *26*(2), 582–599.

Romeiko, X. X., Zhang, X., Pang, Y., Gao, F., Xu, M., Lin, S., & Babbitt, C. (2024). A review of machine learning applications in life cycle assessment studies. *The Science of the Total Environment*, *912*, 168969. PMID: 38036122

Rotter, V., Chancerel, P., & Schill, W. (2009). Implementing individual producer responsibility (IPR) under the European WEEE Directive – Experiences in Germany. In *2009 IEEE International Symposium on Sustainable Systems and Technology* (pp. 1–6). IEEE.

Roy, S., Zhai, L., Kim, H. C., Pham, D. H., Alrobei, H., & Kim, J. (2021). Tannic-acid-cross-linked and TiO2-nanoparticle-reinforced chitosan-based nanocomposite film. *Polymers*, *13*(2), 228. PMID: 33440770

Ruiz Puentes, P., Henao, M. C., Cifuentes, J., Muñoz-Camargo, C., Reyes, L. H., Cruz, J. C., & Arbeláez, P. (2022). Rational Discovery of Antimicrobial Peptides by Means of Artificial Intelligence. *Membranes*, *12*(7), 708. DOI: 10.3390/MEMBRANES12070708/S1 PMID: 35877911

Ruiz, L. G. B., Cuéllar, M. P., Calvo-Flores, M. D., & Jiménez, M. D. C. P. (2016). An application of non-linear autoregressive neural networks to predict energy consumption in public buildings. *Energies*, *9*(9), 684.

Russell, S., & Norvig, P. (2020). *Artificial intelligence: A modern approach* (4th ed.). Pearson.

Ryan-Fogarty, Y., Coughlan, D., & Fitzpatrick, C. (2021). Quantifying WEEE arising in scrap metal collections: Method development and application in Ireland. *Journal of Industrial Ecology*, *25*(4), 1021–1033.

Rydel, T. J., Ravichandran, K. G., Tulinsky, A., Bode, W., Huber, R., Roitsch, C., & Fenton, J. W. (1990). The structure of a complex of recombinant hirudin and human α-thrombin. *Science*, *249*(4966), 277–280. DOI: 10.1126/science.2374926 PMID: 2374926

Saavedra, Y., Iritani, D., Pavan, A., & Ometto, A. (2018). Theoretical contribution of industrial ecology to circular economy. *Journal of Cleaner Production*, *170*, 1514–1522.

Sacha, G. M., & Varona, P. (2013). Artificial intelligence in nanotechnology. *Nanotechnology*, *24*(45), 452002. Advance online publication. DOI: 10.1088/0957-4484/24/45/452002 PMID: 24121558

Saha, A., Pugno, N. M., & Kundanati, L. (2025). Agricultural Waste-based Composites for Electric Vehicle Interiors: A Step Towards Sustainability Using Waste to Value Approach.

Sajid, M., Syed, J. H., Iqbal, M., Abbas, Z., Hussain, I., & Baig, M. A. (2019). Assessing the generation, recycling and disposal practices of electronic/electrical-waste (E-Waste) from major cities in Pakistan. *Waste Management (New York, N.Y.)*, *84*, 394–401.

Sajja, R., Sermet, Y., Cwiertny, D., & Demir, I. (2025). *Integrating AI and Learning Analytics for Data-Driven Pedagogical Decisions and Personalized Interventions in Education*. Technology, Knowledge and Learning., DOI: 10.1007/s10758-025-09897-9

Sakthivel, U., Swaminathan, G., & Anis, J. (2022). Strategies for Quantifying Metal Recovery from Waste Electrical and Electronic Equipment (WEEE/E-waste) Using Mathematical Approach. *Process Integration and Optimization for Sustainability*, *6*(4), 781–790.

Salaverría, R. (2019). Digital journalism: 25 years of research. *EPI*, *28*, 1. DOI: 10.3145/epi.2019.ene.01

Salihoğlu, G., & Kahraman, A. E. (2016). Electrical and Electronical Waste Generation in Turkey: Bursa Case Study. *Uludağ University Journal of the Faculty of Engineering*, *21*(1), 95–106.

Sanders, W. (2018). *Basic Principles of Nanotechnology*. CRC Press., DOI: 10.1201/9781351054423

Santana, R., Zuluaga, R., Gañán, P., Arrasate, S., Onieva, E., & González-Díaz, H. (2019). Designing nanoparticle release systems for drug–vitamin cancer co-therapy with multiplicative perturbation-theory machine learning (PTML) models. *Nanoscale*, *11*(44), 21811–21823. DOI: 10.1039/C9NR05070A PMID: 31691701

Saravanakumar, K., Sathiyaseelan, A., Mariadoss, A. V. A., Xiaowen, H., & Wang, M. H. (2020). Physical and bioactivities of biopolymeric films incorporated with cellulose, sodium alginate and copper oxide nanoparticles for food packaging application. *International Journal of Biological Macromolecules*, *153*, 207–214. PMID: 32105688

Sarı, U., Duygu, E., Şen, Ö. F., & Kırındı, T. (2020). The effect of STEM education on scientific process skills and STEM awareness in simulation-based inquiry learning environment. *Journal of Turkish Science Education*, *17*(3), 387–405. DOI: 10.36681/tused.2020.34

Saxena, A., Prakash Gupta, J., Tiwary, J. K., Kumar, A., Sharma, S., Pandey, G., & Raghav Chaturvedi, K. (2024). Innovative Pathways in Carbon Capture: Advancements and Strategic Approaches for Effective Carbon Capture, Utilization, and Storage. *Sustainability*, *16*(22), 10132.

Schluep, M., Hagelüken, C., Kuehr, R., Magalini, F., Maurer, C., Meskers, C., & Wang, F. (2009). *Recycling: From e-waste to resources*. United Nations Environment Programme & United Nations University.

Schmidhuber, J., & Tubiello, F. N. (2007). Global food security under climate change. *Proceedings of the National Academy of Sciences of the United States of America*, *104*(50), 19703–19708. PMID: 18077404

Schüler, D., Buchert, M., Liu, R., Dittrich, S., & Merz, C. (2011). Study on the review of the list of critical raw materials: Final report. Öko-Institut e.V.

Scruggs, C., Nimpuno, N., & Moore, R. (2016). Improving information flow on chemicals in electronic products and e-waste to minimize negative consequences for health and the environment. *Resources, Conservation and Recycling*, *113*, 149–164.

Şenocak, E., Ozdemir, T., & Yilmaz, F., Tayhan, & Mcnally. (2021). A NANOSCIENCE AND NANOTECHNOLOGY EDUCATIONAL PROGRAM FOR HIGH SCHOOL STUDENTS: EXAMINATION OF PRELIMINARY TRIALS. *The Journal of Materials Education*, *43*, 93–108.

Senthilkumar, A., Ravindran, V., Arthanari, A., & Ramalingam, K. (2024). Evaluation of Forensic Luminol in the Detection of Blood Stains in Instruments Following Dental Treatment. *Cureus*, *16*(4), e57676. DOI: 10.7759/CUREUS.57676 PMID: 38707075

Shahouni, R., Abbasi, M., Dibaj, M., & Akrami, M. (2024). Utilizing Artificial Intelligence to Predict Membrane Behavior in Water Purification and Desalination. *Water (Switzerland)*, *16*(20). Advance online publication. DOI: 10.3390/w16202940

Shah, S. S., Casanova, N., Antuono, G., & Sabatino, D. (2020). Polyamide Backbone Modified Cell Targeting and Penetrating Peptides in Cancer Detection and Treatment. *Frontiers in Chemistry*, *8*, 526066. DOI: 10.3389/FCHEM.2020.00218/XML PMID: 32296681

Shankar, S., Khodaei, D., & Lacroix, M. (2021). Effect of chitosan/essential oils/silver nanoparticles composite films packaging and gamma irradiation on shelf life of strawberries. Food Hydrocolloids, 117, 106750. Khaki, S., & Wang, L. (2019). Crop yield prediction using deep neural networks. *Frontiers in Plant Science*, *10*, 621.

Sharma, A., Sharma, A., Tselykh, A., Bozhenyuk, A., Choudhury, T., Alomar, M. A., & Sánchez-Chero, M. (2023). Artificial intelligence and internet of things oriented sustainable precision farming: Towards modern agriculture. *Open Life Sciences*, *18*(1), 20220713. Advance online publication. DOI: 10.1515/biol-2022-0713 PMID: 37854322

Sharma, N., Naorem, L. D., Jain, S., & Raghava, G. P. S. (2022). ToxinPred2: An improved method for predicting toxicity of proteins. *Briefings in Bioinformatics*, *23*(5), 1–12. DOI: 10.1093/bib/bbac174 PMID: 35595541

Sharma, P., Singh, V., Thakur, A., & Malik, A. K. (2024). Nanotechnology, Artificial Intelligence, and the Future of Manufacturing. In George, S. C., & Tawiah, B. (Eds.), *Nanotechnology in Societal Development* (pp. 307–332). Springer Nature., DOI: 10.1007/978-981-97-6184-5_9

Sharma, R., Shrivastava, S., Kumar Singh, S., Kumar, A., Saxena, S., & Kumar Singh, R. (2021). AniAMPpred: Artificial intelligence guided discovery of novel antimicrobial peptides in animal kingdom. *Briefings in Bioinformatics*, *22*(6), 1–23. DOI: 10.1093/BIB/BBAB242 PMID: 34259329

Sharma, R., Shrivastava, S., Singh, S. K., Kumar, A., Saxena, S., & Singh, R. K. (2022). Deep-AFPpred: Identifying novel antifungal peptides using pretrained embeddings from seq2vec with 1DCNN-BiLSTM. *Briefings in Bioinformatics*, *23*(1), 1–16. DOI: 10.1093/BIB/BBAB422 PMID: 34670278

Sharmile, N., Chowdhury, R. R., & Desai, S. (2025). A Comprehensive Review of Quality Control and Reliability Research in Micro–Nano Technology. *Technologies*, *13*(3), 3. Advance online publication. DOI: 10.3390/technologies13030094

Sharples, M., & Domingue, J. (2016). The blockchain and kudos: A distributed system for educational record, reputation and reward. In *International Conference on Web-Based Learning* (pp. 490–496). Springer.

Sha, W., Guo, Y., Yuan, Q., Tang, S., Zhang, X., Lu, S., Guo, X., Cao, Y.-C., & Cheng, S. (2020). Artificial intelligence to power the future of materials science and engineering. *Advanced Intelligent Systems*, *2*(4), 1900143. Advance online publication. DOI: 10.1002/aisy.201900143

Shende, A. D., Tekade, S. A., Deshmukh, A. A., Tembhurkar, S. P., & Selvakumar, P. (2025). Water Quality Monitoring and Control With AI. In Ruiz-Vanoye, J., & Díaz-Parra, O. (Eds.), *Smart Water Technology for Sustainable Management in Modern Cities* (pp. 51–80). IGI Global Scientific Publishing., DOI: 10.4018/979-8-3693-8074-1.ch003

Shi, G., Kang, X., Dong, F., Liu, Y., Zhu, N., Hu, Y., Xu, H., Lao, X., & Zheng, H. (2022). DRAMP 3.0: An enhanced comprehensive data repository of antimicrobial peptides. *Nucleic Acids Research*, *50*(D1), D488–D496. DOI: 10.1093/nar/gkab651 PMID: 34390348

Shi, S., Wang, W., Liu, L., Wu, S., Wei, Y., & Li, W. (2013). Effect of chitosan/nano-silica coating on the physicochemical characteristics of longan fruit under ambient temperature. *Journal of Food Engineering*, *118*(1), 125–131.

Shi, Y., & Sun, L. (2024). How generative AI is transforming journalism: Development. *Appl. Ethics J. Medi.*, *5*, 582–594. DOI: 10.3390/journalmedia5020039

Shobanke, M., Bhatt, M., & Shittu, E. (2025). Advancements and future outlook of Artificial Intelligence in energy and climate change modeling. *Advances in Applied Energy*, *17*, 100211. DOI: 10.1016/j.adapen.2025.100211

Short, S. (2015). Serology: Blood Identification. *Encyclopedia of Forensic and Legal Medicine: Second Edition*, 239–243. https://doi.org/DOI: 10.1016/B978-0-12-800034-2.00338-4

Shreffler, J., & Huecker, M. R. (2023). Diagnostic Testing Accuracy: Sensitivity, Specificity, Predictive Values and Likelihood Ratios. *StatPearls*. https://www.ncbi.nlm.nih.gov/books/NBK557491/

Shukla, P. S., Mantin, E. G., Adil, M., Bajpai, S., Critchley, A. T., & Prithiviraj, B. (2019). Ascophyllum nodosum-based biostimulants: Sustainable applications in agriculture for the stimulation of plant growth, stress tolerance, and disease management. *Frontiers in Plant Science*, *10*, 462648. PMID: 31191576

Siddique, S., & Chow, J. C. L. (2020). Gold nanoparticles for drug delivery and cancer therapy. *Applied Sciences (Basel, Switzerland)*, *10*(11), 3824. Advance online publication. DOI: 10.3390/app10113824

Sidorenko, A., Gutul, T., Dvornikov, D., Şeker, M. G., Arit, T., Gutul, E., Dimoglo, A., & Vaseashta, A. (2022). Synthesis of nZVI/PVP nanoparticles for bioremediation applications. *Bioremediation Journal*, *26*(2), 162–170. DOI: 10.1080/10889868.2021.1911922

Simonyan, K., & Zisserman, A. (2015). Very deep convolutional networks for large-scale image recognition. *arXiv*. https://arxiv.org/abs/1409.1556

Singh, A. V., Maharjan, R.-S., Kanase, A., Siewert, K., Rosenkranz, D., Singh, R., Laux, P., & Luch, A. (2021). Machine-learning-based approach to decode the influence of nanomaterial properties on their interaction with cells. *ACS Applied Materials & Interfaces*, *13*(1), 1943–1955. DOI: 10.1021/acsami.0c18470 PMID: 33373205

Singh, N., Ogunseitan, O. A., Wong, M. H., & Tang, Y. (2022). Sustainable materials alternative to petrochemical plastics pollution: A review analysis. *Sustainable Horizons*, *2*, 100016.

Singh, R.. (2022). Sustainability in nanomaterials through AI-driven life cycle assessment. *Journal of Cleaner Production*, *356*, 131885. DOI: 10.1016/j.jclepro.2022.131885

Singh, S., Chaudhary, K., Dhanda, S. K., Bhalla, S., Usmani, S. S., Gautam, A., Tuknait, A., Agrawal, P., Mathur, D., & Raghava, G. P. S. (2015). SATPdb: A database of structurally annotated therapeutic peptides. *Nucleic Acids Research, 44*(D1), D1119–D1126. DOI: 10.1093/nar/gkv1114 PMID: 26527728

Soler, P., (2011). "La investigación cualitativa. Un enfoque integrador," in La investigación en comunicación; métodos y técnicas en la era digital, coord. L. Vilches Manterola, 189–236.

Song, Q., Li, J., Liu, L., Dong, Q., Yang, J., Liang, Y., & Zhang, C. (2016). Measuring the generation and management status of waste office equipment in China: A case study of waste printers. *Journal of Cleaner Production, 112*(Part 5), 4461–4468.

Song, X., Gao, X., & Wang, H.. (2025). Artificial intelligence for nanomedicine. *Science China. Chemistry*. Advance online publication. DOI: 10.1007/s11426-025-2942-5

Soori, M., Jough, F. K. G., Dastres, R., & Arezoo, B. (2025). Additive Manufacturing Modification by Artificial Intelligence, Machine Learning, and Deep Learning: A Review. *Additive Manufacturing Frontiers, 200198*. Advance online publication. DOI: 10.1016/j.amf.2025.200198

Sousa, M., Martins, S., Santos, M. J., & Amorim, P. (2024). Dynamic pricing strategy for substitute perishable products. Available at *SSRN* 5019518.

Sowers, A., Wang, G., Xing, M., & Li, B. (2023). Advances in Antimicrobial Peptide Discovery via Machine Learning and Delivery via Nanotechnology. *Microorganisms 2023, Vol. 11, Page 1129, 11*(5), 1129. https://doi.org/DOI: 10.3390/MICROORGANISMS11051129

Spector, J. M. (2013). Emerging Educational Technologies and Research Directions. *Journal of Educational Technology & Society, 16*(2), 21–30.

Sreenivasulu, A., Manimozhi, T., Satheesh Kumar, P., Selvakumar, P., & Singh, P. (2022). "The Model of Smart Sensing Device For Sensitive Nanoclusters Modification in Sensing Properties," 2022 2nd International Conference on Advanced Computing and Innovative Technologies in Engineering (ICACITE), Greater Noida, India, 2022, pp. 1043–1047, doi: DOI: 10.1109/ICACITE53722.2022.9823846

Srikantaiah, K., & Muralidhar, S. (2020). Pedagogical challenges in nanotechnology education in India. *Indian Journal of Engineering Education, 36*(2), 489–498.

Stamatin, I., Berlic, C., & Vaseashta, A. (2006). On the computer-aided modelling of analyte–receptor interactions for an efficient sensor design. *Thin Solid Films, 495*(1–2), 312–315. DOI: 10.1016/j.tsf.2005.08.299

Steubing, B., Boni, H., Schluep, M., Silva, U., & Ludwig, C. (2010). Assessing computer waste generation in Chile using material flow analysis. *Waste Management (New York, N.Y.)*, *30*, 473–482.

Sthiannopkao, S., & Wong, M. (2013). Handling e-waste in developed and developing countries: Initiatives, practices, and consequences. *The Science of the Total Environment*, *463–464*, 1147–1153.

Streicher-Porte, M., Widmer, R., Jain, A., Bader, H.-P., Scheidegger, R., & Kytzia, S. (2005). Key drivers of the e-waste recycling system: Assessing and modelling e-waste processing in the informal sector in Delhi. *Environmental Impact Assessment Review*, *25*, 472–491.

Strokach, A., & Kim, P. M. (2022). Deep generative modeling for protein design. *Current Opinion in Structural Biology*, *72*, 226–236. DOI: 10.1016/j.sbi.2021.11.008 PMID: 34963082

Swain, A., Abdellatif, E., Mousa, A., & Pong, P. W. T. (2022). Sensor Technologies for Transmission and Distribution Systems: A Review of the Latest Developments. *Energies*, *15*(19), 19. Advance online publication. DOI: 10.3390/en15197339

Sweeney, A. E., & Seal, S. (2018). Nanotechnology education: Addressing the issues of workforce preparation and public understanding. *Journal of Nano Education*, *10*(1), 23–32.

Sweeney, A. E., Vaidyanathan, P., & Seal, S. (2006). Undergraduate research and education in nanotechnology: Development, implementation and evaluation of a Research Experiences for Undergraduates program in nanomaterials processing and characterization. *International Journal of Engineering Education*, *22*(1), 157–170.

Szymanski, N. J.. (2023). An autonomous laboratory for the accelerated synthesis of inorganic materials. *Nature*, *614*, 413–420. DOI: 10.1038/s41586-023-06734-w PMID: 38030721

Szymczak, P., Zarzecki, W., Wang, J., Duan, Y., Wang, J., Coelho, L. P., de la Fuente-Nunez, C., & Szczurek, E. (2025). AI-Driven Antimicrobial Peptide Discovery: Mining and Generation. *Accounts of Chemical Research*, *58*(12), 1831–1846. DOI: 10.1021/acs.accounts.0c00594 PMID: 40459283

Tahir, M. A., Hameed, S., Munawar, A., Amin, I., Mansoor, S., Khan, W. S., & Bajwa, S. Z. (2017). Investigating the potential of multiwalled carbon nanotubes based zinc nanocomposite as a recognition interface towards plant pathogen detection. *Journal of Virological Methods*, *249*, 130–136. PMID: 28888402

Tahrani, A. A., Bailey, C. J., Del Prato, S., & Barnett, A. H. (2011). Management of type 2 diabetes: New and future developments in treatment. *Lancet, 378*(9786), 182–197. DOI: 10.1016/S0140-6736(11)60207-9 PMID: 21705062

Tan, X., Liu, Q., Fang, Y., Yang, S., Chen, F., Wang, J., Ouyang, D., Dong, J., & Zeng, W. (2024). Introducing enzymatic cleavage features and transfer learning realizes accurate peptide half-life prediction across species and organs. *Briefings in Bioinformatics, 25*(4). Advance online publication. DOI: 10.1093/BIB/BBAE350 PMID: 39038937

Tasaki, T., Takasuga, T., Osako, M., & Sakai, S. (2004). Substance flow analysis of brominated flame retardants and related compounds in waste TV sets in Japan. *Waste Management (New York, N.Y.), 24*, 571–580.

Tawiah, B., Seidu, R. K., Fobiri, G. K., & Chichi, C. A. (2024). Nanotechnology and education: Preparing the next generation of future leaders. In George, S. C., & Tawiah, B. (Eds.), *Nanotechnology in societal development* (pp. 141–161). Springer., DOI: 10.1007/978-981-97-6184-5_7

The Fluorescein Method of Latent Blood Detection. (n.d.). Retrieved February 23, 2025, from https://www.crime-scene-investigator.net/fluoresceinmethod.html #google_vignette

The Nobel Prize in Chemistry 2023 - Popular information - NobelPrize.org. (n.d.). Retrieved February 23, 2025, from https://www.nobelprize.org/prizes/chemistry/ 2023/popular-information/

Thiébaud-Müller, E., Hilty, L. M., Schluep, M., Widmer, R., & Faulstich, M. (2018). Service Lifetime, Storage Time, and Disposal Pathways of Electronic Equipment: A Swiss Case Study. *Journal of Industrial Ecology, 22*(2), 196–208.

Thimmaiah, S. H., Narayanappa, K., Thyavihalli Girijappa, Y., Gulihonenahali Rajakumara, A., Hemath, M., Thiagamani, S. M. K., & Verma, A. (2023). An artificial neural network and Taguchi prediction on wear characteristics of Kenaf–Kevlar fabric reinforced hybrid polyester composites. *Polymer Composites, 44*(1), 261–273.

Thomas, D. G., Chikkagoudar, S., Heredia-Langner, A., Tardiff, M. F., Xu, Z., Hourcade, D. E., Pham, C. T. N., Lanza, G. M., Weinberger, K. Q., & Baker, N. A. (2014). Physicochemical signatures of nanoparticle-dependent complement activation. *Computational Science & Discovery, 7*(1), 015003. Advance online publication. DOI: 10.1088/1749-4699/7/1/015003 PMID: 25254068

Tillu, R., Muthusubramanian, M., & Periyasamy, V. (2023). From data to compliance: the role of ai/ml in optimizing regulatory reporting processes. Journal of Knowledge Learning and Science Technology ISSN: 2959-6386 (online), 2(3), 381-391.

Tobias, A. V., Yip, J., Rouhanipour, A., & Ziolek, F. (2025). Autonomous 'self-driving' laboratories: A review of technology, challenges, and future directions. *Royal Society Open Science, 12*, 250646. DOI: 10.1098/rsos.250646 PMID: 40852582

Tom, G., Steiner, A., Kwon, S., & Rahman, T. (2024). Self-Driving Laboratories for Chemistry and Materials Science. *Chemical Reviews, 124*(20), 17245–17369. DOI: 10.1021/acs.chemrev.4c00055 PMID: 39137296

Topcu, E., Ridgeway, N. H., & Biggar, K. K. (2022). PeSA 2.0: A software tool for peptide specificity analysis implementing positive and negative motifs and motif-based peptide scoring. *Computational Biology and Chemistry, 101*. Advance online publication. DOI: 10.1016/j.compbiolchem.2022.107753 PMID: 35998543

Touretzky, D. S., Gardner-McCune, C., Martin, F., & Seehorn, D. (2019). Envisioning AI for K-12: What should every child know about AI? *Proceedings of the AAAI Conference on Artificial Intelligence, 33*(1), 9795–9799.

Tran, R., Lan, J., Shuaibi, M., Wood, B. M., Shoghi, N., Sriram, A., Therrien, F., ... (2022). The Open Catalyst 2022 (OC22) dataset and challenges for oxide electrocatalysts. *arXiv preprint*, arXiv:2206.08917. https://arxiv.org/abs/2206.08917

Tran, H. P., Schaubroeck, T., Nguyen, D. Q., Ha, V. H., Huynh, T. H., & Dewulf, J. (2018). Material flow analysis for management of waste TVs from households in urban areas of Vietnam. *Resources, Conservation and Recycling, 139*, 78–89.

Tsydenova, O., & Bengtsson, M. (2011). Chemical hazards associated with treatment of waste electrical and electronic equipment. *Waste Management (New York, N.Y.), 31*(1), 45–58.

Tucs, A., Tran, D. P., Yumoto, A., Ito, Y., Uzawa, T., & Tsuda, K. (2020). Generating ampicillin-level antimicrobial peptides with activity-aware generative adversarial networks. *ACS Omega, 5*(36), 22847–22851. DOI: 10.1021/acsomega.0c02088 PMID: 32954133

Turner, D., Williams, I., & Kemp, S. (2015). Greenhouse gas emission factors for recycling of source-segregated waste materials. *Resources, Conservation and Recycling, 105*, 186–197.

Tyagi, A., Tuknait, A., Anand, P., Gupta, S., Sharma, M., Mathur, D., Joshi, A., Singh, S., Gautam, A., & Raghava, G. P. S. (2015). CancerPPD: A database of anticancer peptides and proteins. *Nucleic Acids Research, 43*(D1), D837–D843. DOI: 10.1093/nar/gku892 PMID: 25270878

Ufarte Ruiz, M. J., & Manfredi Sánchez, J. L. (2019). Algorithms and bots applied to journalism. The case of Narrativa Inteligencia artificial: Structure, production, and informative quality. *Doxa Comun., 2019*, 213–233. DOI: 10.31921/doxacom.n29a11

Ufarte-Ruiz, M. J., Murcia-Verdú, F. J. y Túñez-López, J. M. (2023). Use of artificial intelligence in synthetic media: first newsrooms without journalists. Prof. Inf. 32:2. Doi: .. 2023. mar.03DOI: 10.3145/epi

Ullah, M. F., Khan, Y., Khan, M. I., Abdullaeva, B. S., & Waqas, M. (2024). Exploring nanotechnology in forensic investigations: Techniques, innovations, and prospects. *Sensing and Bio-Sensing Research, 45*, 100674. DOI: 10.1016/J.SBSR.2024.100674

UNESCO. (2022). *Technology in education: A tool on the road to equality.* https://www.unesco.org/en/education/technology

Ünlütabak, B., Ozdil, Z. C. C., & Cirit, E. S. (2023). *Exploring the Use of ChatGPT as a Learning and Teaching Tool in Material Science and Nanotechnology Engineering Education. ChemRxiv.* https://doi.org/DOI: 10.26434/chemrxiv-2023-4d99t

Vamathevan, J., Clark, D., Czodrowski, P., Dunham, I., Ferran, E., Lee, G., Li, B., Madabhushi, A., Shah, P., Spitzer, M., & Zhao, S. (2019). Applications of machine learning in drug discovery and development. *Nature Reviews. Drug Discovery, 18*(6), 463–477. DOI: 10.1038/S41573-019-0024-5 PMID: 30976107

Van Dalen, A. (2024). Revisiting the algorithms behind the headlines. How journalists respond to the professional competition of Generative AI. *J. Prac., 14*, 1–18. DOI: 10.1080/17512786.2024.2389209

Van de Voorde, M., & Fitzpatrick, M. E. (2019). Discovery and Innovation in Science and Technology: Inter- and trans-disciplinary in cognitive learning. *2019 IEEE 18th International Conference on Cognitive Informatics & Cognitive Computing (ICCI*CC)*, 261–264. https://doi.org/DOI: 10.1109/ICCICC46617.2019.9146083

Vaseashta, A. (2022). Nexus of Advanced Technology Platforms for Strengthening Cyber-Defense Capabilities, Vaseashta, DOI: , NATO Science for Peace and Security Series - E: Human and Societal Dynamics. Vol. 155: Practical Applications of Advanced Technologies for Enhancing Security and Defense Capabilities: Perspectives and Challenges for the Western BalkansDOI: 10.3233/NHSDP220003

Vaseashta, A. (Ed.). (2015). *Life cycle analysis of nanoparticles: Risk, assessment, and sustainability*. DEStech Publications.

Vaseashta, A. (2005). Nanostructured materials based next generation devices and sensors. In Vaseashta, A., Dimova-Malinovska, D., & Marshall, J. M. (Eds.), *Nanostructured and advanced materials for applications in sensor, optoelectronic and photovoltaic technology* (pp. 1–30). Springer., DOI: 10.1007/1-4020-3562-4_1

Vaseashta, A. (2006). Carbon nanotubes based devices and sensors. *Materials and Manufacturing Processes, 21*(7), 710–716. DOI: 10.1080/10426910600613595

Vaseashta, A. (2008). Nanoscale Materials, Devices, and Systems for Chem.-Bio Sensors, Photonics, and Energy Generation and Storage. In Vaseashta, A., & Mihailescu, I. N. (Eds.), *Functionalized Nanoscale Materials, Devices and Systems. NATO Science for Peace and Security Series B: Physics and Biophysics*. Springer., DOI: 10.1007/978-1-4020-8903-9_1

Vaseashta, A. (2021). Introduction to Water Safety, Security and Sustainability. In Vaseashta, A., & Maftei, C. (Eds.), *Water Safety, Security and Sustainability. Advanced Sciences and Technologies for Security Applications*. Springer., DOI: 10.1007/978-3-030-76008-3_1

Vaseashta, A., & Batra, A. (2022). Electrospun nanofibers of high-performance electret polymers for tactile sensing and wearable electronics. In Vaseashta, A., Achour, M. E., Mabrouki, M., Fasquelle, D., & Tachafine, A. (Eds.), *Proceedings of the Sixth International Symposium on Dielectric Materials and Applications (ISyDMA'6)* (pp. 31–52). Springer. https://doi.org/DOI: 10.1007/978-3-031-11397-0_3

Vaseashta, A., & Bölgen, N. (Eds.). (2022). *Electrospun nanofibers: Principles, technology and novel applications*. Springer., DOI: 10.1007/978-3-030-99958-2

Vaseashta, A., Demir, D., Sakım, B., Aşık, M., & Bölgen, N. (2022). Hierarchical integration of 3D printing and electrospinning of nanofibers for rapid prototyping. In Vaseashta, A., & Bölgen, N. (Eds.), *Electrospun nanofibers* (pp. 631–655). Springer., DOI: 10.1007/978-3-030-99958-2_24

Vaseashta, A., Duca, G., & Travin, S. (2022). *Handbook of Research on Water Science and Society* (Vols. 1–2). IGI Global publishers., DOI: 10.4018/978-1-7998-7356-3

Vaseashta, A., & Stamatin, I. (2007). Electrospun polymers for controlled release of drugs, vaccine delivery, and system-on-fibers. *Journal of Optoelectronics and Advanced Materials, 9*(6), 1606–1613.

Vázquez-Herrero, J., Negreira-Rey, M. C., & López-García, X. (2020). Let's dance the news! How the news media are adapting to the logic of TikTok. *Journalism, 23*, 1717–1735. DOI: 10.1177/1464884920969092

Vega-Baudrit, J.-R., Camacho, M., Araya, A., & Camacho, M. (2023). Regulating Nanotechnology: Ensuring Responsible and Safe Innovation in the Advancement of Science and Technology. *Ciencia. Tecnología y Salud, 10*(2), 177–191. DOI: 10.36829/63CTS.v10i2.1581

Velis, C., & Mavropoulos, A. (2016). Unsound waste management and public health: The neglected link? *Waste Management & Research, 34*(4), 277–279.

Ventura-Pocino, P. (2021). *Algorithms in the newsrooms: Challenges and recommendations for artificial intelligence with the ethical values of journalism.* Catalan Fundació Consell de la Información de Catalunya/Press Council.

Viner, K., & Bateson, A. (2024). The Guardian's approach to generative AI | | The Guardian. The Guardian. Available at: https://www.theguardian.com/help/insideguardian/2023/jun/16/the-guardians-approach-to-generative-ai (Accessed October 2, 2024).

Vishnoi, V., Chaudhary, S., Joshi, N., & Consul, P. (2025). Nanoscale Industry 4.0 Empowers Advanced Nanomaterials and Devices. In *Nanotechnology Applications for Industry 4.0* (pp. 297-317). CRC Press.

Vishnyakov, A. (2025). Machine Learning in Computational Design and Optimization of Disordered Nanoporous Materials. *Materials (Basel), 18*(3). Advance online publication. DOI: 10.3390/ma18030534 PMID: 39942200

Vladár, A. E., & Hodoroaba, V.-D. (2020). Characterization of nanoparticles by scanning electron microscopy. In *Characterization of nanoparticles* (pp. 7–27). Elsevier., DOI: 10.1016/B978-0-12-814182-3.00002-6

Volk, A. A., Epps, R. W., Yonemoto, D. T., Masters, B. S., Castellano, F. N., Reyes, K. G., Abolhasani, M., et al. (2023). AlphaFlow: Autonomous discovery and optimization of multi-step chemistry using a self-driven fluidic lab guided by reinforcement learning. *Nature Communications, 14*, Article. https://doi.org/DOI: 10.1038/s41467-023-37139-y

Vora, L. K., Gholap, A. D., Jetha, K., Thakur, R. R. S., Solanki, H. K., & Chavda, V. P. (2023). Artificial Intelligence in Pharmaceutical Technology and Drug Delivery Design. *Pharmaceutics, 15*(7), 1916. DOI: 10.3390/pharmaceutics15071916

Voulodimos, A., Doulamis, N., Doulamis, A., & Protopapadakis, E. (2018). Deep learning for computer vision: A brief review. *Computational Intelligence and Neuroscience, 2018*, 7068349. Advance online publication. DOI: 10.1155/2018/7068349 PMID: 29487619

Vyas, M., Thakur, S., Riyaz, B., Bansal, K. K., Tomar, B., & Mishra, V. (2018). Artificial intelligence: The beginning of a new era in the pharmacy profession. *Asian Journal of Pharmaceutics, 12*(2), 72–76. DOI: 10.22377/AJP.V12I02.2317

Waghu, F. H., Gopi, L., Barai, R. S., Ramteke, P., Nizami, B., & Idicula-Thomas, S. (2014). CAMP: Collection of sequences and structures of antimicrobial peptides. *Nucleic Acids Research, 42*(D1), D1154–D1158. DOI: 10.1093/nar/gkt1157 PMID: 24265220

Wagner, A. M., Gran, M. P., & Peppas, N. A. (2018). Designing the new generation of intelligent biocompatible carriers for protein and peptide delivery. *Acta Pharmaceutica Sinica. B, 8*(2), 147–164. DOI: 10.1016/J.APSB.2018.01.013 PMID: 29719776

Walk, W. (2009). Forecasting quantities of disused household CRT appliances—A regional case study approach and its application to Baden-Wurttemberg. *Waste Management (New York, N.Y.), 29*, 945–951.

Walther, G., Steinborn, J., Spengler, T., Luger, T., & Herrmann, C. (2009). Implementation of the WEEE–Directive – Economic effects and improvement potentials for reuse and recycling in Germany. *International Journal of Advanced Manufacturing Technology, 47*, 461–474.

Wan, F., Torres, M. D. T., Peng, J., & de la Fuente-Nunez, C. (2024). Deep-learning-enabled antibiotic discovery through molecular de-extinction. *Nature Biomedical Engineering, 8*(7), 854–871. DOI: 10.1038/s41551-024-01201-x PMID: 38862735

Wan, F., Wong, F., Collins, J. J., & de la Fuente-Nunez, C. (2024). Machine learning for antimicrobial peptide identification and design. *Nature Reviews Bioengineering, 2*(5), 392–407. DOI: 10.1038/s44222-024-00152-x PMID: 39850516

Wang, J., Chang, K., De, R. O., Costa, A., Passos, T. S., Mikaelly, E., Silva, S., Santos, N. C. S. Dos, Heloneida, A., & Morais, A. (2023). Encapsulated Peptides and Proteins with an Effect on Satiety. *Nanomaterials 2023, Vol. 13, Page 1166, 13*(7), 1166. https://doi.org/DOI: 10.3390/NANO13071166

Wang, A. J., Li, H., He, Z., Tao, Y., Wang, H., Yang, M., Savic, D., Daigger, G. T., & Ren, N. (2024). Digital Twins for Wastewater Treatment: A Technical Review. *Engineering, 36*, 21–35. DOI: 10.1016/j.eng.2024.04.012

Wang, F., Huisman, J., Stevels, A., & Balde, C. P. (2013). Enhancing e-waste estimates: Improving data quality by multivariate Input-Output Analysis. *Waste Management (New York, N.Y.)*, *33*, 2397–2407.

Wang, F., Yu, L., & Wu, A. (2021). Forecasting the electronic waste quantity with a decomposition-ensemble approach. *Waste Management (New York, N.Y.)*, *120*, 828–838.

Wang, G., Li, X., & Wang, Z. (2016). APD3: The antimicrobial peptide database as a tool for research and education. *Nucleic Acids Research*, *44*(D1), D1087–D1093. DOI: 10.1093/nar/gkv1278 PMID: 26602694

Wang, M., Li, M., Yu, A., Zhu, Y., Yang, M., & Mao, C. (2017). Fluorescent Nanomaterials for the Development of Latent Fingerprints in Forensic Sciences. *Advanced Functional Materials*, *27*(14), 1606243. DOI: 10.1002/ADFM.201606243 PMID: 29657570

Wang, M., Tan, Q., Chiang, J., & Li, J. (2017). Recovery of rare and precious metals from urban mines—A review. *Frontiers of Environmental Science & Engineering*, *11*(5), 1–17.

Wang, M., Wang, T., Cai, P., & Chen, X. (2019). Nanomaterials discovery and design through machine learning. *Small Methods*, *3*(8), 1900025. Advance online publication. DOI: 10.1002/smtd.201900025

Wang, M., You, X., Li, X., & Liu, G. (2018). Watch more, waste more? A stock-driven dynamic material flow analysis of metals and plastics in TV sets in China. *Journal of Cleaner Production*, *187*, 730–739.

Wang, N., & Lester, J. C. (2023). K-12 education in the age of AI: A call to action for K-12 AI literacy. *International Journal of Artificial Intelligence in Education*, *33*(2), 228–260. DOI: 10.1007/s40593-023-00358-x

Wang, Y. M., & Fang, C. (2024). Cycle-ESM: Generation-assisted classification of antifungal peptides using ESM protein language model. *Computational Biology and Chemistry*, *113*, 108240. DOI: 10.1016/j.compbiolchem.2024.108240 PMID: 39437594

Wang, Y., Zhang, L., Liu, C., Luo, Y., & Chen, D. (2024). Peptide-Mediated Nanocarriers for Targeted Drug Delivery: Developments and Strategies. *Pharmaceutics*, *16*(2), 240. DOI: 10.3390/PHARMACEUTICS16020240 PMID: 38399294

Wang, Z., Pal, D., Pilechi, A., & Ariya, P. A. (2024). Nanoplastics in Water: Artificial Intelligence-Assisted 4D Physicochemical Characterization and Rapid In Situ Detection. *Environmental Science & Technology*, *58*(20), 8919–8931. DOI: 10.1021/acs.est.3c10408 PMID: 38709668

Watson, J. L., Juergens, D., Bennett, N. R., Trippe, B. L., Yim, J., Eisenach, H. E., Ahern, W., Borst, A. J., Ragotte, R. J., Milles, L. F., Wicky, B. I. M., Hanikel, N., Pellock, S. J., Courbet, A., Sheffler, W., Wang, J., Venkatesh, P., Sappington, I., Torres, S. V., & Baker, D. (2023). De novo design of protein structure and function with RFdiffusion. *Nature*, *620*(7976), 1089–1100. DOI: 10.1038/s41586-023-06415-8 PMID: 37433327

Wei, L., Ye, X., Sakurai, T., Mu, Z., & Wei, L. (2022). ToxIBTL: Prediction of peptide toxicity based on information bottleneck and transfer learning. *Bioinformatics (Oxford, England)*, *38*(6), 1514–1524. DOI: 10.1093/bioinformatics/btac006 PMID: 34999757

Wei, L., Ye, X., Xue, Y., Sakurai, T., & Wei, L. (2021). Atse: A peptide toxicity predictor by exploiting structural and evolutionary information based on graph neural network and attention mechanism. *Briefings in Bioinformatics*, *22*(5), 1–13. DOI: 10.1093/bib/bbab041 PMID: 33822870

Wei, T., Han, Z., Zhong, X., Xiao, Q., Liu, T., & Xiang, D. (2022). Two-dimensional semiconducting materials for ultimately scaled transistors. *iScience*, *25*(10), 105160. DOI: 10.1016/j.isci.2022.105160

Wilburn, D. (2012). Byproduct metals and rare-earth elements used in the production of light-emitting diodes—Overview of principal sources of supply and material requirements for selected markets. *U.S. Geological Survey*.

Winnifrith, A., Outeiral, C., & Hie, B. L. (2024). Generative artificial intelligence for de novo protein design. *Current Opinion in Structural Biology*, *86*, 102794. DOI: 10.1016/j.sbi.2024.102794 PMID: 38663170

Women, U. N. (2021). *Gender equality in STEM education*. https://www.unwomen.org

Wong, M. H., Misra, R. P., Giraldo, J. P., Kwak, S. Y., Son, Y., Landry, M. P., & Strano, M. S. (2016). Lipid exchange envelope penetration (LEEP) of nanoparticles for plant engineering: A universal localization mechanism. *Nano Letters*, *16*(2), 1161–1172. PMID: 26760228

Woolf, B. P. (2010). *Building intelligent interactive tutors: Student-centered strategies for revolutionizing e-learning*. Morgan Kaufmann.

Wu, X., Lin, H., Bai, R., & Duan, H. (2024). Deep learning for advancing peptide drug development: Tools and methods in structure prediction and design. *European Journal of Medicinal Chemistry*, *268*, 116262. DOI: 10.1016/J.EJMECH.2024.116262 PMID: 38387334

Wu, Y., Duan, H., & Xi, H. (2020). Machine learning-driven insights into defects of zirconium metal–organic frameworks for enhanced ethane–ethylene separation. *Chemistry of Materials*, *32*(7), 2986–2997. DOI: 10.1021/acs.chemmater.9b05322

Wu, Z., He, D., Cui, B., & Jin, Z. (2018). A bimodal (SERS and colorimetric) aptasensor for the detection of Pseudomonas aeruginosa. *Microchimica Acta*, *185*, 1–7. PMID: 30382404

Xiao, X., Shao, Y. T., Cheng, X., & Stamatovic, B. (2021). iAMP-CA2L: A new CNN-BiLSTM-SVM classifier based on cellular automata image for identifying antimicrobial peptides and their functional types. *Briefings in Bioinformatics*, *22*(6), 1–10. DOI: 10.1093/BIB/BBAB209 PMID: 34086856

Xu, C., Deng, X., & Yu, P. (2024). High-throughput computational study and machine learning prediction of electronic properties in transition metal dichalcogenide/two-dimensional layered halide perovskite heterostructures. *ACS Applied Materials & Interfaces*, *16*(39), 52351–52362. DOI: 10.1021/acsami.4c11973 PMID: 39361426

Xu, D., Ma, Y., Jin, G., & Cao, L. (2025). Intelligent Photonics: A Disruptive Technology to Shape the Present and Redefine the Future. *Engineering*, *46*, 186–213. DOI: 10.1016/j.eng.2024.08.016

Xu, W., & Ouyang, F. (2022). The application of AI technologies in STEM education: A systematic review from 2011 to 2021. *International Journal of STEM Education*, *9*(1), 59. Advance online publication. DOI: 10.1186/s40594-022-00377-5

Yamankurt, G., Berns, E. J., Xue, A., Lee, A., Bagheri, N., Mrksich, M., & Mirkin, C. A. (2019). Exploration of the nanomedicine-design space with high-throughput screening and machine learning. *Nature Biomedical Engineering*, *3*(4), 318–327. DOI: 10.1038/s41551-019-0351-1 PMID: 30952978

Yang, C., Tan, Q., Liu, L., Dong, Q., & Li, J. (2017). Recycling tin from electronic waste: A problem that needs more attention. *ACS Sustainable Chemistry & Engineering*, *5*(11), 9586–9598.

Yang, H., Zhang, Z., Zhang, J., & Zeng, X. C. (2018). Machine learning and artificial neural network prediction of interfacial thermal resistance between graphene and hexagonal boron nitride. *Nanoscale*, *10*(40), 19092–19099. DOI: 10.1039/C8NR05703F PMID: 30288513

Yang, L., Wang, H., Leng, D., Fang, S., Yang, Y., & Du, Y. (2024). Machine learning applications in nanomaterials: Recent advances and future perspectives. *Chemical Engineering Journal*, *500*, 156687. Advance online publication. DOI: 10.1016/j.cej.2024.156687

Yang, R. X., McCandler, C. A., Andriuc, O., Siron, M., Woods-Robinson, R., Horton, M. K., & Persson, K. A. (2022). Big data in a nano world: A review on computational, data-driven design of nanomaterials structures, properties, and synthesis. *ACS Nano*, *16*(12), 19873–19891. DOI: 10.1021/acsnano.2c08411 PMID: 36378904

Yang, Y., & Williams, E. (2009). Logistic model-based forecast of sales and generation of obsolete computers in the U.S. *Technological Forecasting and Social Change*, *76*, 1105–1114.

Yan, J., Cai, J., Zhang, B., Wang, Y., Wong, D. F., & Siu, S. W. I. (2022). Recent Progress in the Discovery and Design of Antimicrobial Peptides Using Traditional Machine Learning and Deep Learning. *Antibiotics (Basel, Switzerland)*, *11*(10), 1451. DOI: 10.3390/antibiotics11101451 PMID: 36290108

Yan, X., Sedykh, A., Wang, W., Yan, B., & Zhu, H. (2020). Construction of a web-based nanomaterial database by big data curation and modeling friendly nanostructure annotations. *Nature Communications*, *11*(1), 2519. Advance online publication. DOI: 10.1038/s41467-020-16413-3 PMID: 32433469

Yao, J., & Wang, X. (2025). Artificial intelligence in de novo protein design. *Medicine in Novel Technology and Devices*, *26*, 100366. DOI: 10.1016/j.medntd.2025.100366

Yao, L., & Chen, Q. (2023). Machine learning in nanomaterial electron microscopy data analysis. In *Intelligent nanotechnology* (pp. 279–305). Elsevier., DOI: 10.1016/B978-0-323-85796-3.00009-0

Yao, Z., Sánchez-Lengeling, B., Bobbitt, N. S., Bucior, B. J., Kumar, S. G. H., Collins, S. P., Burns, T., Woo, T. K., Farha, O. K., Snurr, R. Q., & Aspuru-Guzik, A. (2021). Inverse design of nanoporous crystalline reticular materials with deep generative models. *Nature Machine Intelligence*, *3*(1), 76–86. DOI: 10.1038/s42256-020-00271-1

Yaseen, H., Mohammad, A. S., Ashal, N., Abusaimeh, H., Ali, A., & Sharabati, A.-A. A. (2025). The Impact of Adaptive Learning Technologies, Personalized Feedback, and Interactive AI Tools on Student Engagement: The Moderating Role of Digital Literacy. *Sustainability*, *17*(3), 1133. DOI: 10.3390/su17031133

Yawson, R. M. (2010). Skill needs and human resources development in the emerging field of nanotechnology. *Journal of Vocational Education and Training*, *62*(3), 285–296. DOI: 10.1080/13636820.2010.499474

Yawson, R. M. (2012). An epistemological framework for nanoscience and nanotechnology literacy. *International Journal of Technology and Design Education*, *22*(3), 297–310. DOI: 10.1007/s10798-010-9145-1

Yawson, R. M., & Greiman, B. C. (2017a). Strategic flexibility analysis of agrifood nanotechnology skill needs identification. *Technological Forecasting and Social Change*, *118*, 184–194. DOI: 10.1016/j.techfore.2017.02.019

Yazici, B., Can, Z. S., & Calli, B. (2014). Prediction of future disposal of end-of-life refrigerators containing CFC-11. *Waste Management (New York, N.Y.)*, *34*, 162–166.

Yeh, Y. C., Creran, B., & Rotello, V. M. (2011). Gold Nanoparticles: Preparation, Properties, and Applications in Bionanotechnology. *Nanoscale*, *4*(6), 1871. DOI: 10.1039/C1NR11188D PMID: 22076024

Yen, J. H., Chang, C. C., Wu, T. Y., Yang, C. H., Hsu, H. J., & Liou, J. W. (2025). Therapeutic peptides and their delivery using lipid-based nanoparticles. *Tzu-Chi Medical Journal*, *37*(3), 223. DOI: 10.4103/TCMJ.TCMJ_321_24 PMID: 40741609

Yıldız, M., Kaltakçı Gürel, D., Salmankurt, B., & Gürel, H. H. (2025). Exploring the Evolution of Nanotechnology Education: Insights from Bibliometric Analysis. *Journal of Chemical Education*, *102*(1), 253–269. DOI: 10.1021/acs.jchemed.4c01120

Yin, H., Xu, M., Luo, Z., Bi, X., Li, J., Zhang, S., & Wang, X. (2024). Machine learning for membrane design and discovery. *Green Energy & Environment*, *9*(1), 54–70. DOI: 10.1016/j.gee.2022.12.001

Yin, P., Niu, X., Li, S.-B., Chen, K., Zhang, X., Zuo, M., Zhang, L., & Liang, H.-W. (2024). Machine-learning-accelerated design of high-performance platinum intermetallic nanoparticle fuel cell catalysts. *Nature Communications*, *15*(1), 415. Advance online publication. DOI: 10.1038/s41467-023-44674-1 PMID: 38195668

Yoshihara, N., Tahara, Y., & Noda, M. (2023). Machine learning method for determining chemical vapor deposition conditions for large-area graphene growth. *Asia-Pacific Journal of Chemical Engineering*, *18*(4), e2911. Advance online publication. DOI: 10.1002/apj.2911

Yuan, L., Ni, Y. Q., Deng, X. Y., & Hao, S. (2022). A-PINN: Auxiliary physics informed neural networks for forward and inverse problems of nonlinear integro-differential equations. *Journal of Computational Physics*, *462*, 111260.

Yuan, L., Richardson, C. J., Ho, M., Willis, C. W., Colman, B. P., & Wiesner, M. R. (2018). Stress responses of aquatic plants to silver nanoparticles. *Environmental Science & Technology*, *52*(5), 2558–2565. PMID: 29381864

Yuan, Q., Chen, K., Yu, Y., Le, N. Q. K., & Chua, M. C. H. (2023). Prediction of anticancer peptides based on an ensemble model of deep learning and machine learning using ordinal positional encoding. *Briefings in Bioinformatics*, *24*(1), 1–10. DOI: 10.1093/BIB/BBAC630 PMID: 36642410

Yuan, Y., Wu, Y., Cheng, J., Yang, K., Xia, Y., Wu, H., & Pan, X. (2024). Applications of artificial intelligence to lipid nanoparticle delivery. *Particuology*, *90*, 88–97. DOI: 10.1016/j.partic.2023.11.014

Yüksel, N., Börklü, H. R., Sezer, H. K., & Canyurt, O. E. (2023). Review of artificial intelligence applications from an engineering design perspective. *Engineering Applications of Artificial Intelligence*, *118*, 105697. DOI: 10.1016/j.engappai.2022.105697

Yu, T., Su, S., Hu, J., Zhang, J., & Xianyu, Y. (2022). A new strategy for microbial taxonomic identification through micro-biosynthetic gold nanoparticles and machine learning. *Advanced Materials*, *34*(24), 2109365. Advance online publication. DOI: 10.1002/adma.202109365 PMID: 34989446

Zafar, S., & Rana, N. (2025). *The Convergence of Nanotechnology and Artificial Intelligence: Unlocking Future Innovations*. https://doi.org/DOI: 10.2174/0124055204359215250127071256

Zare, H., Meshkat, Z., Hatamluyi, B., Rezayi, M., Ghazvini, K., Derakhshan, M., & Aryan, E. (2022). The first diagnostic test for specific detection of Mycobacterium simiae using an electrochemical label-free DNA nanobiosensor. *Talanta*, *238*, 123049. PMID: 34801906

Zawacki-Richter, O., Marín, V. I., Bond, M., & Gouverneur, F. (2019). Systematic review of research on artificial intelligence applications in higher education – where are the educators? *International Journal of Educational Technology in Higher Education*, *16*(1), 39. Advance online publication. DOI: 10.1186/s41239-019-0171-0

Zeinali, S., Dehghani, M., & Talebbeydokhti, N. (2021). Artificial neural network for the prediction of shoreline changes in Narrabeen, Australia. *Applied Ocean Research*, *107*, 102362.

Zeng, X., Ali, S. H., & Li, J. (2021). Estimation of waste outflows for multiple product types in China from 2010–2050. *Scientific Data*, *8*(1), 15.

Zeng, X., Yang, C., Chiang, J., & Li, J. (2017). Innovating e-waste management: From macroscopic to microscopic scales. *The Science of the Total Environment*, *575*, 1–5.

Zhai, X., & Krajcik, J. (2024). Artificial intelligence-based STEM education. In Zhai, X., & Krajcik, J. (Eds.), *Uses of artificial intelligence in STEM education* (pp. 1–24). Oxford University Press., DOI: 10.1093/oso/9780198882077.003.0001

Zhang, Y., ... (2022). Interdisciplinary discovery of nanomaterials based on convolutional neural networks. *arXiv preprint*, arXiv:2212.02805. https://arxiv.org/abs/2212.02805

Zhang, K., Yang, X., Wang, Y., Yu, Y., Huang, N., Li, G., Li, X., Wu, J. C., & Yang, S. (2025). Artificial intelligence in drug development. *Nature Medicine*, *31*(1), 45–59. DOI: 10.1038/S41591-024-03434-4;SUBJMETA PMID: 39833407

Zhang, P., Mao, Y., Yuan, W., Zheng, J., Hu, S., & Wang, K. (2024). A critical review on modeling and prediction on properties of fresh and hardened geopolymer composites. *Journal of Building Engineering*, •••, 109184.

Zhang, S., Gu, Y., Tang, A., Li, B., Li, B., Pan, D., & Wu, Y. (2021). Forecast of future yield for printed circuit board resin waste generated from major household electrical and electronic equipment in China. *Journal of Cleaner Production*, *283*, 124575.

Zhao, M., Zhao, C., Yu, L., Li, G., Huang, J., Zhu, H., & He, W. (2016). Prediction and Analysis of WEEE in China Based on the Gray Model. *Procedia Environmental Sciences*, *31*, 925–934.

Zheng, Z., Zhang, O., Nguyen, H. L., Rampal, N., Alawadhi, A. H., Rong, Z., Head-Gordon, T., Borgs, C., Chayes, J. T., & Yaghi, O. M. (2023). ChatGPT Research Group for Optimizing the Crystallinity of MOFs and COFs. *ACS Central Science*, *9*(11), 2161–2170. DOI: 10.1021/acscentsci.3c01087 PMID: 38033801

Zhong, X., et al. (2022). Explainable machine learning in materials science. *npj Computational Materials*. https://doi.org/DOI: 10.1038/s41524-022-00884-7

Zhou, S., Luo, Y., & Lovell, J. F. (2023). Vaccine approaches for antigen capture by liposomes. *Expert Review of Vaccines*, *22*(1), 1022. DOI: 10.1080/14760584.2023.2274479 PMID: 37878481

Zhou, W., Fang, J., Tang, S., Wu, Z., & Wang, X. (2021). 3D-printed nanocellulose-based cushioning–antibacterial dual-function food packaging aerogel. *Molecules (Basel, Switzerland)*, *26*(12), 3543. PMID: 34200653

Zhu, Y., Xu, W., Zhang, J., Liu, Q., Wu, S., & Wang, L. (2021). Deep graph structure learning for robust representations: A survey. arXiv preprint arXiv:2103.03036, 14, 1-1.

Zhu, B., Fan, C., Cheng, C., Lan, T., Li, L., & Qin, Y. (2021). Migration kinetic of silver from polylactic acid nanocomposite film into acidic food simulant after different high-pressure food processing. *Journal of Food Science*, *86*(6), 2481–2490. PMID: 33948960

Zhu, C., Bamidele, E. A., Shen, X., Zhu, G., & Li, B. (2024). Machine learning aided design and optimization of thermal metamaterials. *Chemical Reviews*, *124*(7), 4258–4331. DOI: 10.1021/acs.chemrev.3c00708 PMID: 38546632

Zhu, Z., Dong, S., Zhang, H., Parker, W., Yin, R., Bai, X., Yu, Z., Wang, J., Gao, Y., & Ren, H. (2025). Bayesian Optimization-Enhanced Reinforcement learning for Self-adaptive and multi-objective control of wastewater treatment. *Bioresource Technology*, *421*(February), 132210. DOI: 10.1016/j.biortech.2025.132210 PMID: 39933666

About the Contributors

Ashok Vaseashta received a Ph.D. from the Virginia Polytechnic Institute and State University, Blacksburg, VA, in 1990, followed by Kobe's post-doctoral fellowship. He was awarded an Honoris Causa doctorate from Riga Technical University, Honorary membership of the Academy of Sciences of Moldova, and a Gold Medal for his leadership in Nanotechnology from the National Polytechnic University of Armenia. Currently, he is Executive Director for strategic research with the International Clean Water Institute in Virginia, USA. He served as a Professor at Virginia Tech and Marshall University, Director of Research at the Institute for Advanced Sciences Convergence and International Clean Water Institute for Norwich University Applied Research Institutes, Vice-Provost (Rector) for Research in South Carolina, and Executive Director and Chair of the Institutional Review Board at a State University in New Jersey. He held visiting positions as a Professor at the Riga Technical University, Latvia, 3 Nano-SAE Research Centre, University of Bucharest, Romania, Transilvania University of Brasov, Romania; member of CIRET, France, and a scientist at the Weizmann Institute of Science, Israel. He served the U.S. Department of State in two rotations as a strategic S&T advisor in the Bureau of International Security and Nonproliferation (ISN), Office of Weapons of Mass Destruction and Terrorism. He served as NATO project director (NPD) for 7 activities funded under NATO's Science for Peace and Security program. Inspired by nature and guided by societal necessities, he strives for technological innovations to address the global challenges of the 21st century. He specializes in dual-use research to identify embedded hybrid threats, using research tools such as foresight, heuristics, artificial intelligence, and complexity science. His research interests include chem.-bio sensors, nanotechnology, environmental/ecotoxicology science, critical infrastructure protection, and the environmental impact of micro/nano plastics - all using the nexus of advanced technological solution platforms. He is a Fulbright Specialist and serves as a research fellow at the Center for Disruptive Technologies and Future Warfare in the Institute for National Strategic Studies at the

National Defense University. He is the author/editor of 27 books and has published over 350 articles in scientific journals, book chapters, and conferences. He serves on the editorial boards of several international journals and is an active member of various professional organizations.

Ioan Stamatin is a Professor of Physics at the University of Bucharest, where he earned his doctorate in 1998. He is a supervisor in the Doctoral School of Physics, specializing in Renewable and Alternative Energy Sources. He is also the coordinator of the 3nano-SAE Research Center (Nanosciences & Alternative Energy Sources) and founder of the Fuel Cells & Hydrogen Storage branch of the University of Bucharest's Postdoctoral School. Recently in 2024, as a fervent supporter, he implemented the new idea by starting "innoversity" – a concept based on AI & IoT. Prof. Stamatin's research interests span a wide range of topics, including Energy: Fuel cells, hydrogen storage, sodium-ion batteries, heat storage nanomaterials, and radioisotope thermal generator applications. Nanomaterials: Synthesis, characterization, and applications of nanomaterials in electrocatalysis, photocatalysis, CO2 reduction, and enhancing plant pest resistance. Biotechnology: Microbial fuel cells, microorganism-based metal recovery, and algal biotechnologies. Agriculture: Nanopesticides for seed revitalization, green permaculture, and electroculture. He has authored numerous publications, including books, book chapters, and articles in high-impact journals. His work has led to over 20 national patents. Prof. Stamatin is a recognized expert evaluator in EU research programs and a member of Romanian standard associations (MRS, E-MRS). He has supervised numerous doctoral theses for both Romanian and international students, contributing to the development of highly skilled researchers worldwide. Some of the key highlights include his extensive experience in renewable energy and nanomaterials research; proven track record of securing patents and publishing in reputable journals; dedication to fostering research excellence and mentoring future generations of scientists; and an active contributor to national and international research communities

***

Umut Akduğan, has his BA degree in the field of Mathematics Education at Marmara University in 2007. Subsequently, he received his MA degree in the field of Econometrics at Trakya University and PhD degrees in the field of Economics at Istanbul University. The author currently continues his career at Trakya University, Faculty of Economics and Administrative Sciences, Department of Public Finance, since 2017. Dr. Akduğan is interested in Econometrics of Macroeconomic Modelling, Environmental Economics and Fiscal Policy.

Gagik Ayvazyan received his Ph.D. degree in Physics and Technology of Semiconductor Materials and Devices from the St. Petersburg State University of Information Technologies, Mechanics and Optics (Russia) in 1989. He currently holds the position of Associate Professor at the National Polytechnic University of Armenia, Yerevan. His research interests are focused on materials and technological processes for solar cell applications. Dr. Ayvazyan has authored more than 40 scientific publications in international journals and conference proceedings. He is a Corresponding Member of the International Academy of Engineering

Manjunatha Badiger is an Assistant Professor in the Department of VLSI Design and Technology at NMAM Institute of Technology, Nitte. He has over 13 years of experience in academics, research, and administration. Dr. Badiger holds a Ph.D. from Visvesvaraya Technological University, an M.Tech in VLSI & Embedded System Design from Reva Institute of Technology and Management, and a B.E. in Electronics & Communication Engineering from NMAM Institute of Technology, Nitte. His research interests span Machine Learning, Deep Learning, Image Processing, and VLSI. Dr. Badiger has made significant contributions to national and international journals, book chapters, and has completed various NPTEL courses. Previously, he served as an Assistant Professor in the Department of Electronics and Communication Engineering at Sahyadri College of Engineering & Management and Shree Devi Institute of Technology. Dr. Badiger is accessible via email at manjunatha.badiger@nitte.edu.in or badiger_manju@yahoo.com.

Adriana Balan Associate professor at University of Bucharest, Ph.D. since 2011 in Solid State Physics, University of Bucharest. Professional experience: Composite polymer membranes based on nanoparticles, oxides, intrinsic conducting polymers (polypyrrole, polyaniline), carbon materials (nanocarbons, graphenes) and other additives– syntheses and characterization, electrochemical device engineering-fuel cells, batteries and hydrogen storage, ion exchange membranes, characterization techniques: Atomic Force Microscopy, Differential Scanning Calorimetry, Thermogravimetric analysis, fuel cell specific tests. She is co-founder of 3Nano-SAE Research Center and she is supervising student research activity of the Master Program Renewable and Alternative Energy Sources (R&AES) University of Bucharest, Faculty of Physics. She led as coordinator or was responsible for 5 national projects: project no. 205TE/2021- coordinator, project no 76PCCDI/2018- responsible for component project 1, STAR project no 134/2017-coordinator, STAR project no 68/2013- partner, PNCDI II project no 46/2014- coordinator; and participated in over 15 projects as team member. She is co-author of 6 patents and 5 pending patent.

Cătălin Berlic is an Associate Professor at the Faculty of Physics, University of Bucharest, where he teaches mechanics, computational simulation, and polymer physics, and coordinates the master's program Environmental Physics and Eco-Friendly Materials. He holds a Ph.D. in Physics and has published in international journals and conference proceedings, as well as co-authoring several university textbooks. His research focuses on computational modeling of polymers and liquid crystals, polymer–liquid crystal interfaces, plasma-polymerized materials, and polymer-based biosensors, and he also works on advancing physics education by integrating Artificial Intelligence and Machine Learning methods in both research and teaching.

Shikha Choudhary is an Assistant Professor with over 8 years of teaching experience. She is currently pursuing her Ph.D. and specializes in subjects such as C, C++, Data Communication, and Software Engineering. With a strong academic foundation and a passion for research, she has authored 6 research papers in reputed journals and conferences. Her primary research interests lie in the domains of Artificial Intelligence (AI) and Machine Learning (ML), where she actively explores innovative applications and emerging trends. Shikha is known for her engaging teaching methodology, commitment to academic excellence, and continuous pursuit of knowledge. She aims to bridge the gap between theoretical learning and practical implementation through her teaching and research work.

Prakhar Consul brings more than 7 years of expertise in teaching, and Research. He is currently working as an Assistant Professor in School of Computer Science and Engineering, IILM University, Greater Noida. He completed his Ph.D. in Computer Science Engineering from Bennett University, School of Computer Science Engineering and Technology, Greater Noida, India. He received M.Tech. and B. Tech degree in 2015 and 2013, respectively. His research interest includes Edge Computing, Cyber Physical Systems, Machine Learning, Digital Twin, Internet of things, non-orthogonal multiple access and the development of micro-strip patch antenna for various wireless application. He is an active member of IEEE and International Association of Engineers (IAENG). He published 8 SCI research papers (4 in IEEE Transactions, 1 in IEEE Internet of Things Journal, 1 in Alexandria Engineering Journal, 1 in Vehicular Communications Journal and 1 in Ad Hoc Networks Journal). Also, he published 03 Patents and 18 research papers in IEEE and Springer reputed conferences such as INFOCOM, MILCom, UCC GLOBECOM, ICC, IWCNC, ACM, FTNCT, IACC, RTIP2R, and ICIICT.

D. Sai Chaitanya Kishore has about 17 years of teaching and research experience. Presently he is Professor of Mechanical Engineering at Srinivasa

Ramanujan Institute of Technology, Rotarypuram, Anantapur District, A.P. He has filed 5 patents and has published several research papers in reputed Journals indexed by Scopus, Science Citation Index Expanded and Materials Science Citation Index. He is a member of professional bodies like IE, ISTE, SAE and IAENG. His areas of interest include advanced machining processes, additive manufacturing, optimization of machining processes, ANN and composite materials.

Bogdan Dobrică is an AI Engineer and Software Architect who combines theory and industry. His role as Technical Lead for Bolt's Data Platform's Model Lifecycle team is to build and deploy scalable machine learning infrastructures. Machine learning operations (MLOps), distributed computing, data-driven system automation, and AI integration into real-time service platforms are his research and work interests. Before joining Bolt, Dr. Dobrică led multidisciplinary teams in AI consulting projects at Cognizant Softvision, including financial trading anomaly detection, reinforcement learning for autonomous decision systems, and automotive cloud-based model optimization. His previous work as a Machine Learning Architect at Pathwwway centered on predictive modeling and large-scale data analysis in marketing and gaming, focusing on translating mathematical models into high-performance production settings. Dr. Dobrică studied computational modeling and complex systems for his PhD, Doctoral school of Physics, at the University of Bucharest. He then coordinated the development of biomedical monitoring and therapy ISFET nano-sensors and micro-embedded neuro-electrostimulation devices at the 3Nano-SAE Research Center. This research is part of his ongoing interest in AI, embedded systems, and intelligent sensing, key subjects in IoT research. Dr Dobrica is founder of AI & IoT -course and laboratories based on new concept "innoversity", developed at 3NanoSae res center, University of Bucharest. Besides his academic and professional work, Dr. Dobrică is President of the Pro-Youth NGO, which promotes AI and robotics education for youth. He is a Visiting Lecturer at IE University in Madrid, where he promotes applied AI methods and multidisciplinary technology education.

Kiranbhai Ramabhai Dodiya is an experienced Cyber Forensic Expert and Assistant Professor specializing in cyber security and digital forensics. He has taught at prestigious institutions and is currently pursuing a PhD in Biochemistry and Forensic Science. With a strong background in ethical hacking and malware analysis, Kiranbhai is passionate about educating students and contributing to the field of cyber forensics. His expertise includes malware forensic, computer forensics, and data recovery software. Kiranbhai's commitment to professional growth is evident through his participation in workshops, seminars, and research publications. He is dedicated to shaping the next generation of professionals in the field and combatting cybercrimes.

Ajesh F is a Professor and Dean of Administration in the Department of Computer Science and Engineering at TKM Institute of Technology, Kollam. He is recognized for his leadership in academia and his expertise in Artificial Intelligence and Machine Learning. Dr. Ajesh holds a Bachelor of Technology in Computer Science and Engineering from Cochin University of Science and Technology, a Master of Engineering, and a PhD in Computer Science and Engineering from Anna University, Chennai.

F Shabina Fred Rishma completed hes BE, in Anna University and ME from Sathyabama University, Chennai. B.Lib.Sc. from Alagappa University. She is in the teaching profession for the past 3 years. Her area of interest in Machine Learning & Artificial Intelligence.

Gor Gevorgyan was born in Yerevan, Armenia, in 1984. He received his B.S. and M.S. degrees in Biology from Yerevan State University in 2006 and 2008, respectively. In 2011, he was awarded a Ph.D. in Biology by the decision of the Council of the National Academy of Sciences of Armenia. Dr. Gevorgyan currently serves as Deputy Director for Science and Head of the Applied Hydroecology Laboratory at the Scientific Center of Zoology and Hydroecology of the National Academy of Sciences of the Republic of Armenia. He is the author of over 60 scientific publications. His research interests focus on aquatic ecology and biodiversity, water quality, water resources management, limnology, environmental impact assessment, and the effects of climate change on freshwater ecosystems.

Chandra Has is currently serving as an Assistant Professor in the Department of Chemical Engineering at GSFC University, Vadodara, India. He earned his M.Tech–Ph.D. dual degree in Chemical Engineering from the Indian Institute of Technology Bombay (IIT Bombay) in 2018, followed by two years of postdoctoral research at IIT Palakkad. His research expertise lies in membrane biophysics and colloid science. Dr. Has has published several high-quality research articles in reputed journals and actively contributes to the academic community as a reviewer for leading international journals, including Frontiers in Chemistry, Colloid and Interface Science Communications, and Chemical Papers.

Safad Ismail is a dedicated researcher, lawyer, and academician with a strong interdisciplinary background. He holds an M.Tech in Computer Science and Engineering from Government Engineering College (Rajiv Gandhi Institute of Technology & Science), Kottayam, Kerala, India, and a B.Tech in Computer Science and Engineering from Cochin University of Science and Technology. Additionally, he earned a Bachelor's degree in Legislative Law from Karnataka State Law University, Bangalore, India. He has qualified the University Grants Commission-NET examination and the All India Bar Examination. His journey with IEEE began in 2012 as a student member. With 5.5 years of teaching experience in various engineering colleges across India, three years of research experience, and industry exposure, he has published papers in reputed IEEE journals and conferences. His academic interests lie at the intersection of Computer Science, Artificial Intelligence, and Machine Learning, with a particular focus on their applications in the legal domain. He remains passionate about research and teaching, continuously striving to contribute to these evolving fields.

Surik Khudaverdyan received the Ph.D. degree in physics from Moscow Institute of Electronic Technology, Russia, in 1980, and the Doctor of Technical Sciences degree in micro- and nanoelectronics from the National Polytechnic University of Armenia, Yerevan, in 2005. He is currently a Distinguished Professor and the Head of the Photoelectronic Devices Laboratory, National Polytechnic University of Armenia. He has an extensive academic and administrative background, having supervised numerous Ph.D. students and authored a substantial number of publications in leading scientific journals. His research activities are centered on the development of solar cells and photodetectors. He has been honored with several prestigious awards, including the President's Award in Technical Sciences and Information Technologies.

Varuna Kumara a Research Scholar in the department of Electronics Engineering at JAIN (Deemed to be University), Bengaluru, India. He also received his B. E and M. Tech from Visvesvaraya Technological University, Belagavi, India in 2009 and 2012 respectively. He is currently Assistant Professor at Electronics and Communication Engineering in Moodlakatte Institute of Technology, Kundapura, India. His research interests are in Artificial Intelligence, Signal Processing, Control Systems.

M. R. Arun completed his BE, ME and PhD from Anna University, Chennai. He completed PG Diploma (PGDEEM) from Annamalai University, M.Sc. Counselling Psychology from TNOU University, M.A. Public Administration from Madras University and B.Lib.Sc. from Alagappa University. He is in the teaching profession for the past 17 years, published around 79 papers in Journals, Patents and Conferences, Active participant crossing more than 130 events and internationally networked through various professional bodies. He is skilled and technically proficient in various simulation software associated to Research in

Engineering. He organized and coordinated around 15 events in working institution. Authored 9 Books, 3 Book Chapters and he has been technically integrated in various international journals as an Editorial Board Member and Reviewer. His area of interest includes Video Processing, Remote Automation, Wireless Communication, Quantum Computing Machine Learning & Artificial Intelligence.

T.C. Manjunath was born on 6th February 1967. He did his B.E. from RV College of Engg. in 1989 (B'lore Univ), M.E. in Robotics & Automation from Govt.'s LD College of Engg., (Gujarat Univ., Ahmedabad) in 1992 with FCD-First Rank & PhD in Systems & Control Engg. from IIT Bombay in the year 2007, got best research scholar award for the entire IITB. He is currently working as the Dean Research (R & D) and as Professor in Dept. of Computer Science & Engg. with specialization in IoT, Cyber Security, Blockchain Technology in Rajarajeswari College of Engineering, Bengaluru, Karnataka. He has got a teaching (academic), research & administrative experience of more than 35+ years (10 years as Principal & 8 years as HoD, 1 year as Dean Research & rest as faculty) and has published more than 1500+ research papers in various national & international journals & conferences, with 175+ patents, 100+ book-chapters, 25 awards, 100s of Keynote talks and also published 25 text books in various field of science, engineering & technology, completed projects worth Rs. 9 Crores till date & guided 10 PhD scholars in VTU. He is a member of 30 Professional Society Chapters and is a fellow of IE, IETE, AMIEE, Sr. Member of IEEE and a chartered engineer.

Jose Alex Mathew, B.E(ECE), M-Tech (BM), Ph.D., F.I. E, M.I.S.T. E and Senior Member IEEE, received B. E (KVGCE, Sullia) and MTech(MIT Manipal) degrees in Electronics and Communication Engineering from Mangalore University, India in 1992 and 1997 respectively. He received his PhD in Medical Image Processing and System Development from Mangalore University, India in 2012. He Completed many NPTEL courses in the Computer Science field. He worked at MESCE Kuttippuram, SJEC Mangalore, VJEC Kannur, PACE Mangalore, SCME Mangaluru and SIT Mangaluru and held different positions like PG Coordinator, HOD, Director etc. Currently, he is working as a Professor of Data Science at Alva's Institute of Engineering & Technology. His research interests include Artificial Intelligence, Machine Learning, Signal and Image Processing, VLSI, Communication and Control and Renewable Energy. He has contributed many National and International journals, and book chapters to various reputed journals and Conferences. He can be contacted at email: aymanamkuzhy@gmail.com

Tanya Mendez received her B.Tech degree in Electronics and Instrumentation Engineering in 2007 from Mahatma Gandhi University, Kerala, and her M.Tech degree in VLSI and Embedded Systems from CUSAT, Kerala in 2011. She has 8 years of teaching, 3.5 years of research and 2 years of industry experience. She is currently working as an Assistant Professor, Grade 3 in the Department of Robotics and Artificial Intelligence, NMAM Institute of Technology, Nitte. Her research interests include digital system design, low-power VLSI design, electronic design automation and VLSI architecture design

P. Selvakumar is currently an Associate Professor in the Department of Science and Humanities at Nehru Institute of Technology, Coimbatore, Tamil Nadu, India. Prior to joining Nehru Institute of Technology, he held positions as an Assistant Professor in the Department of Science and Humanities various engineering colleges in Coimbatore, Tamil Nadu, India, spanning from 2011 to 2023. Dr.P. Selvakumar's academic and research interests lie primarily in the field of phytochemistry. He has made significant contributions to his field, with an impressive record that includes 30 journal publications, 12 patents, and 20 conference papers. Additionally, he has organized 5 seminars and conferences, further demonstrating his commitment to academic and scientific exchange. Throughout his career, Dr. P.Selvakumar has been recognized for his dedication and excellence in teaching, having been honored with several awards, including the Best Teacher Awards. He holds a prominent position in the academic community, contributing actively to research.

Mikail Pehlivan, has his BA degree in the fields of Public Administration and Public Finance (double major) at Afyon Kocatepe University in 2015. Subsequently, he received his MA and PhD degrees in the field of Public Finance at Istanbul University. The author currently continues his career at Trakya University, Faculty of Economics and Administrative Sciences, Department of Public Finance, since 2017. Dr. Pehlivan is interested in Public Economics, Environmental Economics and Regulation Economics.

Shibu K. R is an Associate Professor in the Department of Computer Science and Engineering at Saintgits College of Engineering, Kottayam, India. He received his Ph.D in Computer Science and Engineering from Noorul Islam University in the year 2021. His research interests include wireless networks, mobile ad hoc networks (MANETs), network security, and machine learning applications in communication systems. He has published papers in reputed journals and conferences and is actively engaged in guiding undergraduate and postgraduate students in research projects.

Renuka Mahesh Raut is an Assistant Professor in the Department of Computer Science and Engineering at St. Vincent Pallotti College of Engineering and Technology, Nagpur, with over 8 years of teaching experience. I secured First Merit in M.Tech. (CSE) with First Class with Distinction and is currently pursuing her Ph.D. in Computer Science and Engineering from IIIT Nagpur. My research interests include Machine Learning, Deep Learning, and Artificial Intelligence. I had published 3 international journals papers, along with 5 copyrights and 2 patents, and continues to contribute actively to innovation and research in emerging technologies.

Ramadevi Vitthal Salunkhe is an Assistant Professor in the Department of Artificial Intelligence at G. H. Raisoni College of Engineering, Nagpur, and a Ph.D. scholar in Computer Science at Visvesvaraya National Institute of Technology (VNIT), Nagpur. She has over 11 years of diverse experience in academia, government, and industry, with research interests in artificial intelligence, quantum computing, and nanomaterials discovery.

Rushikesh M. Shete (b. 16th December 1987) is an Assistant Professor at St. Vincent Pallotti College of Engineering and Technology, Nagpur. He has been in academia since 2010 and is currently pursuing a Ph.D. in Computer Science and Engineering from Bhagwant University, Ajmer. A cloud computing enthusiast, he has presented research in 5 conferences and published 15 papers in reputed journals. His academic interests lie in cloud computing, emerging technologies, and advanced computing systems.

Sreekumar G, presently working as the Principal, College of Applied Science, (under Institute of Human Resources Development (IHRD), Kerala) Perissery, Chengannur, received his Bachelor's in Physics from Sree Narayana College, Punalur, in 1990, M.Sc in Physics from Mahatma Gandhi College, Thiruvananthapuram in 1992 and B.Ed. from Mount Tabor Training College, Pathanapuram in 1993, affiliated to University of Kerala, Thiruvananthapuram, Kerala, India. He has been awarded Ph.D. degree for his work "Preparation and Characterization of Thin/Thick Solid Dye Films Applicable to Various Optoelectronic Purposes" by the University of Kerala in 2010. He started his career in IHRD as Lecturer in Physics in 1995, and later has been working as principal of different streams (Technical Higher Secondary Schools & College of Applied Sciences) under IHRD, since 2005.

Nidhi Tyagi Professor in School of Computational Sciences and Engineering at Shobhit Institute of Engineering & Technology (Deemed to-be-University), Meerut. She completed her Bachelors of Engineering in Computers from Pune University. Dr. Tyagi obtained her M. Tech. & Ph. D. in Computer Engineering from Shobhit University Meerut. She has a teaching experience of 25 years. Her areas of interest include Data Mining, Internet of Things, Databases, Robotic Process Automation, Machine Learning, Information Retrieval System and Big Data Analytics. She has published about 50 research papers in reputed journals and has authored 3 books, patents, monographs & articles. She has guided 2 Ph.D. scholar, 25 M.Tech. students and, number of Ph.D. scholars are doing research under her supervision. She has delivered expert talks in different domains & platforms. She has organized many conferences, coordinated government funded Faculty Development Program and has been a reviewer in international conferences. She is a member of technical professional societies like IEEE, ISTE & Editor-in-Chief of IJISSE and Editor of IJWNS. Dr. Tyagi is an Innovation Ambassador certified by Ministry of Education.

Hemant K. Upadhyay (Ph.D. pursuing) is an accomplished academic leader with over 20 years of experience in higher education, research, and administration. Currently serving as Assistant Professor at Poornima University, Jaipur, he holds B.Tech, M.Tech, and MBA degrees. His research focuses on multi-objective optimization and decision-making models for electric vehicle adoption in India. He has published extensively in SCIE/Scopus-indexed journals, authored international book chapters, and presented papers at global conferences. With eight patents, numerous awards, and memberships in leading professional societies, he is dedicated to advancing sustainable engineering and fostering academic excellence.

Lavanya V M.Sc.,MPhil., is serving as an Assistant Professor at CMR University, Bengaluru for the past two years. She has around 11 years of teaching experience and 1 year of Industry experience. She has done her Undergraduate in B.Sc Computer Science from Sri Ramakrishna College of arts & science, Coimbatore, postgraduate in M.Sc Computer technology from Anna University, Coimbatore. She has published two books on the title "Mobile Application Development" and "Data Science with R: An Integrated Approach to programming and Analytics" and has published papers in 5 International journals. She has also presented papers at 2 IEEE conferences. She has participated in many national and international conferences, workshops, seminars and FDP programs. The Areas of Interest are Data Mining, Deep Learning and Artificial Intelligence. She has presented and participated in many national and international conferences, workshops, seminars and FDP programs.

Vineet Vishnoi earned his Ph.D. in Computer Science Engineering from the Bennett University, Greater Noida, India in 2023. He received M.Tech. and B. Tech degree in Electronics and Communication Engineering from Uttar Pradesh Technical University, Lucknow, India, in 2012 and IET, Barielly, in 2008, respectively Some of his research findings are published in top-cited journals, such as IEEE Transactions on Vehicular Technology, Science-Direct, IEEE Internet of Things, and various international top-tiered conferences, such as IEEE GLOBECOM, IEEE ICC, and IEEE Infocom. His research interests include device-to-device communications, the Internet of Things, Non-orthogonal multiple access, deep reinforcement learning, and microstrip patch antenna.

Index

A

B

C

D

E

F

P

Q

R

S